T0319263

Storing Energy
with Special Reference to Renewable Energy Sources

Storing Energy

with Special Reference to Renewable Energy Sources

Trevor M. Letcher
Emeritus Professor, Department of Chemistry
University of KwaZulu-Natal
Durban, South Africa

AMSTERDAM • BOSTON • HEIDELBERG • LONDON • NEW YORK • OXFORD • PARIS
SAN DIEGO • SAN FRANCISCO • SINGAPORE • SYDNEY • TOKYO

Elsevier
Radarweg 29, PO Box 211, 1000 AE Amsterdam, Netherlands
The Boulevard, Langford Lane, Kidlington, Oxford OX5 1GB, UK
50 Hampshire Street, 5th Floor, Cambridge, MA 02139, USA

British Library Cataloguing-in-Publication Data
A catalogue record for this book is available from the British Library

Library of Congress Cataloging-in-Publication Data
A catalog record for this book is available from the Library of Congress

ISBN: 978-0-12-803440-8

For information on all Elsevier publications
visit our website at http://www.elsevier.com/

Typeset by Thomson Digital

Contents

Part B
Electrical Energy Storage Techniques Gravitational/Mechanical/Thermomechanical

4. Advanced Rail Energy Storage: Green Energy Storage for Green Energy

Francesca Cava, James Kelly, William Peitzke, Matt Brown, Steve Sullivan

5. Compressed Air Energy Storage

Seamus D. Garvey, Andrew Pimm

Part C
Electrochemical

Part D
Thermal

List of Contributors

Aliakbar Akbarzadeh, School of Aerospace, Mechanical and Manufacturing Engineering, RMIT University, Australia
aliakbar.akbarzadeh@rmit.edu.au

Christopher Baldwin, Department of Mechanical and Aerospace Engineering, Carleton University, Ottawa, ON, Canada
Christopher.Baldwin@carleton.ca

Stephan Bauer, Innovation and Development, RAG Oil Exploration Company, Vienna, Austria
Stephan.bauer@rag-austria.at

Donald Bender, System Surety Engineering, Sandia National Laboratories, California, United States of America
dbende@sandia.gov

Pierrick Bouffaron, MINES ParisTech, PSL Research University, Centre de mathématiques appliquées, France; Berkeley Energy & Climate Institute, University of California, Berkeley, United States of America
pbouffaron@gmail.com

Matt Brown, ARES, Santa Barbara, CA, United States of America
matt@aresnorthamerica.com

Jens Burfeind, Fraunhofer Institute for Environmental, Safety, and Energy Technology UMSICHT, Oberhausen, Germany
jens.burfeind@umsicht.fraunhofer.de

Francesca Cava, ARES, Santa Barbara, CA, United States of America
francesca@aresnorthamerica.com

Haisheng Chen, Institute of Engineering Thermophysics, Chinese Academy of Sciences, Beijing, China
chen_hs@iet.cn

Greig Chisholm, School of Chemistry, University of Glasgow, Glasgow, United Kingdom
greig.chisholm@glasgow.ac.uk; valandgreig@gmail.com

José Luis Cortina, Departament d'Enginyeria Química, Universitat Politècnica de Catalunya; Water Technology Center CETaqua, Barcelona, Spain
jose.luis.cortina@upc.edu

Leroy Cronin, School of Chemistry, University of Glasgow, Glasgow, United Kingdom
lee.cronin@glasgow.ac.uk

Fritz Crotogino, R&D Department, KBB Underground Technologies GmbH, Hannover, Germany
frcrotogino@kbbnet.de

Cynthia Ann Cruickshank, Department of Mechanical and Aerospace Engineering, Carleton University, Ottawa, ON, Canada
Cynthia.Cruickshank@carleton.ca

Louis Desgrosseilliers, Dalhousie University, Halifax, Nova Scotia, Canada
louis.d@dal.ca.

Yulong Ding, Birmingham Centre for Cryogenic Energy Storage, School of Chemical Engineering, University of Birmingham, Edgbaston, Birmingham, UK
y.ding@bham.ac.uk

Paul E. Dodds, UCL Energy Institute, University College London, London, UK
p.dodds@ucl.ac.uk

Christian Doetsch, Fraunhofer Institute for Environmental, Safety, and Energy Technology UMSICHT, Oberhausen, Germany
christian.doetsch@umsicht.fraunhofer.de

Sabine Donadei, KBB Underground Technologies GmbH, Hannover, Germany
donadei@kbbnet.de

Frank Escombe, EscoVale Consultancy Services, Reigate, Surrey, United Kingdom
frank.escombe@escovale.com

Leuserina Garniati, Centre for Understanding Sustainable Practice (CUSP), Robert Gordon University, Aberdeen, Scotland, United Kingdom
L.garniati@rgu.ac.uk

Seamus D. Garvey, Department of Mechanical, Materials and Manufacturing Engineering; Faculty of Engineering, University of Nottingham, Nottingham, United Kingdom
seamus.garvey@nottingham.ac.uk

David Greenwood, School of Electrical and Electronic Engineering, Newcastle University, Newcastle upon Tyne, United Kingdom
David.Greenwood@ncl.ac.uk

Dominic Groulx, Dalhousie University, Halifax, Nova Scotia, Canada
Dominic.Groulx@dal.ca

Fengjuan He, Institute of Engineering Thermophysics, Chinese Academy of Sciences, Beijing, China
hefengjuan@iet.cn

Shan Hu, Institute of Engineering Thermophysics, Chinese Academy of Sciences, Beijing, China
hushan@iet.cn

Samer Kahwaji, Dalhousie University, Halifax, Nova Scotia, Canada
sam@dal.ca

James Kelly, ARES, Santa Barbara, CA, United States of America
jim@aresnorthamerica.com

Henner Kerskes, Research and Testing Centre for Solar Thermal Systems (TZS), Institute for Thermodynamics and Thermal Engineering (ITW), University of Stuttgart, Germany
kerskes@itw.uni-stugarttgart.de

Trevor M. Letcher, Emeritus Professor, Department of Chemistry, University of KwaZulu-Natal, Durban, South Africa; Laurel House, FosseWay, Stratton on the Fosse, United Kingdom
trevor@letcher.eclipse.co.uk

Yongliang Li, Birmingham Centre for Cryogenic Energy Storage, School of Chemical Engineering, University of Birmingham, Edgbaston, Birmingham, UK
y.li.1@bham.ac.uk

Chang Liu, Institute of Engineering Thermophysics, Chinese Academy of Sciences, Beijing, China
liuchang@iet.cn

Stephan Lux, Fraunhofer Institute for Solar Energy Systems ISE, Freiburg, Germany
stephan.lux@ise.fraunhofer.de

John A. Noël, Dalhousie University, Halifax, Nova Scotia, Canada
John.noel@dal.ca

Alan Owen, Centre for Understanding Sustainable Practice (CUSP), Robert Gordon University, Aberdeen, Scotland, United Kingdom
a.owen@rgu.ac.uk

Charalampos Patsios, School of Electrical and Electronic Engineering, Newcastle University, Newcastle upon Tyne, United Kingdom
Haris.Patsios@ncl.ac.uk

William Peitzke, ARES, Santa Barbara, CA, United States of America
bill@aresnorthamerica.com

Andrew Pimm, Faculty of Engineering, University of Nottingham, Nottingham, United Kingdom
andrew.pimm@nottingham.ac.uk

Jonathan Radcliffe, Birmingham Centre for Cryogenic Energy Storage, School of Chemical Engineering, University of Birmingham, Edgbaston, Birmingham, UK
J.radcliffe@bham.ac.uk

Gregor-Sönke Schneider, KBB Underground Technologies GmbH, Hannover, Germany
schneider@kbbnet.de

Catalina Spataru, Energy Institute, University College London, United Kingdom
c.spataru@ucl.ac.uk

Steve Sullivan, ARES, Santa Barbara, CA, United States of America
steve@aresnorthamerica.com

Trevor Sweetnam, Energy Institute, University College London, United Kingdom
trevor.sweetnam.09@ucl.ac.uk

Philip Taylor, Institute for Sustainability, Newcastle University, Newcastle upon Tyne, United Kingdom
Phil.Taylor@ncl.ac.uk

Robert Tichler, Department of Energy Economics, Energy Institute, Johannes Kepler University Linz, Linz, Austria
tichler@energieinstitut-linz.at

Lige Tong, School of Mechanical Engineering, University of Science & Technology Beijing, Beijing, China
tonglige@me.ustb.edu.cn

César Valderrama, Departament d'Enginyeria Química, Universitat Politècnica de Catalunya, Spain.
cesar.alberto.valderrama@upc.edu

Stalin Munoz Vaca, School of Electrical and Electronic Engineering, Newcastle University, Newcastle upon Tyne, United Kingdom
S.E.Munoz-Vaca@ncl.ac.uk

Matthias Vetter, Fraunhofer Institute for Solar Energy Systems ISE, Freiburg, Germany.
matthias.vetter@ise.fraunhofer.de

Neal Wade, School of Electrical and Electronic Engineering, Newcastle University, Newcastle upon Tyne, United Kingdom
Neal.Wade@ncl.ac.uk

Huanran Wang, School of Energy and Power Engineering, Xi'an Jiaotong University, Xi'an, China
huanran@mail.xjtu.edu.cn

Li Wang, School of Mechanical Engineering, University of Science & Technology Beijing, Beijing, China
lwang@me.ustb.edu.cn

Mary Anne White, Dalhousie University, Halifax, Nova Scotia, Canada
mary.anne.white@dal.edu

Guang Xi, School of Energy and Power Engineering, Xi'an Jiaotong University, Xi'an, China
xiguang@mail.xjtu.edu.cn

Yujie Xu, Institute of Engineering Thermophysics, Chinese Academy of Sciences, Beijing, China
xuyujie@iet.cn

Chi-Jen Yang, Center on Global Change, Duke University, Durham, NC, United States
Cj.y@duke.edu

Erren Yao, School of Energy and Power Engineering, Xi'an Jiaotong University, Xi'an, China
yao.erren@stu.xjtu.edu.cn

Peikuan Zhang, School of Mechanical Engineering, University of Science & Technology Beijing, Beijing, China
pkzhang@me.ustb.edu.cn

Preface

Renewable energy sources such as wind turbines and solar panels for electricity generation have become commonplace in our society. Their aim is to supply energy that is free from carbon dioxide production while sustainable and not dependent on a finite energy supply. Unfortunately their full potential is reduced by their intermittency. For these and other developing renewable technologies, such as tidal current energy and wave energy, to make a real difference we need to find effective ways to store this energy. This book is a showcase for the current state of the different methods that are being explored to store energy and make it available not only when the Sun is shining, the wind is blowing, the tides flowing, the sea currents moving or when the waves are breaking. These new storage methods will also be useful in times when demand for electricity is low and electricity can be bought cheaply and stored until demand rises and the stored energy can be used. At present the chief way of storing energy is through pumped hydroelectric storage. Most countries have now exhausted the places where large reservoirs can be built so this new focus on storing energy is both timely and necessary as the world moves towards sustainable carbon-free energy.

Some chapters in the book are concerned with developments of well-known energy storage techniques, others are concerned with new techniques which are being tested and researched for the first time, and a few involve techniques which have yet to leave the drawing board. Unfortunately a few interesting and novel processes are missing as authors were unavailable to write the chapters. One process is that of superconducting magnetic energy storage (SMES) which has recently been reviewed by Weijia Yuan and Min Zhang in *A Handbook of Clean Energy System* published by Wiley (2015) (DOI: 10.1002/9781118991978.hces210). Two other links are: http://link.springer.com/book/10.1007%2F978-0-85729-742-6 and http://onlinelibrary.wiley.com/doi/10.1002/9781118991978.hces210/abstract. Facilities for SMES exist all round the world for use in power quality control and for grid stabilization and units of 1 MW h are not uncommon.

Another technology which is not represented here is that involving supercapacitors which are very effective at relatively small-scale energy storage (thus in competition with batteries) and which is particularly useful in transport vehicles. Its applications are reviewed by Yank, Yeh, and Ramea et al. of the University of California, Davis at: http://www.its.ucdavis.edu/wp-content/themes/ucdavis/pubs/download_pdf (document 2014-UCD-ITS-RR-14-04).

A third process not covered in this volume is the pumped heat electrical energy storage system currently being developed by Isentropic Ltd. in Hampshire, UK. This is a grid scale storage system, and the short term goal is to develop a 1.5 MW unit. The method has great promise but has yet to be commercially available. A good introduction to the topic is the paper by Derues, Ruer, Marty and Fourmigue in Applied Thermal Engineering, 2010; 30:425–432. Yet another explanation of the process is given by staff of Isentropic Ltd. http://www.isentropic.co.uk/our-phes-technology.

Our book *Storing Energy: with Special Reference to Renewable Energy Sources,* is a natural follow-up to *Future Energy: Improved Sustainable and Clean Options for our Planet* (2nd ed.), which was published by Elsevier in 2014. In *Future Energy* the case was made for developing new and sustainable energy sources in the light of climate change and increasing levels of greenhouse gases. In many ways, *Storing Energy* also goes hand in hand with another book we published recently: *Climate Change: Observed Impacts on Planet Earth* (2nd ed.) (Elsevier 2015).

The present book is divided into four sections, namely an Introduction; Electrical Energy Storage Techniques; Integration; and International Issues and the Politics of Introducing Renewable Energy Schemes. The Electrical Energy Storage section is divided into further sections headed: Gravitational, Mechanical, and Thermomechanical; Electrical; Thermal; and Chemical. The Gravitational, Mechanical, and Thermomechanical storage methods include chapters on: pumped hydroelectricity storage (PHES) as well as novel hydroelectricity processes; liquid air (LAES); compressed air (CAES); pumped hydro combined with compressed air; and finally advanced rail energy storage (ARES). The Electrical section has chapters on: rechargeable batteries and vanadium redox flow batteries. The Thermal section has chapters on: phase changes; solar ponds; and sensible thermal energy storage and the Chemical section includes chapters on: hydrogen and water electrolysis; chemical reactions including zeolite–water reactions; power to gas; traditional energy storage (gas oil and coal) and large scale hydrogen storage. The Integration chapters are on network integration, smart grids and off-grid energy. The three chapters in the section on International Issues and the Politics of Introducing Renewable Energy are: on energy storage in China; energy storage worldwide; and on the politics of investing in renewable energy.

Many governments and people of influence throughout the world are supporting the drive to reduce our dependency on fossil fuels with interesting and innovative programmes. One such programme is the Global Apollo Programme, spearheaded by Sir David King, which calls for £15 $\times$ 10^9 (£15 billion) a year to be spent on research, development and demonstration of green energy and energy storage. Significantly this amount is the same, in today's money that the US Apollo programme spent in putting astronauts on the moon. Professor Martin Rees, former head of the Royal Society and another member of the Apollo group, explains the reason for using the name Apollo: "NASA showed how a stupendous goal could be achieved, amazingly fast, if the will and the resources are there."

This book has been produced in order to allow the reader to have an understanding and insight into a vital aspect of our future use of energy—its storage. The final decision as to which option should be developed in a country or region must take into account many factors including: topography, for example, are there suitable sites for reservoirs to tap into PHES?; are there convenient salt caverns available for gas storage?; is the amount of sunlight available sufficient?; is it possible to take advantage of thermal energy storage?; is the chemical industry infrastructure sufficiently mature?; is it possible to install electrolysis plants for hydrogen production or develop chemical reaction storage or install a sophisticated battery system?; is the density of population important and should off-grid technologies be incorporated or can network integration and smart grids be used?

It is also to be hoped that the book will act as a springboard for new developments. One way that this can take place is through contact between readers and authors and to this effect mail addresses of the authors have been included.

The book is supported by IUPAC through its Physical Chemistry Division and both the logos of IUPAC, and our publisher Elsevier, appear on the front cover. The adherence of IUPAC to the International System of Quantities through its Interdivisional Committee for Terminology, Nomenclature and Symbols (ICTNS), is reflected in the book with the use of SI Units throughout. The index notation is used to remove any ambiguities; for example, billion and trillion are written as 10^9 and 10^{12}, respectively. To further remove any ambiguities the concept of the quantity calculus is used. It is based on the equation: physical quantity = number $\times$ unit. To give an example, power = 200 W and hence, 200 = power/W. This is of particular importance in the headings of tables and the labelling of graph axes.

This volume is unique in the genre of books of related interests in that each chapter of *Storing Energy* has been written by an expert scientist or engineer, working in the field. Authors have been chosen for their expertise in their respective fields and come from ten countries: Australia, Austria, Canada, China, France, Germany, South Africa, Spain, United Kingdom, and the United States. Most of the authors come from developed countries as most of the research and development in this fast moving field, presently, come from these countries. We look forward to the future when new approaches to storing energy from scientists and engineers working in developing countries will be developed which focus on their local conditions.

A vital concern related to future energy and storing energy is: what is to be done when it appears that politicians misunderstand or ignore and corporations overlook, the realities of climate change and the importance of renewable energy sources? The solution lies in sound scientific data and education. As educators we believe that only a sustained grassroots movement to educate citizens, politicians and corporate leaders of the world has any hope of success.

This book is part of this aim. It presents options for readers to consider and we hope that not only students, teachers, professors, and researchers of renewable energy, but also politicians, government decision makers, captains of industry, corporate leaders, journalists, editors, and all interested people, will read the book, take heed of its contents and absorb the underlying message that renewable energy sources are our future and storing energy is a vital part of it.

I wish to thank all 59 authors and coauthors for their cooperation, help and especially for writing their chapters. It has been a pleasure working with each and every one of the authors. I thank my wife Valerie for all the help she has given me over these long months of putting the book together. I also wish to thank Elsevier for their professionalism and help in producing this well presented volume. Finally I wish to thank Professor Ron Weir of IUPACs Interdivisional Committee for Terminology, Nomenclature and Symbols for his help and advice.

Trevor M. Letcher
Stratton on the Fosse
Somerset
Sep. 2015

Part A

Introduction

Chapter 1

The Role of Energy Storage in Low-Carbon Energy Systems

Paul E. Dodds*, Seamus D. Garvey**

**UCL Energy Institute, University College London, London, UK; **Department of Mechanical, Materials and Manufacturing Engineering, University of Nottingham, Nottingham, UK*

1 INTRODUCTION

Energy storage makes a vital contribution to energy security in existing energy systems. At present, most energy is stored as raw or processed hydrocarbons, whether in the form of coal heaps or oil and gas reservoirs. Since electricity storage is much more expensive by comparison, precursors to electricity are stored rather than electrical energy, and generation is varied to meet demand. The principal exception to this *modus operandi* is pumped hydroelectric generation, which can generate a large power output for a short period, at very short notice, and is used to increase electricity system stability.

As energy systems gradually evolve toward using low-carbon technologies, the role and type of energy storage is likely to change substantially. Two broad trends are likely to drive this transition. First, as intermittent renewable and fixed output nuclear generation take an ever greater role, it will become increasingly difficult to match electricity supply to demand, with imbalances becoming both larger and more common over time. The move away from fossil generation will mean that it will no longer be possible to store most electricity precursors as hydrocarbons, with the exception perhaps of gas for flexible generation. Second, if low-carbon electricity displaces oil and gas for transport and heat provision, electricity demand patterns could change substantially, with peak demands becoming much more pronounced. Numerous energy storage technologies are under development that store electricity at times of excess supply in order to meet periods of high demand. Other storage technologies could support the energy system in other ways, for example, by storing excess electricity as heat or hydrogen for use in other sectors. The move to a low-carbon economy will cause nothing less than a revolution in how energy storage is used.

This chapter considers how new energy storage technologies can support future low-carbon energy systems in the long term. It introduces a wide range of energy storage technologies, which are explored in this book, and identifies

Storing Energy. http://dx.doi.org/10.1016/B978-0-12-803440-8.00001-4

key characteristics with which to compare the technologies. Finally, it identifies challenges for commercializing and deploying these technologies into existing energy systems in the short to medium term.

2 THE NEED FOR NEW TYPES OF STORAGE

Most new storage technologies are designed to contribute to low-carbon electricity systems. Electricity generation is relatively stable in most countries and intraday variations can often be larger than interseasonal variations. An example is shown in Fig. 1.1 for the United Kingdom. Average consumption through the year is 37 GW, with an average intraday variation over the year of 18 GW and a peak of 59 GW. Yet the difference in average daily consumption between winter and summer is much lower than the intraday variation, at only 11 GW (winter here is defined as Dec. to Feb., and summer as Jun. to Aug.). Larger intraseasonal variations would be expected in countries that use predominantly electric heating, with the peak demand in winter, or in countries with warmer climates, where demand for air conditioning would lead to a peak demand in summer.

A mix of baseload and flexible electricity generation technologies are generally used to meet these demands. Baseload is generally nuclear or coal plants that produce a constant output and have high capital costs and low fuel costs. Flexible generators tend to be gas- and oil-fired power plants with low capital costs and high fuel costs. There is a tradeoff between flexibility, efficiency, and cost between technologies; for example, open-cycle gas turbines (OCGTs) are more flexible and have lower capital costs than combined-cycle gas turbines (CCGTs), but have lower fuel efficiencies. The optimal generation portfolio depends on demand patterns; in the United Kingdom, to meet the demand pattern in Fig. 1.1, there is around 45 GW baseload and 40 GW flexible supply.

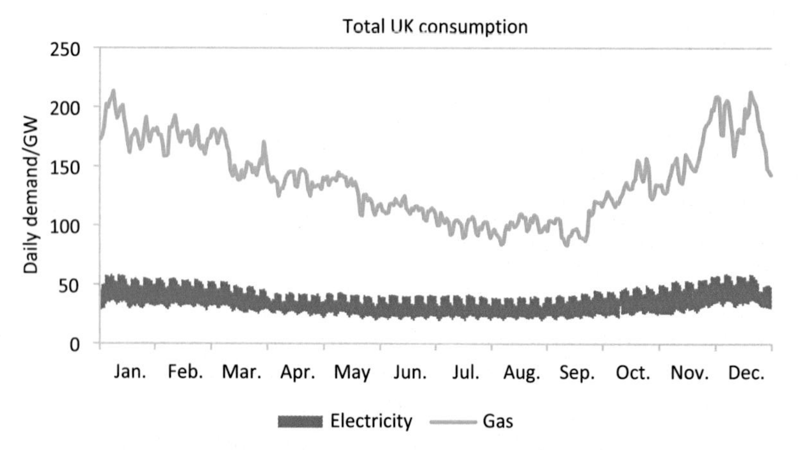

FIGURE 1.1 **Total UK electricity and natural gas consumption in 2010.** For electricity, the width of the line demarks the maximum and minimum for each day. *(Source: Electricity data are half-hourly averages from the National Grid [1] and daily gas data are from the National Grid [2].)*

A new challenge for electricity systems is the increasing penetration of intermittent renewable generation. Low-carbon solar, wind, and wave technologies have high capital costs and negligible operating costs, but the intermittent outputs cannot be easily forecast or controlled. Small penetrations of intermittent generation can be incorporated through standard flexible electricity generation. However, as renewables account for an increasingly large part of the generation portfolio, two issues arise:

1. At times of high demand and low renewable generation, deficits occur that can affect the stability of the electricity system.
2. At times of low demand and high renewable generation, surplus electricity is generated that must be stored or lost.

These electricity system imbalances between generation and demand present an opportunity for new types of energy storage to have an important role in future energy systems.

2.1 Impact of Demands on Generation Imbalances

The deficits and surpluses from renewable generation could be greatly magnified in the future if transport and heat are electrified to reduce greenhouse gas (GHG) emissions. Fig. 1.1 shows that natural gas consumption has much wider intraseasonal variations than electricity consumption in the United Kingdom. This is primarily due to heat demand, as demonstrated by Fig. 1.2 for the UK residential sector. While electrification would not lead to intraseasonal electricity peaks of this magnitude, since heat pumps with a COP of 3 would likely replace gas

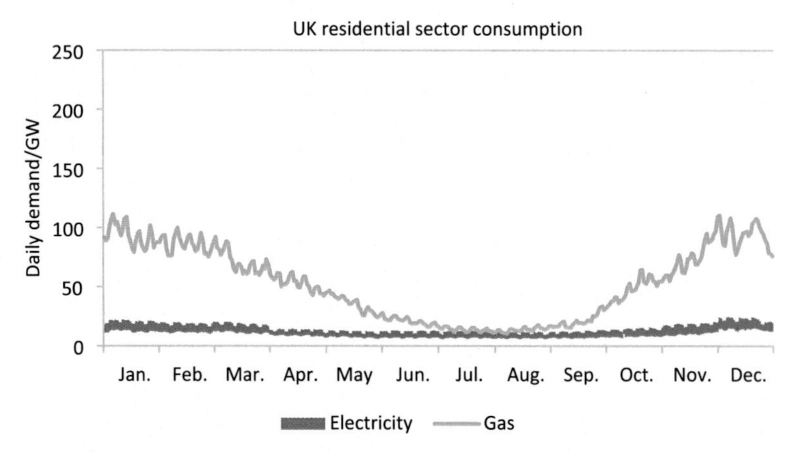

FIGURE 1.2 **Estimated UK electricity and natural gas consumption in 2010 in the residential sector.** For electricity, the width of the line demarks the maximum and minimum for each day. *(Source: Electricity data are half-hourly from the National Grid [1] and are linearly interpolated to the average residential monthly consumption. Residential daily gas consumption is estimated from National Grid [2] statistics of customers with low usage.)*

boilers with a fuel conversion efficiency of up to 90%, the resulting intraseasonal variations would still be much more pronounced and overall electricity demand much higher in this scenario [3] (here COP stands for coefficient of performance, which is a measure of the efficiency of the heat pump; a COP of 3 means that each unit of input electricity produces three units of output heat on average).

The sizes of the deficits and surpluses also depend to some extent on whether renewable outputs are correlated with demands. Wind generation tends to be higher in winter than summer so is better correlated to winter peak demands in high-latitude countries, while solar generation is much higher in summer daytime and is most closely correlated to lower latitude countries, where peak demand occurs in summer daytime for air conditioning. Germany provides a good example of some of the challenges that can emerge. Renewable generation accounted for 23% of German electricity generation in 2012, and a strong expansion of solar photovoltaics in southern Germany in particular has led to some areas already producing more electricity than they consume [4]. The two-way electricity flows in these areas have resulted in some distribution networks already operating at their technical limit, as these have historically been designed to transfer electricity in only one direction, from transmission networks to end-users.

2.2 Strategies to Cope with Electricity System Imbalances

Four principal strategies have been proposed to manage electricity deficits and surpluses [5]:

1. Dispatchable generation. Flexible generators are used to avoid deficits. The main disadvantages of this existing approach to imbalances are the high capital cost of generation capacity and the lack of a strategy to benefit from electricity surpluses.
2. Transmission and distribution network reinforcement. By increasing network capacity, this strategy enables greater movement of electricity *in space*, so supply and demand are averaged over larger areas which is likely to lead to lower imbalances. The proposed European Supergrid is an example of network enhancement that would use Scandinavian hydropower to balance renewable generation across Europe [6].
3. Demand-side management. Agreements with large electricity consumers are already used in some countries to reduce demand at peak times. In the future, demand-side response (DSR) technologies could be used by electricity system operators to shift demand from peak to non-peak periods, for example, by changing refrigerator or water-heating patterns in homes.
4. Energy storage deployment. Some storage technologies, such as power-to-power and power-to-heat with storage, could manage both electricity deficits and surpluses. Others, such as producing hydrogen for use outside these sectors, would only address surpluses.

A schematic of the relationships between these technologies is shown in Fig. 1.3. Most energy storage studies have examined grid-scale storage

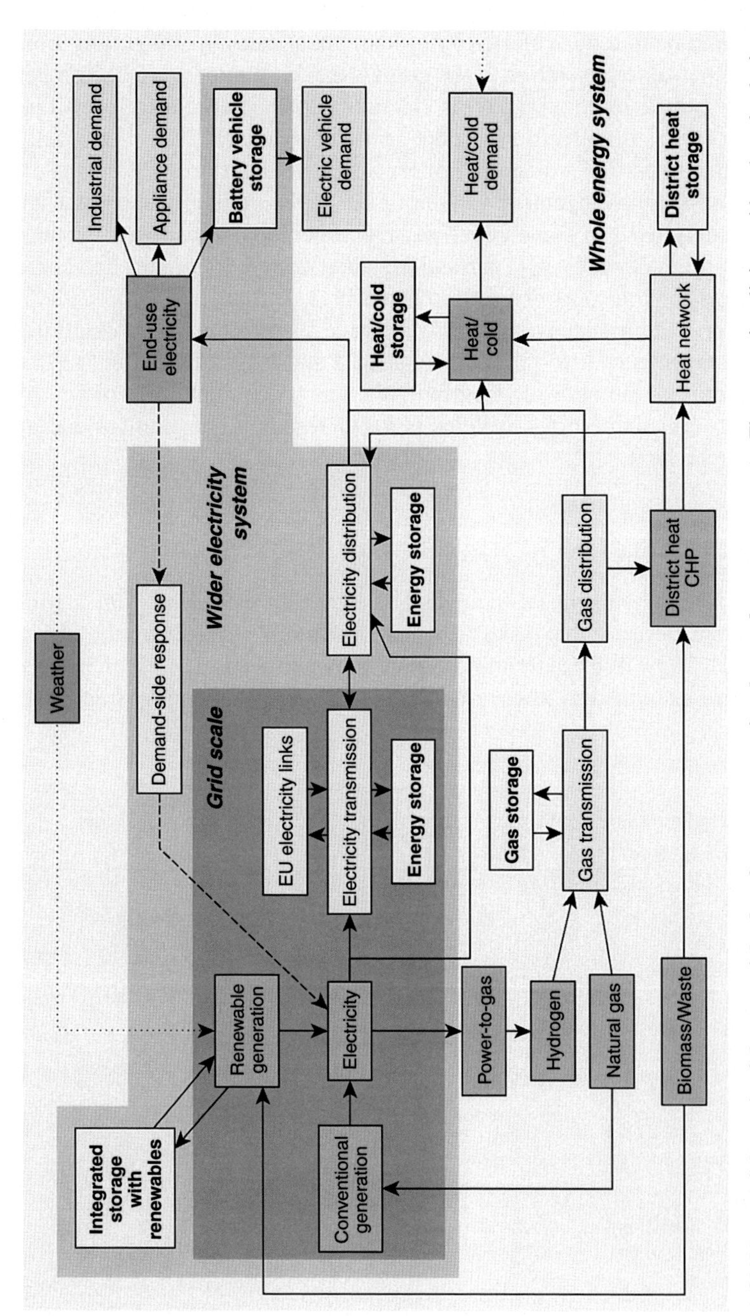

FIGURE 1.3 Schematic of the potential roles of energy storage in a low-carbon energy system. The system is split into grid-scale technologies, the wider electricity system and the whole energy system. Network and storage technologies (denoted with bold text) are integrated throughout the energy system.

or, at most, the electricity system in isolation (e.g., [7]). Yet energy storage could be integrated much more widely across the energy system. One approach would be to reduce electricity system imbalances by integrating storage with renewable generation at the point of generation (e.g., [8]). Another strategy would be to convert excess electricity into other storage fuels that are not used to return electricity to the grid. For example, excess electricity could produce heat for storage in boilers at district heat or building scales, in a demand-side management version of the night storage heaters that are already widely-used in some countries. Hydrogen could be stored for later electricity production [9,10], particularly interseasonal storage, but could also be used for transport or heat provision [11]. Although battery vehicles primarily provide transport services, vehicle-to-grid technologies could use car batteries for power-to-power storage in a smart grid [12]. Finding the most appropriate methods of integrating the many different types of energy storage into existing energy systems is a key research question for energy system researchers.

3 STORAGE TECHNOLOGIES

Numerous energy storage technologies are under development, with a wide range of characteristics that make them suitable for different roles in the energy system [13]. Many of these technologies are shown in Table 1.1, which lists the technologies examined in this book by category. One important characteristic

TABLE 1.1 List of Energy Storage Technologies that are Examined in this Book, by Category

Mechanical/thermomechanical/ gravitational: • Pumped hydro • Ground-breaking energy storage (GBES) • Advance rail energy storage (ARES) • Compressed air energy storage (CAES) • Pumped hydro with compressed air • Flywheels • Liquid air energy storage (LAES) • Heat-pumped temperature difference Thermal: • Sensible thermal energy storage • Latent heat storage • Solar ponds	Electrochemical: • Rechargeable batteries (e.g., lead–acid, lithium–ion, sodium–sulfur) • Flow batteries • Supercapacitors Chemical: • Reversible endothermic chemical reactions • Power-to-gas • Large-scale hydrogen storage • Traditional energy storage (natural gas, oil, and coal)

for comparing systems is the roundtrip efficiency, which is a measure of the overall loss of electricity from storage in power-to-power systems. Other characteristics and metrics for comparing energy storage systems are discussed in Section 4.

3.1 Gravitational/Mechanical/Thermomechanical

The principal power-to-power energy storage technology in operation around the world is pumped hydroelectricity, in which excess electricity is used to pump water to a high reservoir where it is stored as gravitational potential energy (see chapter 2: Pumped Hydroelectric Storage). Pumped hydro can produce a high power output at short notice so is normally used to meet intraday peak demands. A roundtrip efficiency of 80% can be achieved but the number of suitable sites for building schemes is limited. A novel derivative of pumped hydro is ground-breaking energy storage (GBES), in which a large mass (e.g., a concrete disk) is raised or lowered hydraulically during electricity surpluses and deficits. The aim is to create a smaller, cheaper system than pumped hydro, which is not limited to a small number of available sites. Gravitational schemes using solid rather than liquid masses are also under development. These novel hydro schemes are discussed in chapter 3: Novel Hydroelectric Storage Concepts.

Advanced rail energy storage (ARES) uses surplus electricity to power a heavy electric train to a high elevation (see chapter 4: Advance Rail Energy Storage (ARES)). At times of high demand, the train is returned to the lower elevation and generates electricity on the way. The advantages of this technology are that there are no energy losses over time once the train has reached high elevation, and it is particularly suited for dry regions where little water is available and high evaporation would cause hydroelectric reservoir losses.

Mechanical storage converts surplus electrical energy into potential or kinetic energy. The two principal commercial technologies are compressed air energy storage (CAES) and flywheel storage. CAES stores compressed air in constant volume or constant pressure storage vessels (see chapter 5: Compressed Air Energy Storage (CAES)). Storage mediums can include salt caverns, aquifers, or purpose-built pressure vessels. Underground storage (see chapter 6: Compressed Air Energy Storage (CAES) with Underground Storage) and undersea energy bags (see chapter 7: Underwater Compressed Air Energy Storage (CAES)) could be integrated with offshore wind generation. A novel hybrid energy storage system has been proposed that combines CAES and pumped hydro (see chapter 8: Pumped Hydro Combined with Compressed Air). The principal challenge for CAES is to conserve heat energy produced during compression in order to maximize roundtrip energy efficiency.

Novel thermomechanical storage technologies are under development with the aim of improving roundtrip energy efficiency by combining mechanical and thermal approaches. Liquid air energy storage (LAES), also known as cryogenic energy storage, is an alternative to CAES in which surplus electricity is used to cool air until it liquefies (see chapter 9: Liquid Air Energy Storage: (LAES)). The liquid air is stored in a tank and electricity is generated when required by warming the air until it expands into a gaseous state and turns a turbine. Since capacity and energy are decoupled, LAES is particularly suited to long-duration applications. Another thermomechanical scheme is heat-pumped temperature difference, which uses a reversible heat pump to store energy in the form of a temperature difference between two heat stores. For example, Isentropic have developed a system with a hot vessel storing thermal energy at high temperature and high pressure, and a cold vessel storing thermal energy at low temperature and low pressure [14]. Both vessels are filled with crushed rock or gravel, which acts as the heat storage medium. The whole system is filled with argon that is pumped between vessels if there are electricity surpluses or deficits, in a heat pump system, with a claimed roundtrip efficiency of up to 80%.

Flywheels use surplus electricity to accelerate a rotor to very high speeds and to maintain these speeds, maintaining the energy as rotational energy (see chapter 10: Flywheels). Rotational speed is reduced as energy is extracted from the system due to electricity generation. The principal challenge is to minimize friction losses in order to maximize roundtrip energy efficiency, so flywheels are most suited to short-duration applications.

3.2 Electrochemical

Rechargeable batteries are widely used in transport and electronic devices but have had limited deployment for grid-scale energy storage, with the 300 MW sodium–sulfur investment in Abu Dhabi a notable exception [15]. Common battery types include lead–acid, nickel–metal hydride, sodium–sulfur, and lithium–ion, and the latter is discussed in detail in chapter 11: Rechargeable Batteries. Larger megawatt-scale systems use banks of batteries arranged in racks, so cost reductions through economies of scale are lower than for gravitational and mechanical systems. Key issues for batteries are high capital costs, material availability, loss of charge when idle, and loss of capacity over time.

A flow battery is a type of rechargeable battery in which two chemical components are dissolved in liquids separated by a membrane. Ion exchange occurs across the membrane, meaning that the battery is similar to a fuel cell from a technical perspective. Flow batteries have longer durability than conventional batteries but reducing capital costs is a key challenge. Vanadium redox flow batteries are examined in chapter 12: The Vanadium Redox Flow Batteries.

Supercapacitors bridge the gap between conventional capacitors and rechargeable batteries. They use electrostatic double-layer capacitance that can hold up to 10 000 times the charge of a conventional solid dielectric capacitor, but they still have a much lower energy density than batteries. Supercapacitors are most appropriate for applications requiring many rapid charge/discharge cycles, for example, regenerative braking in vehicles or improving power quality for electric grids.

3.3 Thermal

Thermal storage involves the storage or removal of heat for later use. Sensible thermal energy storage heats or cools a liquid or solid storage medium, mostly commonly water (see chapter 15: Sensible Thermal Energy Storage: Diurnal and Seasonal). For example, many buildings have hot-water storage that could in the future be used as a sink for surplus renewable electricity. Larger storage tanks are often used to support district heat schemes. Large-scale seasonal heat storage has also been proposed, using purpose-built plants or aquifers; on long timescales, the rate of heat loss is an important determinant of the value of the plant.

Latent heat storage uses phase change materials to store heat through the reversible conversion from solid to liquid phases (see chapter 13: Phase Changes). The principal advantage of latent heat storage over sensible heat storage is that energy is stored at the temperature of the process application. A range of inorganic, organic, and bio-based materials have been developed with different characteristics.

Solar ponds are saltwater pools that act as solar thermal energy collectors (see chapter 14: Solar Ponds). Pond salinity prevents water from flowing from the bottom to the top of the pond, meaning that temperatures are much higher at the bottom of the pond and the trapped heat can be used for heating buildings or hot water, particularly in industry, or to drive a turbine or Stirling engine to generate electricity.

3.4 Chemical

Hydrogen is a potential storage vector. As a zero-carbon energy carrier, it could have a similar role in a low-carbon energy system to electricity, but has the key advantage that it is much easier and cheaper to store (see chapter 20: Large Scale Hydrogen Storage). Large underground cavern storage of hydrogen is one of the few low-carbon interseasonal energy storage solutions that could support electrification of heat demand [9,11,16]. Power-to-power systems have been proposed [17], with hydrogen produced by electrolysis (see chapter 16: Hydrogen from Water Electrolysis). Power-to-gas uses surplus electricity to produce hydrogen that is then injected into the natural gas network (see chapter 18: Power to Gas), using surplus power but not contributing to meeting deficits. The tight tolerance of gas appliances and the different characteristics of hydrogen compared with natural gas mean that it can only supply (1–6)% of the total gas by energy content [18]. This could be increased by methanating the hydrogen using waste CO_2 from an industrial process, but at a cost and with an energy efficiency penalty. Otherwise, converting the gas networks to deliver hydrogen would avoid this issue [19]. Hydrogen could also provide flexible generation for peak energy demand, for example, by producing hydrogen from fossil fuels that is stored in salt caverns until required [20], but such a system would not utilize surplus power from renewables.

New materials are being developed that store heat using reversible endothermic chemical reactions (see chapter 17: Chemical Reactions (zeolites/water/inorganic oxides)). Aluminosilicate minerals called "zeolites" absorb water in an endothermic reaction. The water is desorbed when the zeolite crystals are heated, and long-term losses of this stored heat are negligible as long as water is not present. Moreover, the energy density is higher than for sensible or latent heat storage.

With the exception of pumped hydroelectricity, most existing energy storage is in the form of fossil fuels (see chapter 19: Traditional Energy Storage: natural gas, oil and coal). Contemporary energy systems store precursors to electricity rather than building power-to-power storage. Coal can be piled while oil is cheap to store in reservoirs. Gas is generally stored in underground caverns or storage holders. All of these methods are cheaper than the other storage technologies discussed in this book.

4 COMPARING STORAGE SYSTEMS

The numerous energy storage technologies reviewed above have a wide range of characteristics that affect their suitability for different roles in low-carbon energy systems. The International Energy Agency (IEA) [13] categorizes technologies according to the range of time periods over which their charge/discharge cycles can operate. Barton and Infield [21] similarly identify storage durations for contributions to the electricity system from 20 s to 4 months, with different technologies able to operate over different time periods. The key characteristics chosen by these studies to compare technologies are listed in Table 1.2, together with the characteristics chosen in a review of power-to-power systems by Luo et al. [22].

This list of characteristics is by no means exhaustive. Numerous other metrics could be used to compare technologies, including minimum natural energy and power scales of a single device, optimum natural energy and power scales of a single device, nominal cost per unit energy and power at optimum scale, marginal cost per unit energy and power at optimum scale, lowest power slew rate at which performance degrades noticeably, and effective turnaround efficiency. Further metrics are required for specific technology types (e.g., operating temperature for latent heat storage) and for energy storage technologies that do not provide a power-to-power service.

The relative importance of each characteristic depends on the technology application. It is necessary to identify uncertainties in each parameter and to consider the relative importance of these uncertainties on technology performance

TABLE 1.2 Key Characteristics Used to Compare Energy Storage Technologies in Studies by the IEA [13], Barton and Infield [21], and Luo et al. [22]

	IEA	B&I	Luo
Storage duration	×	×	×
Typical size	×		×
Charge duration		×	
Discharge duration	×	×	×
Cycles	×		×
Response time	×		×
Roundtrip efficiency		×	×
Discharge efficiency			×
Daily self-discharge			×
Energy and power density			×
Specific energy and power			×
Maturity			×
Energy and power capital costs		×	×
Operating and maintenance costs			×

and value. The studies examined in Table 1.2 provide a valuable comparison of technologies but there is still a need for a more exhaustive comparison using a wider range of metrics. The UK-funded Realising Energy Storage Technologies in Low-Carbon Energy Systems (RESTLESS) project is currently producing such a comparison.

5 CHALLENGES FOR ENERGY STORAGE

Energy storage technologies are among the most complicated and least well–understood low-carbon technologies. They are arguably underresearched compared with other low-carbon technologies. For example, the Global Energy Assessment has little consideration of individual energy storage technologies, yet notes that, "providing integrated and affordable energy storage systems for modern energy carriers is … perhaps the largest and most perplexing part of the energy systems for a sustainable future that is needed for future economic security" [23].

As suggested by this statement, integrating storage into evolving energy systems is a key challenge that is not well understood, partly because there is currently no energy model that can fully represent the benefits of different types of energy storage across an energy system. There are a number of economic, social, and regulatory barriers to energy storage deployment. The capital costs of most energy storage technologies are thought to be too high to justify their deployment at present and there is a need for innovation to reduce these costs. Public acceptance of storage technologies is important but has received little investigation. There is a need to find the most appropriate roles for different storage technologies, based on the value that each technology offers and taking into consideration barriers to each technology. Finally, existing energy markets are not designed to realize the value of energy storage to the energy system and would need to be redesigned to reflect this value. These barriers are explored in this section.

5.1 Integrating Energy Storage into Low-Carbon Energy Systems

The potential importance of novel energy storage technologies for low-carbon energy systems is uncertain for several reasons. First, the optimal amount of storage depends on the amount of flexible generation in the overall electricity generation portfolio and the magnitude of demand peaks. While it is straightforward to show that storage has positive economic benefits for a very inflexible system, for example, [7], it is unlikely that such a portfolio of generation technologies would be intentionally constructed. Second, there are tradeoffs between energy storage and alternatives such as network reinforcement, including connections between national electricity systems, and demand-side management, as described in Section 2.2. Third, the number and diversity of novel energy storage technologies makes it challenging to identify their most appropriate roles in supporting different low-carbon electricity systems. Fourth, the role of storage could change if it is not viewed as being an independent system to the generation technology.

5.1.1 Generation-Integrated Energy Storage

For energy storage that is associated with supporting electricity generation, most assume that this is power-to-power storage that involves converting energy from electricity to some storable form and back again. However, there are two other broad categories. Generation-integrated energy storage (GIES) systems store energy before electricity is generated. Load-integrated energy storage (LIES) systems store energy (or some energy-based service) after electricity has been consumed (e.g., power-to-gas, with hydrogen stored prior to consumption for transport or another end-use). GIES systems have received little attention to date but could have a very important role in the future [24].

As mentioned in Section 1, most countries store precursors to electricity in the form of raw or processed hydrocarbons, whether in coal and biomass heaps or oil and gas reservoirs. These are examples of GIES technologies, and offer two general advantages over other storage systems:

- There may be low (or even zero) marginal costs associated with the extra equipment or infrastructure required to enable energy to be put into storage.
- There may be low (or even zero) marginal losses of energy associated with passing energy through storage.

Many instances of GIES systems are already in existence. All natural hydropower plant with a dam fall into the GIES category. Such plants accumulate energy in the form of gravitational potential energy of water; their energy stores are filled up as a result of rain but they can generate electricity when no rain is falling. It is conceivable that some new nuclear power stations could be equipped with thermal energy storage (as indicated in [25]) so that the reactors would run at a constant power rate, but electricity would be generated to match demand, and these nuclear power stations would be GIES systems. Fourth-generation high-temperature nuclear power plants could dissociate water directly into hydrogen and oxygen [26], and would also share this classification.

Selected concentrated solar power plants such as the Andasol and Gemasol plants [27] are equipped with thermal energy storage, enabling heat to be stored prior to its use in raising steam to drive electrical generators, and these are also GIES systems. Another arrangement for converting solar power into an immediately storable energy form involves photolysis, in which water is split using photons [28,29].

There are several mechanical renewable energy devices for harvesting wind, wave, and tidal power by directly compressing air. Some of these are discussed in chapter 5: Compressed Air Energy Storage (CAES) and all fall within the GIES aegis. Finally, there are several possible configurations of equipment that exploit the interaction between mechanical work and heat to integrate energy storage with devices that collect renewable energy directly [30,31]. Fig. 1.4 outlines a potential GIES system for wind generation.

5.1.2 Analyzing Energy Storage Integration Using Models

One method to better understand the relative benefits of different technologies is to compare them using a wide range of operational, economic, and environmental metrics, as described in Section 4, and this is undoubtedly important. Another is to compare a small number of technologies for a particular energy system, for example, [32]. But the most common method to understand how these technologies might be integrated into a low-carbon energy system is to explore scenarios using models. These can give insights over the transition to a low-carbon system about which technologies are most economically deployed, how the level of deployment might change over time, and where the technologies are best deployed. Several model paradigms can give useful insights:

- Energy system models: market-based economic optimization models that represent all energy flows and GHG emissions in an economy and are used to examine how energy systems might evolve, at least cost, to meet long-term emission targets. Demands are specified for energy services and the contribution of individual fuels to meeting these is optimized, meaning that the demand for electricity evolves over time depending on the relative competitiveness of electricity generation against alternative energy vectors. This practically means that these models can compare different methods of

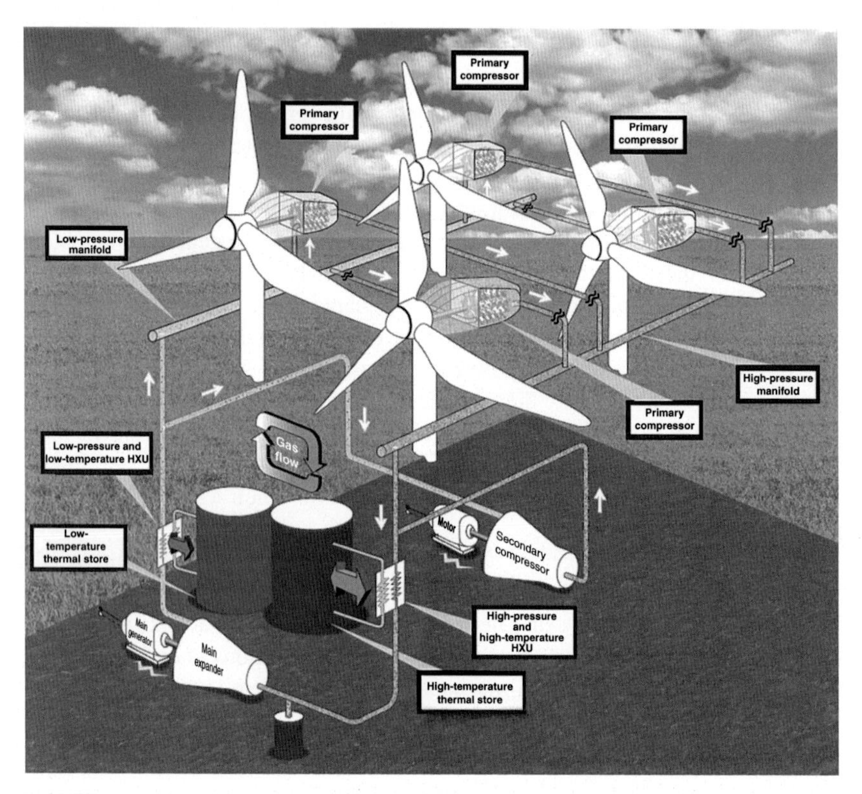

FIGURE 1.4 **Schematic of a GIES system suitable for wind generation [30].**

balancing supply and demand (network reinforcement, energy storage, etc.) or can construct the energy system to minimize imbalances if all of these options are too expensive. For example, one storage option would be to use excess electrical generation to produce hydrogen for transport or heat for buildings, rather than deploying power-to-power storage. Energy system models can be used to compare all types of energy storage, on different timescales, but they tend to have low spatial and temporal resolution, meaning the need for and the value of energy storage is generally underestimated.

- Electricity dispatch models: market-based electricity system models that calculate the merit order of electricity generation in a mixed portfolio to meet evolving demands. Electricity demands are fixed in each time period, meaning that in contrast to energy system models, dispatch models cannot consider options to change the electricity demand by using alternative energy vectors. Energy storage can be simulated by more advanced dispatch models, but principally grid-scale power-to-power storage. DSR has also been simulated, and simplistic network reinforcement can be investigated using multiregion dispatch models. The principal advantages of dispatch models over energy system models are, first, that the temporal resolution is

much higher, typically 30 min. compared with 6 h, so supply/demand imbalances are better represented, and, second, that network reliability requirements such as loss of load can be set by the modeler.
- Electricity network models: detailed spatial models of electricity transmission or distribution networks, to examine the performance of networks in meeting peak loads. These can provide detailed insights into the relative benefits of grid-scale energy storage and network reinforcement, and can be used to identify the most appropriate locations to deploy different energy storage technologies. Chapter 21: Network Integration and Smart Grids examines the integration of energy storage with transmission and distribution networks in smart grids.

Most energy storage integration studies have used only electricity dispatch models, considering grid-scale or very occasionally the wider electricity system shown in Fig. 1.3. Since these models do not consider the whole energy system, these studies necessarily exclude some storage options and make broad assumptions about electricity demand. Yet while energy system models do not have these drawbacks, their low temporal resolution causes them to underestimate the value of energy storage. Moreover, different energy storage technologies work at different scales and it is difficult to construct a model that can consider the relative benefits of in-house, distribution, and grid-scale storage. No single model is capable of holistically assessing the value of all energy storage technologies. There is a need for studies that combine these three types of model to understand the whole system shown in Fig. 1.3, which could then produce more credible scenarios for energy storage within the context of energy system transitions.

5.2 Innovation to Reduce Technology Costs

The Global Energy Assessment identifies affordability as a key challenge for energy storage systems [23]. Although most energy storage research is targeted at the technologies discussed in Section 3, they all currently have high costs per unit energy relative to existing energy storage and none are currently commercially competitive.

Innovation is required to reduce capital costs and improve the performance of key technologies, but the rate of technology cost reduction is generally related to the level of technology deployment [33], so the lack of an existing market is an impediment to energy storage having a wider role in the future. Moreover, as energy storage technologies are so numerous and so diverse, deployment rates could be low even as demand increases, particularly for larger devices, which could inhibit cost reduction. One promising niche area for early deployment of energy storage is in off-grid applications, where storage has a higher value for smaller, more constrained energy systems (see chapter 22: Off-Grid Energy Storage). Various types of energy storage are now being developed and deployed globally (see chapter 23: Energy Storage Worldwide). A particular source of innovation is likely to be China, which has a rapidly-evolving electricity system and strong investment in infrastructure and industrial development (see chapter 24: Energy Storage in China).

5.3 Public Acceptance

Public engagement with energy supply and demand technologies has been identified as a critical issue for the future deployment of innovative and low-carbon energy systems [34], but there is a dearth of knowledge on public attitudes toward energy storage technologies and the roles that they might have in future energy systems. There are difficult conceptual and methodological challenges to such research, including the need to integrate social science methods with appropriate technical descriptions of the technologies, in order to engage with people about technologies for which they may be unfamiliar. Deliberative methods have been used to achieve similar goals for other energy technologies [35] and a similar approach could be used to identify energy storage technologies that are fully responsive to societal views. Public acceptance is particularly important for in-house storage technologies (e.g., heat storage). Since network reinforcement and demand-side management are potential alternatives to energy storage, it would be useful to compare the societal views of these three approaches with balancing flows in the energy system.

5.4 Finding the Most Appropriate Roles for Energy Storage Technologies

The most appropriate roles for each energy storage technology depend foremost on the design of the electricity system, particularly the fraction of inflexible renewable and nuclear generation and the fraction of flexible peak generation. Other important factors include how the technologies are integrated with the electricity system from an engineering perspective, the value of each technology relative to the cost of deploying it, accounting for potential cost reductions in the future, and social acceptance of preferred technologies.

Identifying the most appropriate roles is therefore a difficult challenge. The integrated modeling approach in Section 5.1.2 could provide insights, if cost innovation and social barriers were incorporated into "integrated" scenarios. Yet other than pumped hydro, most technologies are currently at the demonstration stage and there are no broad guidelines available about the suitability of different technologies for particular situations. Some technologies have been tested at large scale (100–1000 MW), but the future performance of most technologies is not well understood. Novel approaches to integration are still under development, for example, the GIES systems discussed in Section 5.1.1. GIES systems tend to perform especially well when a high proportion of all energy passing through these systems also passes through their internal storage [24], so their utility again depends on the wider electricity system configuration. There is a need for R&D programs to further develop such novel approaches and to test a range of technologies at scale so their operating characteristics and costs can be better understood.

A key challenge for finding the most appropriate roles for energy storage is how to cope with future uncertainty, both in terms of the electricity system and the technologies themselves [36]. Roadmaps for energy storage are occasionally published, for example, [13], based on a long-term vision, but these do not generally consider uncertainties in the costs and the value of energy storage technologies, nor whether energy storage investments can be justified now to keep the option open for their use in the future. A real options approach could be used to assess the option value of energy storage in an uncertain future.

5.5 Adapting Energy Markets to Realize the Value of Energy Storage

Even if the value of energy storage could be demonstrated, the technologies would not currently be viable in most countries because their electricity markets are designed for incumbent systems in which supply is varied to meet demand. It would be necessary to adapt market regulation to reflect the technological, economic, and social value of energy storage to an energy system. Some countries have taken first steps in this direction, for example, through the development of an electricity capacity market in the United Kingdom. Chapter 25: Politics of Investing in Renewable Energy Systems examines the politics of investing in renewable energy systems.

As with many emerging technologies, there is a need for incentives to encourage energy innovation in the short term and also investments that reflect the long-term value of energy storage, through the design of both electricity markets and business models. One approach would be to provide incentives from capacity markets and feed-in tariffs in the short term to encourage the deployment of new technologies, and subsequent innovation. Some experimentation would likely be required to identify optimal market structures.

End-users could potentially reduce bills or offer balancing services to the electricity grid through storage provision or demand flexibility. This means that retail as well as wholesale markets could need to be adapted to enable the integration of energy storage at different scales, in order to reward end-users for the value of the balancing services that they provide to the electricity system. Other barriers, such as the deployment of suitable smart meters, might also need to be overcome.

6 CONCLUSIONS

Energy storage makes an important contribution to energy security. Most contemporary storage systems are based around fossil fuels but novel energy storage technologies could make an important contribution to future low-carbon energy systems, particularly in the event of heat and transport electrification or if intermittent renewables and other inflexible low-carbon technologies come to dominate electricity generation. Energy storage would be in competition with existing system control strategies, notably dispatchable generation and network reinforcement, as well as newer strategies such as DSR.

Numerous energy storage technologies have been proposed to store excess electricity, with electrical energy conversion to mechanical, thermal, gravitational, electrochemical, and chemical energy for storage. Pumped hydro is the only widely-used electricity storage technology at the moment. As storage technologies have a wide range of characteristics that affect their suitability for different roles in low-carbon energy systems, a series of energy storage metrics have been proposed by different studies for comparison purposes, but these are by no means comprehensive and there is still a need for a more exhaustive comparison using a wider range of metrics.

Energy storage technologies are complicated and poorly understood relative to most low-carbon technologies. Understanding how to integrate energy storage into low-carbon energy systems is a difficult challenge for several reasons. First, the proportion of inflexible generation in the electricity system affects the value of energy storage to the system; if the cost of energy storage were too high then the proportion of inflexible generation could be reduced. Second, existing energy system, dispatch, and network models are either not broad enough to examine all energy storage and alternative options, or have insufficient temporal resolution to realistically portray the need for and performance of storage technologies. This means that the optimum approaches to integrating novel energy storage technologies into low-carbon energy systems cannot be fully understood using existing models. Third, the costs of most energy storage technologies are currently too high. Innovation is required to reduce costs but the rate of cost reduction generally depends on the level of technology deployment. Fourth, there is a dearth of knowledge on public attitudes toward energy storage technologies that could be particularly important for decentralized systems, particularly in-house systems. Finally, even if the long-term value of energy storage could be demonstrated, existing electricity markets are designed for incumbent systems and market regulation would need to be adapted to reflect the technological, economic, and social value of energy storage to an energy system.

A whole energy system approach to energy storage is necessary to fully understand how different technologies might contribute as innovation reduces costs during transitions toward low-carbon energy systems. Studies have shown that the value of energy storage in some electricity system configurations is substantial. It is likely that the economic value of the difference between good and bad policy decisions relating to the role of energy storage in the transition to low-carbon generation is in the order of billions of dollars. Further R&D and a better understanding of the integration of these technologies is vital to provide information to underpin future market design and regulation to realize the value of energy storage.

REFERENCES

[1] National Grid. Electricity transmission operational data: historical demand data, 2015. Available from: http://www2.nationalgrid.com/UK/Industry-information/Electricity-transmission-operational-data/Data-explorer/

[2] National Grid. Gas seasonal and annual data, 2013 Available from: http://www.nationalgrid.com/uk/Gas/Data/misc

[3] Wilson IAG, Rennie AJR, Ding Y, Eames PC, Hall PJ, Kelly NJ. Energ Policy 2013;61:301–5.

[4] Dodds PE, Fais B. Network infrastructure and energy storage for low-carbon energy systems. In: Ekins P, Bradshaw M, Watson J, editors. Global energy: issues, potentials and policy implications. Oxford, UK: Oxford University Press; 2015. p. 426–51.

[5] Coalition study. Commercialisation of energy storage in Europe. McKinsey & Company, 2015. Available from: http://www.fch.europa.eu/sites/default/files/CommercializationofEnergyStorageFinal_3.pdf

[6] Energy and Climate Change Committee. A European supergrid. UK Parliament, London, UK, 2011. Available from: http://www.publications.parliament.uk/pa/cm201012/cmselect/cmenergy/1040/104002.htm

[7] Strbac G, Aunedi M, Pudjianto D, Djapic P, Teng F, Sturt A, Jackravut D, Sansom R, Yufit V, Brandon N, 2012. Strategic assessment of the role and value of energy storage systems in the UK low carbon energy future. Imperial College London, UK. http://www.carbontrust.com/media/129310/energy-storage-systems-role-value-strategic-assessment.pdf

[8] Pimm AJ, Garvey SD, de Jong M. Energy 2014;66:496–508.

[9] Lohner T, D'Aveni A, Dehouche Z, Johnson P. Int J Hydrogen Energ 2013;38:14638–53.

[10] Anderson D, Leach M. Energ Policy 2004;32:1603–14.

[11] Dodds, PE. A whole systems analysis of the benefits of hydrogen storage to the UK. World Hydrogen Technologies Convention 2013, Shanghai, China, 2013.

[12] Wade NS, Taylor PC, Lang PD, Jones PR. Energ Policy 2010;38:7180–8.

[13] IEA. Energy storage technology roadmap. International Energy Agency, Paris, France, 2014.

[14] Isentropic. How Isentropic PHES works. Isentropic, Fareham, UK, 2015.Available from: http://www.isentropic.co.uk/our-phes-technology.

[15] Decourt, B, Debarre, R. Electricity storage. SBC Energy Institute, The Hague, The Netherlands, 2013. Available from: https://www.sbc.slb.com/SBCInstitute/Publications/ElectricityStorage.aspx

[16] Maton J-P, Zhao L, Brouwer J. Int J Hydrogen Energ 2013;38:7867–80.

[17] Carr S, Premier GC, Guwy AJ, Dinsdale RM, Maddy J. Int J Hydrogen Energ 2014;39:10195–207.

[18] Dodds PE, McDowall W. Energ Policy 2013;60:305–16.

[19] Dodds PE, Demoullin S. Int J Hydrogen Energ 2013;38:7189–200.

[20] ETI. Hydrogen: the role of hydrogen storage in a clean responsive power system. Energy Technologies Institute Loughborough, UK, 2015. Available from: http://www.eti.co.uk/wp-content/uploads/2015/05/3380-ETI-Hydrogen-Insights-paper.pdf

[21] Barton JP, Infield DG. IEEE T Energ Conver 2004;19:441–8.

[22] Luo X, Wang J, Dooner M, Clarke J. Appl Energ 2015;137:511–36.

[23] Johansson TB, Nakicenovic N, Patwardhan A, Gomez-Echeverri L. Summary for policymakers: global energy assessment: toward a sustainable future. International Institute for Applied Systems Analysis, Laxenburg, Austria, 2012.

[24] Garvey SD, Eames PC, Wang JH, Pimm AJ, Waterson M, MacKay RS, Giulietti M, Flatley LC, Thomson M, Barton J, Evans DJ, Busby J, Garvey JE. Energ Policy 2015:86:544–51.

[25] Denholm P, King JC, Kutcher CF, Wilson PPH. Energ Policy 2012;44:301–11.
[26] Verfondern K. Nuclear energy for hydrogen production. Energy Technology vol. 58, Forschungszentrum Jülich, Germany, 2007. Available from: http://juser.fz-juelich.de/record/58871/files/Energietechnik_58.pdf
[27] Dunn RI, Hearps PJ, Wright MN. Proc IEEE 2012;100:504–15.
[28] Kibria MG, Chowdhury FA, Zhao S, AlOtaibi B, Trudeau ML, Guo H, Mi Z. Nat Commun 2015;6.
[29] Ismail AA, Bahnemann DW. Sol Energ Mat Sol C 2014;128:85–101.
[30] Garvey S, Pimm A, Buck J, Woolhead S, Liew K, Kantharaj B, Garvey J, Brewster B. Wind Eng 2015;39:149–74.
[31] Lee JE. On-demand generation of electricity from stored wind energy. USA patent, Application number US2,012,326,445–A1, 2012.
[32] Karellas S, Tzouganatos N. Renew Sust Energ Rev 2014;29:865–82.
[33] Grübler A, Nakićenović N, Victor DG. Energ Policy 1999;27:247–80.
[34] Whitmarsh L, Upham P, Poortinga W, McLachlan C, Darnton A, Devine-Wright P, Demski C, Sherry-Brennan F. Public attitudes, understanding, and engagement in relation to low-carbon energy: a selective review of academic and non-academic literatures, 2011. Available from: http://psych.cf.ac.uk/home2/whitmarsh/Energy%20Synthesis%20FINAL%20(24%20Jan).pdf
[35] Pidgeon N, Demski C, Butler C, Parkhill K, Spence A. Proc Natl Acad Sci USA 2014;111:13606–13.
[36] Taylor PG, Bolton R, Stone D, Upham P. Energ Policy 2013;63:230–43.

Part B

Electrical Energy Storage Techniques Gravitational/Mechanical/ Thermomechanical

Chapter 2

Pumped Hydroelectric Storage

Chi-Jen Yang
Center on Global Change, Duke University, Durham, NC, United States

1 INTRODUCTION

Pumped hydroelectric storage (PHES) is the most widely adopted utility-scale electricity storage technology. Furthermore, PHES provides the most mature and commercially available solution to bulk electricity storage. It serves to stabilize the electricity grid through peak shaving, load balancing, frequency regulation, and reserve generation.

Japan currently has the largest installed PHES capacity in the world [1], followed by China [2] and the United States [3]. China currently has the most aggressive plan to expand PHES installation, with over 27-GW capacity under construction as of December 2015, and much more planned. China is expected to surpass Japan in installed PHES capacity by 2018. Table 2.1 shows the installed PHES capacities in major countries [2–5].

A PHES facility is typically equipped with pumps and generators connecting an upper and a lower reservoir (Fig. 2.1). The pumps utilize relatively cheap electricity from the power grid during off-peak hours to move water from the lower reservoir to the upper one to store energy. During periods of high electricity demand (peak hours and when electricity is expensive), water is released from the upper reservoir to generate power at a higher price.

There are two main types of PHES facilities: (1) pure or off-stream PHES, which rely entirely on water that was previously pumped into an upper reservoir as the source of energy; (2) combined, hybrid, or pumpback PHES, which use both pumped water and natural stream flow water to generate power [4]. Off-stream PHES is sometimes also referred to as "closed-loop" systems. However, some may define closed-loop systems more strictly as being entirely isolated from natural ecosystems. The US Federal Energy Regulatory Commission defines closed-loop pumped storage as projects that are not continuously connected to a naturally flowing water feature [5].

The efficiency of PHES varies quite significantly due to the long history of the technology and the long life of a facility. The roundtrip efficiency (electricity generated divided by the electricity used to pump water) of facilities with older designs may be lower than 60%, while a state-of-the-art PHES system may achieve over 80% efficiency.

Storing Energy. http://dx.doi.org/10.1016/B978-0-12-803440-8.00002-6

TABLE 2.1 Installed Pumped Hydroelectric Storage (PHES) Capacities

Country	Installed PHES capacity/MW
Japan	27 438
China	21 545
United States	20 858
Italy	7 071
Spain	6 889
Germany	6 388
France	5 894
India	5 072
Austria	4 808
Korea, South	4 700
United Kingdom	2 828
Switzerland	2 687
Taiwan	2 608
Australia	2 542
Poland	1 745
Portugal	1 592
South Africa	1 580
Thailand	1 391
Belgium	1 307
Russia	1 246
Czech Republic	1 145
Luxembourg	1 096
Bulgaria	1 052
Iran	1 040
Slovakia	1 017
Argentina	974
Norway	967
Ukraine	905
Lithuania	900
Philippines	709
Greece	699
Serbia	614
Morocco	465
Ireland	292
Croatia	282
Slovenia	185
Canada	174
Romania	53
Chile	31
Brazil	20

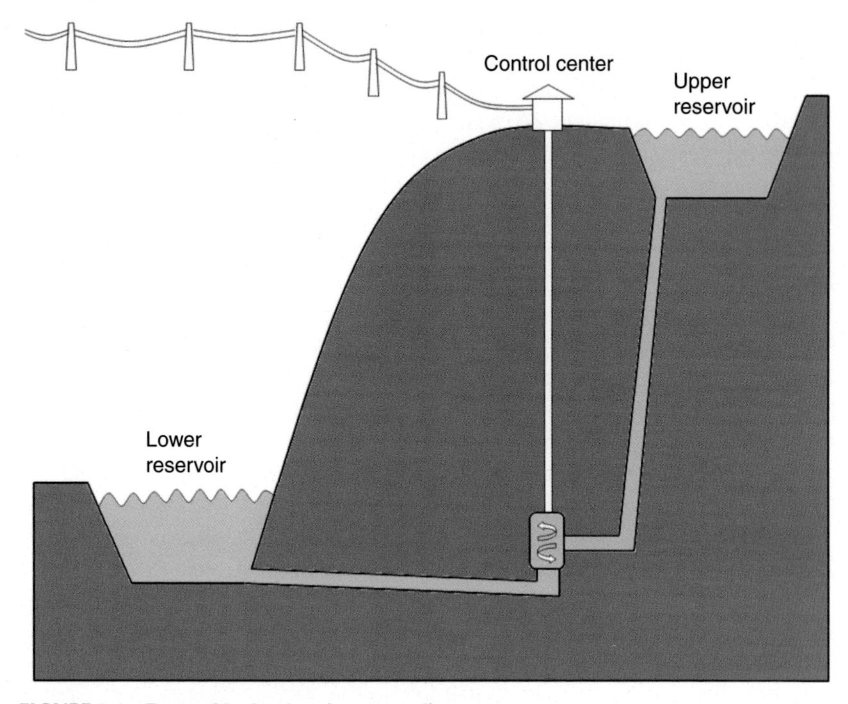

FIGURE 2.1 **Pumped hydroelectric storage diagram.**

2 PROS AND CONS

By storing electricity, PHES facilities can protect the power system from outages. Coupled with advanced power electronics, PHES systems can also reduce harmonic distortions and eliminate voltage sags and surges. Of all kinds of power generators, peak load generators typically produce electricity at much higher costs than base load ones. PHES provides an alternative to peaking power by storing cheap base load electricity and releasing it during peak hours.

PHES facilities provide very large capacities of electricity, with low operation and maintenance cost, and high reliability. The levelized storage cost for electricity using PHES is usually much lower than other electricity storage technologies.

There are several drawbacks to PHES technology. The deployment of PHES requires suitable terrains with a significant elevation difference between the two reservoirs and a significant amount of water resource. Typical PHES facilities have a hydraulic head of (200–300) m with reservoirs of volumes of the order of 10×10^6 m^3. The construction of a PHES station typically takes many years, sometimes over a decade. Although the operation and maintenance cost is very low, there is a high upfront capital investment in civil construction, which can only be recouped over decades.

Environmental impacts are also serious concerns and have caused many cancellations of proposed PHES projects. Conventional PHES construction sometimes involves damming a river to create a reservoir. Blocking natural water flows disrupts the aquatic ecosystem and the flooding of previously dry areas may destroy terrestrial wildlife habitats and significantly change the landscape. Pumping may also increase the water temperature and stir up sediments at the bottom of reservoirs and deteriorate water quality. PHES operation may also trap and kill fish. There are technologies to mitigate ecological impacts. Fish-deterrent systems could be installed to minimize fish entrapment and reduce fish kill. Water intake and outlet could be designed to minimize turbulence. An oxygen injection system could also compensate for potential oxygen loss due to warming of the water because of pumping. In some cases, the PHES system may serve to stabilize water level and maintain water quality [6]. The potential impacts of PHES projects are site specific and must be evaluated on a case-by-case basis. Governments usually require an environmental impact assessment before approving a PHES project.

Most PHES facilities have good safety records. Nevertheless, the upper-reservoir failure of Taum Sauk PHES station in Missouri (United States) should be heeded as a reminder of its potential danger. On Dec. 14, 2005 the upper reservoir of Taum Sauk was overfilled, and the embankment was then overtopped and breached. The suddenly released water washed away over 1 km^2 of forest, uprooted all the trees, and obliterated a house in its path. Although the Taum Sauk PHES station was later repaired and brought back to operation in 2010, the reservoir failure incident should provide important lessons for future PHES design, construction, and operation.

3 HISTORICAL DEVELOPMENT

The earliest PHES system in the world appeared in the alpine regions of Switzerland, Austria, and Italy in the 1890s. The earliest designs use separate pump impellers and turbine generators. Since the 1950s a single reversible pump turbine has become the dominant design for PHES [7]. The development of PHES remained relatively slow until the 1960s, when utilities in many countries began to envision a dominant role for nuclear power. Many PHES facilities were intended to complement nuclear power in providing peaking power.

In the 1990s the development of PHES significantly declined in many countries. Many factors may have contributed to the decline. Low natural gas prices during this period make gas turbines more competitive in providing peaking power than PHES.

Before the 1950s, most of the PHES facilities were located in Europe. The United States completed its first PHES station in 1928. Japan built its first PHES in 1934 and China in 1968. Since the 1950s the adoption of PHES has gradually spread all over the world. As of 2014, the US Department of Energy (DOE)

Global Energy Storage Database recorded over 300 operating PHES stations with a total capacity of 142 GW in 41 countries [8].

The regulations and financing of PHES facilities are greatly influenced by national policies. Japan, China, and the United States have the largest PHES capacities in the world. Tables 2.2–2.4 list the PHES facilities in Japan, China, and the United States. The distinctive policies and regulatory regimes for PHES in Japan, China, and the United States offer interesting contrasts and may reveal useful policy insights.

The buildup of PHES capacities in Japan has been relatively steady over several decades. The Japanese power sector is mainly composed of vertically integrated regional electric power utilities, which build, own, and operate the PHES facilities. The vertically integrated power sector structure has provided a stable and predictable business environment that is favorable to investments in PHES. The path of PHES development in Japan is the epitome of PHES development worldwide. Before the early 1960s, PHES facilities were rare and small, mostly of hybrid design. Deployment started to accelerate in the 1960s and continued throughout the 1990s. Since the 1970s, pure/off-stream PHES has become the dominant design, which is likely a result of increased concerns for the ecological impacts of hybrid systems. In addition to having the world's largest PHES capacities, Japan is also the world leader in employing seawater PHES and variable-speed PHES.

China is a latecomer in worldwide PHES deployment, but it is catching up quickly. With the largest PHES capacities planned and under construction, China will soon overtake Japan as host of the largest PHES capacities in the world. Most PHES facilities in China are relatively new, with large capacity and off-stream designs.

China's regulatory regime for PHES has been through great changes in the past two decades. Before 2004, most of the PHES facilities in China were built by local governments and local grid companies with diverse pricing models. In 2002, China restructured its power sector by separating them into two state-owned grid companies and five power generation corporations. In 2004 the National Development and Reform Commission promulgated a regulation which specified that PHES stations are transmission facilities and should be constructed and managed by the grid companies, and that the construction and operation costs of PHES should be incorporated into the operation costs of the grid companies [9]. The decision to treat PHES as transmission facilities has contributed to the rapid expansion of PHES in China.

Most of the PHES facilities in the United States were built in the 1970s and 1980s. Since the 1990s the construction of PHES slowed down in the United States. Environmental concerns caused the cancellation of several PHES projects and significantly prolonged the permit process. Power sector restructure also contributed to this slowdown. During the 1990s the United States started to restructure the power sector by separating generation from transmission. The nature of energy storage falls into the gray area between generation and

TABLE 2.2 Pumped Hydroelectric Storage Stations in Japan

Plant name (Japanese)	Plant name (English)	Location	Type	Rating/ MW	Commi-ssion year
池尻川	Ikejirigawa	Nagano Prefecture	Hybrid	2	1934
大森川	Omorikawa	Kochi Prefecture	Hybrid	12	1959
諸塚	Morotsuka	Miyazaki Prefecture	Hybrid	50	1961
畑薙第一	Hatakenagi No. 1	Shizuoka Prefecture	Hybrid	137	1962
三尾	Mio	Nagano Prefecture	Hybrid	36	1963
池原	Ikehara	Nara Prefecture	Hybrid	350	1964
穴内川	Ananaigawa	Kochi Prefecture	Hybrid	13	1964
城山	Shiroyama	Kanagawa Prefecture	Pure	250	1965
矢木沢	Yagisawa	Gunma Prefecture	Hybrid	240	1965
新成羽川	Shinnaruhagawa	Okayama Prefecture	Hybrid	303	1968
長野	Nagano	Fukui Prefecture	Hybrid	220	1968
蔭平	Kagetaira	Tokushima Prefecture	Hybrid	47	1968
安曇	Azumi	Nagano Prefecture	Hybrid	623	1969
高根第一	Takane No. 1	Gifu Prefecture	Hybrid	340	1969
水殿	Midono	Nagano Prefecture	Hybrid	245	1969
喜撰山	Kisen'yama	Kyoto	Pure	466	1970
新豊根	Shintoyone	Aichi Prefecture	Hybrid	1125	1972
沼原	Numappara	Tochigi Prefecture	Pure	675	1973
奥多々良木	Okutataragi	Hyogo Prefecture	Pure	1932	1974
新冠	Niikappu	Hokkaido	Hybrid	200	1974
大平	Ohira	Kumamoto Prefecture	Pure	500	1975
南原	Namwon	Hiroshima Prefecture	Pure	620	1976
馬瀬川第一	Mazekawa No. 1	Gifu Prefecture	Hybrid	288	1976
奥清津	Futai Dam	Niigata Prefecture	Pure	1000	1978
新高瀬川	Shin-Takasegawa	Nagano Prefecture	Hybrid	1280	1979
奥吉野	Okuyoshino	Nara Prefecture	Pure	1206	1980
奥矢作第二	Okuyahagi No. 2	Aichi Prefecture	Pure	780	1980
奥矢作第一	Okuyahagi No. 1	Aichi Prefecture	Pure	323	1980

TABLE 2.2 Pumped Hydroelectric Storage Stations in Japan (*cont.*)

Plant name (Japanese)	Plant name (English)	Location	Type	Rating/MW	Commission year
玉原	Tamahara	Gunma Prefecture	Pure	1200	1981
本川	Motokawa	Kochi Prefecture	Pure	615	1982
第二沼沢	Daini Numazawa	Fukushima Prefecture	Pure	460	1982
高見	Takami	Hokkaido	Hybrid	200	1983
俣野川	Matanoagawa	Tottori Prefecture	Pure	1200	1986
天山	Tenzan	Saga Prefecture	Pure	600	1986
今市	Imaichi	Tochigi Prefecture	Pure	1050	1988
下郷	Shimogo	Fukushima Prefecture	Pure	1000	1988
大河内	Okawachi	Hyogo Prefecture	Pure	1280	1992
奥美濃	Okumino	Gifu Prefecture	Pure	1500	1994
塩原	Shiobara	Tochigi Prefecture	Pure	900	1994
奥清津第二	Futai Dam No. 2	Niigata Prefecture	Pure	600	1996
葛野川	Kazunogawa	Yamanashi Prefecture	Pure	1200	1999
沖縄やんばる海水揚水	Okinawa Seawater Pumped Hydro	Okinawa Prefecture	Pure	30	1999
神流川	Kannagawa	Gunma Prefecture	Pure	940	2005
小丸川	Omarugawa	Miyazaki Prefecture	Pure	1200	2007
朱鞠内	Shumarinai	Hokkaido	Hybrid	1	2013
京極	Kyogoku	Hokkaido	Pure	200	2014

transmission [10]. Because the net electricity output of PHES operation is negative, a PHES facility usually cannot qualify as a power generator. Although their crucial load-balancing and ancillary services to the grid reduce the need for transmission upgrades, PHES facilities are not recognized as parts of the transmission infrastructure [11]. This confusion in business models has deterred the development of PHES in the United States.

The diverging paths of PHES development in Japan, China, and the United States have shown that national regulatory and institutional environments have tremendous impacts on the deployment of PHES. PHES facilities are large

TABLE 2.3 Pumped Hydroelectric Storage Stations in China

Plant name (Chinese)	Plant name (English)	Province	Type	Rating/ MW	Commission Year
岗南	Gangnan	Hebei	Hybrid	11	1968
密云	Miyun	Beijing	Hybrid	22	1975
潘家口	Panjiakou	Hebei	Hybrid	270	1992
寸塘口	Cuntangkou	Sichuan	Pure	2	1992
广州一期	Guangzhou Phase 1	Guangdong	Pure	1200	1994
十三陵	Shisanling	Beijing	Pure	800	1997
羊卓雍湖	Yangzhuoyong	Tibet	Pure	90	1997
溪口	Xikou	Zhejiang	Pure	80	1998
天荒坪	Tianhuangping	Zhejiang	Pure	1800	2000
广州二期	Guangzhou Phase 2	Guangdong	Pure	1200	2000
响洪甸	Xianghongdian	Anhui	Hybrid	80	2000
天堂	Tiantang	Hubei	Pure	70	2001
沙河	Shahe	Jiangsu	Pure	100	2002
回龙	Huilong	Henan	Pure	120	2005
桐柏	TONGBAI	Zhejiang	Pure	1200	2006
白山	Hakusan	Jilin	Hybrid	300	2006
泰安	Taian	Shandong	Pure	1000	2007
琅琊山	Langyashan	Anhui	Pure	600	2007
西龙池	Xilongchi	Shanxi	Pure	1200	2008
宜兴	Yixing	Jiangsu	Pure	1000	2008
张河湾	Zhanghewan	Hebei	Pure	1000	2008
惠州	Huizhou	Guangdong	Pure	2400	2009
黑麋峰一期	Heimifeng Phase 1	Hunan	Pure	1200	2009
白莲河	Bailianhe	Hubei	Pure	1200	2010
宝泉	Baoquan	Henan	Pure	1200	2011
蒲石河	Pushihe	Liaoning	Pure	1200	2012
响水涧	Xiangshuijian	Anhui	Pure	1000	2012
仙游	Xianyou	Fujian	Pure	1 200	2013

TABLE 2.4 Pumped Hydroelectric Storage Stations in the United States

Plant name	County	State	Type	Rating/ MW	Commi-ssion year
Rocky River	Litchfield	CT	Hybrid	31	1928
Flatiron	Larimer	CO	Hybrid	9	1954
Hiwassee Dam	Cherokee	NC	Hybrid	95	1956
Lewiston Niagara	Niagara	NY	Hybrid	240	1962
Taum Sauk	Reynolds	MO	Pure	408	1963
Yards Creek	Warren	NJ	Pure	453	1965
Cabin Creek	Clear Creek	CO	Pure	300	1967
W R Gianelli	Merced	CA	Pure	424	1967
Muddy Run	Lancaster	PA	Pure	1072	1967
ONeill	Merced	CA	Pure	25	1968
Thermalito	Butte	CA	Hybrid	83	1968
Edward C Hyatt	Butte	CA	Hybrid	293	1968
Salina	Mayes	OK	Pure	288	1970
FirstEnergy Seneca	Warren	PA	Pure	469	1970
Smith Mountain	Franklin	VA	Hybrid	247	1970
Mormon Flat	Maricopa	AZ	Hybrid	54	1971
Horse Mesa	Maricopa	AZ	Hybrid	100	1972
Degray	Clark	AR	Hybrid	28	1972
Northfield Mountain	Franklin	MA	Pure	940	1973
Ludington	Mason	MI	Pure	1979	1973
Blenheim Gilboa	Schoharie	NY	Pure	1000	1973
Jocassee	Pickens	SC	Pure	612	1974
Bear Swamp	Berkshire	MA	Pure	600	1974
Castaic	Los Angeles	CA	Hybrid	1275	1976
Carters	Murray	GA	Hybrid	250	1977
Fairfield Pumped Storage	Fairfield	SC	Pure	511	1978
Raccoon Mountain	Hamilton	TN	Pure	1714	1979
Wallace Dam	Hancock	GA	Hybrid	209	1980
Grand Coulee	Grant	WA	Hybrid	314	1980
Harry Truman	Benton	MO	Pure	161	1981
Mount Elbert	Lake	CO	Pure	200	1983
Helms Pumped Storage	Fresno	CA	Pure	1053	1984
Clarence Cannon	Ralls	MO	Hybrid	31	1984
Bath County	Bath	VA	Pure	2 862	1985

(*Continued*)

TABLE 2.4 Pumped Hydroelectric Storage Stations in the United States (*cont.*)

Plant name	County	State	Type	Rating/ MW	Commi-ssion year
J S Eastwood	Fresno	CA	Pure	200	1987
Bad Creek	Oconee	SC	Pure	1 065	1991
Waddell	Maricopa	AZ	Pure	40	1993
North Hollywood	Los Angeles	CA	Hybrid	5	1993
Rocky Mountain	Floyd	GA	Pure	848	1995
Richard B Russell	Elbert	GA	Hybrid	328	2002
Lake Hodges	San Diego	CA	Pure	42	2012

facilities that require huge upfront capital investments, and paybacks are spread over many decades. If a government wishes to facilitate the development of PHES, it needs to provide a stable and predictable regulatory environment and a reasonable pricing scheme that allows PHES facilities to be compensated for their services to the transmission grid.

4 PROSPECTS

In recent years, due to increasing concern for global warming and the call to decarbonize electricity, there has been increasing commercial interest in PHES [12]. Developers are actively pursuing new PHES projects around the world.

4.1 Revival of Conventional PHES

More than 100 new PHES plants with about 74 GW capacity worldwide are expected to be in operation by 2020 [13]. China has the most aggressive plan. In 2014 the Chinese government announced its plan to more than quadruple its current PHES installations to a total capacity of 100 GW by 2025 [14]. Driven by the need to accommodate rapid growth of intermittent renewable electricity, Europe is also witnessing a renaissance of PHES, particularly in Spain, Switzerland, and Austria, with 27 GW new PHES capacity expected by 2020 [15]. Although Japan already has the highest density of PHES installations in the world, Japanese power companies are continuing to develop more PHES plants. The United States is also experiencing a revival of PHES development. In 2014 the US Federal Energy Regulatory Commission issued licences to construct and operate two new PHES facilities (1.3 GW Eagle Mountain PHES and 400 MW Iowa River PHES). Another application for a construction and operation licence (for the 1 GW Parker Noll PHES) is currently under review. In addition, there are over 40 proposed PHES projects currently conducting feasibility studies with preliminary permits issued. With the mature technology and high volume of commercial development activities, PHES will certainly remain the most dominant energy storage technology in the foreseeable future.

4.2 Alternative and Novel PHES Designs

There are a number of PHES designs:

- *Variable-speed PHES*: Most of the existing PHES facilities are equipped with fixed-speed pump turbines. While fixed-speed PHES facilities may provide economical bulk electricity storage, they can only provide frequency regulation during generating mode—not in pumping mode. With the increasing adoption of renewable power sources, such as wind and solar, there is increasing demand for frequency regulation. New variable-speed technology allows PHES facilities to regulate frequency during both pumping and generating modes. Japan has pioneered variable-speed PHES technology and has successfully operated such systems at the Okawachi PHES station for over 20 years [16,17]. European countries are actively introducing variable-speed PHES to accommodate more variable renewable electricity in their power portfolios [18].
- *Seawater PHES*: In addition to the worldwide revived interest in developing conventional PHES projects, many developers are also proposing new approaches. Japan has pioneered seawater PHES. The Okinawa seawater PHES station, which commenced operation in 1999, was the world's first seawater PHES system [19]. It uses the open sea as the lower reservoir together with a constructed upper reservoir at 150 m above sea level. New seawater projects have been proposed in Ireland, Greece, Belgium, and the Netherlands [20–23]. The energy islands concept proposed by the Dutch consulting company DNV KEMA is an interesting approach: they plan to use the open sea as the upper reservoir and construct the lower reservoir by dredging and building a ring of dikes at a depth of 50 m below sea level.
- *Underground PHES*: Researchers since the 1970s [24] have proposed the possibility of utilizing underground caverns as lower reservoirs for PHES projects, but so far none have been built [25]. Commercial interest in developing underground PHES has resurfaced in recent years in the United States. Several US developers have received preliminary permits to study the feasibility of building underground PHES facilities at their identified sites. The UK company Quarry Battery is also working on developing underground PHES facilities in abandoned quarries [26].
- *Compressed air PHES*: A promising innovative design (Fig. 2.2) is to replace the upper reservoir in PHES with a pressurized water container [27]. The air within the pressure vessel becomes pressurized when water is pumped into the vessel. Instead of storing potential energy in elevated water, the proposed compressed air pumped hydro system stores the energy in compressed air. This innovative design could potentially free PHES from

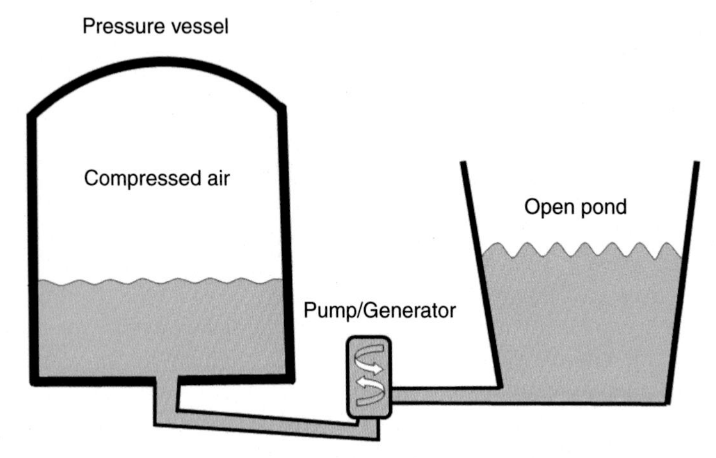

FIGURE 2.2 **Diagram of compressed air pumped hydroelectric storage.**

geographic requirements and make it feasible at almost any location with flexible and scalable capacity. This concept is discussed in more detail in chapter: A Novel Pumped Hydro Combined with Compressed Air Energy.

- *Undersea PHES*: Another innovative concept (Fig. 2.3) is to utilize the water pressure at the bottom of the sea to store electricity from offshore wind turbines [28]. The system places submerged pressure vessels (hollow concrete tanks) on the seafloor. It uses electricity to pump water out of the tank to store energy, and generate electricity when seawater is filling into the tank through the generator. This concept is further discussed in chapter: Underwater Compressed Air Energy Storage.

4.3 Retrofits of Existing PHES and Conventional Hydropower Stations

Many existing PHES facilities were built many decades ago and therefore were equipped with outdated and inefficient technology. There is significant potential in increasing PHES capacity simply by renovating and upgrading existing PHES facilities. Upgrades to old PHES facilities typically include replacing outdated pumps/turbines, impellers, and control systems with new advanced equipment. Many existing PHES station may increase capacity by (15–20)% and efficiency by (5–10)% [7]. In addition, many existing conventional hydropower stations could be reengineered to add a lower reservoir and pumpback units to pump the water back to the upper reservoir during off-peak hours, and become combined PHES stations for use with intermittent energy from renewable sources such as wind turbines and solar panels.

Although PHES may be an essential enabling technology for decarbonizing electricity (replacing fossil fuel with renewable forms of energy such as wind and solar), the political will to mitigate carbon dioxide and to remove regulatory

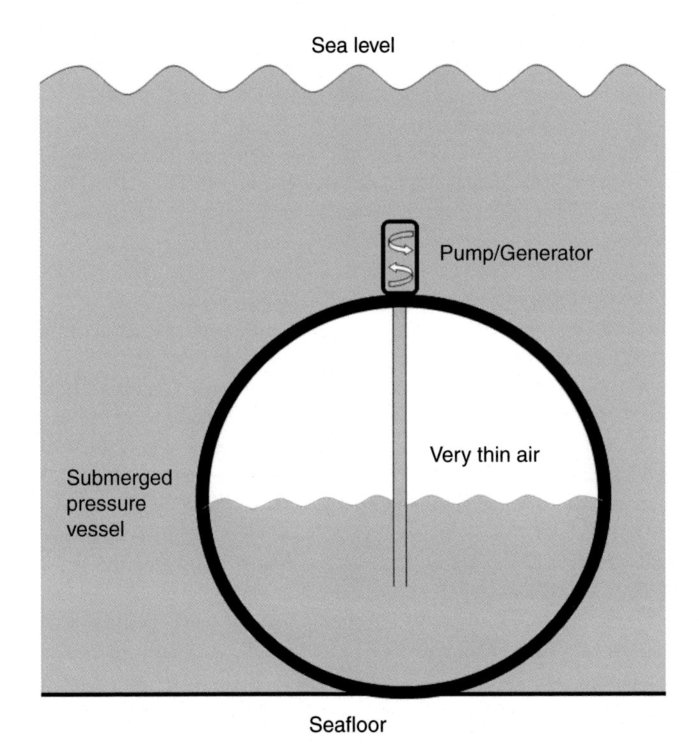

FIGURE 2.3 **Diagram of undersea pumped hydroelectric storage.**

barriers for PHES is far from certain. The price of natural gas is also a key determinant in the future of PHES. Because PHES is essentially a peak load technology, which competes directly with gas-fired power, the low natural gas price may render PHES uncompetitive. The vision of decarbonizing electricity and how PHES fits into it will likely vary from country to country.

REFERENCES

[1] Electrical Japan. 日本全国の揚水発電所ランキング. http://agora.ex.nii.ac.jp/earthquake/201103-eastjapan/energy/electrical-japan/type/5.html.ja

[2] Peng W, Chen D. Some considerations on the development of pumped hydroelectric storage power station in China (In Chinese: 对我国抽水蓄能电站发展的几点思考). Beijing, People's Republic of China: State Electricity Regulatory Commission; 2010. http://www.serc.gov.cn/jgyj/ztbg/201006/t20100621_13195.htm

[3] EIA. Electricity: Form EIA-860 detailed data 2012. Washington, DC: US Department of Energy, Energy Information Administration; 2012. http://www.eia.gov/electricity/data/eia860/

[4] Army Corps. Engineering and design—hydropower, No. 1110-2-1701. Washington, DC: Army Corps of Engineers, Department of the Army; 1985.

[5] US Federal Energy Regulatory Commission. http://www.ferc.gov/industries/hydropower/gen-info/licensing/pump-storage.asp

[6] Yang C-J, Jackson R. Opportunities and barriers to pumped-hydro energy storage in the United States. Renew Sust Energ Rev 2011;15:839–44.
[7] Baxter R. Energy storage: a nontechnical guide. Tulsa, OK: PennWell; 2006.
[8] http://www.energystorageexchange.org/
[9] Zeng M, Zhang K, Liu D. Overall review of pumped-hydro energy storage in China: status quo, operation mechanism and policy barriers. Renew Sust Energ Rev 2013;17:35–43.
[10] APS Panel on Public Affairs Committee on Energy Environment. Challenges of electricity storage technologies. College Park, MD: American Physical Society; 2007.
[11] FERC. FERC encourages transmission grid investment. Docket No. ER06-278-000. Washington, DC: Federal Energy Regulatory Commission; 2008.
[12] Deane JP, Gallachóir BP, McKeogh EJ. Techno-economic review of existing and new pumped hydro energy storage plant. Renew Sust Energ Rev 2010;14:1293–302.
[13] Ecoprog GmbH. 2013. The World Market for Pumped-Storage Power Plants. Köln, Germany.
[14] Xinhuanet, 2014. http://news.xinhuanet.com/energy/2014-11/19/c_127223035.htm
[15] Zuber M. Renaissance for pumped storage in Europe. Hydro Rev World 2011;19. http://www.hydroworld.com/articles/print/volume-19/issue-3/articles/new-development/renaissance-for-pumped-storage-in-europe.html
[16] Nagura O, Higuchi M, Tani K, Oyake T. Hitachi's adjustable-speed pumped-storage system contributing to prevention of global warming. Hitachi Rev 2010;59:99–105.
[17] Henry JM, Maurer F, Drommi J-L, Sautereau T. Converting to variable speed at a pumped-storage plant. Hydro Rev World 2013;21. http://www.hydroworld.com/articles/print/volume-21/issue-5/articles/pumped-storage/converting-to-variable-speed-at-a-pumped-storage-plant.html
[18] Lefebvre N, Tabarin M, Teller O. A solution to intermittent renewable using pumped hydropower. Renew Energ World 2015;:49–57.
[19] http://www.kankeiren.or.jp/kankyou/en/pdf/en108.pdf
[20] McLean E, Kearney D. An evaluation of seawater pumped hydro storage for regulating the export of renewable energy to the national grid. Energ Procedia 2014;46:152–60.
[21] Katsaprakakis DA, Christakis DG, Stefanakis I, Spanos P, Stefanakis N. Technical details regarding the design, the construction and the operation of seawater pumped storage systems. Energy 2013;55:619–30.
[22] LaMonica M. A manmade island to store wind energy. MIT Technol Rev 2013; Feb. 5.
[23] http://earthtechling.com/2013/01/energy-island-for-wind-power-storage-draws-belgiums-interest/
[24] Tam SW, Blomquist CA, Kartsounes GT. Underground pumped hydro storage—an overview. Energ Sources 1979;4:329–51.
[25] Pickard WF. The history, present state, and future prospects of underground pumped hydro for massive energy storage. Proceedings of the IEEE 2012;100:473–83.
[26] http://www.quarrybatterycompany.com/
[27] Wang H, Wang L, Wang X, Yao E. A novel pumped hydro combined with compressed air energy storage system. Energies 2013;6:1554–67.
[28] Slocum AA, Fennell GE, Dundar G, Hodder BG, Meredith JDC, Sager MA. Ocean renewable energy storage (ORES) system: analysis of an undersea energy storage concept. P IEEE 2013;101:906–24.

Chapter 3

Novel Hydroelectric Storage Concepts

Frank Escombe
EscoVale Consultancy Services, Reigate, Surrey United Kingdom

1 INTRODUCTION

1.1 Scope and Purpose

Following on from coverage of conventional pumped hydroelectric storage (PHES), this chapter examines other concepts that share the same principle: using hydroelectric equipment (usually reversible pump-turbines and motor-generators) to convert electrical energy to and from gravitational potential energy. It explores their ability to complement PHES by:

- Extending use into regions where the terrain is unsuitable for conventional PHES or where public opposition/approval difficulties (areas of outstanding natural beauty or villages or agricultural land or conflicting priorities for water resources) make PHES projects impractical.
- Opening up new applications—in particular, to meet future requirements in electricity networks with high concentrations of wind and solar resources that need much longer duration/higher energy solutions than are usually feasible with conventional PHES.

1.2 Constraints

Developers of these novel concepts hope to benefit from the status, technical expertise, and market dominance of PHES, which accounts for >95% of global grid-scale installed capacity. They also have to live within its limitations. The main problem with gravitational storage is gravity, which is incredibly weak compared with the other fundamental forces. This is a good thing for the universe but rather inconvenient for energy storage developers.

As an exercise in the absurd, a record-breaking weightlifter can support an astonishing mass of over 400 kg. If an electricity company could persuade him to store energy by carrying this from sea level to the top of Everest, it could afford to pay him almost $1 for his trouble. A simple time-shifting energy storage

Storing Energy. http://dx.doi.org/10.1016/B978-0-12-803440-8.00003-8

service is not worth much to the average electricity supplier—<$100 (MW h)$^{-1}$. Any form of gravitational storage has to accept some uncompromising facts of physical life.

Kilowatt-hours make a sensible starting point for electrical energy storage calculations. The SI unit of energy is the joule (J), equal to the energy expended when applying one watt of power for one second (1 W s). Therefore, 1 kW h (1000 W for 3600 s) is equal to 3.6×10^6 J (3.6 MJ).

A joule is also equal to the energy expended in moving a force of one newton through one meter (1 N m). A mass of one kilogram exerts a downward force of "g" newtons, where g is the acceleration due to gravity (about 9.81 m s^{-2}):

- In a perfect world, raising/lowering 1 kg by 1 m will store/generate 9.81 J.
- Storing/generating 1 kWh (3.6 MJ) therefore requires 367 tonne-meters (367 t m).
- The world is not perfect. Even with the best turbines, hydroelectric generation needs about 400 t m (tonnes of water × meters of vertical travel) per kilowatt-hour. We might require 450 t m with less efficient turbines.
- For our purposes: each kilowatt-hour from storage requires (400–450) t m (tonnes × meters vertical travel).

It is sometimes more convenient to express hydroelectric energy in terms of its volume/pressure relationship:

- In a perfect world, $E = V \times P$, where E = energy in joules; V is the volume of water passing through the turbine in m^3; and P is the pressure in pascals. One pascal (symbol Pa) = 1 N m^{-2}. One megapascal (MPa) = 10^6 Pa = 1 MPa = 10 bar pressure.
- Storing/generating 1 kW h (3.6 MJ) requires 3.6 m^3 MPa (cubic meters × megapascals). Real world hydroelectric generation has to allow for turbine efficiency.
- For our purposes: each kilowatt-hour from storage requires (4.0–4.5) m^3 MPa or (40–45) m^3 bar.

For comparison, we could retrieve 1 kW h from a battery weighing less than 10 kg. Or we could electrolyze less than 1 kg of water and use the resulting hydrogen to generate 1 kW h in a fuel cell. Or, if only we knew how to use really strong fundamental forces ($E = mc^2$), we could convert our kilowatt-hour to and from 40×10^{-9} g (40 nanograms) of matter—less than 1 mm of fine hair!

PHES is the elephant in the storage room, but it is there for good reasons. Novel hydroelectric storage technologies will also have to offer real advantages if they are to make real progress. This may be in terms of capital and operating costs; cycle life and longevity; or other performance characteristics, including power and energy capability (where PHES has often been seen as the only practical answer).

1.3 How Did We Get Here?

Pumped hydro concepts can trace their origins back to 18th century water towers or to 19th century hydraulic accumulators.

Given enough time, a small steam-driven pump could elevate a considerable quantity of water into an 18th century water tower. The water was then used for intensive operations, by releasing it rapidly through high-power hydraulic machinery. The 20th century saw the adoption of PHES as the dominant method of storing electricity. Conventional PHES uses mountains rather than water towers, but the principle is the same.

Nineteenth century hydraulic accumulators [1] provided attractive alternatives to costly and occasionally fallible water towers. Early systems were vertical cylinders at ground level, sealed at the top by massively weighted pistons. Operating pressures were an order of magnitude higher than available from water towers at that time, with a corresponding reduction in the volume of water required for a given task. The 21st century may see history repeat itself, with the introduction of accumulator-type systems to store electricity in high-pressure, low-level water, without needing mountains and, perhaps, on a scale that eclipses PHES.

1.4 Novel Hydroelectric Storage Categories

The technologies covered in this chapter are grouped in the following sections:

- Piston-in-cylinder storage
- Energy membrane storage
- Novel land-based and seabed pumped hydro configurations
- Offshore lagoon and energy island storage.

The first two categories use pumped hydro techniques to elevate a solid working mass (similar in principle to hydraulic accumulators). The remainder use water (or seawater) as the working mass in configurations that differ in some respects from conventional PHES.

1.5 Future Applications and Markets

Most of the concepts covered here are intended for high-power, high-energy applications: multiple hours of power delivery at ratings from tens of megawatts to multigigawatts. Looking ahead, we can envisage four nominal overlapping segments, discussed further in Section 2.4

- Day-ahead storage (say, (3–20) h capacity). This accounts for almost all of today's high-energy business and will remain extremely important. Such systems can provide many reserve and regulation services, help meet the next day's peak, and tailor the erratic production profile of variable renewables. There is little need for anything more than day-ahead storage at present.

- Week-ahead storage (say, (15–40) h capacity). Such systems will be used by future networks that rely heavily on variable renewables, subject to multi-day extremes in terms of high or low production.
- Strategic storage (say, (>50–250) h capacity). In addition to providing fairly long-term storage and quasigenerator services, the objective is to harvest energy that would otherwise be lost from the electricity supply system.
- Seasonal electricity storage (say, >200 h capacity). A hypothetical seasonal storage system may use summer sunshine to power winter loads. In practice, there usually comes a point where the marginal cost of adding energy capacity cannot be justified by the additional energy throughput.

2 PISTON-IN-CYLINDER ELECTRICAL ENERGY STORAGE

2.1 Background and Operating Principle

The idea of storing energy by using a hydraulic piston to raise a solid working mass has been suggested many times, usually by people who do not realize the miniscule energy released when a few tonnes are lowered by a few meters. Others have proposed constructing large commercial and public buildings on hydraulic jacks for energy storage and earthquake resilience. Buildings could be elevated by (5–10) m using cheap overnight electricity and lowered slowly to provide daytime power. Unfortunately, power levels are unexciting. One would need an Empire State Building (400 000 t) to get close to 1 MW over a working day.

Clearly, we are going to need something much larger if we want to be in the same ballpark as PHES (in fact, a ballpark may be about the right size!). For context, 20 GW h is a useful amount of energy for a network—it might keep a city of a million going for a day in Europe or for half a day in the United States. Using the 400 t m $(\mathrm{kW\ h})^{-1}$ metric, 20 GW h requires some 8×10^9 t m. PHES could store 20 GW h by pumping 20×10^6 t of water 400 m uphill. In the absence of a suitable mountain, we could get the same result by pumping 400×10^6 t of water up a 20 m hill. That would be absurd because of the size and cost of the two reservoirs, each the size of a small country. We would not need reservoirs if we were to raise a 400×10^6 t concrete disk by 20 m, but the disk would be far too expensive (even if we could work out how to lift it).

However, since we live on a rocky planet, we might be able to make a comparable disk cheaply by excavating material from around a disk shape. This is the basis of several storage ideas, including EscoVale's GBES (ground-breaking energy storage), which incorporates the components of a gravitational storage system in a single construction. The concept is straightforward and construction looks simple—ground-breaking in the literal sense! In practice, the method and sequence would differ greatly from this highly simplified outline (Fig. 3.1) but, in essence:

- Excavate and reinforce a deep trench to form the periphery of the piston and the cylinder wall.
- Excavate horizontally to form and reinforce the base of the piston.
- Install a seal between the trench and the working area beneath the base.
- Store energy by pumping water into the working area to raise the piston in the cylinder.
- Recover energy by releasing the water through hydroelectric generators.

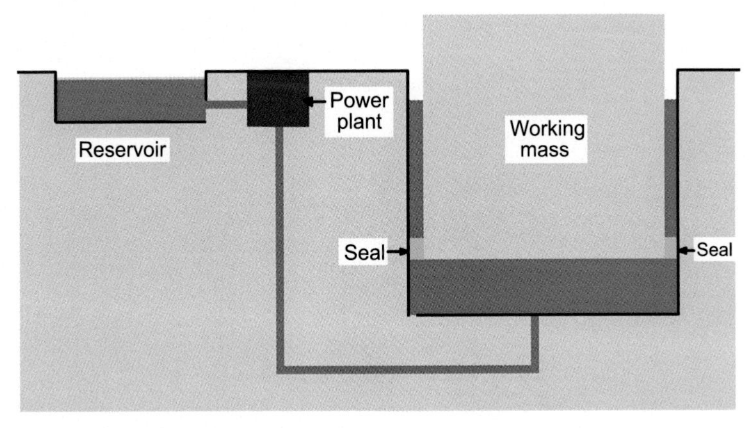

FIGURE 3.1 **Piston-in-cylinder energy storage.**

A GBES system could be constructed as above, but the concept is versatile with many possible configurations of reservoir, piston, and power plant, some of which are outlined in Fig. 3.2:

- The walls need not be vertical above the travel zone, simplifying construction tolerances, reinforcement, and local geology issues.
- The piston can be excavated within the upper reservoir, reducing the footprint and cost.
- It could be submerged to be aesthetically pleasing (the reservoir level remains constant, avoiding the unsightly cyclic drainage issues of PHES).
- A reservoir would not be needed if constructed within a lake or offshore.
- A large piston might accommodate some (perhaps all) of the power plant—short, cheap pressure tunnels and more piston control options.
- GBES can be integrated with other energy infrastructure. It could add energy storage capability to a conventional hydroelectric station without requiring a second reservoir. It could increase substantially the capacity of an existing PHES plant. GBES could be located within tidal lagoons or tidal barrage schemes and would fit within turbine separation distances in offshore wind farms.
- Alternatively, GBES could accommodate a large public water supply reservoir, using excavated material to create a retaining barrier, for example, thousands of hectares inside a 10 m to 20 m barrier.

- A similar arrangement could store surplus seasonal water or be used for flood relief, temporarily using part of the power plant to pump external water into the impoundment area. Energy supply and water management rank among the century's greatest challenges. A technology that tackles two high-value issues could be useful in sharing costs and gaining support (water provision, flood control and coastal protection have tangible and popular local benefits, unlike energy storage).

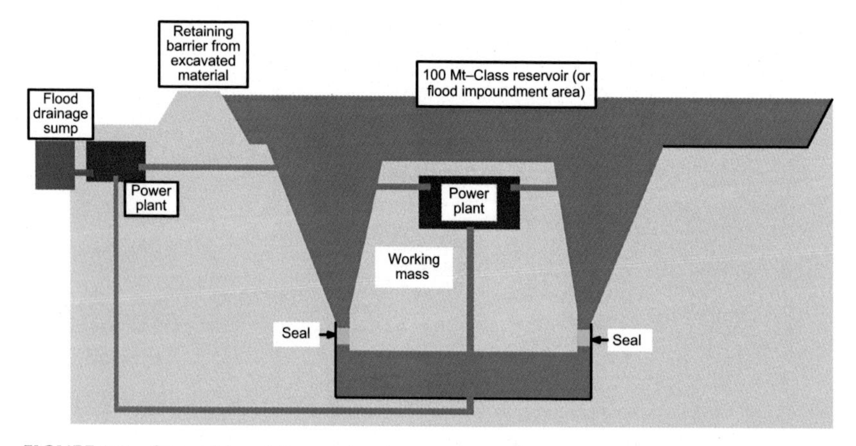

FIGURE 3.2 **Ground-breaking energy storage schematic.**

At least two other programs share the same principle as GBES, starting from different origins within the past ten years or so.

EscoVale [2] commenced work on GBES in 2007, initially as an idea for gigawatt-class seabed storage, using piston diameters of around 500 m within large offshore wind turbine arrays. Onshore applications of interest include PHES lookalikes in flat terrain and future high-energy markets that are beyond the reach of other electrical energy storage systems—even the great majority of PHES plants.

Gravity Power (GP) [3] is believed to be the longest established and furthest advanced firm, initially based on relatively small–diameter pistons fabricated from high-density concrete, moving a considerable vertical distance in very deep shafts. GP's focus is now on somewhat larger units, from about 30 m to 100 m diameter, giving the option of geological or fabricated pistons. The market entry target is in peaking applications, typically providing 4 h of power delivery from about 40 MW/160 MW h for the smallest unit and up to 1.6 GW/6.4 GW h with the 100 m module.

Heindl Energy (HE) [4] based its work on geological pistons from the outset, with proposed applications on the same general scale as large PHES plants—typically, gigawatt power levels and at least tens of gigawatt-hours energy. HE was ahead of EscoVale in making its plans public, including extremely large designs intended for 10^{12} W h (terawatt-hour) seasonal storage.

GP and HE have advanced plans for demonstration projects. EscoVale is completing an analysis of market opportunities for systems with GBES characteristics to provide a rational framework for development choices [5].

2.2 Piston versus PHES?

The job of the piston (and of the high-level PHES reservoir) is to create working pressure at the power plant. This puts us on familiar ground. If PHES and piston storage operate at the same pressure and pump the same amount of water at the same rate for the same time, they will store a similar amount of energy and require hydroelectric equipment of similar rating. The two systems have much in common, but they are not competitors. In any event, PHES would normally win—it works brilliantly, is immortal (near enough), and is backed by many millions of hours in service. It will take decades for piston storage to gain comparable experience. In the meantime, it must seek opportunities that can exploit some of its five major points of difference:

- Location
- Operating pressure
- Energy rating
- Discharge power rating
- Charging power rating.

2.2.1 Location, Location, Location

The most important attribute of piston storage is that it brings pumped hydro down from the mountains, overcoming fundamental limitations of PHES:

- Most networks do not have mountains, making PHES a nonstarter.
- Where there are mountains, there are not many people and so PHES has to serve distant applications.
- Public opinion opposes knocking mountains about to create PHES reservoirs.

2.2.2 Pressure by Design

PHES operating pressure is dictated by the terrain (typically, a few megapascals or low tens of bar). In general, designers prefer higher pressures, enabling physically smaller hydroelectric equipment, reservoirs, and pressure tunnel diameters. Pressure is a design variable for piston systems, dependent on piston height. For example, a piston height of 500 m usually provides about 7.5 MPa (75 bar) pressure at the power plant [6]. Given this freedom, the piston storage industry is likely to opt for pressures toward the top of the range for which there is PHES experience—there are few examples over 8 MPa (80 bar). In principle, piston storage might go well above 10 MPa (100 bar) and use equipment other than the industry norm of reversible Francis pump turbines. In any event, designers will probably adopt a small number of preferred operating pressures, selecting from standard hydroelectric pump motor/turbine generator packages, in much the same way that airlines and aircraft manufacturers select engines.

2.2.3 No Energy Limitations

The energy rating of a PHES system is constrained by the reservoirs. As an approximation, $E \approx 2.5VH \times 10^{-9}$, where E is the stored energy (in units of gigawatt-hours); V is the usable volume of the smaller of the two reservoirs (in units of cubic meters) or water mass in tons; and H is the difference in elevation between reservoirs (expressed in meters).

PHES systems generally serve day-ahead markets where the largest systems require tens of gigawatt-hours. This is difficult, even for PHES. Information on energy ratings is sparse, but there are probably only ~30 PHES systems >10 GW h worldwide. This compares with ~150 medium energy (1–10) GW h plants and 200+ smaller systems of <1 GW h. Most are entirely adequate for their applications, but there is no doubt that sites suitable for PHES of >10 GW h are already hard to come by (see chapter: Pumped Hydroelectric Storage).

This limits the ability of PHES to serve future requirements for gigawatt-scale week-ahead and larger strategic storage markets (see Sections 1.5 and 2.4), with system ratings of tens or hundreds of gigawatt-hours. A 500 GW h PHES would need to transfer water between 500 Mt reservoirs (assuming an elevation difference of 400 m). There are few locations where such systems are feasible, requiring two large natural bodies of water (and even with 100 km^2 lakes, there may be objections to 5 m changes in water levels).

Naturally, 500 GW h (100+ hours) piston storage would also be a huge engineering task, but of a different nature. Prospective sites are almost as common for 100 h as for 10 h systems, and are as likely (in fact, more likely) to be found in flattish terrain or fairly close to load centers, rather than in mountains. The 100 h systems need not require much more space than 10 h schemes and marginal costs are moderate.

We can illustrate this with a 3 GW GBES design (matching the current highest power PHES plant). A piston diameter of 800 m (0.5 km^2 area) and operating pressure of 8 MPa (80 bar) delivers a convenient 1 GW h for every 1 m of travel [6].

A 10 h, 30 GW h day-ahead unit (e.g., Fig. 3.2) requires 30 m vertical travel.

Energy capacity could be doubled at negligible cost by extending the vertical cylinder wall section to allow 60 m travel. There is a limit to the height that the piston can be raised (one reason being that the operating pressure at the power plant increases slowly as the piston rises above the surface). It should be possible to get well into the week-ahead sector—in our example a 40 h, 120 GW h, 120 m travel system would result in a pressure increase of about 10%.

Ultrahigh-energy strategic storage requires a different approach, with a radical increase in the piston diameter, travel distance or operating pressure—the latter seems unlikely. A 600 GW h, 200 h system, with the same travel distance and pressure would require a massive (but not inherently more difficult) piston of ~1800 m diameter. The marginal cost of the additional storage capacity would be low (perhaps 10% of the dollars per kilowatt-hour cost of the 40 h design and <5% of the 10 h unit).

Alternatively, the cylinder depth of the smaller diameter system could be increased from ~600 m to, say, 1200 m by excavating a deeper trench. It would be necessary to take ~600 m off the top of the piston to maintain the design operating pressure. This gives an extra 600 m of travel (600 GW h, 200 h) before the piston reaches the surface. The main issue is the huge increase in excavated material, mostly from the top of the piston—over 500 Mt for our example. The cost of extraction is surprisingly low (a few dollars per kilowatt-hour of energy storage capacity) and so disposal costs are critical. We live in a world that will have to accommodate, feed, and improve the conditions of a few billion additional people while coping with rising sea levels and increasingly extreme weather conditions. This will push up demand for material associated with coastal defense, flood protection, water management, and land reclamation projects. If blocks of rock have enough value to cover extraction and local transport, the marginal cost of deeper cylinders would be even lower than that of larger diameters.

With either approach, stored energy is limited only by system requirements and by economics. The 10th hour of storage has a higher cycle count and more earning opportunities than the 100th, and the 1000th might only be used once a year. Multiterawatt-hour storage systems are technically feasible. At the other end of the scale, GP's modular system starts at little more than 100 MW h.

Another point to note is that piston storage projects are physically small, at least compared with PHES. A deep cylinder system can store >1 MW h m^{-2} in terms of cylinder area and >100 kW h m^{-2} of site area. This is one or two orders of magnitude better than PHES—indeed, the energy storage capacity of the entire global PHES portfolio could be matched by just two of the above 600 GW h piston storage systems.

2.2.4 No Power Limitations

Geological piston storage has no practical upper power limit—the bigger, the better because of strong economic benefits of scale. If required, power ratings far above the present storage maximum of ~3 GW are feasible. Gigawatt-class systems will remain the norm for large-scale storage, although some locations could accommodate >10 GW. The lower threshold will depend on other design features, especially the energy rating, but may be ~100 MW. GP's fabricated piston design is a special case, enabling use of relatively small diameters at powers down to tens of megawatts in applications that are not accessible to HE and GBES.

2.2.5 Asymmetric Charging

Asymmetric charging is an interesting design quirk for slow-speed piston designs such as GBES or HE. It should be one of the most valuable features for future networks with a very high proportion of variable renewables.

Slow speed gives many hours of power delivery from a limited travel distance and more options in terms of seal subsystems. It also simplifies transition from discharge to charge and makes it much easier to handle extreme faults, for example, loss of grid connection when the piston is traveling at its highest descent speed and generating maximum power. Basically, if the turbines are unable to extract energy and send it to the grid, we need to deal with the kinetic energy in the system and the large amount of potential energy that would be released if the piston continued to descend an appreciable distance before it is brought to a halt. That is relatively easy if the highest descent speed is a brisk snail's pace [6].

Piston speed is not an issue during the charging phase, with energy taken from the grid and converted via pumps into potential energy as the piston rises. Gravity assists (rather than counteracts) a switch from charge to discharge mode and aids piston deceleration if grid power is lost. Also, some of the more complex components of seal systems may not be needed when the piston is ascending. Consequently, the storage system can incorporate extra pumping capacity (at low cost), increasing the input power rating during the charge cycle. The charge rating can be several times the power delivery rating, if this is beneficial.

Asymmetric charging has little value at present. Large-scale storage systems usually have plenty of time to gather sufficient overnight energy to provide services during the following day. The situation will be quite different in networks that rely on variable renewables for a substantial part of their energy (see Section 2.4).

2.2.6 Other Performance Characteristics

The above piston storage features (power, energy, footprint, pressure, location) are quite different from PHES. In other respects, performance and longevity should be similar. Roundtrip efficiency may be marginally higher [6] and should exceed 80%.

The construction phase would be as disruptive as it is for PHES and more evident—PHES schemes are usually in very remote locations. Completed piston storage projects can be unobtrusive and may be aesthetically pleasing or locally beneficial, for example, submerged piston designs and storage integrated with public water supply or flood prevention.

Storage has a much better safety record than electricity production, but all energy projects have safety issues. PHES is the main culprit as far as storage is concerned, with its virtual monopoly of grid-scale systems. Risks are largely confined to construction and supply industry personnel. Piston systems carry an extra element of risk because they will be somewhat closer to the general public. However, in the unlikely event that a terawatt-hour of gravitational energy escapes, it will be comforting if this happens hundreds of meters below the general public's feet, rather than above its head.

This is not just a question of up and down. As in a traffic accident, the speed at which it happens is critical. A megaton nuclear explosion can release 1 TWh almost instantaneously, destroying a city. The same energy, released in a Richter Scale 7 earthquake, can cause widespread damage in less than a minute. Catastrophic failure of a terawatt-hour hydroelectric dam might drain a 10^9 t (thousand million ton) reservoir in tens of minutes, with serious consequences along downstream watercourses. In comparison, fragmentation of a 1-TW h GBES piston would be a slow-motion car crash as rock and water jostle to swap places in a confined space—not a disaster, but hugely expensive and best kept a sensible distance from population centers.

2.2.7 What Could Possibly Go Wrong?

Gigawatt-class piston storage systems will face formidable engineering issues. Those raised most frequently relate to construction difficulties; the seal subsystem; structural integrity of the piston; preventing leakage from the high-pressure chamber beneath the piston; safety considerations (see Section 2.2.6, for example); "parking" the piston for maintenance; and preventing piston tilt (GBES only [6]). These are serious challenges, but they are not obvious show-stoppers:

- Construction: In essence, piston storage is much like any other very large construction project and could be undertaken by adapting equipment and techniques that are well understood in the civil engineering, electromechanical, hydroelectric, and extractive industries. It may encourage development of advanced technologies, for example, for rock cutting, but it does not depend on them.
- Seal: The seal subsystem is absolutely critical to the success of piston storage. All the developers believe that they have plausible solutions for evaluation during the crucial technical development stages.
- Piston integrity: There is never a cubic kilometer of seamless basalt around when you want one. It may seem that there is little chance of a real world piston surviving the huge forces exerted by the high-pressure water needed to lift it. In fact, the force on the base of the piston is virtually identical to that which was provided by the underlying rock for millions of years before the piston was separated. This, of course, was just enough to support it. If the exerted force is fractionally more, the piston starts to move up: if fractionally less, it starts to move down.
- Leakage: For the same reason, containment of the water beneath the piston is relatively easy, because it is surrounded by material of very similar lithostatic pressure (except for the seal, of course).

The nonengineering challenges are just as important. Given a choice, potential backers prefer charismatic technologies that can be demonstrated convincingly on a laboratory bench. They should promise short-haul development, with opportunities to gain experience (and early revenue) from niche markets. Modular systems are preferred, with economies of production scale rather than physical scale. Gigawatt-class piston storage ticks all the wrong boxes, especially for most venture capital investors. It will need to make a particularly convincing case, both technically and economically.

2.3 Piston Storage Economic Performance

2.3.1 Capital Cost

Cost comparisons for new technologies are not very meaningful—early estimates are notoriously unreliable and the comparison point is a moving target. For piston systems, PHES is the logical benchmark and current gold standard for high-power storage. PHES specific costs vary widely from project to project, but \$1500 kW^{-1} to \$2000 kW^{-1} (in terms of power) or about \$200 $(kW\ h)^{-1}$ to \$300 $(kW\ h)^{-1}$ (in terms of energy) are reasonable averages. Such figures exclude interest costs during construction, land acquisition, approvals, hookup costs, and dedicated transmission links. Multibillion dollar annual markets suggest that these cost levels are broadly acceptable for day-ahead storage.

The range spans ~\$1000 to >\$5000 kW^{-1}. Below-average costs are more prevalent in developing economies, which account for much of the global activity. High-end costs are sometimes attributed to PHES by those advocating alternative we-can-do-better-than-that technologies. We are not aware of any significant PHES activity toward the top of the cost range.

Capital cost estimates put forward by piston storage companies suggest that parity (or better) with PHES costs is plausible in key markets from hundreds of megawatt to multigigawatt (Table 3.1). One cannot yet put much confidence in such figures, and coverage in terms of project scope is uncertain. However, there is nothing to suggest that dramatic cost reduction will be needed to match PHES.

GP has probably made the most thorough of these cost appraisals, including assessments undertaken by German engineering companies. The GP figures in Table 3.1 are taken from a German 2013 source [7], converted to US dollars at the exchange rate prevailing at that time. These are for the 4 h modular system, targeting peak power markets. This relatively short duration leads to fairly high dollar per kilowatt-hour figures, especially for the 40 MW unit. An important point is that \$3200 kW^{-1} and \$800 $(kW\ h)^{-1}$ are competitive when compared to high-energy battery systems—a more realistic comparison point at this power and energy level. In any event, the 40 MW design is best regarded as a demonstrator since GP's interests are mainly at higher powers, where projected costs compare well with PHES. This is noteworthy as GP's figures are for relatively small diameter pistons—larger diameters should offer further economies of scale.

HE's figures [8] are intended to underline the steep fall in specific costs as size increases, rather than to represent the all-in costs of practical storage systems. They relate to the core mechanical construction costs of the piston/cylinder (excavation, reinforcement, seal system, etc.), where the HE model assigns costs to the various construction tasks [9]. They exclude major items such as hydroelectric equipment, the hydraulic circuit, upper reservoir, and powerhouse. The full system cost will be appreciably higher in dollar per kilowatt terms (although below the PHES average). System costs in dollar per kilowatt-hour terms will remain very low for the larger HE systems—say \$2 (kW h)$^{-1}$ for the 2000 GW h design. This is an extreme example, operating at ultrahigh pressures with variation between 15 and 20 MPa (150 and 200 bar) during the cycle.

GBES features should keep costs low, but preparation of authoritative estimates is not a priority at this stage. The figures in Table 3.1 are tentative (back of envelope plus ~50% contingency) and intended to illustrate the progression in moving to higher levels of stored energy, as discussed in Section 2.2.3. Moving from day-ahead to week-ahead applications is virtually cost free. Longer duration strategic storage incurs considerable additional cost, but the dollar per kilowatt-hour metric continues to fall sharply.

TABLE 3.1 Piston Storage Capital Cost Estimates

Power/ MW	Energy/ (GW h)	Time/h	Power capital cost/ (\$ kW^{-1})	Energy capital cost/ \$ (kW h)$^{-1}$	Comments (see text)
40	0.16	4	3200	800	GP—compare with battery storage
250	0.8	4	1700	425	GP
1000	6.4	4	870	220	GP
20	0.5	24	2400	100	HE—construction costs only
330	8	24	500	21	HE—construction costs only
750	125	168	800	4.7	HE—construction costs only
2750	2000	720	800	1.1	HE—construction costs only
2000	20	10	1300	130	GBES
2000	80	40	1400	35	GBES
2000	400	200	1800	9	GBES

Capital cost estimates require caution, but piston storage economics seem likely to be acceptable in proven day-ahead markets. Indeed, piston storage might outcompete PHES, if necessary. More importantly, piston storage is a potential frontrunner in areas where PHES cannot be used, and where competition is perceived as being much weaker. There are provisos, of course: piston storage has yet to demonstrate its technical and economic prowess; other emerging technologies may set tough new benchmarks in day-ahead markets; and smaller-scale storage is becoming much more prominent.

Piston storage is a promising candidate, not a shoo-in, but it has a couple of extra shots in its locker, if there is value in going beyond the energy needed for day-ahead markets:

1. Even if we are mistaken and piston storage holds no cost advantage over other contenders at 10 h capacity, the marginal cost of adding energy storage is inherently very low and should be far less than for other electricity-in/out storage technologies.
2. Asymmetric charging enables networks with highly variable production to capture large surpluses (far in excess of the power delivery rating of the storage unit).

Capital costs are important, but the overall cost and value of the storage service are crucial. As well as the return required to recover the investment, additional factors include operation and maintenance costs, input energy costs (including roundtrip losses), annual utilization, and the value of the output energy.

2.3.2 Finance Cost

In its simplest form, the finance cost (C_f, in terms of dollars per unit of energy) depends on the specific capital cost (C in terms of dollars per unit of power), utilization (U which is the equivalent annual power delivery time at the rated output), and the required rate of return (r expressed as a percentage) according to:

$$C_f = C\, r / U$$

As an example: assuming $C = \$2000\ (\text{kW})^{-1}$; $r = 6\%\ \text{a}^{-1}$ (where a is annum) and $U = 3000\ \text{h a}^{-1}$ then $C_f = \$40\ (\text{MW h})^{-1}$

In practice, adjustments may result in a lower figure to take account of the tax position or specific incentives for this type of investment, or the fact that storage is said to increase the value of other assets in the network's portfolio.

2.3.3 Operation and Maintenance (O&M) Costs

O&M costs for PHES are probably the lowest of all storage technologies, typically around \$5 $(\text{MW h})^{-1}$ for high-power installations.

Ignoring special issues for the moment, piston storage conditions bring benefits that might yield a 40% reduction to \$3 $(\text{MW h})^{-1}$ (~3000 h utilization, use of variable speed technology as standard, higher average rating).

In practice, this advantage is likely to be offset by seal subsystem maintenance and other deep-level work and so \$5 $(\text{MW h})^{-1}$ is a reasonable target. In the first instance, it would be prudent to budget for a higher figure— say, \$10 $(\text{MW h})^{-1}$.

2.3.4 Energy Costs

For a storage system with 80% roundtrip efficiency, the input energy is 1.25 times the delivered energy. Input energy costs can be calculated in several ways:

- One approach is to operate the storage system as a network asset in which participants can park surplus energy and share revenue from subsequent energy sales, reserve services, etc. The input energy could be treated as having zero value, since it is being parked rather than sold at that stage.

Other essential participants, for example, transmission and distribution operators, government agencies, and those financing the system, would also be included as beneficiaries. An agreement would account for energy and other costs as part of the equitable distribution of revenue.

- Alternatively, the storage system can be regarded as an independent entity, with input costs based on the market price of electricity at the time it is used to charge the system. Storage can take advantage of low off-peak power (there are times when it is near zero or negative in some networks). This is attractive initially, but is not particularly stable. Introducing storage increases demand and narrows the window during which low-cost surplus power is available. The sale of power from storage adds to supply and reduces the price of on-peak electricity. At some point, it becomes uneconomic to invest in further storage capacity (other than for low-energy reserve and regulation services). This will fall short of the optimum capacity in terms of overall benefit to the network (independent storage requires a production surplus to drive down the market price of input electricity and a subsequent supply deficit to drive up the price of peak power). A totally competitive storage market is unlikely to be in the best interests of the sector, or of the wider electricity supply industry. However, if this happens, piston storage could be the biggest bully on the street: it can afford an input energy price that would put some competing storage technologies out of business; asymmetric charging enables it to grab more than its fair share of input energy during shorter intervals of low prices; higher capacity systems (>20 h, say) will be the only buyers left when prolonged periods of high renewables production exceed the charging capacity of day-ahead storage.
- A third option is to base input energy costs on an estimate of the levelized cost of electricity (LCOE) delivered to the storage site. This gives stability, but introduces anomalies regarding the "correct" LCOE for input energy. For example, onshore wind power may be a principal energy input in a renewables-rich network. In some parts of the world, this would be low enough for profitable high-energy storage—perhaps a good deal less than \$50 (MW h)$^{-1}$. At the other extreme, ~\$150 (MW h)$^{-1}$ might be regarded as appropriate and storage is left with the weaker argument that it loses less money than the alternatives (dumping surplus energy or using it for other purposes).

2.3.5 Overall Storage Costs

At its simplest the LCOE delivered from the piston storage service is the sum of the costs of finance, O&M, and input energy.

This might put the output LCOE at around \$100 (MW h)$^{-1}$, comprising:

Finance cost	\$40 (MW h)$^{-1}$
O&M	\$5–\$10 (MW h)$^{-1}$
Input power price	\$40 (MW h)$^{-1}$
Energy losses	\$10 (MW h)$^{-1}$

A storage service costing ~\$60 (MW h)$^{-1}$ (on top of the input energy price) would probably be seen as viable by most networks, especially as it does not take account of other revenue opportunities, such as provision of reserve and regulation services.

Other analyses are more ambitious. An estimate on GP's website [3] gives the output LCOE as \$76 (MW h)$^{-1}$ for their 4 h, gigawatt-class design, based on \$40-(MW h)$^{-1}$ off-peak charging power and a capital cost of ~\$1000 (kW h)$^{-1}$. The storage service cost of \$36 (MW h)$^{-1}$ includes ~\$10-(MW h)$^{-1}$ energy losses. The balance of ~\$26 (MW h)$^{-1}$ for finance and O&M costs probably requires very low cost finance since funding has to be recovered through relatively few annual hours of power delivery with a 4 h system.

HE goes further still, estimating that it would take just four years to recoup the investment in a much higher energy gigawatt-class system [10].

2.4 Markets and Competition for Piston Storage

2.4.1 Size Matters, or Does It?

With few exceptions, piston storage power ratings will be >100 MW and energy ratings will equate to at least four hours at rated delivery power. This puts piston storage at the top of the power/energy spectrum—in fact, it goes well beyond the present storage envelope. There is not much advantage to be gained from output ratings above the present 3 GW achieved by PHES, but ultrahigh-energy/long-duration storage should be very important in future (and difficult for most other storage technologies to achieve).

Competition will also come from much further afield. Storage services are delivered by wire and users requiring high-power, high-energy storage can get it from a large number of low-power, high-energy systems or, for that matter, from an even larger number of low-power, low-energy units. Low-power technologies (<100 MW in a piston storage context) account for less than 5% of grid-connected installed storage capacity at present, but well over 95% of R&D expenditure and press coverage, heralding attractive new challengers.

Equally importantly, delivery by wire means that piston storage can serve low-power or low-energy applications for which a large dedicated unit would be totally inappropriate. The user of storage services does not need to know what is on the other end of the wire.

This assumes that the wires are still there by the time that piston storage becomes commercially available. The popular "utility death spiral" scenario postulates that we are already locked into a total transformation in electricity supply, whereby large central networks will be completely swept away by local energy provision, based on distributed resources. Electricity will be provided by renewables, other generators and cogenerators, integrated with storage in locally connected networks and microgrids. If this were the future, there would be little point in gigawatt-scale concepts, including piston storage. In practice, it is far more likely that we are heading for a future in which distributed resources form a significant component of a fully interconnected system or, failing that, one that maintains the status quo. In a straight fight between the alternatives, it is the small-scale systems that would usually be swept away if they were deprived of the support provided by the central network. The argument should revolve around whether the optimum distributed resource share should be 25% or 50% or 75%—not whether it should be 0% or 100%. Piston storage can live equally comfortably with any of these, other than 100%.

2.4.2 Short-Duration Markets (<4 h Storage Capacity)

There are unlikely to be significant sales of short-duration piston storage systems, but applications in this area represent an important source of revenue.

Piston storage is excluded from one of the most active areas. Modular distributed storage is extremely responsive and can do a far better job than central units in handling local power quality issues and in smoothing the output of highly dispersed wind, solar, and other erratic power sources. Systems can be constructed to any power level (from the smallest behind-the-meter units to ~100 MW thus far).

Piston storage will seldom be associated with a specific resource, unless the site is chosen for its proximity to gigawatt-scale wind farms or solar parks. The main short-duration market opportunity is in broader renewables support, frequency management, and other reserve services, extending into the lower end of time-shifting and peak supply. Large piston storage can ramp up or down at >100 MW min^{-1} (100 MW per minute) and react more quickly than spinning reserves or peaking plant (although not as quickly as batteries or flywheels). High-power storage is also valuable in emergencies, responding to line faults or failure of a large power plant, and in black-starting following network shutdown.

Naturally, these applications require that the storage unit is partially charged and not fully committed to its primary longer duration function at the time it is needed. However, part of the capacity of a large storage system could be ring-fenced for reserve and regulation services (permanently, or on a schedule agreed with the network operator).

2.4.3 Day-Ahead Markets (~3–20 h)

Day-ahead markets have been the mainstay of PHES and so account for the great majority of today's storage portfolio. It will remain extremely important in future. Storage capacity averages (8–10) h at present, but is seldom fully cycled. The average will decrease with the advent of new technologies optimized for somewhat lower stored energy. Annual utilization currently averages about 1000 h (equivalent hours at rated output). There is seldom any problem in gathering sufficient off-peak and surplus power to meet the following day's requirements.

This is a densely populated sector, both in terms of its share of the grid-connected storage market and in development effort. Around 20 storage technologies are vying to serve day-ahead applications and to build a platform from which they can tap into valuable shorter duration revenue streams. Day-ahead targets range from residential storage, through a host of distributed resource applications (extending well beyond 100 MW) and into traditional central network storage systems.

Piston storage should be an effective contender, but there are many high-power alternatives, including PHES, conventional and advanced compressed air energy storage (CAES), several electricity-out storage technologies based on thermal techniques and numerous battery electrochemistries.

Competition will also come from "power-to-X" systems, where surplus electrical energy is used to produce another energy vector such as heat, methane, hydrogen, and electrofuels, which can be used in other energy markets or stored cheaply for subsequent use. The focus is on energy harvesting rather than power delivery, since the roundtrip efficiency of power-to-X-to-power cycles is generally too low to compete in the day-ahead electricity business, if another storage solution is available. However, there is a good case for power-to-X:

- for low cost utilization of prolonged energy surpluses that would otherwise be discarded
- in the many networks that will choose not to use electricity-in/out storage among their techniques for network management
- as a means of using renewables to decarbonize natural gas networks and other fossil fuel usage (especially after decarbonizing electricity production, which is more effective in carbon terms, has run its course).

2.4.4 Week-Ahead Markets (~15–60 h)

There is little demand for systems with tens of hours of storage, although it is a useful feature for PHES or CAES if large lakes or caverns are available. Longer durations will become important in future networks that rely heavily on variable renewables, with prolonged periods when power production is generally in surplus and others when it is in deficit. As well as providing day-ahead services, higher energy units can also act as quasigenerators. When charging prospects are good, an equivalent capacity of fossil plant can be taken offline, in the knowledge that adequate notice can be given when it needs to be brought back into service. Similarly, storage units can be brought to a high state of charge in the days before a forecast period of prolonged low renewable production.

A market for these types of high-energy storage may develop fairly quickly, even though it may take several decades before problems become widespread. As with major road projects, large storage systems should be built to cope with traffic growth over a long working life. That is easier to justify if the cost of adding spare capacity is low or if the design can accommodate a low-cost upgrade, both of which apply to piston storage.

Most other storage technologies have relatively high dollar per kilowatt-hour capital costs. The exception is power-to-X, which is expected to be the main competitor in networks that use high-energy storage as one of their management tools. Power-to-X will be cheaper, but piston storage should represent a better investment under realistic input and output energy values.

Within the piston storage sector itself, moving from day-ahead to week-ahead energy capacity will often be very inexpensive (see Section 2.3.1). In this case, higher energy (tens of hours) systems may become the norm, even if most of the revenue comes from day-ahead and short duration applications.

2.4.5 Long-Duration Markets (>50 h Storage Capacity)

Compared with week-ahead systems, the main additional objective is to harvest energy that would otherwise be lost from the electricity supply system. This is not much of an issue at present, but will become prevalent in networks where variable renewables are the principal energy source.

There is no doubt that strategic storage would be extremely useful and there is unlikely to be much competition from other electricity-in/out technologies at this level. As noted in Section 2.2.3, a 600 GW h strategic store equates to about half the world's electrical energy storage capacity. It is inconceivable that half the present global PHES population would be installed in California or a large European country or a Chinese province, where 600 GW h (or more) strategic storage might be appropriate. This could be an attractive application for piston storage. The key questions are whether the cycle count and cycle value of the additional energy capacity is sufficient to justify its cost, and whether this is a better proposition than alternatives such as long-distance interconnectors or power-to-X systems.

Low marginal cost helps. So does asymmetric charging, which can radically improve utilization—ultramarathon runners would be at a huge disadvantage if their "recharging rates" at feeding stations were restricted to the rate at which they burn energy while running (say 40 kJ min^{-1}), especially if the energy gels are taken off the table every time the wind drops!

Opportunities for strategic (and possibly seasonal) storage will probably be confined to large supply networks. However, piston storage should greatly expand the envelope within which it makes sense to retain electrical energy within the network. It opens prospects for electrical energy storage at much higher energy levels than are envisaged with today's technologies, and in networks that have made little or no use of storage in the past.

3 ENERGY MEMBRANE–UNDERGROUND PUMPED HYDRO STORAGE

3.1 The Energy Membrane Concept

Invented in Denmark by JolTech [11], energy membrane-underground pumped hydro storage (EM-UPHS) stores energy by pumping water into a cavity bounded by two layers of membrane, sealed at the edges. The proposed commercial design is a 30 MW, 200 MW h unit with a working area of 0.2 km^2. The cavity is buried beneath a 25 m layer of soil. The cavity inflates when water (or seawater) is pumped in at about 5 bar, raising the soil until it is approximately 10 m higher than the surrounding ground level. Energy is recovered when the water is released in turbine mode. EM-UPHS is designed for use in areas where the terrain is unsuitable for PHES, but it can take advantage of any difference in elevation between the water source and the cavity installation site. For example, a difference of just 5 m would add about 10% to the operating pressure and energy rating.

The concept has been refined with the help of a pilot project operated over a period of several years. This is a cooperative venture involving EnergiNet.dk, JolTech ApS, GODevelopment ApS, RisøDTU, DTU Byg, SDU/MCI, GEO, Syd Energi, Danfoss A/S, Arkil A/S, Lean Energy Cluster, Sloth-Møller A/S, and Sønderborg Kommune.

The pilot plant (Fig. 3.3) is of appreciable physical size (2500 m^2) but on a small scale compared with the commercial design (~10% linear, 1% area, and 0.01% energy). Table 3.2 compares key design parameters. Early results led to a design revision in which the edges of the membrane are "clamped" within a profiled trench constructed around the perimeter of the membrane area. Together with a geotextile between the upper membrane and the soil in the edge zone, this prevents excessive strain and movement in this vulnerable area.

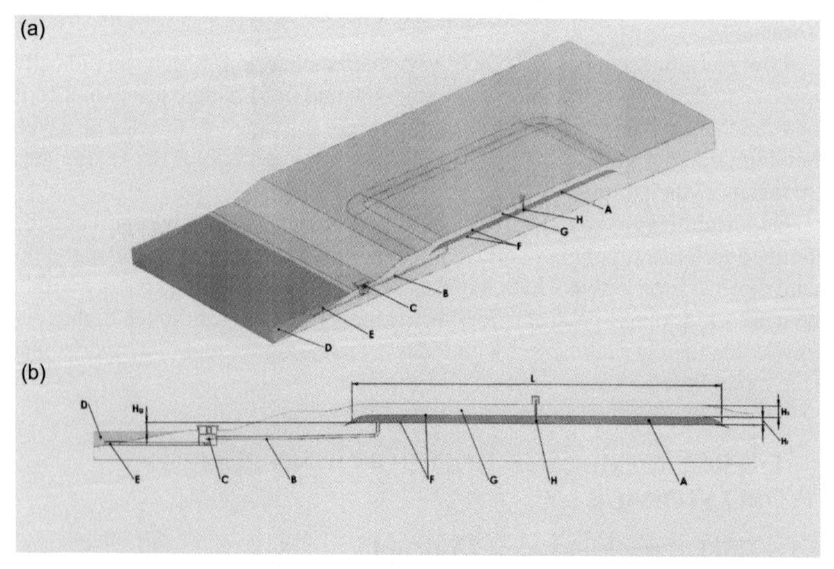

FIGURE 3.3 Energy membrane–underground pumped hydro storage schematic. Schematic representation of the underground energy storage concept. (a) Isometric view of the $L \times L = 50\text{ m} \times 50\text{ m}$ field test plant; (b) Section view showing details of the design. (A) inflatable cavity, (B) connecting pipe, (C) pump system, (D) water reservoir, (E) strainer, (F) lower membrane, (G) top soil, (H) level meter.

TABLE 3.2 EM-UPHS Design Parameters

Design parameters	(5 × 5) m Lab test	(50 × 50) m Field test	(500 × 500) m Plant
Pump/Turbine power/kW	Not measured	11/5.5	30 000
Underground storage area/m^2	25	2 500	250 000
Soil layer lift distance/m	0.1	0.6	10
Soil layer thickness/m	0.5	3	25
Soil layer weight/t	17	15 000	10 000 000
Cavity volume/m^3	1.5	1 500	1 500 000
Water pressure/(100 kPa) or/bar	0.07	0.6	5
Stored energy/kWh	0.003	25	2 900 000
Percentage energy loss in soil layer/%	5	0.4–1.2	<0.1
System efficiency/%	Not measured	30	>80

EM-UPHS topics for investigation include:

- Energy losses due to soil deformation
- Soil migration
- Longevity of the membrane and stability of the water cavity
- Verification of efficiency and economics at commercial scale.

3.2 Energy Losses due to Soil Deformation

Pumping water into the cavity stores energy by raising the soil mass. Energy losses result from relative movement within the soil layer, especially in the zone around the perimeter of the membrane where a ramp develops as the cavity is inflated. One of the main tasks of the trials was to measure these losses during charge/discharge cycles. They were found to average $<1\%$ of the input energy [12] and to correspond well with a finite element model incorporating soil properties. These losses are insignificant compared with the overall roundtrip losses of a commercial system ($\sim$20%). Indeed, the model predicts that scaling benefits should reduce deformation losses to $\sim$0.1% [12] if the model remains valid at higher energy densities (about 1 kW m^{-2} in the commercial design, compared with 10 W m^{-2} in the trials).

3.3 Soil Migration

The surface of the soil layer takes up a similar profile to the cavity, rising and falling during the charge/discharge cycle. When the cavity is inflated, some of the soil tends to migrate horizontally toward the (lower) surrounding ground level. Key factors include soil characteristics, rainfall (rainwater that reaches the membrane must drain laterally), and, most importantly, the cycle history (the number and depth of energy cycles and the gradient profile at different states of charge). It will take longer with clay than with sand but, basically, a block of soil will tend to level out, especially if it is raised and lowered once a day.

The average of the pilot trials appears to be about 1 mm of horizontal movement per meter of vertical travel (combined up and down). The paper reporting the results [12] shows a slowdown in soil movement toward the end of the 200 day trial, suggesting the approach of steady-state conditions. There were few charge/discharge cycles after 50 days and so it is unlikely that a plot of horizontal displacement against number of cycles or aggregate vertical travel would show such a pronounced slowdown. Another caveat is that most of the trials were gentle. The paper shows only two charge/discharge cycles exceeding 50% of the rated volume change/energy capacity.

Soil migration continues after steady-state conditions are reached, slowly reducing the working mass and, therefore, the megawatt-hour energy rating for a given cavity volume. This could be dealt with by undertaking occasional remedial earth movement.

Another issue is that thinning (or thickening) of the soil layer changes the cavity profile and is progressive if not corrected.

3.4 Membrane and Water Cavity

The developers are confident that further trials will confirm membrane longevity and the long-term stability of the water cavity. The surface area of the upper membrane (which is anchored along its edges) increases when the cavity is inflated. The intended shape of the water cavity is roughly similar to the space between two flattened domes. The separation between the upper and lower membranes increases gradually with distance from the clamped edge [12]. Compared with a near cuboid shape, this requires much less membrane expansion during charging and spreads the strain over a larger area. If expansion can be accommodated by membrane stretch, there is no need for excess folded material when deflated (with greater potential for abrasive movement and the possibility of soil accumulation in progressively deeper folds as the cycle count increases). Restricting the depth of discharge (to about 80% in the trial unit) helps membrane management by retaining some water within the cavity, preventing frequent contact between the membrane layers.

The required membrane stretch is small, provided that the cavity shape can be controlled and maintained. The target shape depends on applying the correct pressure profile over a large area of the cavity. This, in turn, requires controlled variation in soil layer depth. Maintaining the correct depth pattern over such a large area may be difficult, since the soil layer is subject to frequent cycling, gradient changes, and substantial movement.

3.5 Efficiency and Economic Performance

Capital costs for the 30 MW, 200 MW h commercial design are estimated [12] at €1111 kW^{-1} and €208 $(kW\ h)^{-1}$—roughly \$1200 kW^{-1} and \$230 $(kW\ h)^{-1}$ at May 2015 exchange rates. Estimates of the cost of the storage function vary from about \$60 $(MW\ h)^{-1}$ [12] to \$100 $(MW\ h)^{-1}$ [13]. Encouragingly, the lower estimate is from the more recent source, with details in terms of cost of finance (7%), energy losses, and utilization.

Energy losses are based on roundtrip efficiency of 80%—a figure generally associated with systems operating under much higher power and pressure conditions, but one which the developers expect to achieve with the 30 MW commercial unit probably using a reversible Francis pump turbine custom-designed for this purpose. Utilization equates to some 1800 h of annual power delivery at rated output.

These are difficult targets and the cost of the storage function will probably edge up with inclusion of O&M costs. However, if further work verifies these general levels of cost and performance, and also confirms the durability of the concept at the commercial scale, EM-UPHS should have a strong competitive position. The majority of the funding is in place [12] for the next planned step: a 100 kW, 400 kW h unit at the trial site, which should clarify these issues.

4 NOVEL LAND-BASED AND SEABED PUMPED HYDRO CONFIGURATIONS

4.1 Background

Chapter: Pumped Hydroelectric Storage focused mainly on conventional PHES systems, where water is pumped up mountains to a high-level lake or reservoir, which provides operating pressure at the low-level pump/turbine plant. The chapter also touched on alternative arrangements, operating on the same principle. The purpose is usually to provide a pumped hydro function in regions that do not have the mountainous terrain required for conventional PHES. Options include systems with:

- Two surface reservoirs
- One surface and one subterranean reservoir
- Seabed systems, where the sea acts as the high-pressure pressure reservoir and water is pumped from a low-pressure chamber anchored to the seabed.

4.2 Surface Reservoir Systems

These are variants of the standard PHES arrangement moving water between two accessible reservoirs. Possible configurations include:

- Use of worked-out open-pit mines as the lower reservoir. Some mines are of impressive depth [14], capable of accommodating large PHES-type reservoirs and operating at typical PHES pressures. The missing components are an equally large surface reservoir and, perhaps, readily accessible water (many mines are in arid areas, where provision of the initial water stock may be a problem, as may the modest ongoing requirement to make up for evaporation and other losses). However, major sites could provide storage systems of more than 10 GW h capacity.

- Use of smaller mining and quarry sites in projects of, say, (10–1000) MW h storage capacity [15,16]. QBC—the Quarry Battery Company—is a leading advocate of this approach, with plans to construct a 600 MW h facility in Wales [16], making use of two former quarry sites with an average elevation difference of about 250 m. This type of project tends to use fairly deep reservoirs of comparatively small area, resulting in substantial pressure variation during the cycle, due to changing water levels.

4.3 Subterranean Reservoir Systems

As discussed in chapter: Pumped Hydroelectric Storage, there is a long history of proposals (with little success to date) for pumped hydro projects in which the low-pressure reservoir and power plant are located underground, fed from a surface reservoir or other water source. The difference in elevation is typically comparable with a PHES plant but there is no particular restriction. Deep mines in South Africa, for example, have been developed over depths of thousands of meters, with workings at different levels. If storage systems were to be developed in deep mines, they would probably operate between worked-out levels, rather than at very high pressure. The main limitations are usually the capacity of underground chambers and the stability of mine workings when subjected to rapid water flows through the shafts and tunnels connecting chambers.

Most proposals envisage adapting old mine works [17], but underground reservoirs could be constructed specifically for the project, as with Riverbank Power's gigawatt-scale Aquabank concept [18]. Riverbank's website seems to have been inaccessible for some time, but a project of similar magnitude has been proposed by Sogecom and O-PAC with their FLES (flat land electricity storage) system [19]. The original plan to utilize a disused coal mine was taken further to propose excavating the lower reservoir within underlying solid rock to develop a 1.4 GW, 8 GW h project.

On a much smaller scale, the University of Colorado, Boulder (United States) [20] has explored underground pumped hydro storage systems operating from aquifers at power levels below 1 MW (down to tens of kilowatts) for use in agricultural and other rural applications.

4.4 Seabed Hydroelectric Storage

More than 99.9% of global hydropower is derived from freshwater lakes and rivers (which account for less than 0.1% of global water). The onshore/offshore balance is not going to change dramatically, but there is considerable interest in power from tidal, wave, and ocean current sources, including gigawatt-scale schemes. This encourages the development of hydroelectric equipment designed for saline conditions, which should also benefit energy storage applications.

Most seabed storage ideas use CAES techniques (see chapter: Underwater Compressed Air Energy Storage). Underwater pumped hydro concepts store energy by pumping water out of concrete pressure vessels anchored to the seabed at depths of 400 m to more than 1000 m. Energy is recovered when water reenters the vessel in turbine mode [21]. The operating pressure at 400 m is about 4 MPa (40 bar)—sufficient to generate a convenient 1 kW h for each cubic meter of water that flows back into the pressure vessel (or 2 kW h at 800 m depth). There are at least two programs investigating this.

The MIT proposal illustrated earlier (see Ref [22] and also Chapter: Pumped Hydroelectric Storage Section 4.2 and Fig. 3.3) envisages use of a large array of concrete spheres of (25–30 m) diameter, each storing around 6 MW h at 400 m depth (an operating volume of 6000 m^3). A depth of 750 m depth is considered optimal at current estimated installed costs [21], aiming for an energy storage cost of about \$60 $(MW\ h)^{-1}$. With a wall thickness of up to 3 m, the design could be used at greater depths. The mass of the concrete is sufficient to offset buoyancy, keeping the structure on the seabed without elaborate anchoring. The tank weight is presumably about 10 000 t, raising some transport issues.

Work in Norway by Subhydro [23] and SINTEF [24] emphasizes advanced concrete technology. The objective is to take weight and cost out of the pressure vessels, which would be buoyant when empty and could be towed to the operating site. These pressure vessels are cylindrical tanks with hemispherical ends, where the design volume depends on the cylinder length. Each hydroelectric unit serves a group of interconnected tanks. A ventilation shaft to the surface provides air at atmospheric pressure above the water in each tank [24] and may have other functions.

A ventilation shaft is a substantial structure, over 400 m in length and capable of withstanding the surrounding water pressure. The MIT approach does not require a ventilation shaft, with a near vacuum above the water, as the sealed tank is pumped out. This also provides a slightly higher operating pressure (one extra bar), but presumably requires a purging mechanism to prevent outgassing products building up as "new" seawater is depressurized during each cycle.

Seabed pumped hydro is intended to be a good match with deep-water floating wind turbine arrays, where it is envisaged that several hundred (perhaps 1000) tanks could be deployed to provide output smoothing and day-ahead storage services. It could also serve other applications in countries where appropriate depths are found reasonably close to shore.

5 OFFSHORE LAGOON AND ISLAND STORAGE SYSTEMS

5.1 Background

Seawater is an attractive resource for energy storage—PHES proposals can fail in areas with good geology because of interference with, or scarcity of, freshwater resources. Disadvantages include the corrosive nature of seawater and the somewhat lower efficiency and high specific cost of low-head hydroelectric plant. Greater use of ocean power resources will encourage development of saline low-head hydroelectric equipment. This will also help with the types of storage covered in this section, the first of which is closely related to tidal lagoon power generation.

5.2 Shallow-Water Lagoon Energy Storage

A simple tidal power generation lagoon is constructed by building a seabed-to-surface enclosure around a substantial area of water at a location with a large tidal range. Water is impounded within the lagoon until low tide, when it is released through low-head hydraulic turbines to generate power. Water is then excluded until high tide, when it flows back into the lagoon, generating further power.

It often makes sense to incorporate a pump storage function in tidal lagoon and barrage systems. It takes very little energy to pump additional water into the lagoon at high tide, because the water is raised only a small distance. The same water can yield much more energy when released through a greater distance at low tide. For example, if it takes 10 MW h of input energy to raise the level of a lagoon by 1 m at high tide, then one might, in principle, recover 100 MW h of additional energy when the water is discharged at low tide (the extra water is raised by an average 0.5 m but falls by more than 5 m in a good tidal area). This trick can be repeated at low tide, by pumping some of the remaining water out of the lagoon, reducing its level by 1 m.

In essence, this adds a PHES function at very low cost. It does not require a larger lagoon or bigger turbines (these are sized to handle the most extreme tidal conditions that the project is expected to encounter, and pumping just makes use of some of the surplus water storage capacity that is nearly always available). It has the intriguing quirk of an intuitively impossible $\gg$100% round-trip efficiency due to the input of "free" external energy (if you ignore the fact that tides slow the planet's rotation!). This makes a nice change from the usual scenario where 60% might be a struggle for a small turbine when most of the water is transferred at less than half the turbine design pressure.

Technically, this qualifies as energy storage, but there is very little control over the timing of charge and discharge cycles. It is best regarded as a means of augmenting tidal generation, rather than as part of the energy storage business. However, there is at least one proposal—by David MacKay of Cambridge University (United Kingdom)—where energy storage takes center stage [25]. This is based on more aggressive pumping regimes between multiple lagoons, at least one of which is built to a height well above the high-tide level. At its simplest, this would firm up intermittent tidal energy and act as an on-demand resource without any input energy from the grid. This flexibility is obtained by self-pumping (using tidal energy to pump between lagoons). Unsurprisingly, annual energy delivery is less than that in intermittent mode (about 70% in the example in [25]).

A more interesting example stores off-peak or surplus grid power (not confined to short periods just after high and low tides), perhaps with additional input from wind turbines built as part of the project (possibly driving pumps rather than generators). This can be regarded as a storage system that time-shifts the grid power used for charging and returns it at a time when it has higher value to the network, much as any other storage system, except that it returns more energy than it takes from the grid. There is some sleight of hand here, in that the extra energy is derived from the tidal power plant (and any integrated wind turbines). However, this is really a matter of accounting. The lagoon system has to recover high capital costs, and the way in which revenue is earned is not very important unless we could get a better result by optimizing the lagoon for power production, with storage services provided by another resource.

Near-shore lagoon storage is characterized by low-energy density and large physical size (maybe 100 km^2 for a system rated at hundreds of megawatts and multiple gigawatt-hours). Operating pressures are low and vary widely during the cycle, generally requiring separate pumps and turbines, rather than reversible machines. Potential sites are located close to shore in shallow water (of the order of 10 m, say), in areas with a wide tidal range.

5.3 Deeper Water Energy Island Storage

A similar technique can be employed in deeper water (of the order of 50 m, say). Suitable locations are fairly close to shore but tidal enhancement is not a factor—or, at least, not of much importance. Much of the work on this approach has been undertaken by KEMA (now within DNV GL) and by Gottlieb Paludan Architects [26,27].

A barrier is constructed to impound a substantial area, ranging from a few square kilometers up to perhaps 100 km^2 to form an island with an internal lagoon. Energy is stored by pumping water out of the lagoon to lower its level by, say, 45 m. Energy is recovered when water is allowed to flow back in through turbines. In principle, the system could be fully discharged, allowing the level to rise to that outside the island. However, little energy would be recovered during the final stages, at much reduced powers and low efficiency. Most designs are intended to operate within a narrower range (perhaps between (45 and 35) m head in our example).

Using the 400 t m $(kW\ h)^{-1}$ metric, an average head of 40 m and a 10 m difference in the charge and discharge levels would store 1 GW h for each square kilometer of lagoon area. Proposals reviewed for this report have energy densities in a range of about (0.3–0.9) GW h km^{-1}. Difficulties with working at higher energy densities in deeper waters (100 m, say) include: the rapid increase in construction costs with barrier height; the scarcity of suitable sites with good seabed geology and reasonably constant depth over a large area fairly close to land; and more demanding sealing issues at higher pressures.

If developed, energy storage lagoons will be huge projects, usually incorporating other infrastructure (which could sometimes be the main motivation for the scheme). Port, airport, marina, and recreational facilities might be integrated with the barrier, and part of the lagoon could be infilled to service these facilities and for residential, commercial, and industrial use. The surrounding barrier is a massive structure and extends well above sea level to prevent overtopping under extreme conditions. More infrastructure can be built on the barrier, including proposals for wind turbines driving mechanical pumps to help charge the system.

6 CONCLUSIONS

Conventional PHES has dominated the grid-connected storage market and, given the opportunity, it remains fully capable of defending its corner (high-power, day-ahead storage). It does not have that opportunity in areas where the terrain is unsuitable, or where appropriate sites have already been developed, or where projects have little chance of overcoming public opposition or approval difficulties. Conventional PHES cannot participate in the rapidly growing distributed storage sector and its role will also be quite limited within any future markets for week-ahead and strategic storage.

The storage sector looks set to become much more important—it is no longer acceptable to take the view that if you cannot have PHES you cannot have storage (or, at least, not of much significance). This book underlines the diversity of technologies that will help build markets around the PHES core and, sometimes, in competition with it.

The novel hydroelectric concepts discussed in this chapter are not yet market proven, but they share many of the characteristics of PHES. If they can get close to their economic and performance targets, they should help take PHES-type solutions into territories that are currently inaccessible. Some of them can also serve emerging markets for longer duration storage, which will be beyond the reach of the great majority of electricity in/out storage technologies, including conventional PHES.

ACKNOWLEDGMENT

The author wishes to thank Jan Olsen of JolTech ApS for helpful advice.

REFERENCES

[1] Gibson JW, Pierce MC. Remnants of early hydraulic power systems. Third Australasian Engineering Heritage Conference; 2009. http://www.ipenz.org.nz/heritage/conference/papers/gibson_j.pdf

[2] EscoVale Consultancy Services. www.escovale.com

[3] Gravity Power, LLC. www.gravitypower.net

[4] Heindl Energy GmbH. www.heindl-energy.com

[5] Escombe F. GBES-03, Proof-of-market study. Available from: escovale@escovale.com
[6] Escombe F. GBES-02, Concept and features. Available from: escovale@escovale.com
[7] Die neue Alternative für eine umweltschonende Energiespeicherung. http://www.bad-toelz.bund-naturschutz.de/fileadmin/kreisgruppen/badtoelz/dokumente/Home/Gravity_Power_Kurzinfo_Januar_2013.pdf
[8] Heindl E. Der Lageenergiespeicher, p. 39–40. www.physik.uni-regensburg.de/forschung/rincke/Allgemeines/Lageenergiespeicher-Lehrertage-2013.pdf
[9] Heindl E. Hydraulic hydro storage, an ecological solution for grid scale storage. www.efzn.de/uploads/media/131121_Heindl_-_Lageenergie-Speicher.pdf
[10] Heindl E. Hydraulic hydro storage system. Energy Storage Forum Europe; 2012.
[11] JolTech ApS. www.joltech.dk
[12] Olsen J, Paasch K, Lassen B, Veje C. A new principle for underground pumped hydroelectric storage. J Energ Storage 2015;2:54–63. http://www.sciencedirect.com/science/article/pii/S2352152X15000365 (with additional information from private communications with Jan Olsen).
[13] Ingeniøren, August 28, 2013. http://ing.dk/artikel/efter-aars-forsoeg-vejen-banet-nedgravet-pumpelager-til-vindmoellestroem-161238
[14] Top 10 deep open-pit mines. http://www.mining-technology.com/features/feature-top-ten-deepest-open-pit-mines-world/
[15] Leyshon Resources completes pumped storage study. International Water Power and Dam Construction; May 29, 2014. http://www.waterpowermagazine.com/news/newsleyshon-resources-completes-pumped-storage-study-4280650
[16] Holmes D. QBC pumped storage. The Engineer 2015; March 31. http://www.theengineer.co.uk/energy/in-depth/pumped-storage-a-new-project-for-wales/1020129.article
[17] Casey T. Old iron mine repurposed for new pumped storage hydroelectricity. http://cleantechnica.com/2013/10/11/new-pumped-hydro-energy-storage-uses-old-iron-mine/
[18] Reynolds P. Investigating Aquabank. Int Water Power Dam Construct 2010; January 4.
[19] FLES—flat-land large-scale electricity storage. http://www.sogecom.nl/energy.html
[20] Martin G. Aquifer underground pumped hydroelectric storage. http://www.colorado.edu/engineering/energystorage/files/EESAT2007/EESAT_AquiferUPHS_Paper.pdf
[21] Almén J, Falk J. Master's thesis T2013-387. Subsea pumped hydro storage: a technology assessment. Gothenburg, Sweden: Chalmers University of Technology. http://publications.lib.chalmers.se/records/fulltext/179756/179756.pdf
[22] Chandler D. MIT News Office. Wind power—even without the wind. Cambridge, MA: Massachusetts Institute of Technology. http://web.mit.edu/newsoffice/2013/wind-power-even-without-the-wind-0425.html
[23] Subhydro AS. http://subhydro.com/
[24] SINTEF. Storage power plant on the seabed. ScienceDaily 2013; May 15. http://www.sciencedaily.com/releases/2013/05/130515085343.htm
[25] MacKay D. Enhancing electrical supply by pumped.storage in tidal lagoons. Cambridge, UK: University of Cambridge. www.inference.eng.cam.ac.uk/sustainable/book/tex/Lagoons.pdf
[26] DNV GL 2000 Group. Energy Island. https://www.dnvgl.com/services/large-scale-electricity-storage-7272
[27] Gottlieb Paludan. Green Power Island. Copenhagen: Gottlieb Paludan; 2000. http://www.greenpowerisland.dk/

Chapter 4

Advanced Rail Energy Storage: Green Energy Storage for Green Energy

Francesca Cava, James Kelly, William Peitzke, Matt Brown, Steve Sullivan
ARES, Santa Barbara, CA, United States of America

1 INTRODUCTION

Advanced Rail Energy Storage (ARES) LLC, based in California, is a technology development firm dedicated to advancing the role of energy storage to improve the resilience, reliability, and environmental performance of the electrical grid. ARES has developed and been granted both domestic and international patents for an alternative method of utility-scale electrical storage that is significantly less expensive, more efficient, and more environmentally responsible than any competing technology. [Patents granted include the following: UTILITY SCALE ELECTRIC ENERGY STORAGE SYSTEM 8593012 (United States) 2529123 (Russia) 2012/01744 (South Africa) 33570 (Morocco); UTILITY SCALE ELECTRIC ENERGY STORAGE SYSTEM 8952563 (United States); COMBINED SYNCHRONOUS AND ASYNCHRONOUS POWER SUPPLY FOR ELECTRICALLY POWERED SHUTTLE TRAINS 9096144 (United States); RAIL BASED POTENTIAL ENERGY STORAGE FOR UTILITY GRID ANCILLARY SERVICES 8674541 (United States).]

ARES is a rail-based energy storage technology that, like pumped storage hydroelectric technology, stores energy by raising the elevation of mass against the force of gravity, and recovers the stored energy as the mass is returned to its original location.

The company has worked closely with key leaders in the railroad, energy, and environmental industries to develop its unique technology. The technology has been tested successfully in a pilot project in Tehachapi, California, and its first commercial deployment is under development in Pahrump, Nevada and will tie into the California electrical grid.

Storing Energy. http://dx.doi.org/10.1016/B978-0-12-803440-8.00004-X

Looking ahead, we see extraordinary economic potential as a result of the following factors:

- The market for electrical storage is expected to experience significant and enduring growth resulting from increased use of renewable sources of power (primarily wind and solar) that are inherently intermittent and do not synchronize with customer energy needs.
- ARES technology is approximately half the capital cost of its major competitor, pumped storage hydro, and is capable of delivering power and energy at an equivalent scale.
- ARES facilities can be constructed over a wide range of power and energy capacities to meet grid requirements. Many locations in hilly or mountainous areas can accommodate ARES with minimal environmental impacts.
- ARES uses no fuel, produces no emissions, requires no hazardous or environmentally troubling substances, and can be decommissioned with no lasting impact on the environment.

2 MARKET FOR UTILITY-SCALE ENERGY STORAGE

According to the US Department of Energy (DOE), "cost-effective grid-scale energy storage technologies are critical for accelerating the adoption of renewable generation and reducing CO_2 emissions from the electricity generation sector" [1]. In simple terms, wind and solar power are intermittent. Without large-scale energy storage, these sources of renewable energy are difficult to synchronize with demand. Energy storage thus plays a vital role in the world economy, a role that will become increasingly important in accommodating the wider use of low-carbon electric power.

Utility-scale energy storage currently represents approximately 3% of world electric power-generating capacity; however, this capability must increase dramatically to accommodate low-carbon energy technologies that cannot flexibly adjust their power output to match fluctuating demand. Energy storage, coupled with renewable energy, provides the system flexibility to adapt to variable electricity generation which otherwise would not be possible.

The electrical grid must continuously balance supply and demand on a moment-to-moment basis. Unlike the fuel supplies for electrical power that can be stored indefinitely, once the fuel is converted into electricity it must be consumed immediately. In the past, it has been difficult to store large amounts of electricity except through the limited availability of pumped storage hydro, discussed in chapter: Pumped Hydroelectric Storage. There are three main barriers to increasing the world storage supply: economic (the options are often too expensive); technological (the options are too inefficient or impractical), and environmental (e.g., construction of dams and impoundment of massive volumes of water have significant environmental impacts). ARES is now poised to provide a large-scale technological solution that removes all three barriers.

ARES offers a major economic opportunity in three main areas. First, ARES can shift bulk energy supplies between hours of greater and lesser demand, reaping the benefit of large price spreads that regularly occur over a 24 h day. Second, by storing energy at the point of intermittent generation ARES can reduce transmission constraint enabling the development of more renewable capacity behind existing transmission facilities. Third, because of its flexibility, ARES can provide a wide range of what are referred to as "ancillary services," including enabling the grid to adjust to momentary changes in demand, stabilizing voltage, and providing emergency capability to restart generators following catastrophic failures.

3 HOW MUCH STORAGE IS NEEDED FOR RENEWABLE ENERGY?

In most areas of the world, the use of renewable resources in power generation is on the increase. However, the electricity available from wind and solar energy is highly intermittent because generation depends on the weather. Technologies that provide the ability to store intermittently generated power can expand the use of these forms of energy by smoothing out the variations in hour-by-hour, minute-by-minute, and second-by-second availability. But these short-term variations are only part of the story. Where wind generates power at night when demand is low, energy storage increases the value of the investments in these generation systems and in transmission lines and other ancillary assets by shifting the delivery of the stored energy to optimal times of day [2].

In the absence of energy storage capacity, it will not possible to harness solar energy produced during the daytime for nighttime consumption regardless of how efficient the production of photovoltaic solar energy becomes. According to the US National Academy of Engineering:

However advanced solar cells become at generating electricity cheaply and efficiently, a major barrier to widespread use of the sun's energy remains: the need for storage.

The US DOE indicates that about 300 GW of storage (compared with current availability of about 25 GW) will be required by 2030 to reach the goal of integrating 20% renewable energy into the nation's electrical grid [1]. The US DOE's conclusions are:

Voltage and frequency support—About (20–50) GW of storage will be required in the seconds-to-minutes time frame for voltage and frequency support depending on the time of day and season of the year.

Smoothing renewable generation—Additional storage capacity of up to 40 GW will be required for periods of minutes-to-hours for smoothing and firming renewable generation.

Energy peak shifting—Storage capacity on the order of 200 GW will be necessary over periods of hours-to-days for daily energy peak shifting.

The Federal Energy Regulatory Commission (FERC) has acknowledged the important role that energy storage will play in the future [2]. In Feb. 2007, the FERC began to reduce barriers to the incorporation of alternative power suppliers in electricity markets [2]. Several independent system operators (ISOs) are developing new market rules that facilitate the incorporation of storage devices, and many are now releasing their first requests for proposals specifically for energy storage [3]. On Dec. 1, 2014, California's three large investor-owned utilities (IOUs)—Pacific Gas & Electric Company (PG&E), Southern California Edison (SCE), and San Diego Gas & Electric Company (SDG&E)—issued requests for offers (RFOs) for a total of approximately 95 MW of energy storage in California. These RFOs were the first competitive solicitation under California's AB 2514 energy storage procurement program, which requires the IOUs to procure a total of 1.325 GW of energy storage resources by 2020 [4].

The world demand for energy is roughly four times that of the United States, and in many places in the world the demand for renewable energy is growing explosively. The most significant growth of energy consumption is currently taking place in China, which has been growing at 5.5% per year over the last 25 years.

Energy storage is essential to the full deployment of renewable energy, but its value is not limited to renewable generation. It also serves important functions in a nuclear and/or carbon capture and storage (CCS)–dominated base load power system. Nuclear and coal with CCS offer large-scale, stable base load output, but cannot adapt well to fluctuating demand as discussed earlier. Building enough new base load units to meet peak demands would be prohibitively expensive. Energy storage mitigates the need to build for peak demand, regardless of the nature of the generation source. ARES technology will make it possible to minimize the increase in nuclear and CCS base load capacity by shifting the power these sources generate at night into daytime availability. In addition, virtually all of the benefits discussed earlier related to renewable energy apply equally to these alternative future energy scenarios.

4 VALUE AND STORAGE MARKET

The next decade is forecasted to be the beginning of a long-term, viable growth opportunity for commercial energy storage, according to Greentech Media (GTM) Research's report of Feb. 6, 2014, "Distributed Energy Storage 2014: Applications and Opportunities for Commercial Energy." Driven in part by the growth of solar photovoltaics, over 720 MW of distributed energy storage will be deployed in the United States between 2014 and 2020 (Fig. 4.1). This represents a 34% cumulative annual growth rate.

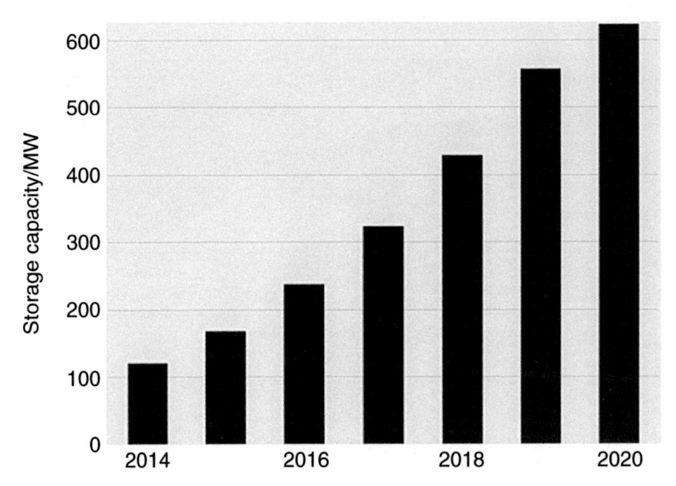

FIGURE 4.1 **Cumulative installed commercial storage capacity, 2014–2020.**

In California alone, the need for energy storage will also be driven by many factors, along with the loss of appetite for use of water in energy generation/storage [3]:

- Greenhouse gas reductions to 1990 levels by 2020
- 33% of load served by renewable generation by 2020
- 12 000 MW of distributed generation by 2020
- Ban on use of once-through cooling in coastal power plants
- California Governor Brown's 2030 goals:
  - Increase to a mandated 50% renewables
  - 50% reduction in petroleum use in cars and trucks
  - Double energy efficiency existing buildings
  - Require cleaner heating fuels

5 COMPETITIVE STORAGE TECHNOLOGIES

Several technologies are available to meet the diverse energy storage requirements of the electrical grid. Fig. 4.2 provides an overview of the scale of these technologies, both in terms of their power *capacity* and discharge *duration* [3]. Note that the figure is plotted on a logarithmic scale that tends to minimize the rather vast differences in scale among the various technologies.

Most of the energy storage technologies shown in Fig. 4.2 are capable of providing very important power quality functions if low power over short durations is the sole requirement. However, aside from ARES, only three energy storage technologies are now available, which offer some realistic potential to meet the storage requirements inherent in the nation's low-carbon energy needs in the near future—pumped hydro storage, compressed air energy storage, and batteries. Aside from ARES, only pumped hydro storage has the robust capacity and technological sophistication to merit wide deployment.

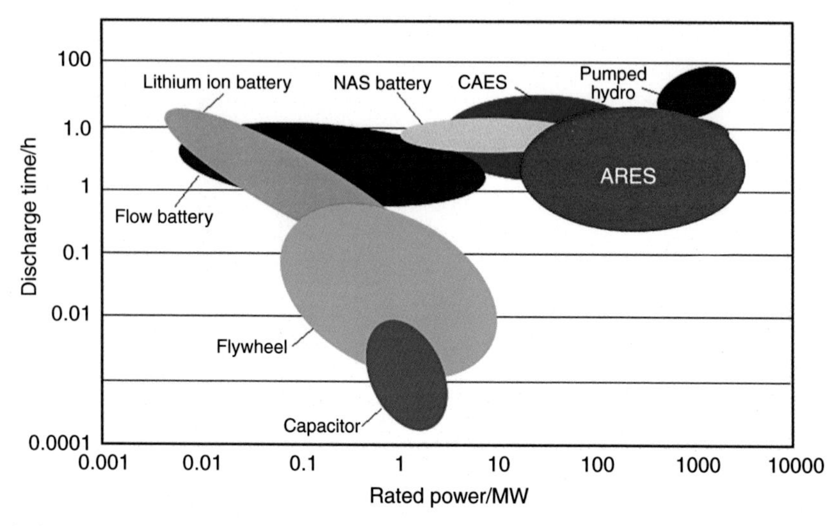

FIGURE 4.2 **Energy storage system ratings.**

6 ADVANCED RAIL ENERGY STORAGE

ARES mission is to deploy its energy storage technology to enable the global grid to integrate unprecedented amounts of clean, environmentally responsible, renewable energy while maintaining reliable electric service necessary for power growth and economic prosperity.

ARES is a rail-based, traction drive technology that offers a highly efficient and economical means of grid-scale electrical energy storage. ARES is a gravitational storage system that shares the robust performance characteristics of pumped hydro with many more siting opportunities, much lower capital costs, much shorter permitting and construction periods, and markedly lower environmental impact.

ARES will use surplus wind/solar energy or other low-cost energy from the grid to move millions of pounds of mass uphill by railroad shuttles, effectively storing thousands of megawatt-hours of potential energy, enough to power a medium-sized city for several hours. Fig. 4.3 shows two large fields of structural masses awaiting transfer from the lower storage yards of an ARES facility.

Once the masses are relocated uphill the transfer process is reversed as the demand for electricity requires. The shuttle units descend under gravity and the motors reverse to operate as generators, releasing the stored potential energy as power that is transmitted back into the electric grid.

ARES employs regenerative traction drive shuttle units operating on a closed, low-friction, automated rail network to shuttle a field of masses between the two storage yards. During periods when excess energy is available on the grid, the masses are transported uphill from a lower storage yard to an upper storage yard. The power lost in the combined storage and generation processes is under 23% more efficient than other large-scale storage technologies.

FIGURE 4.3 **Lower storage areas of an ARES facility.** An animated video of the operation of ARES is available at http://ares-llc.net/ARES_Animation/ARES_Animation.mov

6.1 Shuttle Vehicle

The heart of the ARES system is the shuttle vehicle. The requirements for the vehicle were very specific: it must be capable of reliably carrying and offloading very large loads—up to 45 t (50 short tons) per axle—with daily repetitions up and down a 16 km (10 mile) track at grades of up to 8.5%. In this process, the vehicle must also be highly efficient, resulting in no more than 10% power loss attributable to each roundtrip. The efficiency is discussed later and includes issues such as energy loss due to the motors and friction due to wheel slip. Finally, the vehicle must be designed using state-of-the-art railroad technology to minimize capital and operating costs while incorporating as much standard rail technology as possible to reduce both cost and production time.

The "trucks," or substructure assembly which holds the wheel/axle assemblies and traction motors of the powered units, are based on an Electro-Motive Diesels, Inc. three-axle design. The motors are currently available alternating current (AC) traction motors and are axle hung, which is the method used in modern heavy haul freight diesel–electric locomotives. The truck and motor arrangements were selected to minimize lead time to production, enhance reliability/durability, and minimize wheel wear. The vehicle structure will be adequate to support a large center load and be capable of supporting a mechanism to lift and rotate the masses (Fig. 4.4).

The powered units of an ARES shuttle vehicle consist of a frame, traction motors, underframe structure, electrical components, air compressor, hydraulic pump, and hydraulic cylinders (Fig. 4.5). The total weight of the shuttle vehicle, 70 000 kg (155 000 lbs) for each powered unit and 52 000 kg (115 000 lbs) for each unpowered unit, has a significant impact on system efficiency. The shuttle

FIGURE 4.4 **ARES shuttle vehicle.**

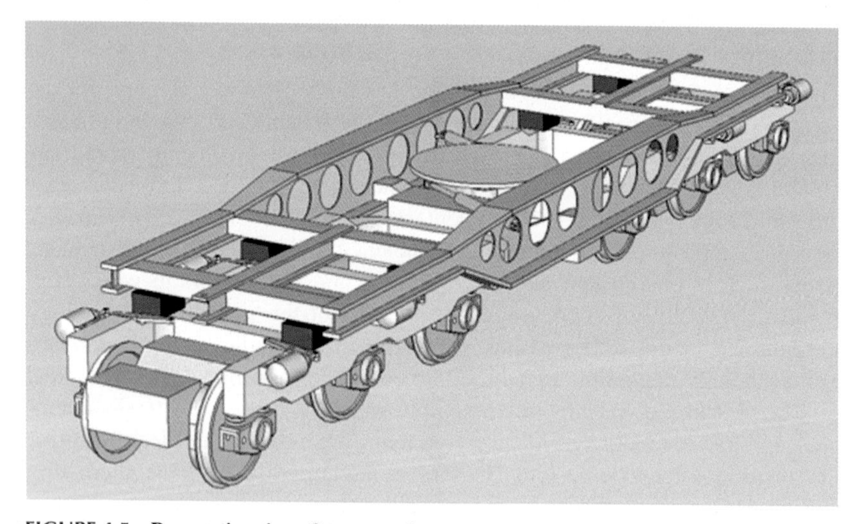

FIGURE 4.5 **Perspective view of a powered car.**

vehicle is currently based on fairly traditional structural design. Advances in this area are under consideration for the detailed design.

The efficiency of the shuttle vehicle is a function of losses due to the motors, inverters, gears, friction, wheel slip, and the third rail shoe. We believe the one-way efficiency of the vehicle as now designed is approximately 90.4%. This refers to the efficiency of the vehicle, excluding line losses, either in charging (transporting mass uphill) or in discharging (regenerative braking coming downhill).

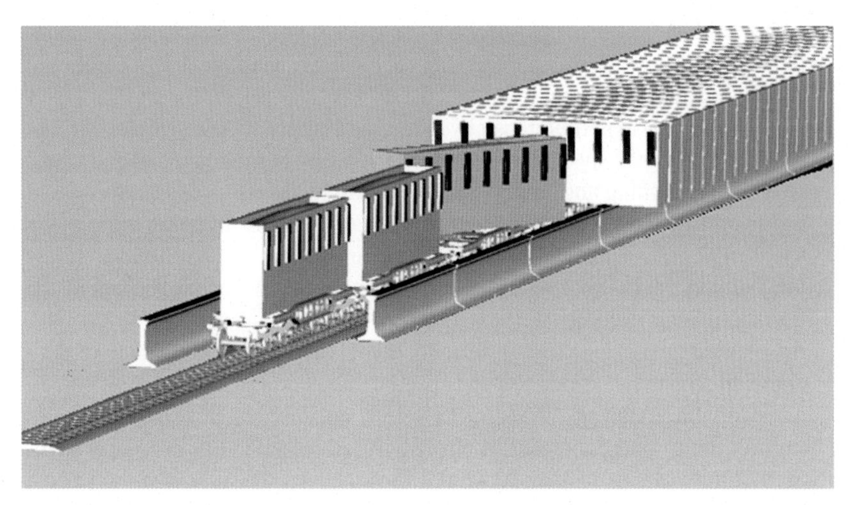

FIGURE 4.6 **Loading process.**

Adjusting for the difference in the mass of the powered and unpowered cars, the powered units weigh 202 000 kg (445 000 lbs) and the unpowered units weigh 220 000 kg (485 000 lbs). In the storage yards the masses rest perpendicular to the track as closely stacked as their 2.44 m (8 ft.) widths allow, on concrete walls, known as K-rails. The shuttle vehicle moves under a line of masses and the first unit stops under the last mass in the line. The unit then lifts the last mass off the K-rails and the vehicle moves so that the second unit is positioned beneath the next mass at the end of the line of masses (Fig. 4.6).

As the second unit lifts its mass, the first unit would be rotating its mass and setting it down on the structural rests. This process continues until all four units are loaded. The shuttle vehicle then moves to the other rail yard, either up or down the hill, depending on whether the system is storing energy or returning energy to the grid. The loaded shuttle vehicle could move at speeds up to 56 km h^{-1} (35 mph).

6.2 Rail

The major considerations in developing the ARES rail and track design are listed and include:

1. The ability to handle the volume of traffic and tonnage envisioned reliably and without interruption.
2. The adaptability to various sites, especially those with problematic ground conditions.
3. The maintainability of the system when subjected to anticipated heavy traffic loads.

Ability to handle the traffic volume—the track layout is simple and nonconflicting, so trains entering and leaving each storage yard are not in conflict in time or location with other trains. All the curves and turnouts have radii of over 610 m (2000 ft.) to permit transit by loaded and empty shuttle vehicles without reducing the transit speed during the entire run, and to minimize rail and wheel wear. Extra running and storage tracks are provided, so that an interruption of traffic on any one track can immediately be corrected by diverting trains to the spare tracks in a matter of seconds. There are set-out tracks provided at key locations in the running tracks so disabled vehicles can be parked out of the traffic stream if need be.

Adaptability to various sites—the track layout scheme is amenable to reasonable adjustment as needed to meet specific site conditions, prevailing grades, and available footprints. Rails/tracks will be anchored against slope creep, as rail fastenings and concrete ties normally used in general railway operations are not adequate to restrain downhill movement of the tracks on the steeper grades ARES will employ. Various methods will be used to reinforce the roadbed such as hot-mix asphalt underlayment, geogrid and/or geocell reinforcement, soil cement, etc., to strengthen low-competence soils to avoid subsidence of the track. These necessary civil design features and their maintenance are incorporated in the design criteria and are covered in both the capital and operations and maintenance cost estimates.

Maintainability and sustainability—The materials used in constructing the track and roadway will be the best and most durable available—rails that have shown excellent strength and wear life in comparable settings, premium turnouts with thick-web switch points, etc., all aimed at using track materials that have demonstrated dependable, trouble-free life in heavy-haul operations. In constructing the ARES rail components, the same basic track materials and design will be used as are currently specified for the heaviest haul railway in the world (FMG in Australia) and UP-BNSF in the United States.

6.3 Power System

The ARES shuttle units receive and deliver power via three-phase 2300 V AC trackside power delivery rails. Each of the powered units in the shuttle vehicles will have a leading and a trailing electrical contactor set maintaining constant contact with the trackside power supply rails. Grounding will be via the connection between the steel wheels, the track, and the ties, which are grounded via copper cables driven into the ground at a prescribed distance along the track.

Overhead 34.5 kV medium-voltage distribution lines parallel to the ARES rail system distribute power to the trackside power delivery rails via transformers spaced at 323.3 m (1060 ft.) intervals along the right of way. These three-phase AC overhead power transmission lines carry trackside power into or out of the ARES substation, which connects the ARES facility into the high-voltage utility grid.

The shuttle units integrate dual high-efficiency insulated gate bipolar transistor (IGBT) inverter/rectifiers onboard each powered shuttle vehicle for power conversion and vehicle speed control. This system converts the 2300 V AC 60 Hz grid frequency trackside power into onboard three-phase AC power at the specific frequencies required by the motor/generators. During periods of charging or discharging the IGBT inverter/rectifier units control the speed of the masses by controlling the frequency of the motor/generators driving the shuttle units and serve to transfer the AC power from the traction motor/generators back into the trackside power delivery rails at 60 Hz grid frequency.

Operationally, the dual inverter/rectifier units are deployed to accelerate the shuttle unit to a set speed at which the frequency of the onboard synchronous motor/generators exactly matches the 60 Hz frequency of the electrical grid. At this point the inverter/rectifiers are bypassed. This allows the motor/generators to lock directly onto the grid at 60 Hz removing onboard AC–DC–AC power conversion losses. The grid synchronization feature has two main benefits. First, it significantly enhances energy storage/redelivery efficiency. Second, it contributes much sought "heavy inertia" to the electrical grid.

An additional benefit of ARES unique grid synchronization system is that, during synchronization, the shuttle units are available to provide up to 100% reactive power for transmission grid volt-ampere reactive (VAR) support with the ARES system idle and 25% at full system load. VAR is a unit used to measure the reactive power of an alternating current. ARES ability to supply this level of VAR support to the grid is unique among large-scale energy storage systems and of significant value to system operators.

7 ARES OPERATIONAL CONTROL SYSTEM

The ARES system is controlled by a supervisory control and data acquisition (SCADA) system which implements, monitors, and controls the performance of broad system commands (such as generation output commands) by converting them into a predetermined set of remote system commands for control of parameters such as individual vehicle acceleration, speed, and braking. Direct commands from the SCADA system are sent to remote terminal units such as the shuttle unit vehicle controllers via multiplex relay over track rail and via wireless signal systems. Additional commands for enabling or modifying programmable control elements such as the switch automation system may be sent to remote programmable logic controllers, which will then automatically sense and monitor traffic to control switch position consistent with the operational mode of the system. Similar remote programmable logic controllers provide oversight to site-specific vehicle commands such as initiation of loading or offloading sequences.

Each shuttle unit may be programmed to respond either to direct control or to a remote programmable logic controller. In the latter case the unit will act as a member of an ad hoc meshed network system, automatically responding to operational requirements sent by the control center, taking into account each shuttle unit's physical location relative to other shuttle units and switch settings.

In general, the ARES SCADA system responds to requirements for storing or releasing energy, the type of ancillary services the system is providing to the grid, and weather conditions. The SCADA system incorporates input from sensors, including individual shuttle unit position, velocity, acceleration, mass loading, wheel speed and slip, electric component amperage draw, electric component voltage, electric component temperature, mechanical component temperature, rail switch position, and others. Depending on the location of the ARES facility, the sensors, remote terminal units, and programmable logic controllers which integrate into the ARES SCADA system may be hardwired, multiplexed through the ARES rail network or wireless with various communications systems and protocols.

Sensors on each shuttle unit will monitor rectifier/inverter functions, backup battery status, motor/generator status, lift mechanism function, brake function, track condition, and hydraulic fluid output through the SCADA system. If a malfunction is detected the shuttle unit is shunted off the main system for repair.

The system receives precise information on vehicle position through a combination of system-wide differential global positioning system (GPS) which tracks the position of each shuttle unit and trackside location tags which are placed every 15.3 m (50 ft.) alongside the track and read by a sensor on the shuttle units. Wheel rotation sensors measure the distance from location tags relative to the shuttle position allowing for granular vehicle position accuracy. Additional location tags placed on the individual masses allow the shuttle units to accurately sense their position relative to a given mass for loading and unloading (Table 4.1).

TABLE 4.1 ARES Performance Statistics

Scalability	(25–2000) MW
Storage duration	(0.25–24) h
Ramp rate (rate of power increase or decrease)	Up to 600 MW min^{-1}
Cycle life (cycled at rated power between charge and discharge)	Unlimited
Roundtrip efficiency	(78–80)%
VAR support capacity (reactive power)	(25–100)% of rated power
Standby storage losses	None
System life	40+ years

8 ADVANTAGES OF ARES

ARES offers several advantages over other technologies.

8.1 Large-Scale Load Shifting and Power Quality Services

In an ARES system the rate of energy input and output may be varied by controlling the speed and quantity of masses in motion (increasing or decreasing the intervals between shuttle units), allowing rapid response to grid power requirements over a wide range of output at a constant efficiency. An ARES facility can be constructed over a wide range of rated power and energy capacities from a small 25 MW facility with 6.25 MW h of storage capacity up to or beyond a 2000 MW facility with 240 000 MW h of storage. Once a site has been selected the basic system may be constructed in such a way as to provide expansion of the storage areas or power components as storage requirements increase in the future. Capital costs associated with expansion are a fraction of original capital costs. For example, for a 1000 MW system, it is anticipated the storage capacity can be doubled for about 13% of original cost because of reduced permit time, ability to use the same geographical footprint, and sharing of many components including the SCADA system and to some degree the main tracks.

The power components of ARES are designed in such a way as to offer the full complement of ancillary services: network frequency and voltage regulation, reserve capacity, black-start capabilities, as well as reactive power production.

8.2 Siting and Permitting

The central drawback with pumped hydro is obviously the requirement to find suitable sites where permits can be obtained in a cost-effective time period.

ARES does require suitable terrain, but there are many locations in hilly or mountainous areas where ARES may be accommodated with minimal environmental impacts. The fact that ARES does not require water or dams and does not involve noxious emissions offers a major advantage and will, in itself, result in significant reductions in capital costs.

8.3 Cost

A large-scale ARES facility will have a capital cost of roughly \$1200 kW^{-1} capacity compared with pumped hydro storage at \$2000–\$4500 kW^{-1} capacity [4], compressed air energy storage at \$1800 kW^{-1} effective capacity, and sodium sulfur batteries at about \$3000 kW^{-1} capacity.

9 POTENTIAL SITES IN THE SOUTHWESTERN UNITED STATES

The ARES team, working with Aspen Environmental Group, has identified over 20 potential ARES project locations in California and Nevada. Aspen excluded potential sites in areas where an ARES facility would likely be prohibited, such as federal Wilderness Areas, Bureau of Land Management (BLM) Areas of Critical Environmental Concern, wildlife refuges, and state parks. Sites within military installations were considered, although the likelihood of military cooperation is not known at this time. The result of this process was a list of the best of the identified sites. Summarized as follows are the initial parameters and criteria that were used to identify potential sites:

- an average gradient of (4–8)% over a minimum distance of between about 13 and 18 km (8 and 11 miles);
- within a reasonable distance of a (220–500) kV substation or transmission line—no more than 48 km (30 miles) away;
- within a reasonable distance of a railroad line (no more than 48 km away); and
- interconnection to the control area of the California ISO.

10 ARES PILOT AND FIRST COMMERCIAL PROJECT

To prove its technology and prepare for its first commercial project, ARES permitted, constructed, and demonstrated its technology with a pilot project in Tehachapi, California, completed in 2013 (Fig. 4.7). This project demonstrated ARES technology to well over 100 utility executives, policy makers, regional transmission system operators, and experts in the field of energy storage. The photograph in Fig. 4.7 depicts the ARES pilot vehicle as it traverses 275 m (900 ft.) of track at a 9% grade. An excellent overview of this pilot project has been

FIGURE 4.7 **ARES Tehachapi Pilot Project Vehicle.**

captured in video news coverage and can be found at http://www.bloomberg.com/news/videos/b/e2169a61-d5f3-485a-91a4-d22f68a55370

10.1 ARES Nevada Project—System Description

ARES is in the process of developing its first commercial project ARES Nevada – that is scheduled for operation in 2017. This project is situated in Nye County, Nevada, on approximately 25 ha (62 acres) of BLM-managed land 13.7 km (8.5 miles) southeast of Pahrump, Nevada (Fig. 4.8).

The ARES Nevada Project is a 50 MW gravity-based rail energy storage system which employs a fleet of seven heavy regenerative traction drive shuttle trains, operating on a high-grade closed low-friction automated steel rail network, to shift mass between alternate elevations, converting electricity into potential energy and back into electric power, as needed.

It will provide 12.5 MW h of fast-response energy storage necessary to assist in the balancing of electrical daily and seasonal supply and demand, as well as assist in balancing the highly and unpredictably variable renewable energy expected to be connected to the transmission grid, increasing renewable energy penetration while maintaining grid reliability. This project thus provides a form of energy storage known as ancillary services to keep supply

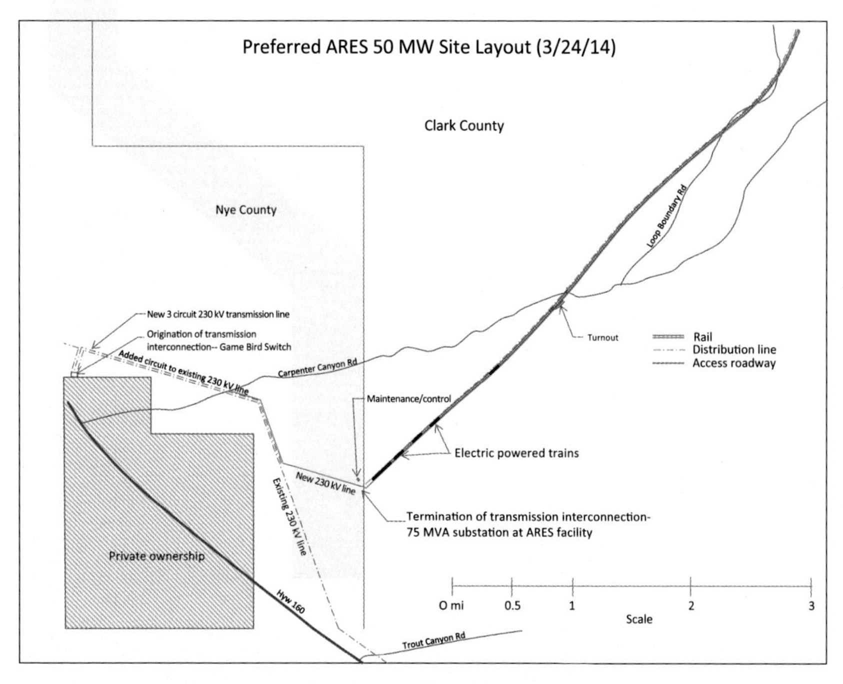

FIGURE 4.8 Location of the ARES Nevada Regulation Energy Management Facility.

and demand balanced. In the past, coal, oil, and nuclear generation have depended on large amounts of contingency reserves. However, energy storage and other demand-side resources can often react faster and provide more flexibility to the system.

During periods where excess energy is available on the grid (Reg-Down), ARES shuttle trains draw electricity from the grid which powers their drive motors to move the trains uphill against the force of gravity—efficiently converting electrical energy into gravitational potential energy. When the grid requires energy (Reg-Up) this process is reversed and the shuttle trains transit downhill with their motors operating as generators, converting the potential energy of the trains' elevation back into electricity in a highly efficient roundtrip process.

In the ARES Nevada Project the seven shuttle trains, each weighing 1223 t (1350 US short tons), operate on a 9 km (5.6 mile) long standard gauge track laid on an 8% average grade that provides 610 m (2000 ft) of overall elevation differential. The shuttle trains are electrified via a 2.5 kV, 3 degree overhead conductor system (OCS/catenary) interconnected into the California ISO (CAISO) transmission system through ARES 230 kV substation.

The ARES Nevada Project employs a three-phase catenary system, AC transmission lines, and AC power that is generally three phase. Three-phase power refers to three-wire AC power circuits. Typically, there are three (Phase A, Phase B, Phase C) power wires (120 degrees out of phase with one another) and one neutral wire. For our purposes let us consider a three-phase four-wire 208 V/120 V power circuit. This arrangement provides three 120 V single-phase power circuits and one 208 V three-phase power circuit. There is 120 V (alternating) potential between any power wire (through the load) and the neutral wire and 208 V (alternating) potential between the three power wires through the load.

A summary of the technical characteristics of the ARES Nevada Project (Fig. 4.8) include:

Rail Line

- 9.3 km (5.75 mile) fully automated electrified railroad; standard gauge track, 136-grade head hardened rail on concrete ties, crushed granite ballast and subballast
- maximum grade 8%; average grade 7.05%; overall elevation differential is 610 m (2000 ft.)
- 25 kV three-phase overhead conductor system (catenary);

Shuttle Trains

- seven heavy-mass shuttle trains consisting of two powered and seven unpowered units (Fig. 4.9)
    - gross mass of all vehicles combined—860 t (9 470 US tons)
    - gross nominal power at wheels 50 MW (67 000 horsepower);

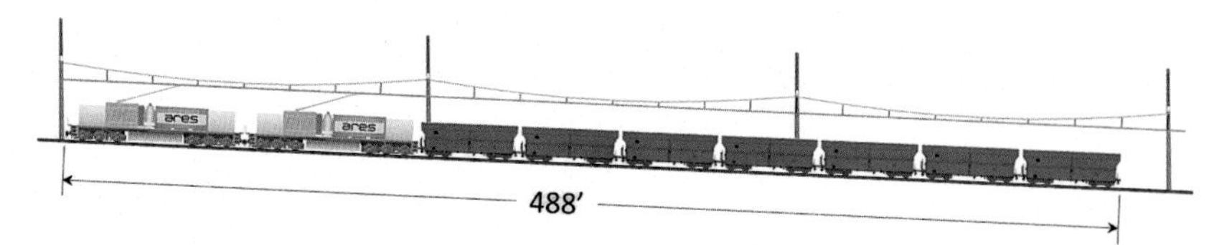

FIGURE 4.9 **The shuttle train.**

Operations, Maintenance, and Control Facilities

- onsite vehicle maintenance facilities—capable of performing scheduled vehicle maintenance
- control facility—monitors and controls vehicle response to Californian ISO automatic generation control (CAISO AGC) signal and provides site right of way (ROW) security monitoring;

Interconnection and Substation

- interconnection and substation supplied by Valley Electric Association (VEA) consisting of:
  - the addition of a 230 kV transmission circuit to the VEA Gamebird switch;
  - addition of the 230 kV circuit from the Gamebird switch to ARES site;
  - addition of 230 kV/25 kV ARES substation at ARES facility.

The new 230 kV transmission line will be constructed and operated by the local utility VEA, connecting the ARES Nevada Project to the regional CAISO electrical grid. This interconnection will include a substation at the southwestern end of the rail line, a gen-tie connecting the substation to the existing VEA 230 kV transmission line, upgrades to the existing VEA 230 kV transmission line to support the facility, two new transmission lines to connect the existing VEA 230 kV transmission line to the Gamebird switch station, removal of the section of the VEA transmission line that bypasses the Gamebird switch station, and a new switch yard at the existing VEA Gamebird switch station with a 230 kV ring bus for ARES, and a 2.1 km (1.3 mile) section of the existing VEA 230 kV transmission line will require tower upgrades to support the addition of the new power distribution line. The project is scheduled to begin operation in 2017.

11 CONCLUSIONS

ARES technology provides a ready alternative to pumped storage hydroelectric facilities for a wide and flexible range of larger scale energy storage applications. No scientific breakthroughs are required to make ARES work today—the durability and efficiency of railroad technology has undergone over a century of refinement and has been proven over tens of millions of safe miles. ARES rail-based storage can be configured to operate in a variety of modes that bring value to the grid (i.e., high power, high-duration ancillary services, etc.) and can move between these configurations. Unlike pumped storage, ARES has minimal environmental impacts, either in construction or operation, and can be decommissioned quickly and with virtually no residual footprint. Unlike batteries, ARES technology can retain stored energy for days, weeks, or months with no degradation and can operate for 40 years with no replacement and only normal maintenance. Further, ARES produces no emissions, uses no water, requires no hazardous or environmentally troubling materials (like lithium) and can be almost entirely repurposed or recycled upon decommissioning. Finally, ARES technology can be deployed quickly and cost-effectively in any site with acceptable topography, and can be scaled at decreasing unit costs as utility and customer needs dictate.

ARES technology represents a significant and valuable advance in the field of clean and environmentally responsible energy storage and should see wide deployment in the years to come.

ACKNOWLEDGMENT

This chapter is dedicated to John Robinson (1938–2011), cofounder of ARES.

REFERENCES

[1] Department of Energy. Grid-scale rampable intermittent dispatchable storage. Washington, DC: Department of Energy; 2010.

[2] California Independent System Operator. Integration of energy storage technology in power systems. Folsom, CA: California Independent System Operator; 2008. p. 12. http://www.caiso.com/1fd6/1fd676263d3d0.pdf

[3] California Independent System Operator. Energy and Aggregated Distributed Energy Education Forum. Folsom, CA: California Independent System Operator; 2015.

[4] FERC. Order 784. Washington, DC: Federal Energy Regulatory Commission.

Chapter 5

Compressed Air Energy Storage

Seamus D. Garvey, Andrew Pimm
Faculty of Engineering, University of Nottingham, Nottingham, United Kingdom

1 INTRODUCTION

Energy is required to compress air—and a very substantial proportion of all electricity generated worldwide is presently used for precisely this purpose. In the United Kingdom, for example, over 2.5% of all electricity is used in air compression [1,2], and this is not unusual in the developed world. Many varieties of air compressors are now highly developed to the point where they are both cost-effective and fairly efficient. Since energy is used to compress air, it follows that energy can be recovered when that air is expanded again. One simple illustration is familiar to most people—pumping up the tires of bicycles or cars with a hand pump or foot pump. The person doing the pumping has a very direct experience of putting energy into air compression, and the tire itself acts as an energy store. One could recover some work by allowing air from the tire into a sealed containment (such as a wine bag) and seeing that bag lift a weight placed on top of it. Though obviously simple, this illustration has most of the important features of compressed air energy storage (CAES). There is a compressor (the pump in this case), an air store (the tire), and an expander of some description (the wine bag that transforms some of the compressed air energy back into work).

The illustration glosses over one of the major concerns with all CAES systems: how to manage thermal aspects of the system. If a person were experimenting with CAES using the system illustrated previously, he or she would probably notice that the air pump became warm while it was being used and might also notice a slight cooling of the tire as air was removed. The set of possible CAES systems is very broad—mainly because of the large variety of ways in which heat can be managed. A section of this chapter has been devoted to the management of heat. The diversity of different possible CAES systems is further increased by the number of different ways in which pressurized air itself is stored. Two separate chapters (following this one) address some viable methods of storing large quantities of compressed air in affordable ways. Obviously,

Storing Energy. http://dx.doi.org/10.1016/B978-0-12-803440-8.00005-1

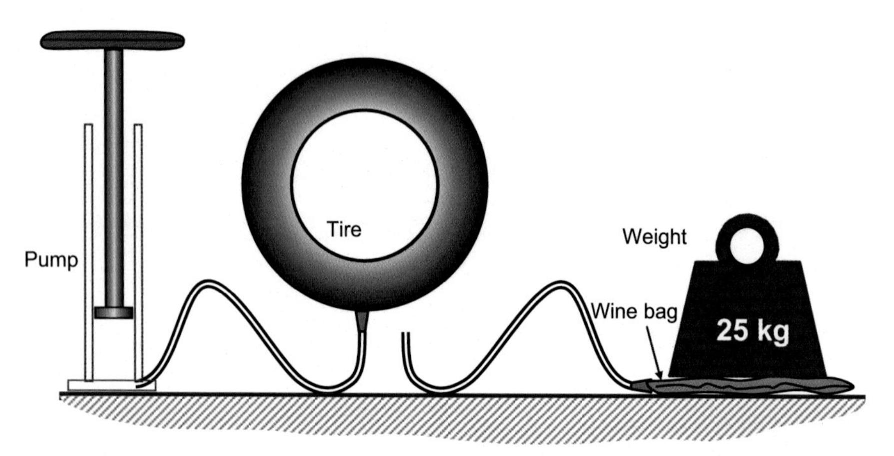

FIGURE 5.1 A simple notional CAES system.

because air is drawn directly from the environment and exhausted directly back into the atmosphere, no store is required for the "ambient air."

There are still more varieties of CAES systems because of the large range of possible machines for air compression and expansion (Fig. 5.1). New machines are under development constantly—especially in the domain of expanders (turbines) since the market for expanders has always been far smaller than that for compressors. A review of many varieties of compressors and expanders is beyond the scope of this chapter. It is generally accepted that small-scale compressors tend to be positive displacement, that is, reciprocating, screw type, vane type, or scroll type, that centrifugal compressors take over at medium-size levels, and the only choices for very large–scale compressors are axial flow dynamic machines. Positive displacement–type compressors can theoretically be run backward as expanders if provisions are made to open the valves at appropriate times (compressors using one-way valves can clearly never be run in reverse as expanders). Dynamic, that is, nonpositive displacement, compression machines cannot be run backward to expand air but there are near-equivalent machines in all cases that can extract work from air as that air expands.

A literal interpretation of CAES admits all systems containing compressed air—including every tire, football, airbed, and inflatable structure created. However, most uses of this term implicitly refer to systems for storing energy to support electrical generation, and we focus on this interpretation in this chapter. As noted in chapter: "The Role of Energy Storage in Low-Carbon Energy Systems", there is no affordable method of storing large quantities of electrical energy directly and all energy stores supporting electricity generation necessarily store the energy in a form other than electricity. CAES has potential to deliver high performance "electrical" energy storage at costs that are proportionately very low—especially at large scales. The achievable performance of CAES is discussed later in this chapter, and the chapter concludes with a discussion on how this energy storage format can be integrated closely with specific renewable energy harvesters and how it can be integrated with specific loads to great advantage.

2 CAES: MODES OF OPERATION AND BASIC PRINCIPLES

In the usual understanding of CAES systems, electrical energy is drawn from the power grid at certain times to charge up the store when the energy value is relatively low and is released back into the power grid (discharging the store) at other times when the value is higher.

When the store is being charged, electricity drives a compressor to inject air under high pressure into a storage facility. As air is compressed, its temperature rises and some deliberate decision is made concerning the heat. For several practical reasons, the high pressure air is never stored hot. Quite apart from the technical difficulty of preventing air from cooling while in store, the mass of air that can be stored in a given containment is inversely proportional to the absolute temperature of that air.

When the store is being discharged, high-pressure air from the storage facility is heated and passed through an expander under high pressure into a storage facility. As air is expanded and work is extracted from it the temperature of this air falls. As discussed later in this chapter, if there is some direct use for cool air, it may be appropriate to allow the air to emerge at a temperature lower than ambient. However, normally it is preferable to achieve exhaust air temperatures similar to ambient temperature to maximize the work output from the compressed air. Note that if an ideal gas is "expanded" through a nozzle, its temperature does not fall because no work has been removed from it. (Air that is not very cold at the start of expansion behaves as an ideal gas. If air is already very cold prior to expansion, it does cool further when throttled. This effect—the Joule–Thomson effect—is a key step in the manufacture of liquid oxygen and nitrogen. This effect is not relevant for most CAES processes.) Also note that if air is allowed to pass through an expander and to emerge with a temperature below 0 °C, there is a real danger that any moisture in the air will freeze onto parts of the expansion machine, and the designer must guard against this.

2.1 The Basic Equations Governing CAES

The ideal gas law is adequately representative of dry air at temperatures of 0 °C (273.15 K) and above and forms a good basis for all calculations on normal CAES:

$$pV = mR_{\mathrm{air}}T \tag{5.1}$$

In this, p represents the air pressure; V represents some closed volume of air; m denotes the mass of air in volume V; T signifies the temperature of the air (K); and R_{air} is the specific gas constant for air. For dry air:

$$R_{air} = 287.058 \text{ J K}^{-1} \text{ kg}^{-1} \tag{5.2}$$

Atmospheric pressure varies slightly with time and strongly with height above sea level for obvious reasons. The average atmospheric pressure at sea level is:

$$p_0 = 101\,325 \text{ Pa} \tag{5.3}$$

and it is common for system design purposes to assume that this is the pressure of air entering a compressor. For systems that are to be located significantly above sea level the value of p_0 is lowered by approximately 12.3 Pa m^{-1}. Thus, for a CAES installation at 500 m above sea level, ambient pressure would be reduced to $p_0 = 95\ 165$ Pa.

A compressor raises the pressure from the ambient pressure p_0 to some higher pressure p_0. The pressure ratio, r is defined as:

$$r := \frac{p_1}{p_0} \tag{5.4}$$

and for most CAES systems that have been considered seriously, r is set between about 20 and 200. When air is compressed, it tends to become warmer. If no heat is allowed to enter or leave the air during compression the temperature of the air after compression is related to that before compression according to:

$$\left(\frac{T_1}{T_0}\right) = \left(\frac{p_1}{p_0}\right)^{\chi} \tag{5.5}$$

where

$$\chi := \left(\gamma - \frac{1}{\gamma}\right) \tag{5.6}$$

and where γ is the ratio of the specific heats of air $\left(c_p/c_v\right)$ and this value is invariably very close to 1.4. Thus, χ is normally 0.2857 for dry air. The temperatures achieved by air compression can be quite high. In thermodynamic calculations such as those done for CAES, we use the absolute (Kelvin) temperature scale in which 0 K corresponds to –273.15 °C, 273.15 K = 0 °C, 373.15 K = 100 °C, etc. Table 5.1 gives the temperatures associated with adiabatic air compression when air has been inducted at 300 K (16.85 °C). In adiabatic compression, negligible heat transfers occur either into or from the air as it passes through the compressor. The word adiabatic will be used later in a slightly different sense to describe a complete system.

TABLE 5.1 Temperature Increases Due to Adiabatic Compression

Pressure ratio, r	1	2	5	10	20	50
Final temperature, T_1/K	300	365.7	475.1	579.2	706.1	917.4
Temperature rise, $T_1 - T_0$/K	0	65.7	175.1	279.2	406.1	617.4

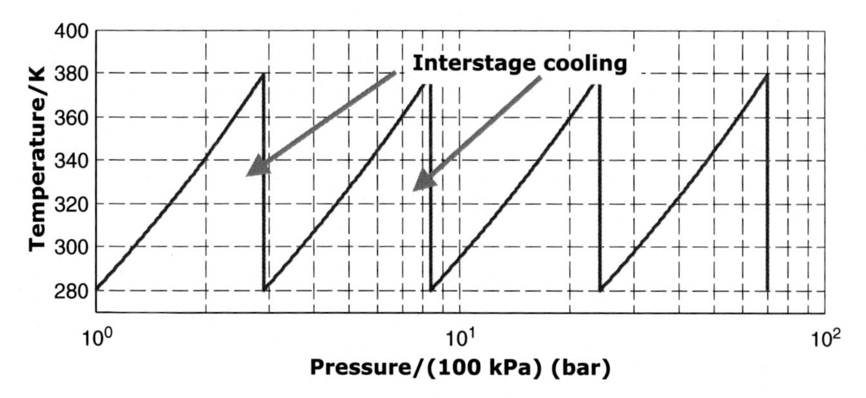

FIGURE 5.2 Temperature–pressure plot for a four-stage compression process.

The data in Table 5.1 do not depend on the pressure of the incoming air. Although CAES invariably sucks in air from the environment at pressure p_0, the overall compression might take place in multiple adiabatic stages with some cooling (called "intercooling") between the stages. For example, a 100:1 pressure ratio, that is, $r=100$, can be achieved by two-stage compression with $r=10$ in each stage and with intercooling being introduced after the primary compression stage and again after the second stage. Temperatures would rise only to 579.2 K and the same work is done in each stage. Fig. 5.2 illustrates the trajectory of pressure and temperature in a four-stage compression process taking air from atmospheric pressure—very close to 10^5 Pa (1 bar)—up to 7 MPa (70 bar) with equal pressure ratios in each stage and assuming that air enters each compression stage at 280 K.

The number of stages used in a compression process is arbitrary. In general, using more compression stages reduces the total work required to compress a given quantity of air and reduces the amount (and grade) of the heat rejected from the process provided that the air is cooled again between stages. In the hypothetical extreme case where air is compressed in a very large number of stages and cooled back to its original temperature after each stage, the work required to compress some initial volume, V_0 of air from its original pressure p_0 up to an increased pressure $p_1 := rp_0$ is given by:

$$W_{\text{isoth}} = p_0 V_0 \log(r) \tag{5.7}$$

This compression process is called "isothermal" because the air temperature remains constant throughout the process. Since the temperature remains the same between intake and exhaust, the product $(p_0 V_0)$ is identical to $(p_1 V_1)$ [see Eq. (5.1)]. The former is the work done by the atmosphere pushing ambient air into the compressor intake and the latter is the work done against air in the high-pressure reservoir pushing the pressurized air out of the compressor. For an isothermal compression process, there is no distinction between the total work done by a closed process where the same gas always remains in place (such as a piston with an airtight seal running within a cylinder) and the total work done by an open process where the gas passes through the compressor and out the other side.

The net work required to induct some initial volume V_0 of air from its original pressure p_0 and then compress it up to an increased pressure $p_1 = rp_0$ and discharge this air into a repository at pressure p_1 with an open adiabatic compression process (an open compression process is one in which the gas passes through the compressor—for adiabatic compression, temperature rises and hence more work is done by the compressor discharging the gas into the high-pressure manifold than is recovered from the atmosphere in pushing gas into the compressor) is given by:

$$W_{\text{adiab}} = p_0 V_0 \frac{(r^{\chi} - 1)}{\chi} \tag{5.8}$$

W_{adiab} approaches W_{isoth} from the previous equation, if the pressure ratio, r falls toward 1 or if χ is reduced toward 0. Eqs. (5.7) and (5.8) enable us to calculate how much air must be compressed to absorb a given amount of energy. Eqs. (5.1) and (5.2) can be used to express the quantity of this air in terms of its mass. For reasons that become clear later, it is useful to present alternative forms of Eq. (5.7) and Eq. (5.8) in terms of the mass of air compressed:

$$W_{\text{isoth}} = mR_{\text{air}} T_0 \log(r) \tag{5.9}$$

$$W_{\text{adiab}} = mR_{\text{air}} T_0 \frac{(r^{\chi} - 1)}{\chi} \tag{5.10}$$

Table 5.2 shows how much air can be compressed by 1 kW h of work (3.6 MJ) in both cases and expresses this in terms of the required intake volume—assuming that p_0 is given by Eq. (5.3)—and in terms of mass of air (taking $T_0 = 300$ K).The data in Table 5.2 show several features, of which two

TABLE 5.2 Quantities of Air that can be Compressed with 1 kW h of Work

	Isothermal compression		Adiabatic compression	
Pressure ratio, *r*	Mass air/kg	Volume air/(m^3)	Mass air/kg	Volume air/(m^3)
2	60.31	51.26	54.53	46.35
5	25.97	22.08	20.46	17.39
10	18.51	15.43	12.83	10.91
20	13.95	11.86	8.824	7.500
50	10.69	9.082	5.804	4.933
100	9.078	7.715	4.379	3.722

are very worthwhile: (a) the work per kilogram or per cubic meter of intake air initially rises linearly with the pressure ratio but rapidly becomes insensitive to the pressure ratio and (b) the distinction between isothermal and adiabatic compression is negligible at small pressure ratios but becomes very significant at larger ratios—exceeding 2:1 for pressure ratios of 100:1.

Table 5.2 can be used directly in several ways. First, it facilitates assessment of the intake flow rates of a compressor for a given capacity. If a given CAES plant is required to have a 50 MW inlet capacity (= 50 000 kW) and if it used isothermal compression with pressure ratio $r = 100$, then we can immediately realize that an air mass flow rate of at least $50\,000 \times 9.078$ kg h^{-1} (= 126.1 kg s^{-1}) would be needed. Conversely, if it used adiabatic compression the air mass flow rate would be $50\,000 \times 4.379$ kg h^{-1} (= 60.8 kg s^{-1}). Table 5.2 also enables us to quantify the size of air store required for a given stored energy. This calculation depends on the nature of the pressurized air store, so the detail of this is reserved for later.

2.2 Electrical Energy, Work, and Heat in CAES

All gas compression processes of relevance to CAES use a mechanical machine to convert mechanical work into energy within a gas of raised pressure. Similarly, all gas expansion processes relevant to CAES utilize a mechanical machine to do the reverse conversion. When a CAES system is installed as a standalone plant in an electricity grid, an electric motor is run to drive the compressor and a generator is turned by the expander. Proportionately, very small amounts of energy are lost in the motor and generator as they convert energy between the forms of electricity and mechanical work. Conversion efficiencies above 98% are quite common for large electrical machines [3] and still higher efficiencies are possible. For simplicity in this chapter, we treat electrical energy and mechanical work as being interchangeable—recognizing that in reality each time that energy is exchanged between the two forms, there will be a small fractional loss.

Despite the apparent complexity of some of the equations, there are some very simple and fundamental principles underpinning CAES. We mention three such principles here:

1. If you take some quantity of air, compress it to raise its pressure, and then cool it such that it returns to its original temperature (maintaining the pressure), then the total heat that has been drawn out of that air is identical to the net work that has been invested in the compression process. (This applies to ideal gases. For temperatures above 200 K, air can be considered to behave as an ideal gas.)
2. Any thermodynamic process is an ideal process if it is "reversible." Any effect that causes a process to be irreversible is bad for system performance. Heat transfer across a finite temperature difference, mixing, and fluid flow across a finite pressure drop, are all instances of irreversibility occurring. Most processes used in CAES plants are intended to be as reversible as possible and any irreversibility present in a final design is either a mistake or a deliberate engineering tradeoff between system performance and increased system cost. By assuming full reversibility of all processes, we can usually reach a very quick first-order understanding of how much electrical energy we may get back out of a body of compressed air by considering how much electrical energy was required to develop that compressed air in the first instance.
3. Heat is a lower grade of energy than mechanical work (or electrical energy). In formal thermodynamics, the term "exergy" is used to indicate how much work (or electric energy) can be extracted from a system or body of material by allowing it to come back into equilibrium with its environment. Exergy is conserved except where irreversibilities destroy some of it. Exergy cannot be created. It is beyond the scope of this chapter to make a formal introduction to this most important thermodynamic quantity but we will use it nevertheless with the informal understanding given here. Note that 100 J of heat energy used to raise the temperature of some material above ambient corresponds to $<$100 J of exergy. Interestingly, it may sometimes require $>$100 J of exergy to remove 100 J of heat from some body and cause it, as a result, to become extremely cold—even with complete reversibility of all processes. However, this is no more than an interesting observation here.

With these three principles established, we can make some highly relevant calculations and reasoning. Rather than speak in generalities, we will now consider some aspects of the design of a possible CAES system that should store 1 GW h (3.6×10^{12} J) of electrical energy, that is, 3.6 TJ of exergy. Applying Eq. (5.1), we realize that when we charge the system up completely from empty, we will have to remove exactly 3.6 TJ of heat energy from the compressed air if, as is normal, the air is to be stored at ambient temperature. Applying Eq. (5.2), we find that the nearest-to-ideal energy storage system will be one in which the expansion process should be almost an exact reverse of the compression process with heat being reinjected into the air during expansion at the same pressures and temperatures as it was removed during compression. Applying Eq. (5.3), we discover that we can make some decision about what fraction of the exergy is stored in the form of heat. If all of the heat rejected is at ambient temperature, the compression is isothermal, and the heat carries zero exergy. There is no point in storing it since there is an infinite quantity of heat at the same temperature all around. Alternatively, if the compression takes place in a single adiabatic stage, then some of that heat may emerge at very high temperatures and this heat may carry a substantial quantity of exergy. Table 5.1 showed the temperatures achieved by single-stage compression with different pressure ratios. If the pressure ratio is relatively low, the highest temperature will not be high and therefore almost no exergy can be stored as heat. If the pressure ratio is high the highest temperature will be substantial and we may store a substantial amount of exergy in the form of heat—possibly as much as one-third of total stored exergy.

This subsection brings out one of the main features of CAES—namely, that it is not simply about how to compress air and store high-pressure air. The management of heat is also important. We close by noting that if a small amount of heat, ΔQ is placed into storage at temperature, T_1 the associated amount of exergy that has been stored is ΔB given by:

$$\Delta B = \Delta Q\left(1 - \frac{T_0}{T_1}\right) \tag{5.11}$$

The previous equation is (effectively) due to Carnot [4] and represents the maximum amount of work that a heat engine could extract from a quantity of heat ΔQ taken from upper temperature T_1 and allowing that low-grade heat will be rejected into the ambient at temperature T_0.

3 AIR CONTAINMENT FOR CAES

Obviously, the heart of every CAES system is a store for high-pressure air. Clearly, this air store determines how much energy the system will be able to retain and it will normally comprise also the most expensive component of the system. One major attraction of CAES systems is that it may often be possible to exploit some natural phenomenon to develop stores for pressurized air that can retain very large quantities of energy at relatively low cost.

There are two profoundly different extremes of high-pressure air stores, and all practicable systems deploy stores that represent either one or the other. Fig. 5.3 illustrates the contrast. In this a vertical *red* line indicates that for all levels of fill, the volume occupied by the air in the store remains constant. This is termed "isochoric" (constant volume) air storage. In reality, no air store is perfectly isochoric since the volume naturally increases with rising

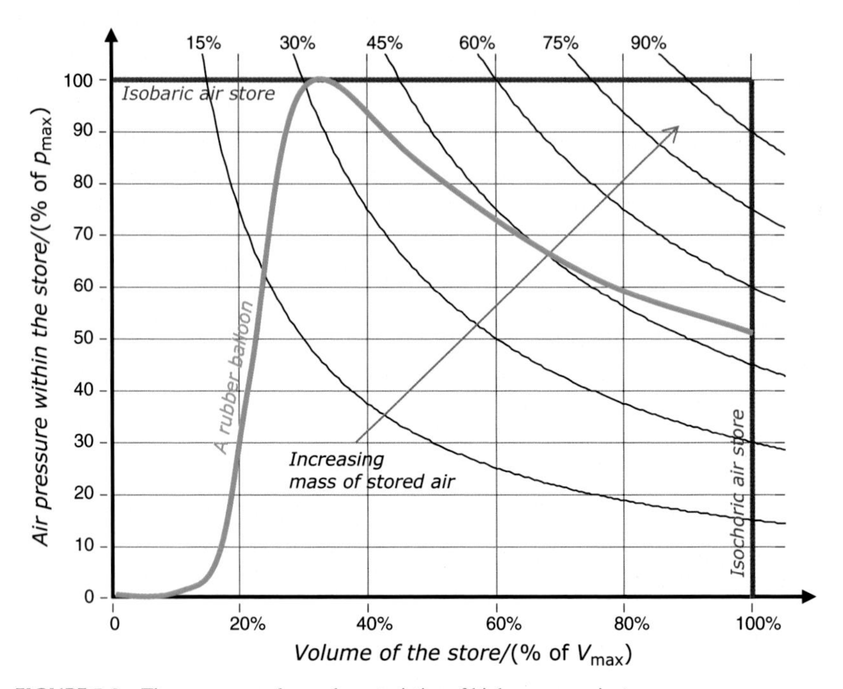

FIGURE 5.3 The pressure–volume characteristics of high-pressure air stores.

internal pressure. However, the change in volume between minimum and maximum pressures may often be very small.

In Fig. 5.3 the horizontal *blue* line indicates that for all levels of fill, the pressure of the air in the store remains constant. This is termed "isobaric" (constant pressure) air storage. Again, no real air store will ever be perfectly isobaric but some can approximate isobaric storage quite well—in that the variation of pressure may be small over the range of fill of the store.

The *green* curve in Fig. 5.3 indicates approximately what happens with an inflatable rubber balloon. This resembles neither isochoric nor isobaric air storage and is presented simply for illustration since most individuals are familiar with the experience of blowing up party balloons.

3.1 Isobaric Air Containment

Isobaric containments for pressurized air have the very strong engineering attraction that the compressor and expander work between (almost exactly) the same two pressures and thus they can be optimized for one single design point. Typically, all of the air present in an isobaric air store can be drawn out and we shall see that this contrasts strongly with isochoric storage. Moreover, isobaric air stores often experience lower variations in stress within the vessel since the air pressure remains essentially constant.

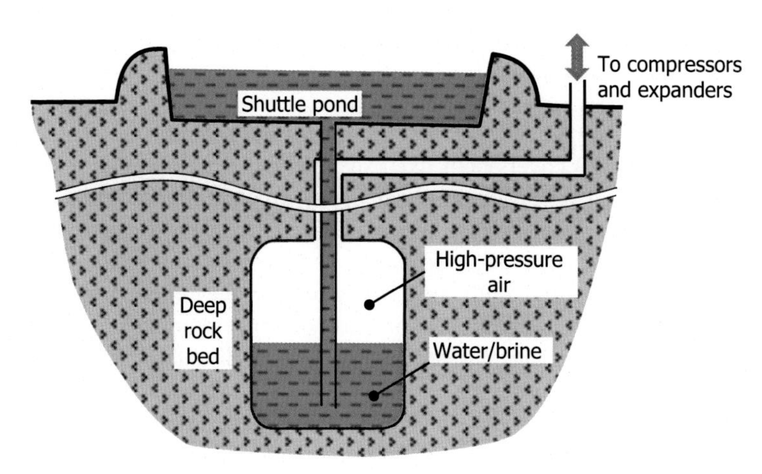

FIGURE 5.4 Fixed volume high-pressure air store uses liquid injection for isobaric characteristic.

Isobaric air stores are the easiest ones to analyze. Consider a store with volume V_1 and pressure $p_1 = rp_0$. The exergy stored in this air store is equal to the work that would be done by expanding this air isothermally at ambient temperature T_0, which may be slightly different from the temperature of the stored air T_1:

$$B_{\text{isobaric\_cavern}} = p_1 V_{1,\max} \frac{T_0}{T_1} \log(r) \tag{5.12}$$

We shall observe in a later section that we can often recover more work from air stored in such a cavern by exploiting other exergy that is stored thermally.

There are several possible ways to achieve near-isobaric air storage. The simplest way is to employ a fixed volume containment such as an underground (or underwater) cavern of fixed shape and to allow (or force) some liquid into the cavern to displace air that has been removed (see Fig. 5.4). In this case, work is done pushing the liquid into the cavern. In some system designs, a shuttle-pond containing water or brine at the surface is connected to the bottom of a cavern and the hydrostatic head between the water surface of the shuttle-pond and the water surface within the cavern provides the pressure to drive the liquid into the cavern. Arguably, such arrangements are a combination of pumped hydro and CAES. Eq. (5.12) captures the total exergy stored, but it should be noted that some of that exergy is not actually contained within the air and lies external to the air containment. In other system designs, there may be a shuttle-pond, but the hydrostatic head may be smaller than p_1. Then a pump would be needed to add additional pressure to the water/brine to drive this into the cavern, but note from Eq. (5.12) that, so long as $(T_0/T_1)\log(r)$ is greater than 1, the work put in by the pump is much lower than the exergy extracted from the pressurized air. Note also that (most of) this work can also be recovered as the cavern is filled with air again.

There are other ways to achieve near-isobaric containment. A separate chapter on underwater storage of compressed air (chapter: "Underwater Compressed Air Energy Storage") covers both fixed shape underwater caverns, as well as deformable shapes. Both can approximate isobaric containment quite well. Another possibility [5] is to separate the volume within a fixed shape cavern into two discrete parts such that the proportion of volume occupied by either part can vary over a wide range and to fill one of those parts with a fluid such as CO_2 that liquefies/evaporates at a suitable combination of temperature and pressure. As air is removed from such an arrangement, pressure tends to fall and more liquid CO_2 evaporates in response to any fall in pressure—soaking in some heat from the surroundings. Then, as air is injected back into the cavern, pressure tends to rise again and CO_2 gas condenses into liquid and rejects some heat into the surroundings.

3.2 Isochoric Air Containment

Isochoric containments for pressurized air are the simplest ones conceptually. They comprise one or more fixed shape volumes coupled by some arrangement of ducts, as Fig. 5.5 illustrates. An isochoric store is full when the pressure of the contained air is at its maximum allowable value $p_{1,\max}$, and the isobaric store is empty when the pressure has fallen to its lowest allowable value $p_{1,\min}$. The ratio between these two is typically less than 2—indicating immediately that when an isochoric store transitions between "full" and "empty" more than one-half of the air present in the "full" state remains present in the "empty" state.

Since the pressure within the store varies with fill level, some care is required to determine the amount of exergy stored in isochoric containment. The

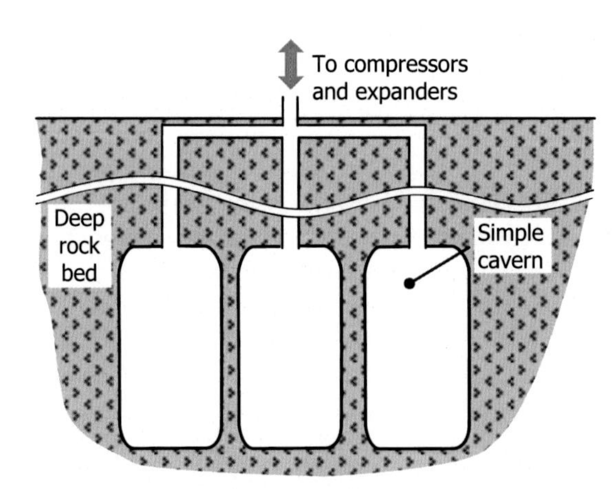

FIGURE 5.5 Simple isochoric containment for pressurized air.

temperature of the stored air will remain constant if the cavern is emptied or filled slowly. We assume that this is the case:

$$B_{\text{isochoric\_cavern}} = V_1 \frac{T_0}{T_1} \left\{ p_{1,\max} \left[\log\left(r_{\max} \right) - 1 \right] - p_{1,\min} \left[\log\left(r_{\min} \right) - 1 \right] \right\} \tag{5.13}$$

As noted in the case of isobaric containment, the total amount of work extracted from air stored in an isochoric store may be greater than $B_{\text{isochoric\_cavern}}$ above if there is some storage of exergy in the form of heat.

For reasons of simplicity and minimizing capital cost of equipment, some plant designs with isochoric air stores employ a simple throttle to reduce pressure from whatever pressure is present in the cavern at one time to a constant value suitable for the expansion equipment. This measure causes some loss of exergy that could be avoided by more sophisticated (but more expensive) machinery. If all air from a cavern will be throttled down from instantaneous cavern pressure to the pressure $p_{1,\min}$ before being passed into an expander, we obtain a lower value for the recoverable exergy:

$$B_{\text{isochoric\_cavern\_thr}} = \left(p_{1,\max} - p_{1,\min} \right) V_1 \frac{T_0}{T_1} \log\left(r_{\min} \right) \tag{5.14}$$

Table 5.3 provides some insight into the contrast between quantities of exergy stored in isobaric containments, isochoric containments with variable pressure ratio machinery and isochoric containments employing throttling. In this table we assume $T_0 = T_1$, $p_0 = 101\ 325$ Pa

TABLE 5.3 Quantities of Exergy Recoverable from Air Stores

Pressure ratio, *r*	Exergy (isobaric)/(MJ m^{-3})	Exergy (isochoric)/(MJ m^{-3})	Exergy (isochoric throttled)/(MJ m^{-3})
10	2.333	1.163	0.8248
20	6.071	3.169	2.528
35	12.61	6.737	5.615
50	19.82	10.71	9.106
70	30.13	16.42	14.18
90	41.04	22.49	19.61
115	55.29	30.45	26.77
140	70.10	38.75	34.26
170	88.47	49.06	43.09
200	107.4	59.69	53.28

and for isochoric arrangements:

$$p_{1.\min} = (p_{1.\max})/2$$

3.3 Air Containment in Tanks

Of course it is possible to store compressed air in tanks above sea level or ground level, and many people have proposed tanks for this purpose. One obvious concern with the use of such tanks is that if a tank failure should occur, a large amount of energy could be released very suddenly as an explosion and the natural shielding that is provided by storing pressurized air underground or underwater largely removes that danger.

The main concern with the use of tanks either at the surface or close to it is the cost [6]. High-pressure air tanks must be built with care, and they use valuable structural material. The design of pressurized containments is a highly developed discipline, and it is beyond the scope of this chapter to explore that in depth but we can develop very good approximate insight by considering that the tanks in question will have thin walls. If the maximum pressure of interest to us is 20 MPa (200 bar) and if the material used to construct the tank has allowable stress, $\sigma_{\max}$ much higher than 20 MPa, the thin-walled assumption is valid.

From simple mechanics (and Fig. 5.6a) a thin-walled sphere of radius R and internal pressure p_1 must have wall thickness t obeying:

$$t > \frac{(p_1 - p_0)}{\sigma_{\max}} \times \frac{R}{2} \tag{5.15}$$

The minimum volume of material required for the sphere wall is then found to be proportional to the volume contained within the sphere by:

$$V_{\text{wall}} > V_1 \times \frac{(p_1 - p_0)}{\sigma_{\max}} \times \frac{3}{2} \tag{5.16}$$

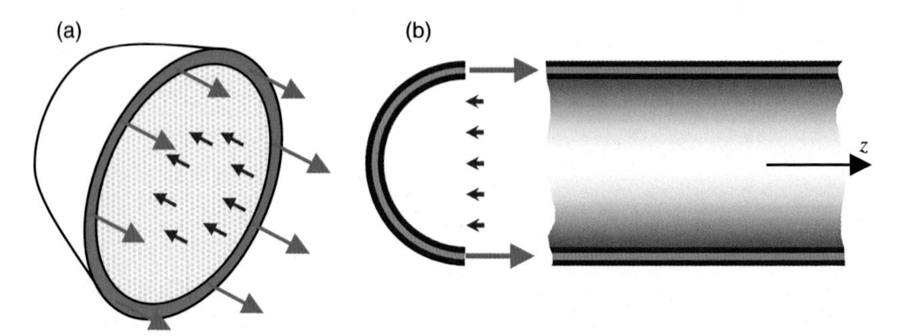

FIGURE 5.6 Stress calculations for thin-walled tanks: (a) spherical and (b) cylindrical.

A similar analysis can be conducted for a long cylindrical thin-walled tank radius R and internal pressure p_1. From Fig. 5.6b, this must have a wall thickness t obeying:

$$t > \frac{(p_1 - p_0)}{\sigma_{max}} \times R \tag{5.17}$$

and substituting this thickness back to assess the volume of material required for the wall returns:

$$V_{wall} > V_1 \times \frac{(p_1 - p_0)}{\sigma_{max}} \times 2 \tag{5.18}$$

Not surprisingly, comparison of Eqs. (5.16) and (5.18) reveals that the spherical tank makes best use of material. Since the two equations do not differ by much, it is conservative and reasonable to apply Eq. (5.18) as a general lower bound and to remove any constraint on the shape of the tank. It could comprise, for example, a long coil of very thin–walled tube.

Having knowledge of the volume of material required to construct a tank is a first step in estimating how much it might cost, and it also provides a very good estimation of the weight of the tank.

A brief calculation provides some interesting insight. Suppose we arrange a compressed air tank to contain 1 m^3 of air at 200 bar and we allow the internal pressure to fall to 2 bar. Consider that this tank is to be constructed from steel with maximum allowable stress of 1000 MPa and density 7800 kg m^{-3}.

From Eq. (5.13) (with $T_0 = T_1$), the exergy stored in this tank is 86 MJ—about 24 kW h. For an ambient temperature of T_0 = 300 K the mass of air in the full tank is 232 kg and the mass of the tank wall itself is 155 kg bringing the complete tank mass to 387 kg. A very good battery might presently achieve an energy density of 200 W h kg^{-1}, and to store 23.9 kW h in that battery would demand a mass of around 120 kg. Evidently, compressed air stored in tanks delivers an energy density that is lower than that of present-day batteries—but not an order of magnitude lower.

Based on a present-day rough assessment of \$300 $(kW\ h)^{-1}$, this energy store might justify spending ~\$7200. This would correspond to ~\$46 kg^{-1} of steel. With an appropriate manufacturing route and tank design, such a value is easily achievable.

3.4 The Case for Underground or Underwater Storage

Section 3 of this chapter makes the point that the cost of storage in pressurized tanks at or near the surface is almost directly proportional to the volume of the storage and to the pressure difference between storage pressure and ambient pressure. This is not the case for stores developed either underground or underwater. In both cases there may be a relatively high fixed cost for causing any

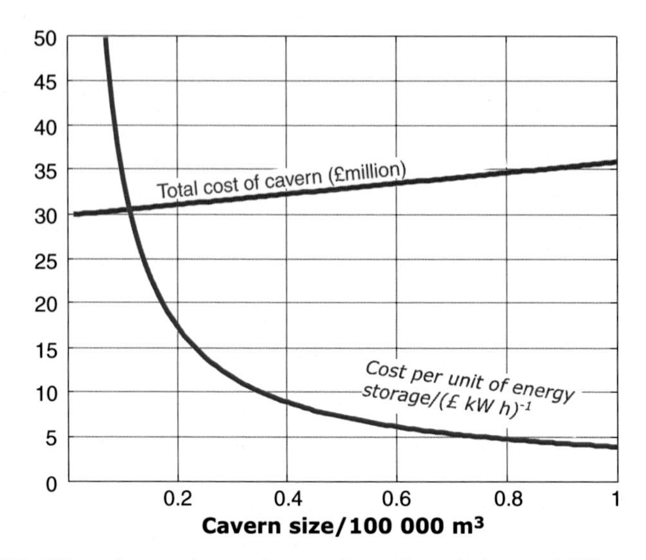

FIGURE 5.7 Why underground storage becomes interesting at the large scale? The *y*-axis is either cost/£10^6 or cost per unit/£ (kW h) $^{-1}$.

such air store to exist, but the marginal costs of increasing the capacity of that store can be very small in both cases. Fig. 5.7 illustrates this point with notional values for the case of a (fictional) cavern. A similar phenomenon happens in the context of underwater stores even though the cost per unit of energy stored tends to plateau at smaller values of stored volume.

4 SYSTEM CONFIGURATIONS AND PLANT CONCEPTS

CAES actually represents a broad family of technologies embracing:

- numerous varieties of air compression machinery
- numerous varieties of air expansion machinery
- heat exchangers for cooling air between/within/after compression stages
- heat exchangers for heating air between/within/before expansion stages
- several options for storage of pressurized air
- several options for thermal storage to complement the stored pressurized air.

Obviously there is a broad set of possible configurations that would all fall under the aegis of CAES systems. Before highlighting some of these, it is useful to discuss one of the most commonly used classifications within CAES systems. This comprises:

- adiabatic CAES systems
- diabatic CAES systems.

The two classes are distinguished by the fact that an adiabatic CAES system does not use any external source of heat while a diabatic CAES system does

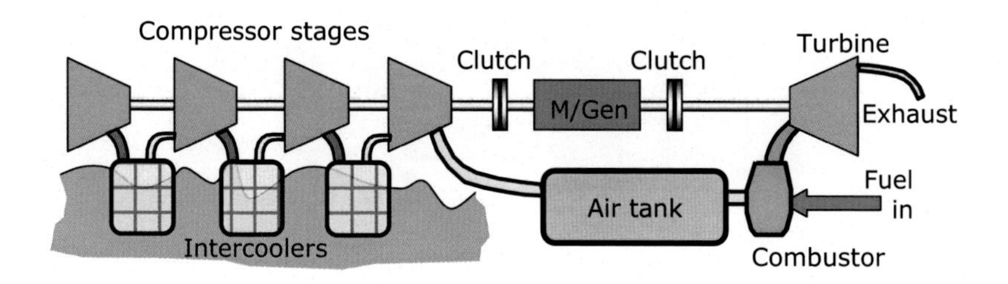

FIGURE 5.8 A simple diabatic CAES system.

make use of some external heat source to extract additional work (electricity) from the stored high-pressure air. Strictly, a diabatic CAES system is a combination of an energy storage system and a generation system since it is normal in these systems that the total amount of electricity generated from such systems exceeds the total amount of electricity consumed.

4.1 Diabatic Concepts

There are only two large-scale CAES plants in the world in operation at present: a 321 MW plant belonging to E.ON Kraftwerke, Huntorf (Germany) and the 110 MW plant of PowerSouth Energy Cooperative in Alabama (United States). Both plants use underground salt caverns for storing the air. Both are diabatic insofar as they do not store heat [7] but burn a fuel to heat the air before expansion. Fig. 5.8 presents the simplest format of a diabatic CAES plant.

Here the same electrical machine (labeled *M/Gen* to indicate that it can act as either motor or generator) can be coupled either to the set of compressor stages or to the expander via the clutches shown in Fig. 5.8. An alternative to having these two clutches is to duplicate the electrical machine so that one machine always acts as a motor when it is in operation and the other always acts as a generator.

The exhaust gas in Fig. 5.8 may exit at a temperature substantially above ambient, and in this case it is clear that exergy is being wasted. An alternative system that does not waste any exergy is shown in Fig. 5.9. Here, a recuperator absorbs heat that is left in the exhaust gas leaving the (final stage of the) expander and transfers this heat to air coming from the high-pressure air store before it reaches the (first stage of) expansion.

It is not essential that the expansion process should comprise only a single stage of expansion. By having more than one stage (together with a recuperator), it is possible to extract substantially more work from the stored air. Fig. 5.10 shows a schematic for such an arrangement.

Fig. 5.11 shows the temperature–pressure profile for the air in the case of Fig. 5.10. In this the recuperator raises the temperature of the air leaving the tank to 800 K by cooling the air exhausted from the second expansion stage back down to ambient temperature. A combustor then adds further heat to the air

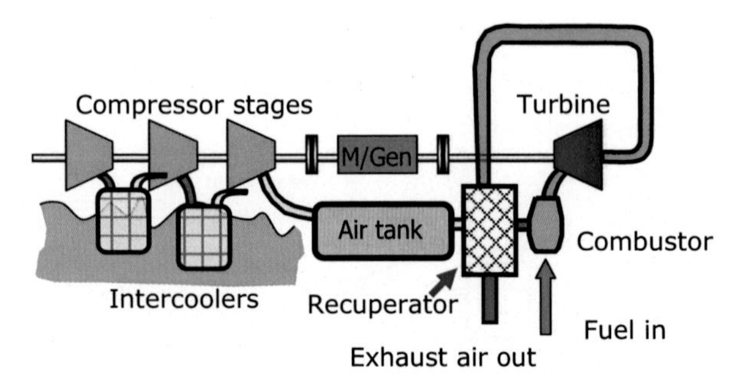

FIGURE 5.9 A simple diabatic CAES system with recuperator.

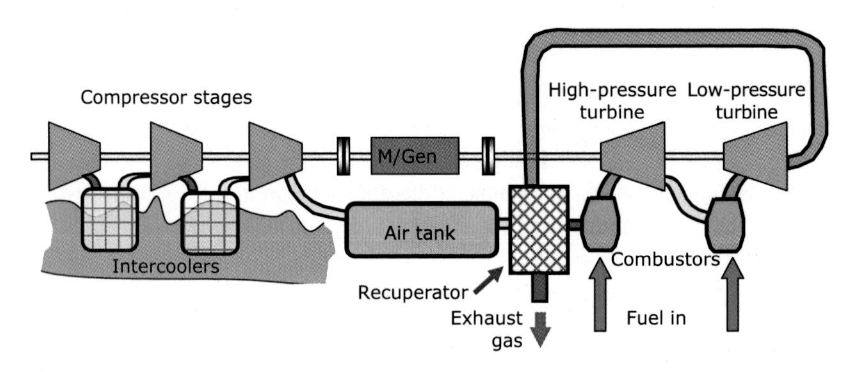

FIGURE 5.10 A diabatic CAES system with recuperator and two-stage expansion.

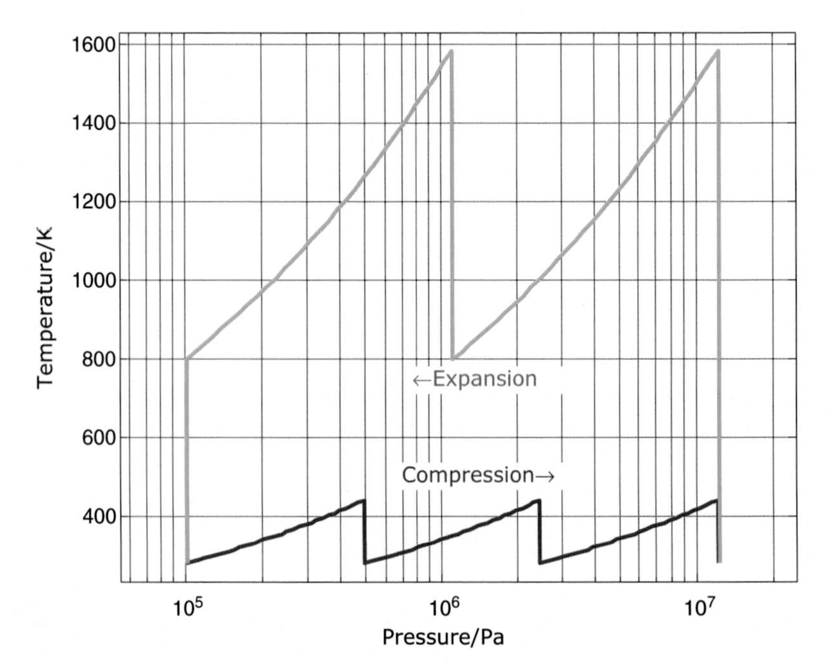

FIGURE 5.11 Temperature–pressure plot for a diabatic CAES system with recuperator and two-stage expansion.

entering the first stage of expansion to raise its temperature to 1585 K. After the first expansion stage, a second combustor raises the temperature again to 1585 K before the second expansion stage, which again reduces the temperature to 800 K.

4.2 Adiabatic Concepts

The term "adiabatic" suggests that no heat is drawn into a process or expelled from that process. Strictly, the term is misused in the context of adiabatic CAES systems. The real meaning is that no net external heat source is used. However, heat may be exchanged with the environment.

Fig. 5.12 shows one of the simplest possible adiabatic CAES system structures. In this, multiple stages of compression with interstage cooling approximate an isothermal compression process. Similarly, multiple stages of expansion with interstage reheating from the environment approximate isothermal expansion.

Fig. 5.13 depicts another possible adiabatic CAES system structure. In this, one single compression stage raises the temperature and pressure of air. Heat is drawn from the air and stored in a thermal store before the air is fed into storage. The same heat is injected back into the air when the air is withdrawn from storage prior to expansion.

If the same size and type of air store is used for the systems of Figs. 5.12 and 5.13, the total storage capacity of the latter system will be much larger since exergy is stored in the thermal store as well as in the compressed air store. It is commonly stated that the purpose of introducing thermal storage into CAES is to improve efficiency. This is quite incorrect. Systems such as that depicted in Fig. 5.12 can be made arbitrarily efficient by using a sufficient number of high-efficiency compression and expansion stages and by demanding high effectiveness of the heat exchanger units—though they would become very expensive. The main motivation for introducing thermal storage to a CAES plant

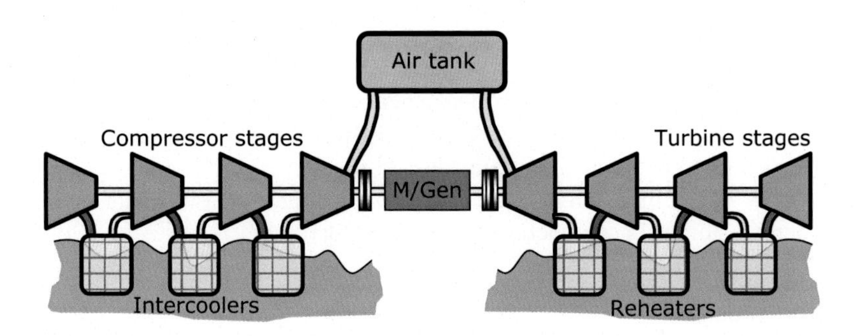

FIGURE 5.12 A straightforward adiabatic CAES plant.

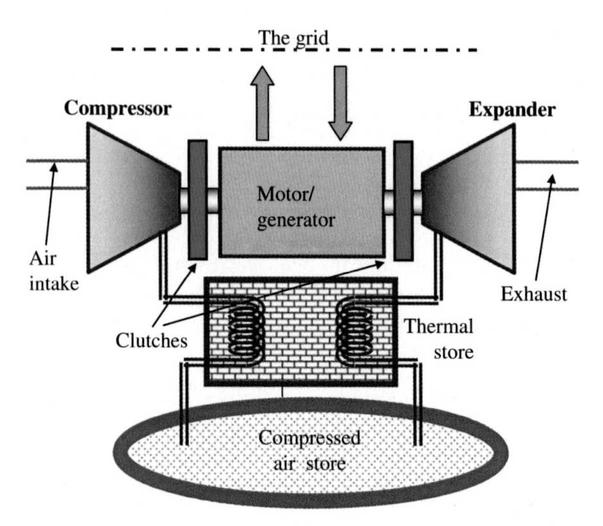

FIGURE 5.13 An adiabatic CAES plant with thermal storage.

is to increase the total quantity of exergy that can be stored for a given size of pressurized air store.

We have seen in the previous section that CAES installations may be either diabatic or adiabatic. In both cases, air is preheated prior to the expansion process—largely for the purposes of extracting more work from the same quantity of compressed air, but partly motivated by avoiding very low temperatures that could cause problems of water or lubricant freezing. In the case of adiabatic CAES plants the heat is stored from the air compression and used again to support expansion.

Thermal energy storage is itself a very large discipline, but its importance is so great in the context of CAES systems that it demands at least some discussion here. The requirement is to be able to transfer heat in and out of a pressurized air stream. If compression and expansion are both single-stage processes, then there is only one pressure to consider for thermal storage. If they both take place in, say, three stages, then there are three separate pressures to be handled by the thermal storage. For reasons connected with minimizing the destruction of exergy by forcing heat transfer to take place across large temperature differences, a well-designed adiabatic CAES plant will always use the same number of stages for compression and expansion.

It takes only a little consideration to realize that there are two major options for thermal storage in conjunction with CAES [6]:

1. Heat is stored within the pressurized system and the high-pressure air circulates around it.
2. Heat is stored outside the pressurized system and transferred across the walls of that system.

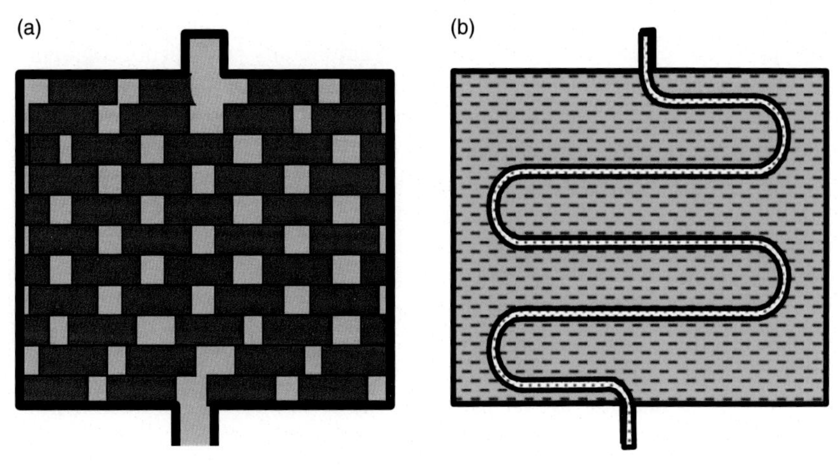

FIGURE 5.14 (a) Heat stored within pressurized system; (b) heat stored outside.

Fig. 5.14 makes the contrast clear. In Fig. 5.14a, a pressurized containment contains the thermal storage medium—shown as bricks. In such cases, heat transfer between the pressurized air and the thermal storage medium may be extremely good and one may achieve very good thermocline behavior (as shown in Fig. 5.15) where a sharp thermal gradient exists between a portion of the thermal store that is hot and another part that remains cool. In Fig. 5.14b, by contrast, a pipe containing pressurized air is shown running through a thermal storage medium that is not pressurized.

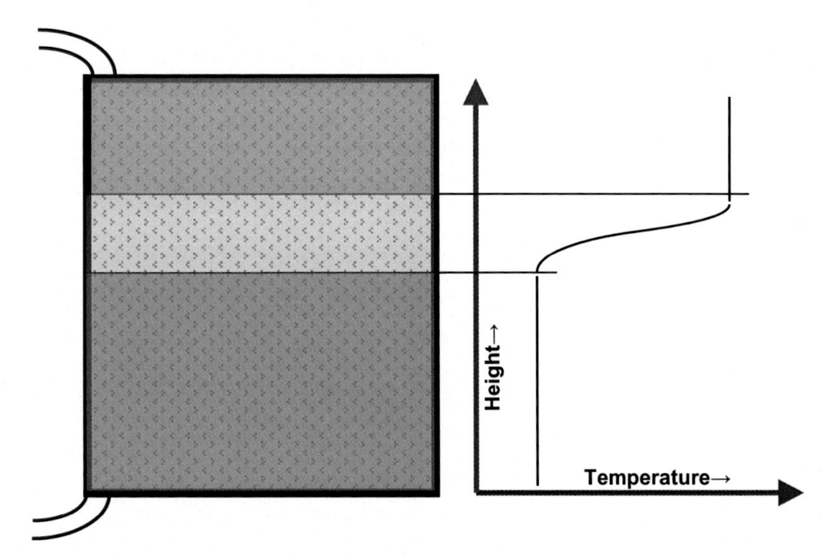

FIGURE 5.15 A thermocline-type packed bed heat store.

The main disadvantage of the first option (Fig. 5.14a) is that one requires a large pressure vessel that may not be able to exploit any geological/geographical features for strength. One rather elegant solution is noted in [8] where thermal energy is stored within the same cavern used for the pressurized air, but this is not applicable in most situations.

If the material within was some rock with specific heat 850 J $(kg\ K)^{-1}$ (typical of a wide variety of rocks) and if the temperature swing in that rock was, say 500 K, then each 1 m^3 volume of space within the containment would account for ~640 MJ of heat (assuming that the mass of rock in each 1 m^3 volume is 1500 kg). Note that this value is independent of pressure. Comparing this quantity of energy with the values in Table 5.3 shows that energy density is quite good relative to the energy stored in the compressed air itself for all realistic storage pressures. However, pressure containment is still expensive and is made more complex by the fact that this containment may have to operate at quite high temperatures (further weakening the containment material).

The main shortcoming of the nonpressurized thermal storage of Fig. 5.14b is that heat transfer is much more indirect between the pressurized air and the thermal storage medium. Numerous creative solutions are being considered for how best to achieve reasonable heat transfer in such contexts.

5 PERFORMANCE METRICS

It is commonly asserted that CAES does not deliver a high-performance energy storage solution. In fact, this position is mainly derived from an erroneous assessment of the performance of the two large-scale diabatic CAES plants—at Huntorf (Germany) and McIntosh, Alabama (United States). Published data, for example reference [7], reveal that to achieve 1 kW h of electrical energy output, 0.8 kW h of electricity is drawn into the plant at Huntorf together with 1.6 kW h of gas. Similarly, to achieve 1 kW h of electrical energy output from the CAES plant at McIntosh, 0.69 kW h of electricity is drawn in together with 1.17 kW h of gas. Performing a straight "output-divided-by-input" calculation suggests that:

$$\eta_{\text{Huntorf}} = \frac{1}{0.8+1.6} = 41.7\% \ ✗ \text{ and } \eta_{\text{McIntosh}} = \frac{1}{0.69+1.17} = 53.8\% \ ✗$$

These values are unrepresentative because they fail to recognize that every diabatic CAES plant is really a combination of a pure energy storage plant and a generation plant. The aforementioned calculations add electrical energy to thermal energy as though 1 kW h of heat was equivalent to 1 kW h of electricity. This is wrong. In fact, the best performing combined cycle generation plants return 60% of the calorific value of the fuel consumed. On this basis, we can obtain much more representative values for the performance of the two long-extant CAES plants:

$$\eta_{\text{Huntorf}} = \frac{1}{0.8 + 0.6 \times 1.6} = 56.8\% \checkmark \text{ and } \eta_{\text{McIntosh}} = \frac{1}{0.69 + 0.6 \times 1.17} = 71.8\% \checkmark$$

The reason that McIntosh performs rather better than Huntorf is connected mainly with the recuperator. Note also that both Huntorf and McIntosh presently use throttles to deliver constant pressure air to the expanders. The alternative would be to use a small high-pressure reciprocating machine or a number of dynamic machines in series to extract exergy as the air pressure is reduced.

Employing Eqs. (5.13) and (5.14) and the knowledge obtained from reference [7], shows that Huntorf caverns operate between 5 MPa (50 bar) and 7 MPa (70 bar). this indicates Huntorf loses 4.4% of the exergy available from the cavern in each cycle. Since that exergy comprises around 43% of total exergy including that from fuel, the improvement that could be achieved by replacing the throttle at Huntorf is around 1.9%, that is, raising its effective turnaround efficiency to 58.7%.

The plant at McIntosh runs from 4.5 MPa (45 bar) to 7.6 MPa (76 bar) and thus loses 6.6% of the available cavern exergy in the nozzle. Moreover, the cavern exergy there comprises around 47% of total exergy, and hence 3.1% improvement in effective turnaround efficiency is possible—raising this to 74.9%.

Other measures can be taken to achieve still higher performance, but, of course, all have associated costs.

6 INTEGRATING CAES WITH GENERATION OR CONSUMPTION

Since CAES intrinsically involves both heat and mechanical work, there are strong opportunities for integrating this with either electricity generation or consumption. The basic arguments for doing this are straightforward. Standalone energy storage for supporting the electricity system requires that energy is converted from one form of electricity to another form compatible with storage and back again, causing two additional sets of energy losses in transformation and requiring additional machinery to effect the transformations. By integrating energy storage with generation, these additional losses and costs may be reduced or avoided [9]. Various proposals have come forward for integrating compression with wind turbines [10,11] and for integrating compression with wave energy [12] and tidal energy [13,14]. For generation from wind, wave, and tides, the conversion efficiency from input energy to electricity is virtually irrelevant and what matters is the combined capital and operational costs of the energy harvester devices per kilowatt-hour of electricity produced. For sound engineering reasons, outlined in [11], exploiting air compression as the primary means of carrying away power from the device can account for significant cost reduction compared with direct generation of electricity. It is to be expected that as more CAES plant will emerge in the near future, there will be an increase in interest in capturing compressed air directly from renewables.

CAES also provides for integration on the consumption side. We noted at the outset that substantial fractions of the electricity supply are presently used to compress air [1], and it follows that if suitable receivers are in place at those locations that use compressed air, these may act as excellent energy stores—obviating the requirement for any new power conversion machinery and avoiding virtually all losses normally associated with storage. One other possibility deserves particular mention: data centers are progressively consuming more and more of all electricity generated and these have very particular requirements for both cooling and power that are especially well suited to CAES. If high-pressure air is stored at near-ambient temperature and then expanded in several stages, it can deliver both cooling and electrical power in equal measure—fitting exactly the requirements of a data center.

7 CONCLUDING REMARKS

The application of air compression to decouple energy absorption from the grid and energy consumption is known and has been practiced for decades. Two large-scale CAES power plants are in operation today, using salt caverns as storage, and both of these burn a fuel to maximize the energy recovered from the stored air. There are many different possible configurations of the CAES system and most of those presently proposed do not combust a fuel. CAES plants can be highly cost-effective and they can deliver very respectable turnaround efficiencies. The containment for high-pressure air is at the heart of every CAES system. Storage of air in man-made tanks has been discussed in this chapter, but this comes at a cost per unit of energy stored that may be two orders of magnitude higher than what may be achieved with underground or underwater storage of the air. For this reason, separate chapters are devoted to these two possibilities.

REFERENCES

[1] Carbon Trust. Compressed air: opportunities for businesses; 2000. https://www.carbontrust.com/media/20267/ctv050_compressed_air.pdf

[2] DECC. Energy consumption in the UK, 1. London: Department of Energy and Climate Change; 2015. https://www.gov.uk/government/statistics/energy-consumption-in-the-uk

[3] Auinger H. Determination and designation of the efficiency of electrical machines. Power Eng J 1999;13(1):15–23.

[4] Carnot S. Réflexions sur la puissance motrice du feu et sur les machines propres à développer cette puissance (in French). Paris: Bachelier; 1st ed. 1824, reissue 1878.

[5] Barbour E. A novel concept for isobaric CAES. Offshore Energy Storage Conference OSES 2014. Windsor, UK: University of Windsor; 2014.

[6] Garvey SD. Compressed air energy storage: performance and affordability. Half-day workshop given at Marcus Evans Biannual Energy Storage Conference, Amsterdam; Dec. 2010. https://app.box.com/s/afdfc8e2bc451647f8be

[7] BINI Informationsdienst. Compressed air energy storage power plants. ISSN 0937-8367. ProjectInfo; 2007.

[8] http://www.alacaes.com/aa-caes-technology/

[9] Garvey SD, Eames PC, Wang J, Pimm AJ, Waterson M, MacKay RS, Giulietti M, Flatley LC, Thomson M, Barton J, Evans DJ, Busby J, Garvey JE. Energ Policy; 2015.

[10] Ingersoll E. Wind turbine system. Patent Application US20080050234-A1; 2008.

[11] Garvey SD. Proceedings of the Institution of Mechanical Engineers, Part A. J Power Energ 2010;224(5):1027–43.

[12] Sieber J. Wave energy accumulator. Patent Application US2009226331-A1; 2009.

[13] Fumio O. Transducer for the conversion of tidal current energy. Patent Application US4071305-A; 1978.

[14] Southcombe AG. Wave or tidal power harnessing apparatus. Patent Application GB2267128-A; 1992.

Chapter 6

Compressed Air Energy Storage in Underground Formations

Sabine Donadei, Gregor-Sönke Schneider
KBB Underground Technologies GmbH, Hannover, Germany

1 INTRODUCTION

Unlike fossil fuels, renewable energy sources such as wind and solar are characterized by short- and long-term seasonal fluctuations and cannot deliver energy on demand. Moreover, only a very small amount of the electrical energy they generate has so far been stored. Compared with the storage of fossil fuels, the storage of electrical energy at the grid scale has only played a very subordinate role so far. This is highlighted by the fact that although the storage capacities in the European Union for liquid and gaseous hydrocarbons guarantee a statutorily regulated supply security for weeks and months, the storage capacity for electrical energy currently corresponds to 5% of total installed generation capacity. In Germany, for instance, electrical storage can only maintain the power in the whole of the German grid for less than one hour. Storage of electrical energy at the grid scale is gaining in significance because of the increased use of fluctuating renewables. In addition to pumped hydro technology, which has proven its worth over many decades, and future hydrogen systems (power-to-gas), attention is again being focused on a storage technology which was developed over 50 years ago: compressed air energy storage (CAES) [1–4].

The use of compressed air to store energy is currently deployed in applications ranging from very small outputs up to triple-figure megawatt installations. In this chapter the focus is on underground energy storage at the grid scale comparable to conventional pumped hydro power plants, which means that the following chapters are restricted to investigation of the megawatt class.

The concept of large-scale compressed air storage was developed in the middle of the last century. The first patent for compressed air storage in artificially constructed cavities deep underground, as a means of storing electrical energy, was issued in the United States in 1948. Frazer W. Gay, the patent holder, described his invention as follows: "In the present invention, I propose to provide equivalent storage space for gas relatively close to the earth's surface and, furthermore, to make this storage space available for the storing of compressed air

Storing Energy. http://dx.doi.org/10.1016/B978-0-12-803440-8.00006-3

to be used for power generation purposes during periods of heavy power load, as well as for natural gas or manufactured gas, butane, propane or other fluids. The invention in general comprises the construction of huge caverns located comparatively close to the earth's surface." [5]

In Germany, a patent for the storage of electrical energy via compressed air was issued in 1956 whereby "energy is used for the isothermal compression of air; the compressed air is stored and transmitted long distances to generate mechanical energy at remote locations by converting heat energy into mechanical energy" [6]. The patent holder, Bozidar Djordjevitch, is sometimes quoted as the inventor of CAES technology. His concept was supplemented by the former Federal Institute for Geology—today's BGR (Bundesanstalt für Geowissenschaften und Rohstoffe), which chose cavities in underground rock salt as the storage facilities for the air because these were considered to be the most economical and the safest [7,8].

CAES power plants attracted interest in the United States at the end of the 1960s with the focus in this case on air storage. This led to the discussion and patenting of CAES systems with salt caverns and aquifer structures [9,10].

Several studies and projects on CAES arose in Europe in the 1970s. Salt caverns, aquifer structures, and mines were investigated and taken into consideration as potential storage spaces.

The world's first underground CAES power plant was constructed in Huntorf (Germany) in the middle of the 1970s and was primarily aimed at storing the electrical energy produced by less flexible coal and nuclear power plants during low periods of demand, and to feed this energy back into the grid again during periods of high demand. Other motives for constructing the plant were the cold-start capacity, its ability to regulate the grid frequency, and the phase shifter operation [11]. Another CAES power plant was constructed in McIntosh, Alabama (United States) in 1991.

After the expected demand for additional CAES power plants evaporated as a result of the merger of smaller grids to form larger shared grids, interest was reawakened at the beginning of this century by the transition from fossil fuels to renewable energy. This was initially stimulated by the growing demand for "minute and hour" reserves in the power grid to balance out deviations between forecast and actual wind energy generation. Growing demand for flexibility again focused attention on underground CAES power plants because of their analogous properties to pumped hydro plants. Attention now concentrates particularly on the development of new power plant components.

2 MODE OF OPERATION

In underground CAES power plants, electrical energy from the power grid drives a compressor to inject large volumes of air under high pressure into a storage facility. When electricity is required, this air can be released from the storage and passed through a turbine and generator to regenerate electrical power, which can be fed back into the grid. The heat energy generated by compression is either lost to the environment or made available to other users, or stored for later deployment. Before being fed back into the turbine and generator installation the air is warmed back up again by using heat from this thermal storage, or a different source, or from combustion gases (for more on mode of operation see chapter: "Compressed Air Energy Storage").

A CAES power plant consists of a storage space for the air and a power plant with motor compressor and turbine generator units. Although the storage of compressed air on the surface is possible, for example, in spherical and pipe storage systems, or in gasometers, these have much lower storage capacities than underground storage systems. Installation concepts at the grid scale therefore usually depend on the underground storage system. Since these underground geological storage systems must be injected with cool air (<10 °C), the air has to be cooled down before injection. In the case of diabatic CAES storage (Fig. 6.1) the extracted heat is not stored. During the generation phase, it is therefore necessary to reheat the cold air flowing out of the storage with the help of combustion gas to prevent the turbine icing up and to compensate for the energy lost during the preceding cooling process. This results in a relatively low level of efficiency of only (42–54)%. This is significantly lower than that obtained from a pumped hydro power plant [12].

In adiabatic CAES systems (Fig. 6.2) the heat of compression is stored in one or more separate storage facilities so that it can be reused to heat up the air when it is withdrawn from the storage. Since this dispenses with the addition of combustion gas, this can be considered a pure power-to-power storage system.

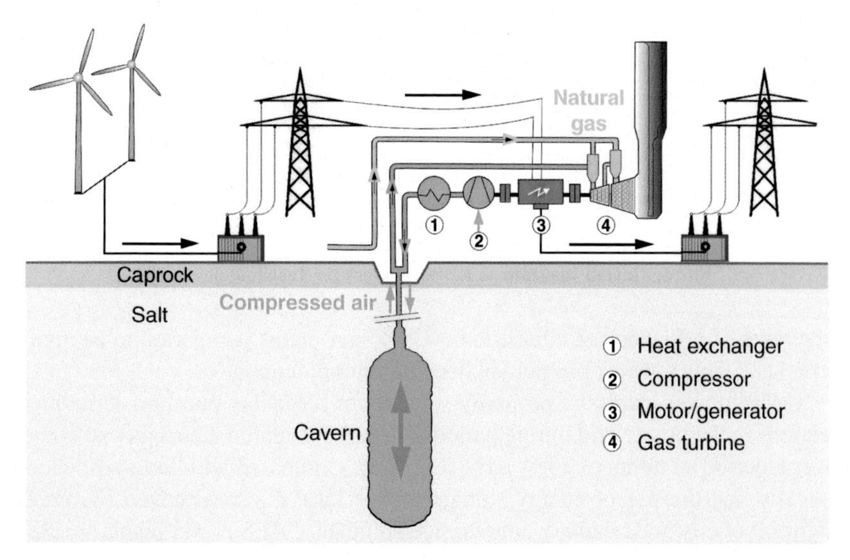

FIGURE 6.1 **Diagram of a diabatic CAES power plant.** © KBB Underground Technologies GmbH.

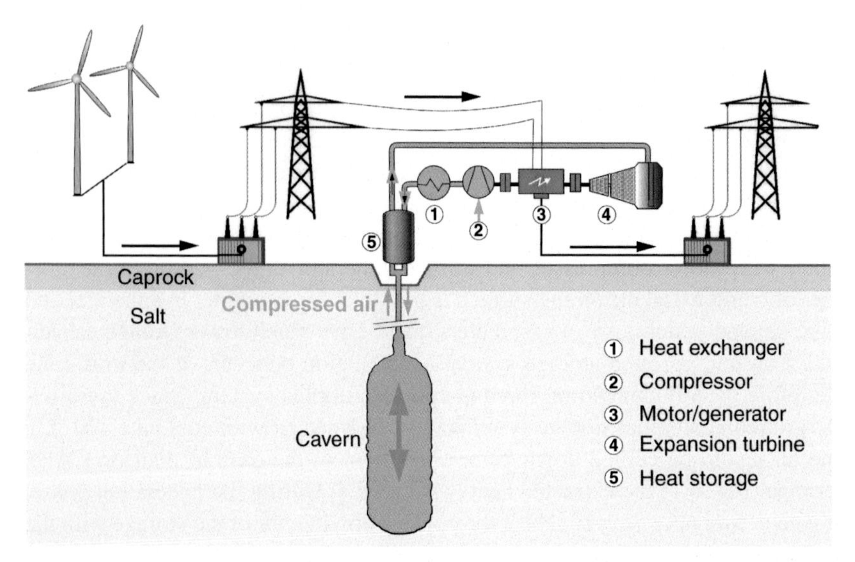

FIGURE 6.2 **Diagram of an adiabatic CAES power plant.** © KBB Underground Technologies GmbH.

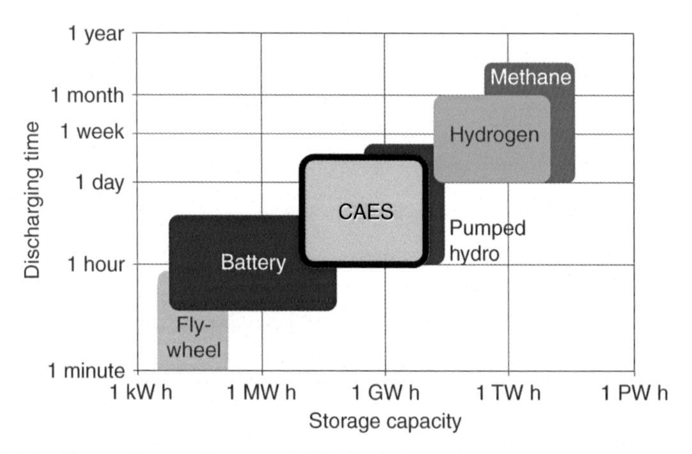

FIGURE 6.3 **Storage layout diagram.** © KBB Underground Technologies GmbH.

The level of efficiency of adiabatic CAES power plants is reported to be up to 70% [12]. This concept has not yet been put into practice.

CAES power plants are primarily suitable for balancing out short-term fluctuations in the power grid during periods of peak demand in the megawatt range over a period of hours to a few days (Fig. 6.3). Compared with long-term storage systems the use of energy storage of this kind is characterized by much higher cyclicity and relatively low storage volumes. CAES power plants are mechanical storage facilities and therefore have a relatively low volumetric energy density—depending on the dimensions—of around twice the value of pumped hydro storage. Long-term storage therefore requires a large amount of storage volume and thus high total costs for a similar output. In contrast, the high efficiency in part of CAES power plants has a positive effect on the economics, particularly when frequent cycles are involved.

3 PLANT CONCEPT

3.1 Diabatic Concept

The basic idea of diabatic CAES was to transfer off-peak energy produced by base nuclear or coal-fired units for high-demand periods, using only a fraction of the gas or oil that would be used by a standard peaking machine, such as a conventional gas turbine. There are only two CAES plants in the world in operation at present: the 321 MW plant belonging to E.ON Kraftwerke, Huntorf (Germany) (Fig. 6.4), and the 110 MW plant of PowerSouth Energy Cooperative in Alabama (United States). Both use underground salt caverns for storing the air (Section 4.7).

During low-cost off-peak load periods a motor uses power to compress and store air in an underground facility. During peak load periods, the process is reversed: the compressed air is returned to the surface to burn natural gas in the combustion chambers. The resulting combustion gas is then expanded in the two-stage gas turbine to spin the generator and produce electricity.

In a pure gas turbine power station, around two-thirds of the output are needed for compressing the combustion air (100 MW net output + 200 MW compressor consumption equal to 300 MW gross output). In a CAES power station

FIGURE 6.4 **Huntorf CAES power plant.** © KBB Underground Technologies GmbH.

no compression is needed during turbine operation because the required energy is already included in the compressed air. This has two advantages:

- using cheaper (excess) power for compression during off-peak periods;
- generation of all the output instead of one-third by the gas turbine [11].

In addition, the compressor and generator operations can be operated at different times. The diabatic CAES concept is not a "pure" energy storage system, but rather a gas turbine power plant in which compression of the combustion air and depressurization of the heated gases take place at different places and at different times. The major advantage of this technology is that cheap (excess) energy can be used for compression.

The efficiency of the diabatic storage plant is determined by the deployment of energy in the form of electricity, as well as the natural gas, and the generated output. Taking all this into account, the efficiency of the Huntorf plant is 42%:

$$\frac{1\ \text{kW h produced electrical energy}}{0.83\ \text{kW h of deployed electrical energy} + 1.56\ \text{kW h of fossil energy}} = 42\,\%$$

If the waste heat of the gas turbine exhaust gas is recovered in a recuperator used to preheat the combustion air for the gas turbine, it is possible to increase the efficiency and reduce the amount of fossil energy required. This results in an enhanced efficiency of 54%:

$$\frac{1\ \text{kW h produced electrical energy}}{0.69\ \text{kW h of deployed electrical energy} + 1.17\ \text{kW h of fossil energy}} = 54\,\%$$

3.2 Single-Stage Adiabatic Concept

Due to the low efficiency compared with modern pumped hydro plants and the desire to construct a "pure" power storage without any additional combustion gas, an increasing number of concepts for CAES power plants have been developed since the beginning of the millennium to incorporate the storage and deployment of compression heat. In these concepts a key element is dimensioning of the heat storage in combination with the compression process. The single-stage adiabatic concept corresponds to compression from atmospheric pressure up to storage pressure in one step, without any interim cooling, and subsequent cooling down of the compressed storage gas by several hundred degrees in one heat storage phase. The basis for this concept was developed in the EU project "Advanced Adiabatic Compressed Air Energy Storage" [13]. This was followed up by establishment of a consortium operated by RWE AG and realization of the ADELE project comprising additional research and development aimed at developing the components and/or the plant to such a level that it would lead to the construction and operation of a pilot plant (Section 4.7).

When air is compressed to pressures of well over 50×10^5 Pa (50 bar) the temperature of the air increases by several hundred degrees centigrade. Since the material properties of underground geological storage are not designed to handle such temperatures the compressed air has to be cooled down considerably prior to injection. If this heat of compression is to be used to heat up the stored air again immediately before it is used to regenerate electricity, it is necessary to cool down the compressed air in a thermal energy storage (TES) unit until it reaches ambient temperature. When the air in the storage is withdrawn the cold compressed air is heated back up again in the TES so that no natural gas needs to be added in the turbine.

The TES has to satisfy a number of challenging specifications:

- The capacity to store heat in the high-megawatt range.
- Ability to rapidly cool down air with temperatures of over 600 °C to ambient temperature and vice versa.
- Resistance to pressures of up to 100×10^5 Pa (100 bar) and temperatures exceeding 600 °C.
- Low-pressure losses with very high mass flow in the several 100 kg s^{-1} range.
- Minor heat losses when on standby.

The compressor also has to satisfy very high specifications, not only because of high input temperatures, but also due to the high temperature–time gradient—because the installation has to generate full capacity within only a few minutes.

An efficiency of 70% is targeted for future adiabatic CAES power plants.

3.3 Multistage Adiabatic Concept

Due to the high technical specifications and the associated costs of realizing the single-stage adiabatic CAES concept, a great deal of interest has been shown in recent years in isothermal or quasiisothermal CAES concepts [14,15]. The focus here is on subdividing compression and decompression into several stages so that each stage is only associated with minor temperature increases (Fig. 6.5).

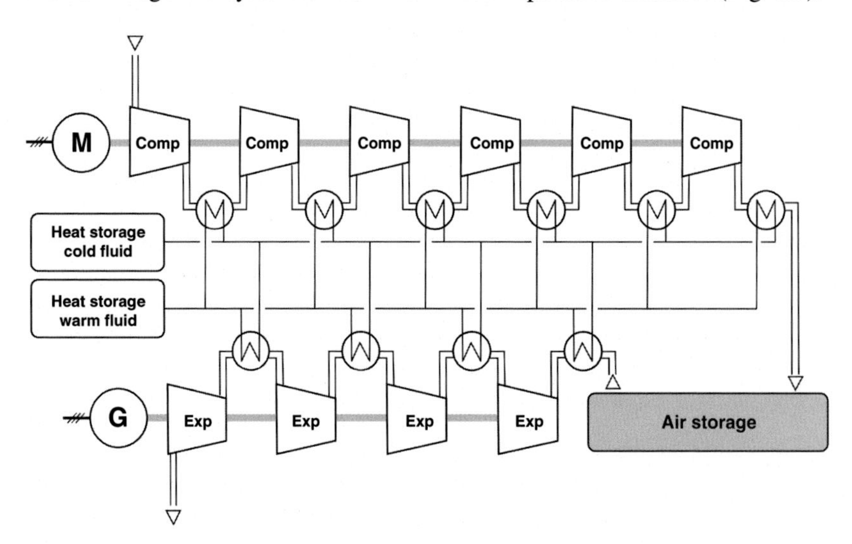

FIGURE 6.5 Sketch of the principle behind the isothermal concept [17]. © KBB Underground Technologies GmbH.

Each stage is followed by a heat exchanger. Due to the low temperatures, heat can be stored in either conventional storage facilities or will only require much less development input. The major requirement here is the size of the heat exchanger surface because only minor temperature differences are available for the heat exchange process, at the same time as involving very large volumes of heat. No project has yet been implemented to apply this concept.

3.4 Performance Metrics

Typical performance metrics of various underground CAES concepts are summarized in Table 6.1 [16–19].

4 UNDERGROUND STORAGE

A great deal of experience has been gained worldwide over several decades with the storage of gas in deep underground geological formations—primarily involving natural gas. Underground storage generally has a number of crucial advantages over surface storage in tanks:

- Achieving high storage capacity as a result of large geometrical volumes and high operating pressures of up to 200×10^5 Pa (200 bar).
- Considerable protection against external influences since the only surface features are the connection valves.
- Very low footprint compared with surface pressure tanks.
- Low specific costs with respect to storage capacity.

There are several options available for the storage of compressed air in underground geological formations at the grid scale: in natural pore storages such as depleted oil and gas fields, aquifer formations, artificially constructed cavities such as salt and rock caverns, and abandoned mines. The following specific aspects have to be taken into consideration when storing compressed air in underground geological formations:

- The high reactivity of oxygen in compressed air, for example, forming compounds with the mineral constituents of the storage rock, and thus leading to oxygen depletion.
- Suitability/dimensioning of the storage for frequent, rapid operation cycles, and high injection and withdrawal rates, because CAES power plants are typically operated in an extremely fluctuating mode.
- Possibility of operating the storage for a short period of time at atmospheric pressure, for example, during repairs and maintenance measures.

The following describes the options available for the underground storage of compressed air, and their suitability in light of today's standard engineering practice. It also evaluates compressed air storage projects currently in operation, construction, or at the planning stage.

TABLE 6.1 Performance Metrics of Various CAES Concepts

	E.ON Kraftwerke/ Huntorf	PowerSouth Energy Cooperative/McIntosh	RWE/ADELE	Fraunhofer/ LTA-CAES	u&í—umwelt-technik und ingenieure GmbH
Concept	Diabatic	Diabatic with waste heat utilization	Single-stage adiabatic	Multistage adiabatic	Multistage adiabatic
Status	In operation	In operation	In planning	Concept	Concept
$p_{min}/10^5$ Pa	46	46	–	–	–
$p_{max}/10^5$ Pa	72	75	65	150	70
$m_{air}/kg\ s^{-1}$	455	154	300	–	840
Power/Generator/ MW	321	110	260	30	290
Full load after time/min	6–9	14	–	<5	–
Efficiency factor/%	42	54	c. 70	52– 60	72.5
Heat output/MW	–	–	–	–	70
V_{cavity}/m^3	310 000	538 000	–	–	–
W_{cavity}/MWh	200–642	240–2 640	1 040	–	–

LTA, low-temperature approach.

4.1 Depleted Oil and Gas Fields

Depleted oil fields, and particularly gas fields, can be used for the storage of gases because their imperviousness over geological time periods has already been proven. Another advantage is that these reservoirs have already been very well researched as part of the preceding exploration and production activities. Moreover, it may also be possible to use existing wells for storage operations after relevant conversion activities have been carried out. However, not all depleted oil and gas fields are suitable for conversion into a gas storage facility. The suitability criteria include depth and adequate permeability and porosity of the reservoir rock. Due to the reservoir engineering properties of depleted oil and gas fields, they tend to have low flexibility for injection and withdrawal, and the gas storage facilities constructed using them therefore tend to be best used for seasonal applications [20].

Natural gas has been stored for many decades in depleted oil and gas fields. A depleted gas field was used for the storage of gas for the first time 100 years ago in Welland County, Ontario (Canada) [21]. They are the main storage option for underground geological storage worldwide primarily because of their capacity. Of the natural gas currently stored underground worldwide, 81% is held in depleted oil and gas fields [22,23].

Even if depleted oil and gas fields play the most important role in the storage of natural gas worldwide, there are problems with their use for storing compressed air because they always contain residual quantities of hydrocarbons, which can lead to the formation of ignitable and explosive mixtures when coming into contact with injected compressed air. The suitability of depleted oil and gas fields for the storage of compressed air is currently being looked at in scientific studies [24–26]. No depleted oil and gas fields have been used so far for compressed air storage.

4.2 Aquifers

Aquifers are porous, permeable, water-bearing underground rock layers, which are in principle suitable for the storage of gas. The injected gas displaces the water to more distant parts of the reservoir to create a gas cap, which acts as the storage. Crucial conditions for the suitability of an aquifer as gas storage are the adequate porosity and permeability of the reservoir rock, and the presence of an upper seal to the aquifer structure formed by a gas-tight rock layer. This is why a large amount of exploration effort is required before constructing a gas storage facility in an aquifer, unlike the situation with depleted oil and gas fields. Aquifers are also less flexible with respect to injection and withdrawal and are therefore more suitable for seasonal storage. Several wells need to be drilled to be able to achieve the rates required for the storage operations.

The first aquifer storage world-wide was constructed in the United States in 1931 [21]. The use of aquifers for the storage of natural gas is now standard engineering practice worldwide and accounts for 13% of the natural gas stored underground globally [22,23].

The positive experience gained from underground storage of natural gas cannot be directly extrapolated to compressed air storages because of the risk of reactions between the oxygen in the air and the minerals and microorganisms in the reservoir rock. This can lead to the loss of oxygen as a result of oxygen depletion, or blockage of flow paths—which can make storage unusable.

Industry has been looking at the use of aquifer structures for the storage of compressed air for many decades [27,28]. Only one pilot plant for the storage of compressed air has been carried out in an aquifer in the megawatt range (Section 4.7). Furthermore, the injection of compressed air into an aquifer was tested over a period of a few years and then withdrawn to investigate the feasibility. Although the feasibility was confirmed from a reservoir engineering point of view, considerable oxygen depletion was a serious problem [29,30].

4.3 Salt Caverns

Salt caverns are artificial cavities in underground salt formations, which are created by the controlled dissolution of rock salt by injection of water during the solution mining process. Geometrical volumes of a few 100 000 m^3, and up to 1 000 000 m^3 and more in individual cases, can be achieved depending on technical specifications and geological conditions. In Germany the caverns lie at depths of around (500–2000) m, with cavern heights of up to 400 m. Depending on the depth, these caverns can be operated with a pressure of up to 200×10^5 Pa (200 bar) and thus allow the storage of very high volumes of gas. The favorable mechanical properties of the salt enable the construction and operation of extremely large cavities stable for long periods of time, which are also completely impervious to gases. In addition, salt is inert with respect to gases and liquid hydrocarbons. The amount of exploration work required is also usually much lower than the aforementioned aquifer storages because many salt structures are already known from oil and gas exploration and the investigation of salt as a raw material itself. Salt caverns are primarily used for the storage of seasonal reserves, trading storages, and as strategic reserves. Moreover, because they are very flexible with respect to injection and withdrawal cycles, they can also be used to cover daily demand peaks.

Artificially constructed salt caverns have been used for the storage of energy carriers for over 50 years—primarily to store fossil fuels such as natural gas, oil, and petroleum products (refined fuels, liquefied gas), but also for the storage of hydrogen and compressed air. Liquefied petroleum gas (LPG) as well as oil were stored in the first caverns in the United States and Europe in the 1950s. The first natural gas cavern was constructed in Marysville, Michigan (United States) in 1961 [31]. The first hydrogen cavern was constructed on Teesside

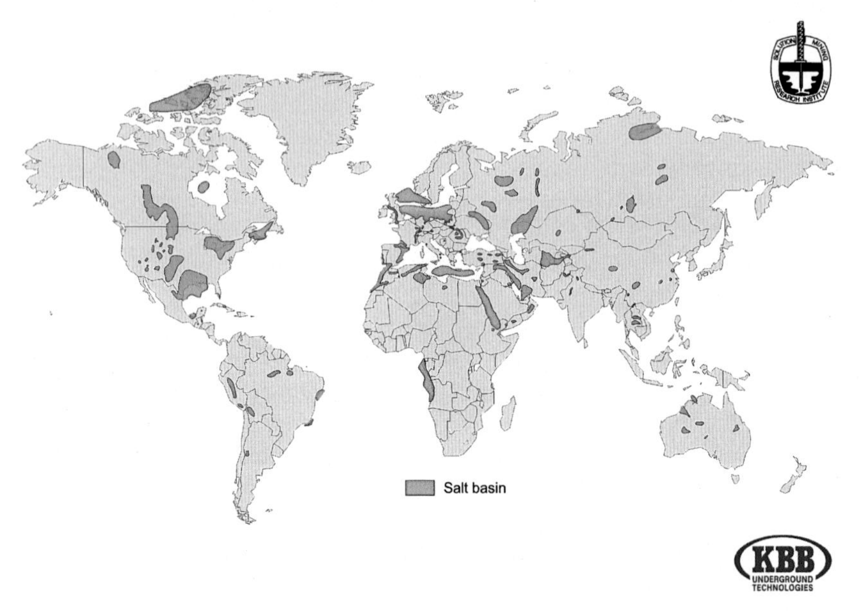

FIGURE 6.6 **Map of underground salt deposits worldwide.** © KBB Underground Technologies GmbH.

in the United Kingdom in 1971–72 and is still in operation [32]. Today, there are more than 2000 salt caverns in North America and over 300 salt caverns in Germany used to store energy carriers [33,23].

The distribution of salt deposits worldwide is very localized (Fig. 6.6). In Europe, for instance, countries such as Germany, Denmark, and the Netherlands have a large amount of salt, with a great deal of additional expansion and enlargement potential for salt caverns, while other countries have very minor or no potential at all. However, it is feasible that countries with large potential for the construction of salt caverns can create capacities beyond their own national needs as part of an international storage system and thus make them available to other countries.

4.4 Rock Caverns

Rock caverns are mined underground using conventional mining techniques and consist of a system of shafts or ramps and drifts, forming cavities in solid rock deep underground, for example, in granite. Although just as stable, unlike rock salt, solid rock is not impervious to liquids, and especially gases, because of fractures within the rock. Sealing therefore has to be achieved by engineering measures.

The most widespread technology to create impervious rock caverns is hydrodynamic sealing, which is primarily used for liquid storage. Since the caverns are located deep underneath aquifers, small amounts of water flow into the storage space. The pressure of the water column prevents the stored medium from leaking out of the caverns through the fractures. However, the water has to be continuously pumped out of the lower part of unlined rock caverns. Since hydrodynamic technology can only build up limited operational pressures at economically acceptable depths for engineering the caverns, a concept using lined rock caverns was developed. Put simply, the rock caverns here are lined with thin sheets of stainless steel [20]. The rock then provides the structural stability of the storage against the gas pressure, while the metal lining ensures that the caverns are sealed.

Unlined rock caverns have been used to store liquid hydrocarbons, mainly in the United States and in Europe since the 1950s. Most of these have been realized in Scandinavia because of the presence of suitable geological formations in these countries. Other rock caverns have been constructed in the United States, Saudi Arabia, and East Asia. The feasibility of the lined rock cavern concept was verified in Skallen (Sweden) in 2004 in a relatively small pilot cavern [20].

Unlined and lined rock caverns have not been used so far for the storage of compressed air. They have, however, been the subject of scientific analysis for a long time analogous to other storage options [34–36]. A pilot plant for the adiabatic storage of compressed air is currently being constructed in Switzerland (Section 4.7).

Compressed air storage in rock caverns—particularly in lined rock caverns—could be interesting in future for countries which are not able to construct salt caverns but have adequate hard-rock potential. In a similar way to salt caverns, no reactions between the oxygen and the in situ constituents of the rock are expected in rock caverns, and flexible operations at high rates are also considered feasible.

4.5 Abandoned Mines

Abandoned mines, which were previously used for the extraction of commodities such as salt, ores, coal, or limestone can sometimes be used for storage of gases and liquids, depending on the local geological situation. Numerous abandoned mines with appropriate volumes and suitable depths exist worldwide. However, whether a mine is actually suitable for the construction of a storage facility largely depends on the imperviousness of the surrounding rock mass and the expense of creating a technical seal. The position and quality of the mined minerals are the priorities in conventional mining so that the resulting underground workings often only have limited suitability later on as compressed air storage facilities.

An abandoned mine was converted for the first time in Sweden at the end of the 1940s for storing liquid hydrocarbons. The storage of gaseous hydrocarbons followed in the 1960s. Compared with other storage options in deep underground geological formations, mines have only been used very rarely for the storage of gas. For instance, only one former mine in Europe is currently used for the storage of natural gas—the former Burggraf-Bernsdorf salt mine in Germany [20].

No experience has been gained to date in the use of abandoned mines for compressed air storage, but this technology has been looked at in some scientific investigations [37–39].

4.6 Conclusions

Unlike the storage space available in depleted oil and gas field, or an aquifer, which consists of a large number of microscopic interconnected pore spaces, caverns, and mines consist of one large open space. This means that caverns can be filled or emptied without having to counteract the resistance of a pore matrix to the flow of gases and are therefore particularly suitable for flexible storage operations with high flow rates and frequent cycles. Combined with lower specific costs a high level of imperviousness and large realizable volumes, this aspect makes salt caverns the best choice for the construction of future CAESs to balance out fluctuating wind and solar power. Existing CAES power plants in Germany and the United States therefore use salt caverns for the storage space. However, because suitable salt formations are highly localized [40] further research is needed to look at other options for the geological storage of compressed air.

4.7 Existing and Proposed Plants

Although only two CAES power plants are currently operated (discussed later), a great deal of work is being done on the realization of additional CAES power plants. The following list provides an overview of some of the plants in operation, in the planning stage, and those that have not come to fruition.

4.7.1 Huntorf (Germany)—in Operation

The world's first CAES power plant began operations in Huntorf (Germany), approximately 40 km northwest of Bremen, in 1978—and is still in operation today. The company E.ON Kraftwerke GmbH controls and monitors the fully automatic plant from the Wilhelmshaven coal-fired power plant 50 km away [41]. When it first began operations the diabatic power plant had an output of 290 MW. This was raised to 321 MW in 2006 after comprehensive upgrading [41]. Two salt caverns with a total geometrical volume of 310 000 m^3, lying at depths between (650 and 800) m, are used to store the compressed air [11]. A third salt cavern with a volume of 300 000 m^3 is used to store the natural gas, which is required to heat up the air when it is released from the storage caverns [11]. The turbine can operate at full power within (6–9) min [41]. The CAES power plant in Huntorf has an efficiency of 42% [12].

4.7.2 McIntosh, Alabama (United States)—in Operation

The world's second diabatic CAES power plant was commissioned in McIntosh, Alabama (United States) in 1991. One salt cavern with a volume of 538 000 m^3 is used to store the compressed air [42,43]. An output of 110 MW is achievable within 14 min, adequate to provide electricity to approximately 110 000 homes [44]. By using the waste heat from the turbine the power plant has an efficiency of 54% [12].

4.7.3 Sesta (Italy)—Shut Down

A 25 MW pilot plant for CAES in an aquifer structure existed in Sesta (Italy) in the 1990s. The plant was shut down after a few years of operation [45–47].

4.7.4 Larne (United Kingdom)—under Construction

A diabatic CAES power plant with an output of (140–270) MW is being planned in Larne (United Kingdom). The current phase of the project involves geological exploration of the salt deposit [48].

4.7.5 Pollegio (Switzerland)—under Construction

ALACAES in Switzerland is currently constructing an adiabatic CAES pilot power plant in an abandoned tunnel in the Alps. Tests are scheduled to begin in the third-quarter of 2015 [49].

4.7.6 Bakersfield, California (United States)—in Planning

The project in Bakersfield, California (United States) involves storing compressed air in a depleted gas field. The aim is to construct a power plant with a capacity of 300 MW by 2020–21. A feasibility study will be completed by the end of 2015 and will form the basis for a decision on whether to continue the project [50].

4.7.7 Tennessee Colony, Texas (United States)—in Planning

The construction of a diabatic CAES power plant with an initial output of 317 MW is being planned in Tennessee Colony, Texas (United States). The power plant will have capacity to provide electricity to 300 000 households. Construction of the power plant, which uses a salt cavern to store the compressed air, is scheduled to begin in 2015. Commissioning is scheduled in 2018 [51].

4.7.8 *Alberta (Canada)—in Planning*

A CAES power plant with an output of (125–150) MW is being planned in Canada as part of the Alberta Saskatchewan Intertie Storage project (ASISt). Compressed air is scheduled to be stored in salt caverns [52].

4.7.9 *Wesel (Germany)—in Planning*

The Energy Storage Niederrhein project involves a CAES power plant in Germany, which is currently at the concept stage. Three salt caverns in Wesel, North Rhine-Westphalia are to be used for the storage of diabatic compressed air in a power plant, which is hoped to achieve an efficiency of (55–60)% [53].

4.7.10 *Cheshire (United Kingdom)—in Planning*

There are plans to erect a CAES power plant in Cheshire (United Kingdom) between 2014 and 2019. The first phase involves a pilot plant with an output of between (25 and 40) MW, which is then scheduled to be expanded to 500 MW for commercial operations. The efficiency of the power plant, which uses salt caverns as storages, is reported to be (70–75)% [54].

4.7.11 *Millard County, Utah (United States)—in Planning*

An ambitious CAES project is planned for commissioning in Millard County, Utah (United States) in 2023. Four salt caverns are planned as the (interim) storages for 2100 MW from a wind farm in the neighboring state of Wyoming. The state of California will be the final user of the energy. Power will be transmitted along power lines several hundreds of kilometers long [55].

4.7.12 *Norton, Ohio (United States)—in Planning*

A project that has been on the cards for several years aims to use a former limestone mine in Norton, Ohio (United States) to store compressed air with a volume of 10×10^6 m^3 and an output of 2700 MW [46,43]. The project was postponed in 2013 according to reports in the media [56]. However, the project still remains unrealized even up to the present time.

4.7.13 *Staßfurt (Germany)—in Planning*

The company RWE AG, together with various partners from industry and science, has been working on the ADELE project since 2010. This involves development of an adiabatic CAES power plant, including the conception and realization of the world's first pilot plant in Germany [57].

4.7.14 Iowa (United States)—Planning Suspended

A project for the construction of a CAES power plant using an aquifer structure to store the air has been pursued in Iowa (United States) for several years as part of the Iowa Stored Energy Park project (ISEP). The project was abandoned in 2011 after 8 years of planning activity when the results of geological investigations revealed that the planned capacity could not be achieved in the selected aquifer formations [58,59].

4.7.15 Donbas (Russia)—Construction Abandoned

Construction of a CAES power plant began in the Donbas region—650 km southwest of Volgograd—shortly before the collapse of the Soviet Union. The plant was scheduled to have an output of 1050 MW and was to use salt caverns to store compressed air. The work was abandoned with the collapse of the Soviet Union [47,60].

4.7.16 Gaines, Texas (United States)—Status Unclear

The Texas Dispatchable Wind Project was commissioned in late 2012. This adiabatic CAES power plant was scheduled to use a salt cavern [61]. The current status of the project remains unclear.

4.7.17 Columbia Hills, Washington (United States)—Status Unclear

A diabatic CAES power plant with a capacity of 207 MW has already been through the conceptual stage and was planned for Columbia Hills in the US state of Washington. Compressed air was scheduled to be stored in four wells drilled into a flood basalt formation. The status of the project is currently unclear [62,63].

4.7.18 Selah, Washington (United States)—Status Unclear

This project in Yakima Canyon in Selah, Washington (United States) brings together compressed air storage and geothermal energy. During the injection process of the adiabatic power plant the heat generated by compression is to be stored in molten salt, which is then used again to heat up the air when it is released from the underground—together with the energy from the geothermal plant, which is part of the overall power plant facility. The total output of the CAES plant is scheduled to be 83 MW [62,63]. The status of this project is also unclear.

5 CONCLUSIONS

The application of air compression to decouple energy absorption from the grid and energy consumption is known and has been practiced for decades. Two underground CAES power plants are in operation today, using salt caverns as storage. With the transition from a fossil fuel–based energy system to a system integrating more and more renewable energies, interest in this technology has been reawakened. Here the main focus is the development of new concept designs and the usage of additional underground formations beside salt caverns. Currently, no further underground CAES power plants are in operation, mainly for economic reasons or due to the need for more investigation into the suitability for proposed underground storage. The technology is available, and with the growth of renewable energy we can expect underground CAES power plants to be developed in suitable situations.

REFERENCES

[1] EU. Council Directive 2009/119/EC of Sep. 14, 2009 imposing an obligation on member states to maintain minimum stocks of crude oil and/or petroleum products. Brussels, Belgium: European Union; 2009.

[2] EU. Regulation (EU) No. 994/2010 of the European Parliament and of the Council of Oct. 20, 2010 concerning measures to safeguard security of gas supply and repealing Council Directive 2004/67/EC. Brussels, Belgium: European Union; 2010.

[3] EC. The future role and challenges of energy storage. DG ENER Working Paper. Brussels, Belgium: European Commission; 2013. http://ec.europa.eu/energy/sites/ener/files/energy_storage.pdf

[4] Sachverständigenrat für Umweltfragen. 100% erneuerbare Stromversorgung bis 2050: klimaverträglich, sicher, bezahlbar. Berlin: Stellungnahme; 2010, pp. 59.

[5] Gay FW. Means for storing fluids for power generation. US Patent 2,433,896; 1948.

[6] Djordjevitch BD. Gasturbinenanlage. German Patent DE940683; 1956.

[7] Allen RD, Doherty TJ, Thomas RL. Geotechnical factors and guidelines for storage of compressed air in solution mined salt cavities. Richland, WA: Pacific Northwest Laboratory; May 1982.

[8] Martinez JD. Role of salt domes in energy productions. In: Coogan AH, editor. Fourth Symposium on Salt, vol. 2. Cleveland; 1974. pp. 259–266.

[9] Lang WJ. Method and apparatus for generating power. US Patent 3,538,340, 1968/1970.

[10] Lang WJ. Method and apparatus for increasing the efficiency of electric generation plants. US Patent 3,523,192, 1968/1970.

[11] Crotogino F, Mohmeyer K-U, Scharf R. Huntorf CAES: more than 20 years of successful operation. SMRI Spring Meeting. Orlando; Apr. 23–24, 2001. pp. 351–362.

[12] Crotogino F, Hübner S. Zukünftige bedeutung der energiespeicherung in salzkavernen. Erdöl Erdgas Kohle 2009;125(2):74–8.

[13] European Commission. Community Research and Development Information Service (CORDIS). Project & Results Service. Advanced adiabatic compressed air energy storage (AA-CAES), http://cordis.europa.eu/project/rcn/67580_en.html.

[14] Wolf D, Budt M, Prümper H-J. LTA-CAES: low-temperature adiabatic compressed air energy storage. 6th International Renewable Energy Storage Conference (IRES). Berlin; Nov. 28–30, 2011.

[15] Nielsen L. GuD-Druckluftspeicherkraftwerk mit Wärmespeicher. Schriftenreihe des Energie-Forschungszentrums Niedersachsen (EFZN). Göttingen: Cuvillier; 2013.

[16] Sterner M, Stadler I. Energiespeicher: Bedarf, Technologien, Integration. Berlin Heidelberg: Springer; 2014.

[17] Moser P. RWE AG, Status der Entwicklung des adiabaten Druckluftspeichers ADELE. Leopoldina-Symposium "Energiespeicher—Der fehlende Baustein der Energiewende?" Halle; Feb 6, 2014.
[18] Wolf D, Budt M. LTA-CAES—a low-temperature approach to adiabatic compressed air. Appl Energ 2014;125:158–64.
[19] Oldhafer N. et al. Druckluftspeicherkraftwerk als Kurzzeitspeicher-, Langzeitspeicher- und Schattenkraftwerk; 2014. http://www.uigmbh.de/images/downloads/KTK2014_Tagungsbeitrag_Druckluftspeicherkraftwerk%20als%20Kurzzeitspeicher-,%20Langzeitspeicher-%20und%20Schattenkraftwerk.pdf
[20] Kruck O, Crotogino F, Prelicz R, Rudolph T. Overview on all known underground storage technologies for hydrogen. HyUnder Deliverable No. 3.1; 2013.
[21] Griesbach H, Heinze F. Untergrundspeicherung: Exploration, Errichtung, Betrieb. Landsberg/Lech: Verlag Neue Industrie; 1996.
[22] International Gas Union. Natural gas facts and figures; Oct. 2014.
[23] LBEG. Erdöl und Erdgas in der Bundesrepublik Deutschland. Hannover: Landesamt für Bergbau, Energie und Geologie; 2014.
[24] Grubelich MC, Bauer SJ, Cooper PW. Potential hazards of compressed air energy storage in depleted natural gas reservoirs. SAND2011-5930. Albuquerque, NM: Sandia National Laboratory; 2011.
[25] Webb SW. Borehole and formation analyses to support compressed air energy storage development in reservoirs. Proceedings TOUGH Symposium 2012, Lawrence Berkeley National Laboratory. Berkeley, CA; Sep. 17–19, 2012.
[26] Gardner WP. Preliminary formation analysis for compressed air energy storage in depleted natural gas reservoirs. A study for the DOE Energy Storage Systems Program (SAND2013-4323). Albuquerque, NM: Sandia National Laboratory; 2013.
[27] Ahluwalia RK, Sharma A, Ahrens FW. Design of optimum aquifer reservoirs for CAES plant. Compressed Air Energy Storage Symposium Proceedings. Pacific Northwest Laboratory, CA; vol. 1; May 15–18, 1978: pp. 327–367.
[28] Kushnir R, Ullmann A, Dayan A. Compressed air flow within aquifer reservoirs of CAES plants. Transport Porous Med 2010;81(2):219–40.
[29] EPRI. Compressed-air energy storage: Pittsfield Aquifer Field Test—test data: engineering analysis and evaluation. Final Report, EPRI GS-6688. Palo Alto, CA: Electric Power Research Institute; 1990.
[30] EPRI/PB-KBB. Compressed-air energy storage field test using the aquifer at Pittsfield, Illinois. Final Report, EPRI GS-6671. Palo Alto, CA: Electric Power Research Institute; 1990.
[31] Thomas RL, Gehle RM. A brief history of salt cavern use. Proceedings of the 8th World Salt Symposium, vol. 1. The Hague; May 2000: pp. 207–14.
[32] Acht A, Donadei S. Hydrogen storage in salt caverns. State of the art, new developments and R&D projects. SMRI Fall 2012 Technical Conference. Bremen, Germany; Oct. 1–2, 2012.
[33] http://www.geostockus.com/what-we-do/leached-salt-caverns
[34] Schaub PE, Hobson MJ. PEPCO/DOE/EPRI Hard Rock Caverns CAES Project status report. Pacific Northwest Laboratory: Compressed Air Energy Storage Symposium Proceedings; May 15–18, 1978. Pacific Grove, CA, vol. 1: pp. 79–94.
[35] EPRI. Compressed-air energy storage using hard-rock geology. Test facility and results, vols. 1 and 2. Palo Alto, CA: Electric Power Research Institute; 1990.
[36] Bauer SJ, Gaither KN, Webb SW, Nelson C. Compressed air energy storage in hard rock. Feasibility study, SAND2012-0540. Albuquerque, NM: Sandia National Laboratory; 2012.
[37] Pape JT. Kombi-Kraftwerke als Mittellast- und Spitzenlastkraftwerke durch Druckluftspeicher. Die Führungskraft; Mar 28–31, 1989: pp. 56.

[38] Väätäinen A, Särkkä P, Sipilä K, Wistbacka M. Compressed air energy storage in mine: prefeasibility study. International Society for Rock Mechanics, Proceedings of 8th International Congress on Rock Mechanics. Tokyo, vol. 2; 1995: pp. 485–489.

[39] K-UTEC Salt Technologies, Evonik, TU Bergakademie. Druckluftspeicher zur Energiespeicherung in stillgelegten Salzbergwerken und Stabilisierung der Grubenhohlräume: Abschlussbericht zum Verbundprojekt. Freiberg, Germany: K-UTEC Salt Technologies, Evonik, TU Bergakademie; 2009.

[40] Gillhaus A, Crotogino F, Albes D et al. Compilation and evaluation of bedded salt deposit and bedded salt cavern characteristics important to successful cavern sealing and abandonment. Research Project Report. Clarks Summit, PA: Solution Mining Reseach Institute; 2006.

[41] http://www.eon.com/content/eon-com/de/about-us/structure/asset-finder/huntorf-power-station.html

[42] Crotogino F. Kavernen als Energiespeicher. Kali und Steinsalz; 2006; 1: pp. 14–21.

[43] Crampsie S. Full charge ahead. Eng Technol; 2009; 4 (6): pp. 52–55.

[44] http://powersouth.com/mcintosh_power_plant/mcintosh_power_plant/compressed_air_energy

[45] Succar S. Compressed air energy storage. In: Barnes FS, Levine JG, editors. Large energy storage systems handbook. USA; 2011. pp. 111–152.

[46] Greenblatt JB et al. Baseload wind energy: modeling the competition between gas turbines and compressed air energy storage for supplemental generation. Energ Policy; 2007; 35, pp. 1474–1492.

[47] Mehta BR. Compressed air energy storage (CAES) in salt caverns. SMRI Spring Meeting, Tulsa; Apr. 27, 1987.

[48] http://www.gaelectric.ie/wp-content/uploads/Public-Consultation-Story-Boards-2012.pdf

[49] http://www.new.alacaes.com/demonstration-plant/

[50] http://www.pge.com/en/about/environment/pge/cleanenergy/caes/index.page

[51] http://www.apexcaes.com/project

[52] Rocky Mountain Power Inc. ASIST Project (Alberta Saskatchewan Intertie Storage): a CAES study of energy storage for Alberta. Alberta Energy Storage Symposium; Nov. 19, 2013. http://de.scribd.com/doc/191156184/Alberta-Saskatchewan-Intertie-Storage#scribd

[53] www.energiespeicher-niederrhein.de

[54] http://storelectric.com

[55] http://www.deseretnews.com/article/865611571/Massive-green-energy-project-taps-salt-caverns-near-Delta.html

[56] http://www.cleveland.com/business/index.ssf/2013/07/firstenergy_postpones_project.html

[57] RWE AG. ADELE—Der adiabate Druckluftspeicher für die Elektrizitätsversorgung. Essen, Germany: RWE AG; 2010.

[58] Schulte RH, Critelli N Jr, Holst K, Huff G. Lessons from Iowa: development of a 270 megawatt compressed air energy storage project in Midwest independent system operator: a study for the DOE Energy Storage Systems Program. SAND2012-0388. Albuquerque, NM: Sandia National Laboratory; 2012.

[59] Heath JE, Bauer SJ, Broome ST, Drewers TA, Rodriguez MA. Petrologic and petrophysical evaluation of the Dallas Center Structure, Iowa, for compressed air energy storage in the Mount Simon sandstone. A study for the DOE Energy Storage Systems Program. SAND2013-0027. Albuquerque, NM: Sandia National Laboratory. USA; 2013.

[60] Knoke S. Compressed air energy storage (CAES). EPRI Handbook of Energy Storage for Transmission or Distribution Applications. Palo Alto, CA: Electric Power Research Institute. Washington; 2002.

[61] http://www.generalcompression.com/index.php/tdw1
[62] Pacific Northwest National Laboratory. Techno-economic performance evaluation of compressed air energy storage in the Pacific Northwest. USA; 2013. http://caes.pnnl.gov/pdf/PNNL-22235.pdf
[63] Pacific Northwest National Laboratory. Compressed air energy storage: grid-scale technology for renewables integration in the Pacific Northwest. USA; 2013. http://caes.pnnl.gov/pdf/PNNL-22235-FL.pdf

Chapter 7

Underwater Compressed Air Energy Storage

Andrew Pimm, Seamus D. Garvey
Faculty of Engineering, University of Nottingham, Nottingham, United Kingdom

1 INTRODUCTION

Compressed air energy storage (CAES) is an energy storage technology that is centered on the concept of storing energy in the form of high-pressure air. The offshore environment provides several ideal conditions for storage of compressed air. By storing pressurized air in an underwater vessel the pressure in the air can be reacted by the surrounding water, greatly reducing loading at the air/water barrier. This simplifies the design requirements of containment vessels and presents potential for their inexpensive manufacture, relative to the cost of a vessel required to handle full pressure without the support of surrounding water.

The density of seawater is around 1025 kg m^{-3}, less than half that of the Earth's upper crust (around 2700 kg m^{-3}), meaning that the storage pressure under a given depth of water is less than half that under the same depth of rock, assuming that compressed air is stored at a pressure roughly equal to the surrounding hydrostatic/geostatic pressure. However, unlike underground stores that require the mining of a cavern, underwater CAES (UWCAES) stores can be manufactured and installed at potentially low cost, though the method of installing and securing the vessels to the seabed requires some development. UWCAES vessels are hidden from view, and suitably deep water can readily be found close to many coastlines, not only in saltwater oceans/seas but also in some freshwater lakes (e.g., Loch Morar in the UK, which has a maximum depth of 310 m). Compared with geological CAES, UWCAES also has the advantage of isobaric characteristics.

In an underwater vessel, compressed air is stored at approximately the same pressure as the hydrostatic pressure in the surrounding water, so the water provides the reaction to the pressure of the compressed air, and the storage vessel can be very low in cost. The load that must be reacted is primarily that of buoyancy, which is proportional to volume, and as a result the cost of a vessel for UWCAES is roughly independent of depth. While the vessel cost is roughly independent of depth, energy storage capacity increases with depth, so it follows

Storing Energy. http://dx.doi.org/10.1016/B978-0-12-803440-8.00007-5

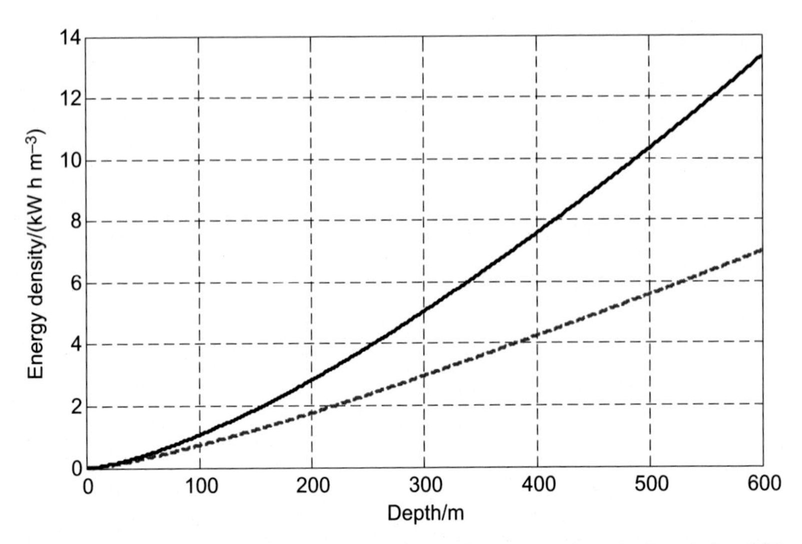

FIGURE 7.1 **Energy density available in a UWCAES vessel against depth.** *Black solid line*, adiabatic expansion with exhaust at ambient temperature; *red dashed line*, isothermal expansion.

that the deeper the vessel is fixed, the lower is the cost per unit of energy storage capacity (in other words the deeper the vessel, the better).

The gauge pressure in seawater at a depth d is given by:

$$p = \rho_{sw} g d \tag{7.1}$$

where ρ_{sw} is the density of seawater (typically 1025 kg m^{-3}) and g is acceleration due to gravity (9.81 m s^{-2}). Using equations from chapter: Compressed Air Energy Storage, it is possible to obtain curves of energy density against depth for an underwater compressed air store, assuming the air is stored at a pressure equal to that of the surrounding seawater. These curves are shown in Fig. 7.1. In both cases (isothermal and adiabatic expansion) the energy density increases more than linearly with depth because of the compressibility of air.

At 500 m depth the energy density is between 5.6 kW h m^{-3} and 10.3 kW h m^{-3}, depending upon how the air is reheated before/during expansion. The lower limit on energy density at this depth is over three times the energy density in the 600 m high upper reservoir at Dinorwig pumped storage plant in the United Kingdom. At depths of the order of hundreds of meters, wave action decays to a very low level and the seabed is typically bare. There is a lack of information on biofouling patterns with depth, and systematic studies are commonly restricted to man-made structures within 150 m depth [1].

2 STORAGE VESSELS FOR UWCAES

Storage vessels for UWCAES can be split into two categories based on their walls: flexible and rigid. Both have advantages and disadvantages, as explained later.

2.1 Flexible Vessels

Flexible vessels (also known as "energy bags") are made of coated fabric serving as the air/water barrier, with reinforcing straps to carry the main buoyancy loads. They can be manufactured in a variety of shapes, though development has focused on vessels with a single point of anchorage (Fig. 7.2) whose inflated shape is very similar to that of hot-air balloons [2]. Such vessels are very similar in design to lift bags used for underwater salvage operations and must be anchored to the seabed using gravity base anchors, seabed piles, or other anchorage means. Similar to lift bags a cluster of several balloon-shaped vessels can be attached to a single anchor point [3].

With careful manufacture, including welded seams, flexible vessels can be watertight and reliable over long periods; fabric air lift bags have been kept underwater for periods of several years and still remained functional [4]. While leakage can be kept to levels so low that the effect on roundtrip efficiency is

FIGURE 7.2 **A 1.8 m-diameter energy bag prototype undergoing tank testing at the University of Nottingham, 2011.** The base of the bag can be seen attached to the base of the tank; in a real-world offshore installation the base would be attached to a gravity base anchor, that is, a large weight, or a pile installed in the seabed.

negligible, it must be possible to prevent any buildup of water at the bottom of both flexible and rigid vessels, which could otherwise cause the vessel to fill up with water and reduce the storage volume. Therefore, both flexible and rigid vessels must have some kind of port in the base to insure that water can be forced out by internal pressure. In flexible vessels, this can take the form of a pressure relief valve, a nonreturn valve, or simply an open port; with the latter, water is allowed to enter the inside of the vessel during discharge, though this is not necessarily a bad thing as the water will be forced out of the vessel again during charge. All of these will also act as pressure relief devices to combat unintended overpressurization, (e.g., due to a valve in the airline that is stuck in the open position, or a pressure regulator accidentally set to a dangerously high pressure), though very small ports may not be large enough to allow air to flow out of the vessel at the rate necessary to prevent a dangerous buildup of pressure.

The airline that connects a vessel to the compression and expansion machinery must be attached to the top of the vessel. If the pipe were connected to the base of the vessel then any water ingress into the vessel (either because of leakage or because of an open base, as mentioned previously) would enter the pipe and block it, potentially rendering the vessel unusable until the water can be removed from the pipe.

Ignoring the mass and volume of materials (which are both relatively small for a fabric vessel) the buoyancy force on a flexible vessel is given by:

$$F_b = (\rho_{sw} - \rho_a) gV \tag{7.2}$$

where ρ_{sw} is the density of seawater, ρ_a is the density of air, and V is the volume of the inflated vessel. This buoyancy must be resisted by some type of anchorage the possible nature of which is discussed later.

The buoyancy load can be carried through meridional straps which run from the top of the vessel down to the anchor, so that the fabric only needs to provide an airtight barrier between the air and surrounding water and react to the small differential pressure load. This rises linearly with height above the base of the vessel; the differential pressure across the air–water barrier at height h above the base of the vessel is given by:

$$p = (\rho_{sw} - \rho_a) gh + p_0 \tag{7.3}$$

where p_0 is the differential pressure at the base of the vessel. In an open-based vessel, p_0 can be no greater than zero and h must be modified to be the height above the water level inside the vessel rather than above the base of the vessel; the differential pressure across the air–water barrier at all points below the water level is zero [5].

The differential pressure load is reacted by curvature in the fabric. Assuming a thin wall, which is quite reasonable for a fabric structure, the differential pressure at a point is related to the tension and curvature in the fabric at that point according to the Young–Laplace equation, which is:

$$p = \frac{F_x}{r_x} + \frac{F_y}{r_y} \tag{7.4}$$

where p is the differential pressure across the fabric; F_x, F_y are the tensions (per unit length) in the fabric in the directions of the principal radii of curvature; and r_x, r_y are the principal radii of curvature.

This equation shows us that the tensions in the fabric of an energy bag are proportional to the local curvature of the fabric. Therefore, it is advantageous to allow the fabric to form lobes between the meridional straps.

It is reasonable to assume that the cost of the anchorage C_a is directly proportional to inflated volume (itself proportional to some nominal geometric dimension, for example, diameter, D^3).

$$C_a = aD^3 \tag{7.5}$$

It is also fair to assume that the cost of the meridional straps C_m is directly proportional to D^4, since a force proportional to D^3 must be carried a distance proportional to D.

$$C_m = bD^4 \tag{7.6}$$

We can assume that the cost of surface, C_s, is directly proportional to D^2.

$$C_s = cD^2 \tag{7.7}$$

Finally, like the anchorage cost the energy storage capacity E is proportional to D^3:

$$E = kD^3 \tag{7.8}$$

Looking at the ratio of vessel and anchorage cost to storage capacity, we arrive at an interesting result. The ratio of cost to storage capacity is given by:

$$\frac{C}{E} = \frac{aD^3 + bD^4 + cD^2}{kD^3} \tag{7.9}$$

We can differentiate this with respect to D and set the result equal to zero to find the vessel diameter that minimizes cost per unit of storage capacity, D^*.

$$\frac{\partial(C/E)}{\partial D} = \frac{1}{k}\left(b - \frac{c}{D^2}\right) = 0 \tag{7.10}$$

$$D^* = \sqrt{c/b} \tag{7.11}$$

Substituting D^* into the equations for C_m and C_s shows that the optimal size of a vessel that minimizes cost per unit of energy storage is that which has equal costs of meridional straps and surface. With realistic materials costs, it was shown in [2] that the optimal vessel can cost less than £10 $(\text{kW h})^{-1}$ of storage capacity. This is further reinforced by the costs of air lift bags, giving costs for 50 t (metric tonne) vessels of around £15 $(\text{kW h})^{-1}$ (not including anchorage).

2.2 Rigid Vessels

Rigid vessels can be built with ballast in mind so that the vessels are heavy enough to remain on the seabed without any additional ballast material or connection to piles in the seabed. In 2005, Davies [6] proposed building modular steel caissons for UWCAES into the base of large offshore wind turbines, and in 2014 Hydrostor installed multiple rigid caissons at a pilot plant in Lake Ontario (Fig. 7.3).

Rigid vessels resist loads in bending because they cannot change shape in the same way as flexible vessels, and because of this it is generally considered that flexible vessels can be of lower cost than rigid vessels (Fig. 7.4). However, flexible vessels are more vulnerable to damage due to handling (though this danger can largely be removed with proper handling procedures and durable fabric), and ballast must be provided separately for a flexible vessel. Also, if

FIGURE 7.3 **Rigid vessels in dry dock, ready for installation at Hydrostor in Lake Ontario, 2014 [7].**

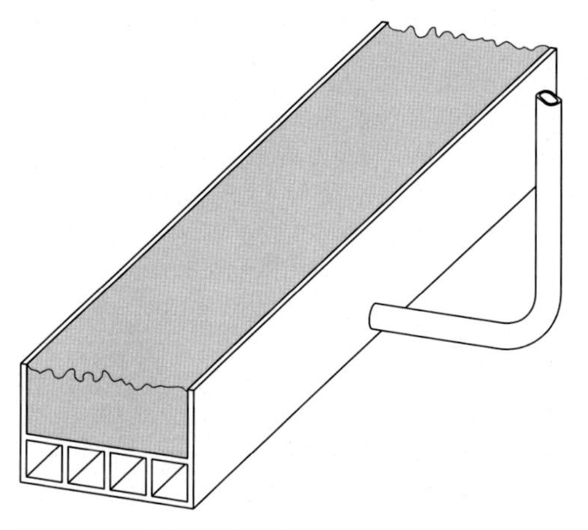

FIGURE 7.4 **Rigid storage vessel incorporating ballast.**

some of the heat of compression was to be stored for later use inside the compressed air, rigid vessels would lose heat to the environment at a slower rate than flexible vessels because their walls provide more insulation.

To avoid large negative differential pressure, it is necessary to ensure that a rigid vessel is open based. As with an open-based flexible vessel, this allows water to fill the vessel from the base upward as the store is discharged and requires that the air pipe is connected to the top of the vessel.

It makes sense to design a rigid vessel as a gravity base structure, so that its resistance to buoyancy comes from its weight and no piles are required. The volume of concrete required in the structure depends upon the anchorage ratio γ (the ratio of anchorage capacity to buoyancy), and is given by:

$$V_c = \frac{\rho_{sw} - \gamma \rho_a}{\gamma \rho_c - \rho_{sw}} V_a \tag{7.12}$$

where ρ_c is the density of concrete, or the average density of the walls.

If the structure were made of concrete with a density of 2400 kg m^{-3} and submerged in seawater with a density of 1025 kg m^{-3}, and if an anchorage ratio of 1.5 is used, then for every cubic meter of stored air, 0.37 m^3 of concrete is required. This means that in a storage vessel at 500 m depth, every megawatt-hour of energy storage capacity would require between 36 m^3 and 66 m^3 of concrete (depending on whether the system was adiabatic or isothermal).

It is possible to incorporate ballast into the design of rigid vessels in other ways. One proposal by Seymour was to create a vessel comprising an air storage unit at the base, above which is an open container filled with rocks or other ballast material [8].

3 ANCHORAGE AND INSTALLATION

Vessels for UWCAES require anchorage capacity in proportion to their storage capacity, and useful plant sizes require significant amounts of anchorage. Economically attractive anchorage and installation techniques therefore provide the biggest challenges facing UWCAES. As mentioned previously, rigid vessels can have mass built into the design and so may not require other anchorage means, but this is not the case for flexible vessels.

The simplest method of anchoring a buoyant object to the seabed is using gravity base anchors (GBAs, also known as "deadweights"). A GBA is simply a large weight that holds the vessel on the seabed. To maximize the lifetime of a GBA underwater, metals should be avoided where possible and so concrete, sand, and rocks make an obvious choice. One interesting idea for GBAs is to design them so that they can be prepared elsewhere then towed to site, using the air storage vessel as support. This approach would require ballast to be added to the GBA before it is sunk; the simplest way to do this would be to tow the sealed GBA to site with air inside, and then open a port in the top of the GBA once it is in position above the site, thus allowing it to fill with water. The GBA/bag combination would then start to sink, and descent must be controlled using a cable attached to a crane or winch on a workboat while also blowing air into the bag (e.g., using a compressor on the workboat), allowing the excess air to flow out of the base of the bag as bubbles. Air must be blown into the bag because as the GBA/bag combination is lowered the surrounding water pressure increases, and thus the bag will tend to reduce in volume and buoyancy unless more air is added. This approach is laid out in Fig. 7.5.

Such a GBA could then be floated at a later date (e.g., for recovery) by connecting an air pipe to the base of the GBA and filling it with air. It would also be necessary to inflate the energy bag and insure that there are holes at the base of the GBA and energy bag to allow air to flow out as the combination rises into lower pressure water.

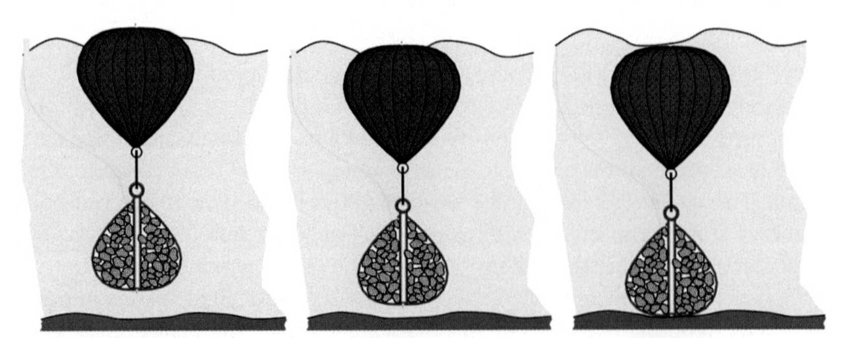

FIGURE 7.5 **Installation technique with gravity base anchor allowing ballast to be floated to site within a container [9].**

Alternatively, anchorage can be provided through piles installed in the seabed. These can be hammered, screwed (screw, or helical, piles), dropped (torpedo piles), or sucked (suction cup) into the seabed. Screw piles and suction cup anchors are of particular interest for UWCAES since they both minimize reliance upon friction and screw pile, in particular, are under consideration for offshore wind foundation applications. They comprise a number of helical steel plates attached to a steel shaft and they are installed into the seabed by applying a torque to the shaft. The anchorage capacity of a screw pile increases with the number of helices, the helix diameter, and the sand compactness [10]. Research into their large-scale application in the seabed is ongoing.

4 SYSTEM CONFIGURATIONS

Discussion in the chapter has so far mostly covered the nature of compressed air stores underwater. We now look at the machinery and infrastructure required for a UWCAES plant and how this might be configured in terms of location and technology. We first consider heat management.

As discussed in chapter: Compressed Air Energy Storage, as well as storing energy as compressed air, CAES plants may also store energy as heat. The reason for this is simple: when a gas is compressed, its temperature increases because thermal energy is generated. Part of the work that went into compressing the air has been converted to thermal energy (i.e., heat) and ideally this energy should not be wasted. In existing underground CAES plants the heat of compression is simply taken out of the compressed air and vented to atmosphere as waste heat.

When a gas is expanded, its temperature drops, and so when the compressed air is allowed out of storage it must be heated before entering the turbine (otherwise moisture in the air would freeze, blocking the pipes). In existing CAES plants, compressed air is mixed with natural gas and burned in a gas turbine, emitting greenhouse gases.An alternative to venting the heat of compression to atmosphere and later burning gas to heat the compressed air is to store the heat of compression in a dedicated thermal energy store and later use it to reheat the air as it enters the turbine. Known as adiabatic CAES (A-CAES), this approach improves utilization of the store and reduces (or removes) the need to burn gas. The Adiabatic CAES Storage for Electricity Supply (ADELE) project, currently under construction in Germany, will be the world's first commercial A-CAES plant [11].

Another choice of system configuration that relates to heat management is the number of stages of compression and expansion, and what intercooling/reheating takes place between compression/expansion stages. UWCAES systems can take many different configurations, including those considered or used for underground CAES (such as single/multistage and adiabatic/diabatic), as effectively the main difference between an underwater CAES system and a fixed volume underground CAES system is the pressure characteristic, which only serves to improve the roundtrip efficiency of an underwater system over the underground equivalent.

In a UWCAES plant the ideal choice of location for the conversion machinery and any heat storage depends upon the distance between the vessels and shore and the relative costs of transmitting power as electricity in a cable and as high-pressure air in a pipe. For nearshore plants, it makes sense to site the machinery onshore, making it easily accessible and removing the need for any underwater power transmission. In such cases an air pipe runs between the machinery and the storage vessels. In some parts of the world there are limited locations where deep water, for example, >400 m, is found within a few kilometers of shore, so in those locations it would make more financial sense to site the machinery on a floating offshore platform moored above the storage vessels, with an air pipe running between the floating platform and the vessels, and a power transmission line running between the floating platform and an onshore electrical substation. Locating the compression and expansion machinery on the seabed could reduce the air transmission distance and hence increase roundtrip efficiency, but it has not been seriously considered because of the sealing requirements and, in particular, the immense difficulty this would introduce to access and maintenance (Fig. 7.6).

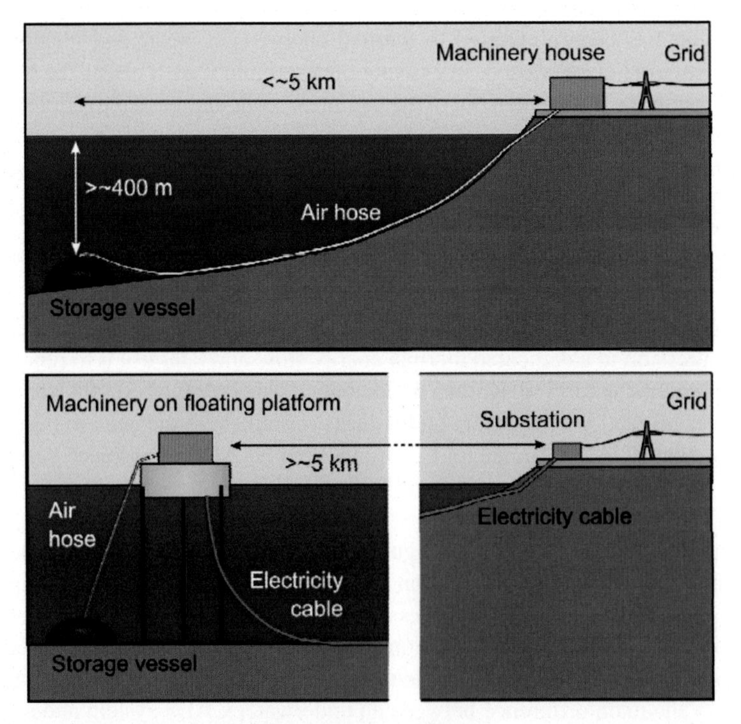

FIGURE 7.6 **Shore-based and platform-based machinery.**

In a near-shore UWCAES plant with onshore conversion machinery a thermal energy storage unit could take exactly the same form as the thermal energy storage unit for underground CAES plants. In a far-offshore plant a thermal energy store would likely be located on the same platform as the conversion machinery. In such cases the size and weight of the required thermal energy stores must be given serious consideration. The weight of thermal energy storage material required for a given storage capacity can easily be calculated: assuming that granite ($c_p \approx 800$ J K^{-1} kg^{-1}) is used to store heat at 500 °C above ambient, 9 kg of rock would be required for a kilowatt-hour of thermal energy storage capacity. For 100 MW h of thermal storage capacity, 900 t of rock would be required, taking up about 550 m^3 (assuming a packing factor of 60%). While the weight of machinery and containment would also need to be taken into account, this kind of size and mass is not excessively large in terms of the capacity of seafaring vessels. In deep water a tension leg platform could be used for this task, similar to those used for offshore oil/gas platforms (Fig. 7.7).

Various aspects of the offshore environment can be advantageous for thermal energy storage. One interesting proposal for offshore thermal energy storage made by Ruer [12] is to locate a pressurized heat store underwater. This could be used in a system incorporating direct contact between the hot compressed air and the thermal storage medium and could use the pressure of the surrounding water to provide resistance to the pressure of the air passing through the heat store. The concept was originally developed for coastal underground CAES but would be equally applicable to underwater CAES (Fig. 7.8).

One other aspect of the offshore environment that could be beneficial to UWCAES is the fact that the sea effectively provides an infinite thermal sink to aid near-isothermal compression, with water also having much higher thermal conductivity than air [0.58 W $(m\ K)^{-1}$ compared with 0.024 W $(m\ K)^{-1}$].

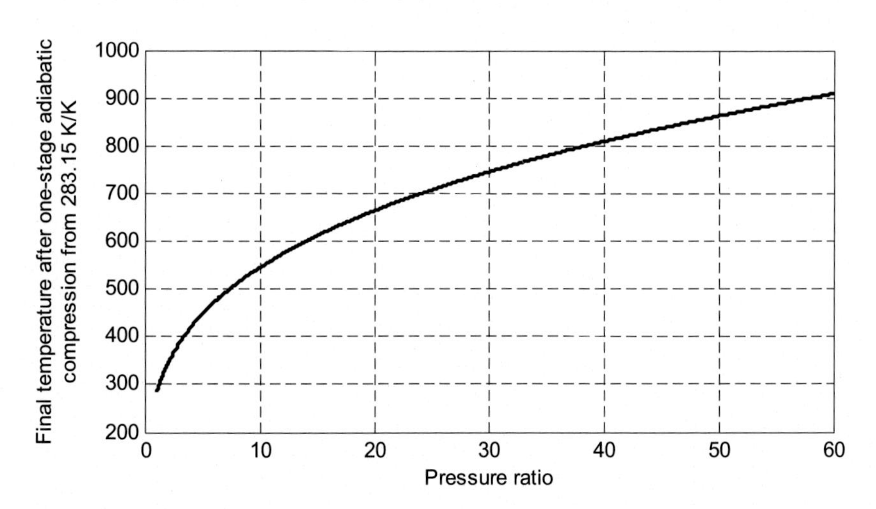

FIGURE 7.7 **Temperature after single-stage adiabatic compression from 283.15 K.**

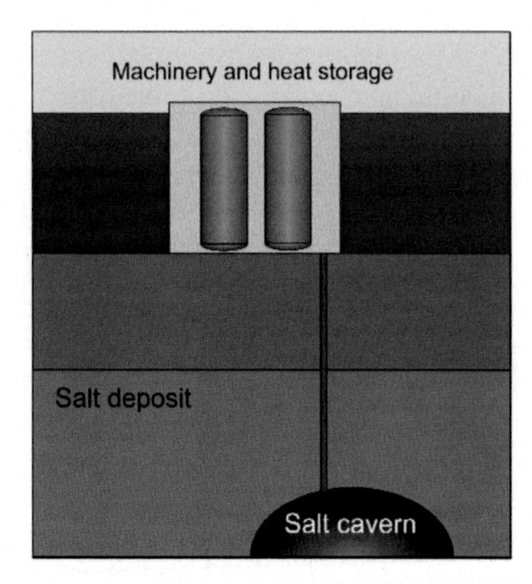

FIGURE 7.8 **Proposed offshore adiabatic CAES system with underwater thermal storage [12].**

5 LOCATIONS

Suitable locations for UWCAES are those with deep water close to shore. In a study of coastal waters with depth greater than 400 m around Europe and North America, it was found that many locations have such depths within reasonable distance of shore, particularly islands [13]. An example of suitable locations in California is shown in Fig. 7.9. A minimum depth of 400 m was used because the water pressure at this depth is about 4 MPa (~40 bar), similar to the inlet pressure of the high pressure turbines at the existing underground CAES plants [14,15]. California is particularly interesting for energy storage because of the mandate approved by the California Public Utilities Commission in late 2013 for 1325 MW of energy storage to be operational by the end of 2024.

Most of the suitable locations for UWCAES in Europe are found within the Mediterranean Sea and the fjords of Norway. The bathymetry of fjords is particularly interesting because they are typically characterized by a phenomenon known as overdeepening, whereby glacial erosion has caused them to become deeper than the surrounding sea. Within the Mediterranean, heavily populated areas with deep water close to shore include the Côte d'Azur and the Amalfi Coast. Fig. 7.10 shows the areas in the central Mediterranean Sea which are within 5 km of land and have a depth greater than 400 m.

Table 7.1 shows the area of water with depth >400 m within the territorial waters of each country in Europe, categorized by distance from shore. Almost all of Norway's vast areas of near-shore deep water are found within the fjords. Aside from Norway, it is typically the case that island nations have the largest areas of deep water close to shore.

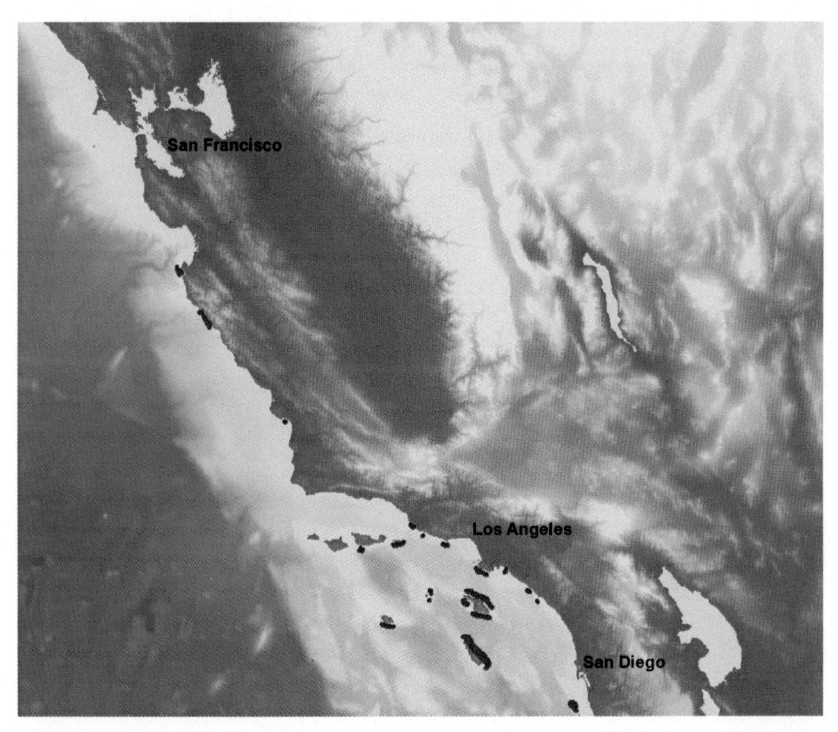

FIGURE 7.9 **Map of the central and southern California coastline, showing areas within 5 km of land which have a water depth greater than 400 m.**

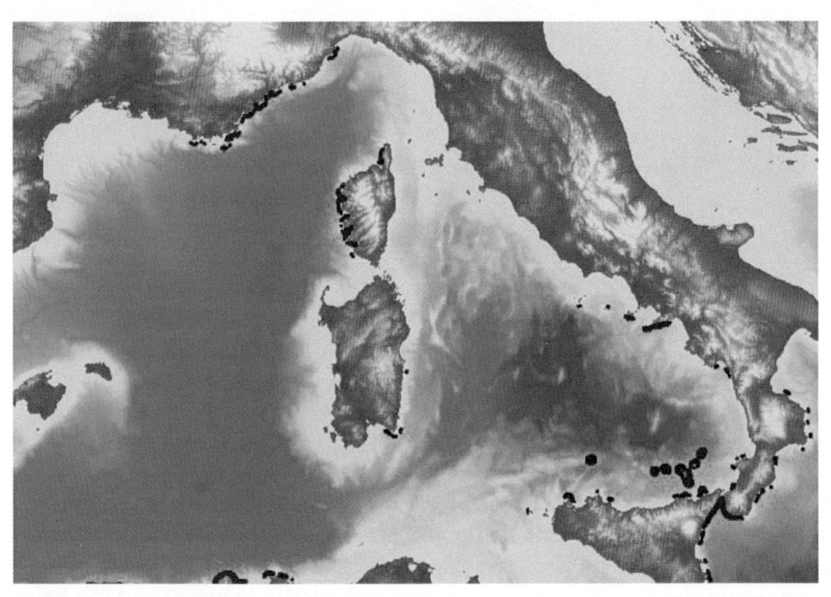

FIGURE 7.10 **Map of the central Mediterranean Sea, showing areas within 5 km of land which have a water depth greater than 400 m.**

TABLE 7.1 Areas of Water Within European Territorial Waters with a Depth Greater Than 400 m, Categorized by Distance from Shore

	Area with depth >400 m/km^2				
Country	**0–22.2 km**	**0–10 km**	**0–5 km**	**0–2 km**	**0–1 km**
Norway	4 787.19	2 692.07	1 810.26	733.01	149.43
Greece	76 346.16	22 473.24	5 289.61	440.02	28.07
Azores	19 561.23	5 746.84	1 472.01	168.61	27.92
Canary Islands	23 340.65	5 746.81	1 171.44	51.42	16.70
Italy	37 577.86	8 460.96	1 663.27	139.94	15.40
Turkey	22 813.23	5 385.08	725.03	27.48	2.26
France	7 934.15	1 887.44	430.85	25.65	1.23
Madeira	9 141.76	2 444.39	513.05	19.91	0.49
Georgia	2 123.36	512.82	99.98	3.83	
Gibraltar	231.24	75.90	17.90	0.95	
Portugal	1 583.32	159.82	21.83	0.45	
Russia (to 65°E)	3 173.24	269.55	23.20	0.43	
Cyprus	10 173.09	2 593.82	445.98		
Spain	12 645.34	1 039.71	67.38		
Monaco	54.33	14.07	1.02		
Albania	585.17	51.44			
Ukraine	1 182.22	48.80			
Malta	970.83	41.95			
Iceland	411.76				
Croatia	107.13				
Totals	235 783.68	59 934.84	13 823.85	1 612.52	241.51

6 COST AND EFFICIENCY

With suitable pipework the roundtrip efficiency of an underwater CAES plant can be very similar to that of an underground CAES plant. The effective roundtrip efficiency of a CAES plant was discussed in chapter: Compressed Air Energy Storage. It is not unusual for compressors to achieve adiabatic efficiencies of 88% and for turbines to achieve efficiencies above 92%, so the basic elements of the system point to a roundtrip efficiency in excess of 80% being achievable.

Assuming that compressed air is stored at a similar temperature to the surroundings—as is the case at Huntorf and at McIntosh), the additional losses introduced by underwater storage are those associated with leakage and pressure drop. With a well-manufactured vessel, leakage losses should be small. The leakage rate in the first prototype offshore flexible vessel was estimated at 1.2% per day at the end of a 3-month deployment [16], with air loss mainly found around stitching of seams and areas of repaired trauma due to handling. It is anticipated that future deployments would benefit from improved handling procedures and more heavy-duty material and so would not be damaged on deck.

The pressure drop in the pipework is likely to be the main source of loss in a UWCAES system. Table 7.1 shows the pipe diameter required for various percentage pressure drops per kilometer of pipe in a UWCAES system taking cool compressed air at 100 MW. Two points should be noted here: (1) the mass flow rate of air is determined here by pessimistically assuming that the air will be expanded isothermally, and (2) the actual losses in output power are much lower than the losses in pressure. As an example of the latter point, at 2.0 MPa (20 bar) a 1% drop in pressure causes only a 0.33% drop in output power, and at 7.0 MPa (70 bar) a 1% drop in pressure causes only a 0.24% drop in output power.

Table 7.3 gives the pipe masses corresponding to the diameters given in Table 7.2, assuming a 10 km length of cylindrical steel pipe with density of 7800 kg m^{-3}, working stress of 100 MPa, and a constant wall thickness. Associated costs are also given, assuming a cost of £2000 per tonne of steel (about five times the cost of the raw material to account for welding and other working). The pressure drop has not been taken into account, as to fully account for its effects on air density (and hence net buoyancy) would require knowledge of the seabed profile along the length of the pipe, which varies by location. In reality

TABLE 7.2 Pipe Diameters Required to Transmit 100 MW of Cool Air for Various Percentage Pressure Drops

		Required pipe diameter *D*/m			
Pressure (10^5 Pa) (bar)	Mass flow rate/(kg s^{-1})	0.05% loss per kilometer	0.2% loss per kilometer	1% loss per kilometer	4% loss per kilometer
20	415.2	2.182	1.648	1.192	0.904
30	365.7	1.761	1.331	0.962	0.730
40	337.2	1.518	1.147	0.830	0.630
50	317.9	1.356	1.025	0.741	0.562
60	303.8	1.237	0.935	0.676	0.513
70	292.8	1.145	0.866	0.626	0.475

Data from Ref. [17].

TABLE 7.3 Pipe Mass *m* and Approximate Costs Assuming 10 km Length of Steel Pipe, Calculated Assuming Constant Wall Thickness from Surface Down to Storage Vessels

	Pipe mass for 10 km length/t			
Pressure (10^5 Pa) (bar)	0.05% loss per kilometer	0.2% loss per kilometer	1% loss per kilometer	4% loss per kilometer
20	11 792	6 709	3 502	2 011
30	11 463	6 540	3 412	1 963
40	11 334	6 465	3 383	1 948
50	11 293	6 449	3 368	1 937
60	11 270	6 436	3 363	1 936
70	11 261	6 439	3 364	1 936
Approx. cost/£	23×10^6 (23 m)	13×10^6 (13 m)	7×10^6 (7 m)	4×10^6 (4 m)

because the highest differential pressure across the pipe would be seen by the parts closest to the surface of the sea it would be possible to reduce the total pipe mass by using a pipe whose wall thickness reduces in stages as it reaches greater depths.

Interestingly, the pipe masses shown are almost independent of internal pressure because the required pipe diameter for a given pressure drop decreases with greater internal pressure. The only reason pipe mass is not constant for all pressures is because air density (and hence differential pressure across the pipe wall) increases with pressure inside the pipe. While the pipe masses shown are large, it should be remembered that the pipe connecting the vessels to the machinery is also a store of compressed air.

The underwater pipes would need to rest on the seabed, so it is necessary to take into account the pipe buoyancy and pipe masses to calculate the net buoyancy of the pipes, which must be counteracted with ballast. The net buoyancy of the 10 km-long pipes referred to in Tables 7.2 and 7.3 is shown in Table 7.4. Clearly higher storage pressures used in deeper water allow narrower pipes to be used for a given pressure drop, hence reducing the net buoyancy of the pipe.

The existing underground CAES plant at Huntorf has the machinery house located almost directly above the caverns, so for each cavern there is only 300 m of surface-based pipework and ~500 m of downpipe. UWCAES plants with offshore machinery could have the machinery located on a platform directly above the storage vessels, and so would only require a pipe directly down to the seabed. Near-shore plants would clearly need to have a longer pipe run than offshore plants and as such would have greater transmission losses.

TABLE 7.4 Net Buoyancy m^* in tonnes for 10 km of Pipe, Using Pipe Diameters from Table 7.2 and Pipe Masses from Table 7.3

	Net buoyancy, m^*, for 10 km length/t			
Pressure/ (10^5 Pa) (bar)	0.05% loss per kilometer	0.2% loss per kilometer	1% loss per kilometer	4% loss per kilometer
20	25 590	14 615	7 654	4 405
30	12 576	7 193	3 762	2 168
40	6 300	3 602	1 889	1 089
50	2 595	1 487	779	449
60	135	80	43	25
70	−1 620	−924	−482	−277

UWCAES has an advantage over underground CAES in that the pressure characteristic in an underwater CAES vessel is roughly isobaric. This contrasts with a fixed volume containment (such as a cavern), in which the air pressure varies strongly with the amount of stored energy. As such, CAES plants that use underground caverns must incorporate pressure regulators to ensure that the turbine inlet pressure is below a certain level. Pressure regulators have energy losses associated with them, losses which are not seen by an underwater system (which needs not use pressure regulation, or certainly does not need such a severe pressure drop). Fig. 7.11 shows a curve of pressure against stored energy for the air in an underwater CAES plant at 400 m depth and the curve for an equivalent plant using a fixed volume container (e.g., an underground salt cavern). The isobaric characteristic of UWCAES is clear. The equivalent underground cavern (of the Huntorf/McIntosh type, with 4 MPa (40 bar) maximum turbine inlet pressure) requires the air to be throttled down to 4 MPa (40 bar) during discharge until roughly two-thirds of its energy content have been removed, and then at lower states of charge the turbine inlet pressure reduces below 4 MPa (40 bar). Both of these characteristics of fixed volume containment serve to reduce efficiency.

Using reasonable costs for materials, it was shown in [2] that total costs of flexible vessels and ballast for UWCAES can be less than £10 $(\text{kW h})^{-1}$ (assuming storage at 500 m depth). Installation and machinery will increase the total cost. Through discussions with a major lift bag manufacturer, it has been found that currently the costs of flexible vessels could be less than £15 $(\text{kW h})^{-1}$ at 500 m depth.

The design life of a UWCAES system will be at least 20 years. Concrete structures underwater have lasted much longer than this (concrete reacts well with seawater), and inflatable vessels have been used underwater for significant periods of time without malfunction.

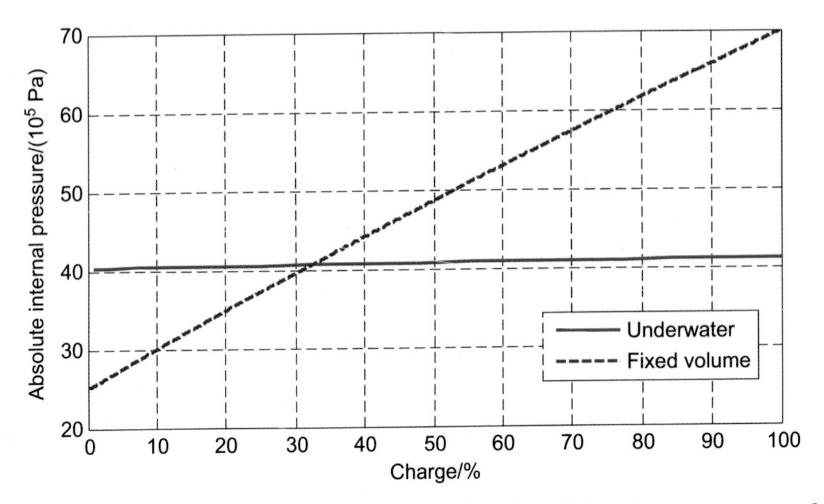

FIGURE 7.11 **Pressure against percentage charge for a 10 m-high underwater compressed air store at 400 m depth and an equivalent underground store with maximum and minimum cavern pressures of 7 MPa (70 bar) and 2.5 MPa (25 bar), respectively.**

7 STATE OF DEVELOPMENT

After early development at the University of Nottingham, Hydrostor Inc. of Toronto have pushed forward commercialisation of UWCAES technology, working with researchers at the University of Windsor. A 750 kW h/1.5 MW h pilot plant was installed at a depth of 80 m in Lake Ontario in 2014, and the first commercial plant will be installed off the coast of Aruba in 2015/2016 (Aruba is one of the Lesser Antilles located in the southern Caribbean Sea and is close to the coast of Venezuela). The pilot plant combines flexible and rigid storage vessels.

Other companies working on or considering UWCAES include Bright Energy Storage, Brayton Energy, Exquadrum, AGNES, DNV, and Moffatt-Nichol. One company working on a novel method of moving the air down to the seabed is OC Energy, who propose the use of an inclined Archimedes screw coupled to a surface-mounted electrical machine [18].

The areas that are in particular need of development are installation procedures and the securing of vessels to the seabed. Rigid vessels are designed such that they have ballast mass built into the structure and, to date, flexible vessels have been held to the seabed using gravity base anchors (i.e., large weights, such as concrete blocks on a steel tray) that were transported to site from shore. However, the use of deadweights could become prohibitively time consuming and costly if carried out for commercial-scale installations of large amounts of storage capacity. By way of example, let us consider a 1 GW h store at 500 m depth. This would require up to 179 000 m^3 of storage capacity, and hence it would be necessary to react 172 000 t of buoyancy force. With a 2:1 ratio of holding-down force to buoyancy force, 344 000 t of holding-down force would be required. Using concrete as a deadweight, approximately 250 000 m^3 would be required to provide this ballast force underwater. To put this in perspective the largest concrete dam in the United Kingdom (at Clywedog reservoir) has a height of 72 m, a length of 230 m, and contains about 200 000 m^3 of concrete [19].

Other than using deadweights, alternative methods of securing vessels to the seabed include driven piles, suction piles, suction-embedded anchors, torpedo piles, and screw (or helical) piles. Of these, screw piles are of particular interest because they provide some amount of positive drive. To minimize cost per unit of anchorage capacity, anchorage which can be installed without use of remotely operated underwater vehicles (ROVs) is of particular interest.

8 CONCLUDING REMARKS

UWCAES is a developing storage technology which is a natural extension of CAES for coastal environments. It is very similar to underground CAES in all aspects but the energy store. Compared with a fixed volume underground store an underwater store brings the benefit of isobaric containment, raising the system's roundtrip efficiency. Around the world there are many coastal locations suitable for UWCAES, particularly around islands, and UWCAES plants could either have the machinery based on shore or on an offshore platform; decisions over the number of compression/expansion stages and whether to use heat storage or burn gas are largely based on the same factors as underground CAES. The main challenges currently facing UWCAES are cost-effective access (including deployment, maintenance, and recovery) and anchorage, and a number of companies and organizations are working to develop solutions to both.

REFERENCES

[1] Cowie PR. Biofouling patterns with depth. Biofouling. New York: Wiley; 2009.
[2] Pimm AJ, Garvey SD. Analysis of flexible fabric structures for large-scale subsea compressed air energy storage. 7th International Conference on Modern Practice in Stress and Vibration Analysis, Cambridge, 2009.
[3] Vasel-Be-Hagh A, Carriveau R, Ting D-K. Flow over submerged energy storage balloons in closely and widely spaced floral configurations. Ocean Eng 2015;95:59–77.
[4] Brading G. Private communication from Seaflex Ltd. to A. J. Pimm relating to performance of lift bags still in use after 7 years to support a pontoon at Brighton; July 2015.
[5] Pimm AJ. Analysis of flexible fabric structures. PhD. thesis, University of Nottingham; 2011.
[6] Davies CM. Energy storage in offshore windfarms. Report. Millbank, London: Corus Group; 2005.
[7] VanWalleghem C. Concept to construction: the world's first grid connected UWCAES facility. Offshore Energy and Storage Symposium (OSES) 2014, Windsor, ON, Canada; 2014.
[8] Seymour RJ. Ocean energy on-demand using underocean compressed air storage. 26th International Conference on Offshore Mechanics and Arctic Engineering, San Diego; 2007.
[9] Garvey S. Leveraging energy bags as a cost effective energy storage solution. Dufresne Energy Storage Forum, Rome; 2012.

[10] Tsuha C, Aoki N, Rault G, Thorel L, Garnier J. Evaluation of the efficiencies of helical anchor plates in sand by centrifuge model tests. Can Geotech J 2012;49:1102–14.
[11] RWE Power. ADELE—adiabatic compressed air energy storage for electricity supply. https://www.rwe.com/web/cms/mediablob/en/391748/data/364260/1/rwe-power-ag/innovations/Brochure-ADELE.pdf
[12] Ruer J. Energy storage for offshore wind power. Conference Proceedings REM2012, Ravenna; 2012.
[13] Pimm AJ, Garvey SD. Potential locations for underwater compressed air energy storage in Europe and North America. Offshore Energy and Storage Symposium 2015 (OSES2015), Edinburgh; 2015.
[14] Crotogino F, Mohmeyer K-U, Scharf R. Huntorf CAES: more than 20 years of successful operation. Solution Mining Research Institute Spring 2001 Meeting, Orlando; 2001.
[15] Meyer F. Compressed air energy storage power plants. Projekt-Info 05/07. Karlsruhe, Germany: BINE; 2007.
[16] Pimm AJ, Garvey SD, de Jong M. Design and testing of energy bags for underwater compressed air energy storage. Energy 2014;66:496–508.
[17] Garvey SD. Compressed air energy storage: performance and affordability. Half-day workshop given at Marcus Evans Biannual Energy Storage Conference, Amsterdam; 2010.
[18] Agrawal P, Nourai A, Markel L, Fioravanti R, Gordon P, Tong N, Huff G. Characterization and assessment of bovel bulk storage technologies: a study for the DOE Energy Storage Systems Program. Sandia National Laboratory, Albuquerque, NM; 2011.
[19] Severn Trent Water. Llyn Clywedog. http://www.stwater.co.uk/leisure-and-learning/reservoir-locations/llyn-clywedog/*/tab/about/

Chapter 8

A Novel Pumped Hydro Combined with Compressed Air Energy

Erren Yao, Huanran Wang, Guang Xi
School of Energy and Power Engineering, Xi'an Jiaotong University, Xi'an, China

1 INTRODUCTION

With the increasing depletion of traditional fossil energy sources, which account for a large proportion of total energy demands, it is becoming more and more urgent to find alternate types of energy [1]. Many countries are doing research on renewable energy such as wind energy, solar energy, bio-energy, etc. However, the intermittent nature of renewable energy is a large barrier to its development. Only by solving this problem can we embrace the future of renewable energy [2,3].

Unfortunately, a considerable proportion of wind and solar power fails to reach the power grid due to base level overloading. Renewable energy is only required if there is a need. It can be expected that this loss of renewable energy will increase with further expansion of wind and solar power, see chapter: The Role of Energy Storage in Low-Carbon Energy Systems.

Two solutions to this problem are to build new pumped hydro energy storage (PHES) facilities (see chapter: Pumped Hydroelectric Storage) and to build compressed air energy storage (CAES) facilities (see chapters: Compressed Air Energy Storage, Compressed Air Energy Storage in Underground Formations, and Underwater Compressed Air Energy Storage) [4,5]. In this way excess energy can be stored when power demand is low and released when required during power peaks.

PHES is currently the most practical and mature energy storage technology available [6–8]. According to data from the Electric Power Research Institute (EPRI), PHES ranks first in the global energy storage market and accounts for more than 99% of total stored energy. However, owing to the shortcomings of this technology, including large investment requirements, long construction periods, dependence on topography, and its influence on regional ecology and

Storing Energy. http://dx.doi.org/10.1016/B978-0-12-803440-8.00008-7

geology, the development of new PHES facilities is slow and is limited to only certain areas [9–11].

On the other hand, CAES technology also has its own problems which include the stability and sealing problems of underground air storage caverns and auxiliary heating problems.

In the chapter a combined PHES and CAES system is introduced which not only integrates the advantages but also overcomes the disadvantages of both the PHES and the CAES systems. The new pumped hydro combined with compressed air (PHCA) system is an attractive solution to the large-scale storage of intermittent renewable energy.

2 STORAGE SYSTEM

Consider a pressure vessel containing compressed air with a certain internal pressure connected to a pump by a pipeline and valve (Fig. 8.1). During off-peak time the pump delivers water to the vessel (left-hand side of Fig. 8.1), and a virtual dam is built between the lower water tank and the upper pressure vessel. For example, when the internal pressure in the pressure vessel is 5 MPa, the bottom height of the virtual dam h is equivalent to 500 m (right-hand side of Fig. 8.1). During peak time, high-pressure water enters the hydro-turbine which generates electricity. The scale of the constant pressure PHCA mainly depends on the model selection of water pump and hydro-turbine, as well as the system's operating time.

Based on this principle, Wang et al. [12] proposed an improved model for a PHES by using compressed air to artificially build a virtual dam. By adjusting the air pressure in the vessel the fall (upstream and downstream) of the virtual dam is changed. The system consists of a storage vessel, a closed pond connected to the atmosphere, a set pump driven by electricity, and a hydro-turbine connected to a generator (Fig. 8.2).

Initially, valves 2, 3, and 4 (Fig. 8.2) are closed while valve 1 is open; this starts the compressor and pumps the air into the vessel, thereby setting up a

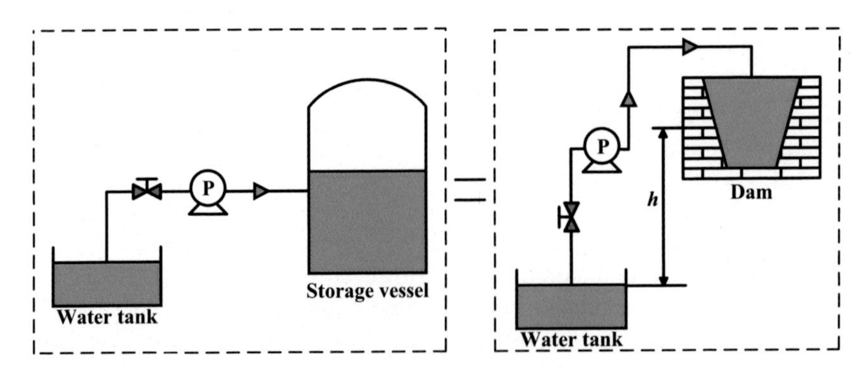

FIGURE 8.1 **Physical model of pumped hydro combined with compressed air energy storage system.**

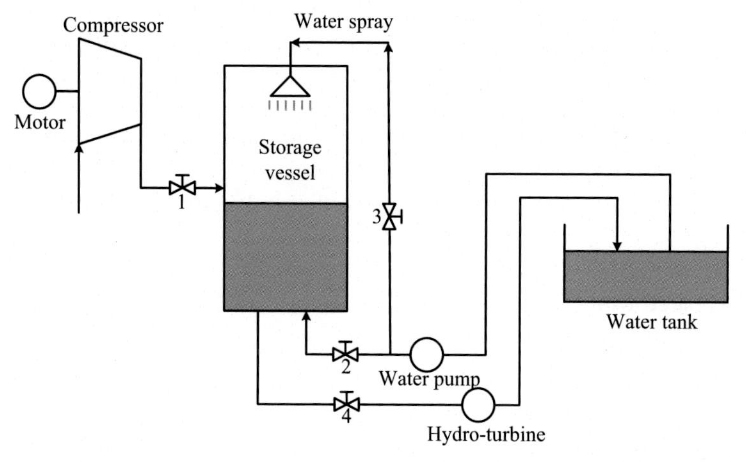

FIGURE 8.2 **Schematic diagram of the pumped hydro combined with compressed air energy storage system.**

preset pressure in the storage vessel. The compressor and valve 1 are then closed once the preset pressure has been reached. When storing energy, valve 2 is opened and the water pump switched on and water is injected into the storage vessel. The storage vessel is further pressurized when the water level rises during this injection period. As the air temperature of the compressed air in the storage vessel increases the pumping process becomes more and more difficult. This is overcome by opening valve 3 and spraying a small amount of water flow into the compressed air vessel in order to decrease its temperature through evaporative cooling. When air pressure in the storage vessel reaches its target value, valves 2 and 3 are closed and the energy storage process is complete. When energy generation is required, valve 4 is opened and high-pressure water drives the hydro-turbine which powers the electricity generator. Exhaust water then goes back to the water tank, which is connected to the atmosphere. No gas turbine or gas combustion chamber is used in this process although they are quite often used in CAES systems.

3 CHARACTERISTICS OF A PHCA SYSTEM

According to the Second Law of Thermodynamics, thermal energy cannot be completely converted to mechanical energy, and some energy would be transferred to a lower temperature reservoir. However, mechanical energy can be totally converted to thermal energy. This implies that the quality of mechanical energy is better than that of thermal energy [13]. Experience shows that mechanical energy is one of the highest "quality" energies of all forms with an extremely wide range of applications. In the PHCA system, energy is converted from mechanical energy (pressure energy) to electrical energy through the turbine, and as a result the energy involved in the PHCA system is theoretically higher than that of the CAES system.

In essence, the thermal machinery (compressor/expander) of the CAES system is substituted by hydraulic machinery (water pump/hydro-turbine) of the PHCA system in the process of energy storage and power generation.

Due to the high-power density of hydraulic machinery, an excessive high-energy head is not required and hydraulic machinery does not need to be multistaged. Hydro-turbines are basically single-stage units and are thus more convenient to operate than multistage units, and the total efficiency of modern large hydro-turbines is of the order of 95%, the highest efficiency of all the prime movers.

Compared with CAES [14–16] the PHCA system has the following advantages:

- the physical structure is simpler
- an electrical cooling and auxiliary heating system is not required
- the efficiency is higher—the efficiency of the water pump and hydro-turbine are both higher than that of the compressor and expander and thus the potential for improving the efficiency of a PHCA system is greater than that for a CAES system
- the cost is lower—for the same scale of energy storage, the cost of a high-pressure water pump and hydro-turbine are less than those of the compressor and expander, and, furthermore, the PHCA system does involve the cost of a cooler or a heater.

Compared with the PHES system [14] the PHCA system does not need any special geographical and geological conditions. Moreover, a PHCA installation does not require a major water supply as it avoids evaporation problems, and is more flexible in that the water energy storage density is adjustable.

However, the PHCA system does have some drawbacks. First, the high pressure ratio and low mass flow through both the water pump and the hydro-turbine would provide the system with both a high work density and a low water requirement, but it is difficult to implement with existing industrial technology. Second, since the pressure in the pressure vessel fluctuates with changing water level, the power generation process and the energy storage process are both variable. This has a deleterious effect on energy storage efficiency and power generation quality and is a result of the design specifications of the hydro-turbine and water pump. In the next section, we will introduce a novel constant pressure PHCA which will ensure a more stable and efficient operation.

4 A NOVEL CONSTANT PRESSURE PHCA ENERGY STORAGE SYSTEM

The system discussed here is a constant pressure PHCA system [17]. Fig. 8.3 presents the basic idea behind its operation. The process consists of the following three phases.

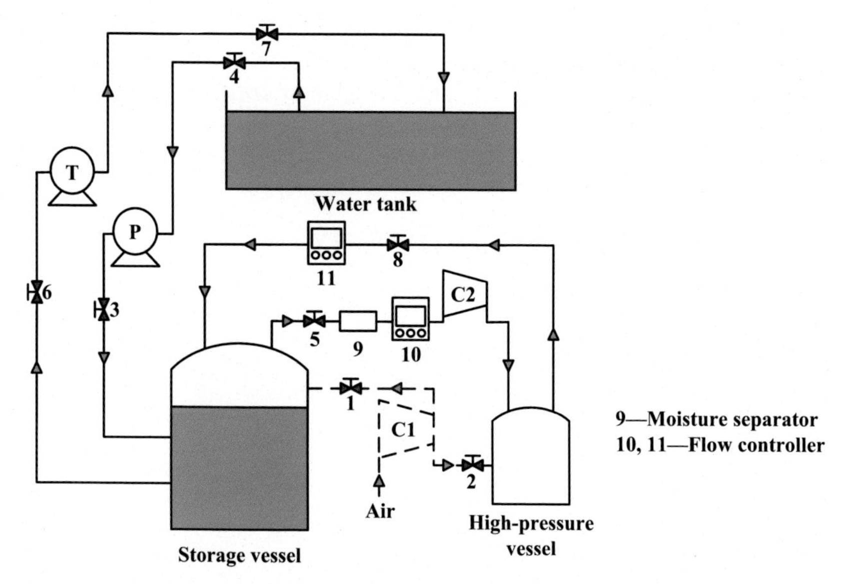

FIGURE 8.3 **Schematic diagram of the constant pressure pumped hydro combined with compressed air energy storage system.**

1. The initial compression process. By controlling the compressor 1 as well as valves 1 and 2, both the storage vessel and high-pressure vessel can be pressurized. This process is to improve working capacity per unit working medium in the same way as water should be pumped to a higher level in a traditional pumped hydro-energy storage station to increase the gravitational potential of upstream water.
2. The water injection process for energy storage. The water in the water tank is pumped into the storage vessel and at the same time air in the storage vessel will be transferred to the high-pressure vessel. This step ensures the pressure in the storage vessel remains the same and the air in the high-pressure vessel is compressed to a predetermined level.
3. The power generation process. When valve 6, valve 7, and throttle valve 8 are opened, the hydro-turbine generates power, driven by the high-pressure water. As the pressure in the high-pressure vessel is higher than that in the storage vessel, air will flow into the storage vessel when the water level decreases, thus keeping the pressure level in the storage vessel constant. Water will flow into the water tank to complete the cycle.

In discussing the thermodynamic performance of the system the following assumptions are made:

1. the gas consists only of nitrogen which is scarcely soluble in water and is considered as an ideal gas
2. in the thermodynamic calculations, the hydraulic water is considered incompressible

3. the high pressure vessel is an adiabatic container
4. it is assumed that there is negligible potential and kinetic energy effects in the gas and liquid flows, and no phase change or chemical reaction takes place
5. there is no pressure drop or loss along the gas and liquid pipelines.

5 THE INFLUENCES OF WORK DENSITY

In the constant pressure PHCA system the storage vessel is one of the key components: it insures normal operation of the whole system and its cost is related to the overall cost of the whole system. The input and output capacities of the pressure vessel per unit volume forms an important basis for optimization of the system.

Work density is defined by:

$$E_p = \frac{W_p}{V_h} = \frac{(p_1 - p_0)\varepsilon}{(1+\varepsilon)\eta_p} \tag{8.1}$$

where p_0, p_1 represent the initial pressure and the terminal pressure of the storage vessel, respectively; V_h is the volume of the storage vessel; V_w is the volume of water in the storage vessel; $\varepsilon = V_w/(V_h - V_w)$ is the water–air volume ratio; η_p is the efficiency of the water pump; W_p is total work done by the pump; E_p is the work done by the water per unit volume flowing through the water pump.

Focusing now on the hydro-turbine:

$$E_t = \frac{W_t}{V_h} = \frac{(p_1 - p_0)\varepsilon\eta_t}{(1+\varepsilon)} \tag{8.2}$$

where η_t is the efficiency of the hydro-turbine; W_t is total work output of the hydro-turbine; E_t is the work done by water per unit volume flowing through the hydro-turbine.

According to the Eqs. 8.1 and 8.2, the relationship between work density, preset pressure, and hydrosphere ratio can be obtained. As shown in Figs. 8.4 and 8.5, with increased preset pressure and water–air volume ratio the work density of the storage system increased.

6 ENERGY AND EXERGY ANALYSIS

The First Law of Thermodynamics states that energy in a closed system is conserved, and the second law of thermodynamics establishes the difference in the quality of different forms of energy and explains why some processes can spontaneously occur while others cannot. It is essential to perform a thorough analysis of both quantity and quality of the energy in the constant pressure PHCA system [18–20].

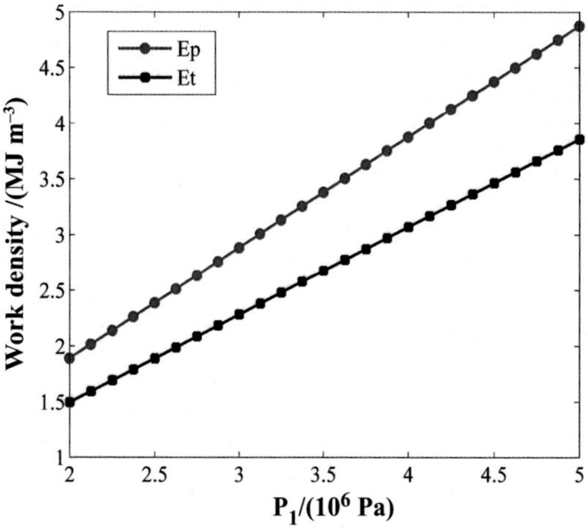

FIGURE 8.4 **Variation of work density with preset pressure.**

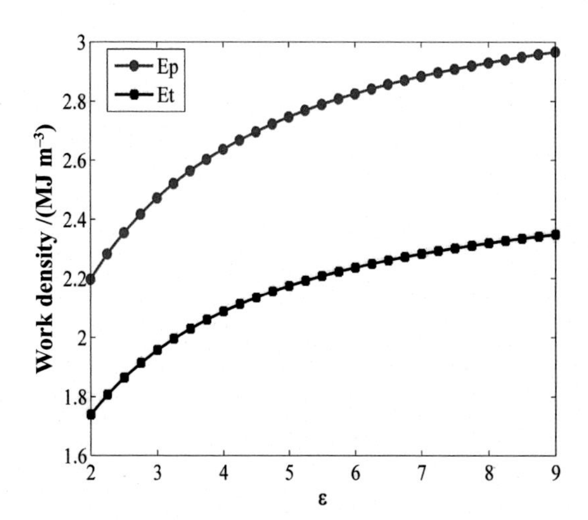

FIGURE 8.5 **Variation of work density with water–air volume ratio.**

6.1 Energy Analysis

Energy analysis is based on the first law of thermodynamics. As suggested by the working process of the constant pressure PHCA, when the air in the storage vessel is compressed to the preset pressure, compressor 1 stops working in subsequent energy storage and power generation processes unless air leakage occurs in the high-pressure vessel. System efficiency η_e can be expressed as:

$$\eta_e = \frac{W_t}{(W_p + W_{c2})} \tag{8.3}$$

where W_{c2} is the total power consumption of compressor 2 to keep the pressure of storage vessel constant.

6.2 Exergy Analysis

Exergy is the maximum theoretical work obtainable from an overall system (consisting of a system and the environment) as the system reaches equilibrium with the environment [20]. Exergy analysis is a method that uses the conservation of mass and conservation of energy principles together with the second law of thermodynamics for the analysis while exergy destruction is the measure of irreversibility which is the source of performance loss [21]. The exergy analysis assessing the magnitude of exergy destruction identifies the equipment, the magnitude, and the source of thermodynamic inefficiencies in an energy storage system. It is felt that exergy analysis would help in producing an efficient system which minimizes exergy destruction in the system [22].

The exergy efficiency of the system η_{ex} is given by:

$$\eta_{ex} = \frac{E_{out}}{E_{in}} \tag{8.4}$$

where E_{out} is the exergy transfer to the system; and E_{in} is the exergy transfer from the system.

7 SIMULATION ANALYSIS

We will show some results based on energy and exergy analysis of the system. When changing the preset pressure in the storage vessel and maintaining pressure P_1 to a level below pressure P_2 in the high-pressure vessel, variations in system efficiency and exergy efficiency with a preset pressure are then obtained. Fig. 8.6 shows that both system efficiency and exergy efficiency increase with increased preset pressure. While both system efficiency and exergy efficiency are not sensitive to the water–air volume ratio the energy intensity of the system could be improved by increasing the water–air volume without influencing system efficiency and exergy efficiency. This must be done within the allowable volume range for the storage vessel.

When changing the pressure in the high-pressure vessel while maintaining pressure P_2 at a level above pressure P_1 in the storage vessel, variations in system efficiency and exergy efficiency with pressure in the high-pressure vessel can be obtained. It can be seen from Fig. 8.7 that both the system efficiency and the exergy efficiency of the system decrease with increased pressure in the high-pressure vessel. This means that when P_1 is constant the smaller the pressure difference between the storage vessel and the high-pressure vessel, the higher the proportion of energy utilization in the constant pressure PHCA.

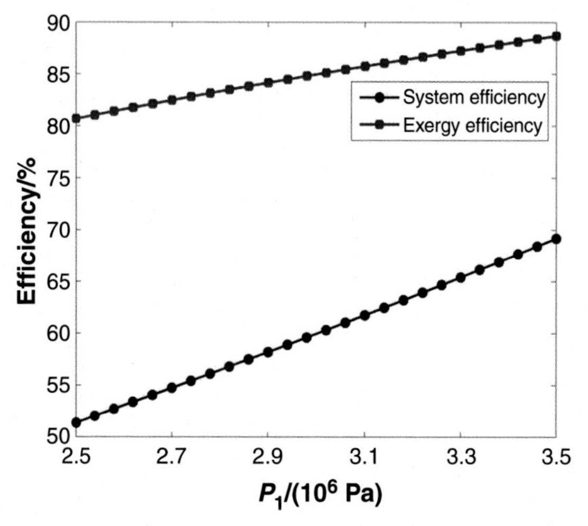

FIGURE 8.6 Variations of system efficiency and exergy efficiency with preset pressure.

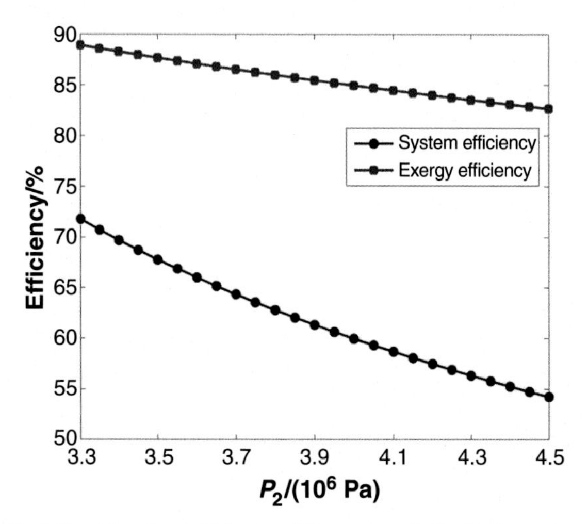

FIGURE 8.7 Variation of system efficiency with pressure in the high-pressure vessel.

Fig. 8.8 shows system efficiency and exergy efficiency versus mechanical efficiency: with increased mechanical efficiency, system efficiency and exergy efficiency increases constantly, but the rate of increasing extent with different factors is different.

Both system efficiency and exergy efficiency increase slowly with increasing compressor efficiency, while system efficiency and exergy efficiency show an obvious upward trend as water pump and hydro-turbine efficiency increases. As far as overall system efficiency is concerned the compressor only increases the pressure of the air; the work done by the water pump and hydro-turbine is

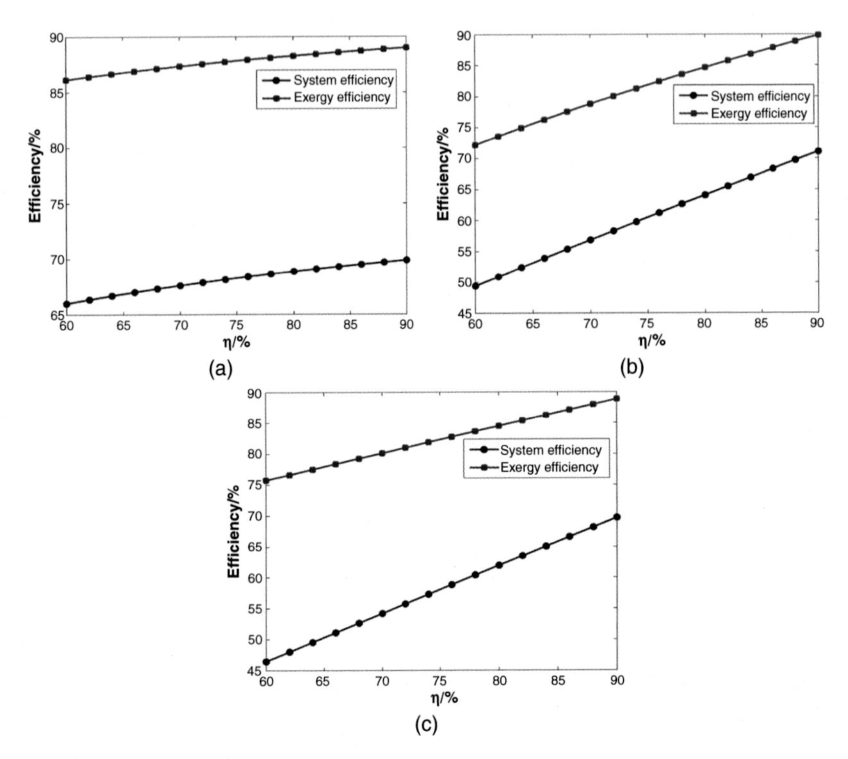

FIGURE 8.8 Variation of system efficiency with mechanical efficiency: (a) variation of compressor efficiency; (b) variation of water pump efficiency; and (c) variation of hydro-turbine efficiency.

much larger than that done by compressor. With respect to exergy efficiency the work done by the compressor accounts for only a small proportion of the total work and, as a result of the irreversible nature of the process, compressor efficiency has little influence on the exergy efficiency of the whole system. Moreover, compared with hydro-turbine efficiency, the pump efficiency exerts a greater influence on the exergy efficiency of the whole system. As the low destruction exergy of the hydro-turbine and the water pump are important issues needed to improve the quality of energy consumption the key to improving system performance is to improve the efficiency of the water pump and hydro-turbine.

The results from Figs. 8.6–8.8 confirm that the system efficiency of the proposed system is 15.6% higher than that of CAES systems and that the exergy efficiency is 29.4% higher than that of CAES systems [23]. This finding is mainly attributed to the following reasons: first, the efficiency of the hydro-turbine and water pump are both higher than that of the compressor and the expander; second, auxiliary heating systems are not required in the proposed system while exergy destruction of the intercooler and external combustion heater is high. In addition, the constant pressure PHCA insures that the hydro-turbine and water pump can operate at their rated working conditions with high efficiency which could improve the energy utilization level.

Our research group is preparing to set up an experimental 7.5 kW constant pressure PHCA system with a 10 m^3 storage vessel to study the actual water injection process for the energy storage and power generation processes. Finally, we believe that the constant pressure PHCA will have broad application prospects in the area of energy storage in the future.

REFERENCES

[1] Bazmi AA, Zahedi G. Sustainable energy systems: role of optimization modelling techniques in power generation and supply—a review. Renew Sustain Energy Rev 2011;15:3480–500.
[2] Liu W, Lund H, Mathiesen BV. Large-scale integration of wind power into the existing Chinese energy system. Energy 2011;36:4753–60.
[3] Blarke MB, Lund H. The effectiveness of storage and relocation options in renewable energy systems. Renew Energy 2008;33:1499–507.
[4] Beaudin M, Zareipour H, Schellenberglabe A, Rosehart W. Energy storage for mitigating the variability of renewable electricity sources: an updated review. Energy Sustain Dev 2010;14:302–14.
[5] Yang C, Jackson RB. Opportunities and barriers to pumped-hydro energy storage in the United States. Renew Sustain. Energy Rev 2011;15:839–44.
[6] Connolly D, Lund H, Finn P, Mathiesen BV, Leahy M. Practical operation strategies for pumped hydroelectric energy storage (PHES) utilising electricity price arbitrage. Energy Policy 2011;39:4189–96.
[7] Ekman CK, Jensen SH. Prospects for large scale electricity storage in Denmark. Energy Convers Manag 2010;51:1140–7.
[8] Deane JP, Ó Gallachóir BP, McKeogh EJ. Techno-economic review of existing and new pumped hydro energy storage plant. Renew Sustain Energy Rev 2010;14:1293–302.
[9] Kapsali M, Kaldellis JK. Combining hydro and variable wind power generation by means of pumped-storage under economically viable terms. Appl Energy 2010;87:3475–85.
[10] Kondoh J, Ishii I, Yamaguchi H, Murata A, Otani K, Sakuta K, Higuchi N, Sekine S, Kamimoto M. Electrical energy storage systems for energy networks. Energy Convers Manag 2000;41:1863–74.
[11] Ibrahim H, Ilinca A, Perron J. Energy storage systems—characteristics and comparisons. Renew Sustain Energy Rev 2008;12:1221–50.
[12] Wang HR, Wang LQ, Wang XB, Yao ER. A novel pumped hydro combined with compressed air energy storage system. Energies 2013;6:1554–67.
[13] Wang, Z. Thermal and power mechanical basis. Beijing, China: China Machine Press; 2000 [in Chinese].
[14] Grazzini G, Milazzo A. A thermodynamic analysis of multistage adiabatic CAES. P IEEE 2012;100:461–72.
[15] Kim YM, Favrat D. Energy and exergy analysis of a micro-compressed air energy storage and air cycle heating and cooling system. Energy 2010;35:213–20.
[16] Kim YM, Shin DG, Favrat D. Operating characteristics of constant-pressure compressed air energy storage (CAES) system combined with pumped hydro storage based on energy and exergy analysis. Energy 2011;36:6220–33.

[17] Yao E, Wang H, Liu L, et al. A novel constant-pressure pumped hydro combined with compressed air energy storage system. Energies 2014;8(1):154–71.
[18] Shekarchian M, Zarifi F, Moghavvemi M, Motasemi F, Mahlia TMI. Energy, exergy, environmental and economic analysis of industrial fired heaters based on heat recovery and preheating techniques. Energy Convers Manag 2013;71:51–61.
[19] Moran MJ, Shapiro HN, Boettner DD, Bailey MB. Fundamentals of engineering thermodynamics. 7th ed. New York, NJ, USA: John Wiley & Sons; 2010.
[20] Dincer I, Rosen MA. Exergy: energy, environment and sustainable development. 2nd ed. Oxford, UK: Elsevier; 2013.
[21] Dincer I. The role of exergy in energy policy making. Energy Policy 2002;30:137–49.
[22] Ranjan KR, Kaushik SC. Energy, exergy and thermo-economic analysis of solar distillation systems: a review. Renew Sustain Energy Rev 2013;27:709–23. 2013.
[23] Kim YM, Favrat D. Energy and exergy analysis of a micro-compressed air energy storage and air cycle heating and cooling system. Energy 2010;35:213–20.

Chapter 9

Liquid Air Energy Storage

Yulong Ding*, Lige Tong, Peikuan Zhang**, Yongliang Li*, Jonathan Radcliffe*, Li Wang****

**Birmingham Centre for Cryogenic Energy Storage, School of Chemical Engineering, University of Birmingham, Edgbaston, Birmingham, UK; **School of Mechanical Engineering, University of Science & Technology Beijing, Beijing, China*

1 INTRODUCTION

Liquid air energy storage (LAES) refers to a technology that uses liquefied air or nitrogen as a storage medium [1]. LAES belongs to the technological category of cryogenic energy storage. The principle of the technology is illustrated schematically in Fig. 9.1. A typical LAES system operates in three steps. Step 1 is the charging process whereby excess (off-peak and cheap) electrical energy is used to clean, compress, and liquefy air. Step 2 is the storing process through which the liquefied air in Step 1 is stored in an insulated tank at ~196 °C and approximately ambient pressure. Step 3 is the discharging process that recovers the energy through pumping, reheating, and expanding to regenerate electricity during peak hours when electrical energy is in high demand and expensive. Step 2 also includes the storage of heat from the air compression process in Step 1 and high-grade cold energy during the reheating process in Step 3. The stored heat and cold energy can be used, respectively, in Step 3 and Step 1 to increase the power output and reduce the energy consumption of the liquefaction process.

The concept of LAES technology was first proposed by researchers at the University of Newcastle-upon-Tyne (United Kingdom) in 1977 for peak shaving of electricity grids [2]. Although the work involved mainly theoretical analyses, it led to subsequent development of the technology, particularly by Mitsubishi Heavy Industries [3] and Hitachi [4–5] in Japan, and by Highview Power Storage in collaboration with the University of Leeds (United Kingdom) [1,6–9]. The work by Mitsubishi Heavy Industries led to a 2.6 MW pilot plant with air liquefaction and power recovery processes operated independently, leading to low roundtrip efficiency [3]. The work by Hitachi looked at integration of the air liquefaction and power recovery processes through a regenerator [4–5]. Such a regenerator stores cold energy released during the power recovery process and reuses the stored cold energy to reduce energy consumption

Storing Energy. http://dx.doi.org/10.1016/B978-0-12-803440-8.00009-9

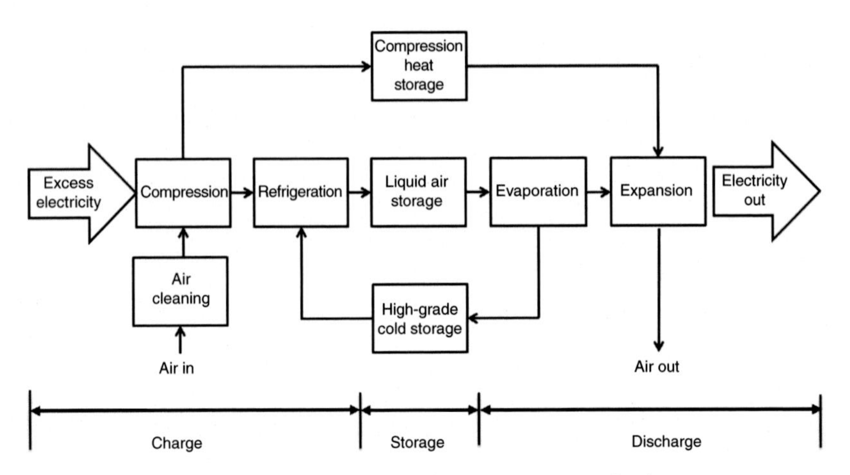

FIGURE 9.1 **Schematic illustration of liquid air energy storage technology.**

of the air liquefaction process. Both simulation and experiments were carried out on the regenerator using solid materials and fluids as cold carriers. Based on the results, Hitachi claimed that the roundtrip efficiency of the cryogenic energy storage system could exceed 70% so long as the regenerator was efficient. However, no fully integrated system demonstration was done. Working with the University of Leeds, Highview Power Storage started to design and build the world's first integrated LAES pilot plant (350 kW, 2.5 MW h) in 2009. The pilot plant was colocated with a Scottish and Southern Energy (SSE) biomass power plant in Slough (United Kingdom) and the whole plant became operational from 2011. The pilot plant has now been relocated to the University of Birmingham (United Kingdom) for further research and development. In collaboration with Virido and with the support of the UK Department of Energy and Climate Change, Highview Power Storage is currently building a 5 MW/15 MW h demonstration plant in Virido's Manchester (United Kingdom) landfill gas power plant.

2 ENERGY AND EXERGY DENSITIES OF LIQUID AIR

Fig. 9.2 shows the exergy density of liquid air as a function of pressure. For comparison the results for compressed air are also included. In the calculation the ambient pressure and temperature are assumed to be 100 kPa (1.0 bar) and 25 °C, respectively. The exergy density of liquid air is independent of the storage pressure because compressibility of liquid is extremely small. Fig. 9.2 indicates that, although the mass exergy density of liquid air is only 1.5–3 times higher than that of compressed air, the volumetric exergy density of liquid air is at least 10 times that of compressed air if the storage pressure of compressed

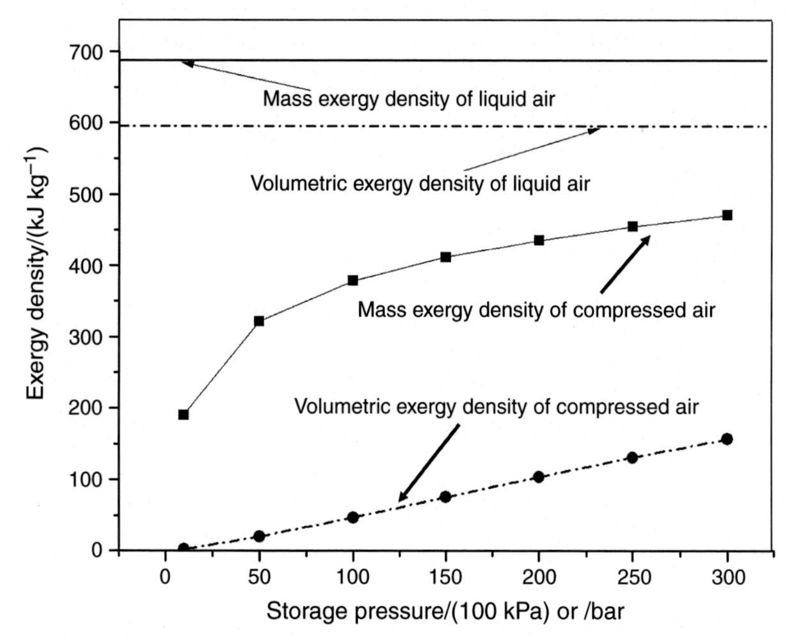

FIGURE 9.2 **Exergy density of liquid air and compression with compressed air.**

air is lower than 10 MPa (100 bar). Such a high volumetric exergy density of liquid air makes it highly competitive even compared with current battery technologies [10].

Assuming the specific heat C_p of a sensible thermal storage material is a constant an increase or a decrease in its temperature by ΔT from the ambient temperature T_a will lead to an amount of thermal energy δQ being charged into or discharged from the material:

$$\delta Q = C_p \Delta T \tag{9.1}$$

Considering a reversibly infinitesimal heat transfer process the exergy change dE of the material can be calculated by:

$$dE = dH - T_a \cdot dS = dH - T_a \cdot \frac{\delta Q}{T} \tag{9.2}$$

where dH is enthalpy change; and dS is entropy change. The exergy ΔE stored in the material is therefore obtained by integrating Eq. (9.2) from T_a to $(T_a + \Delta T)$:

$$\Delta E = C_p \left[\Delta T - T_a \cdot \ln\left(\frac{T_a + \Delta T}{T_a} \right) \right] \tag{9.3}$$

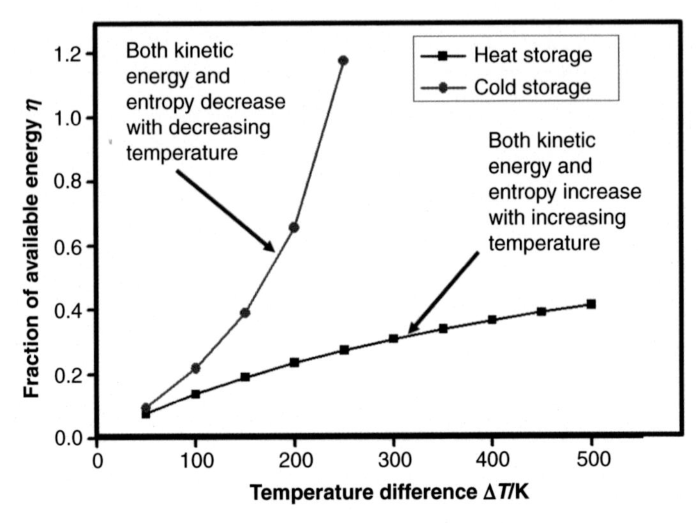

FIGURE 9.3 **Exergy percentage as a function of temperature difference for heat and cold storage.**

Combining Eqs. (9.1) and (9.3) gives the percentage of the available energy stored in the material η as:

$$\eta = \frac{\Delta E}{|\Delta Q|} = \frac{\Delta T - T_a \cdot \ln\left(\frac{T_a + \Delta T}{T_a}\right)}{|\Delta T|} \tag{9.4}$$

Eq. (9.4) is illustrated in Fig. 9.3 where the ambient temperature is assumed to be 25 °C. It can be seen from Fig. 9.3 that, for heat storage, only a significant temperature gap can give a reasonable percentage of available energy. For cold storage, however, the available energy grows far quicker with increasing temperature gap, suggesting that cold storage would be a very attractive option. The physics behind this conclusion rests mainly with the role played by entropy, which increases with increasing temperature.

3 LIQUID AIR AS BOTH A STORAGE MEDIUM AND AN EFFICIENT WORKING FLUID

Currently, low- to medium-grade heat is often recovered by steam cycles with water/steam as the working fluid [11,12]. However, water/steam is not an ideal working fluid for efficient use of low-grade heat due to its high critical temperature of 374 °C compared with the ambient temperature and its extremely high critical pressure of 22.1 MPa (221 bar). It is because of this that a large proportion of heat is consumed to vaporize water during phase change in subcritical or even transcritical cycles. This leads to the loss of a large portion of exergy in heat transfer processes due to temperature glide mismatch between the heat source and the working fluid—the so-called "pinch limitation" [13,14]. The use of liquid air as a storage medium as well as a working fluid in the power recovery step of LAES technology is thermodynamically more efficient than water in terms of recovering low-grade heat, as demonstrated in the next paragraph.

Consider a heat transfer process between a heat source and a working fluid, with the working fluid heated from ambient temperature t_a to t_H = 400 °C, $\{T_a$ to $T_H = 673$ K$\}$, and define normalized heat $\bar{Q}$ as the ratio of heat load at a certain temperature T to total heat exchange amount during the whole process. This gives:

$$\bar{Q}(T) = \frac{H(T) - H(T_a)}{H(T_H) - H(T_a)} \tag{9.5}$$

where H is enthalpy. Fig. 9.4 shows the results of Eq. (5) for liquid nitrogen (the main component of air) and water. For comparison the results for liquid methane and hydrogen are also included. One can see that, given a working pressure, the specific heat (the slope of the lines) is approximately the same for liquid nitrogen, hydrogen, and methane. However, water exhibits very different behavior. If the working pressure is lower than its critical value the specific heat of water changes greatly due to phase change, leading to ineffective use of the heat source considering that the heat sources are mostly provided by fluids with a constant specific heat (e.g., flue gases or hot air). Although water behaves in a similar manner to methane under supercritical conditions (e.g., the case with pressure of 30 MPa (300 bar) in Fig. 9.4), the high working pressure increases technical difficulties in bringing the process about.

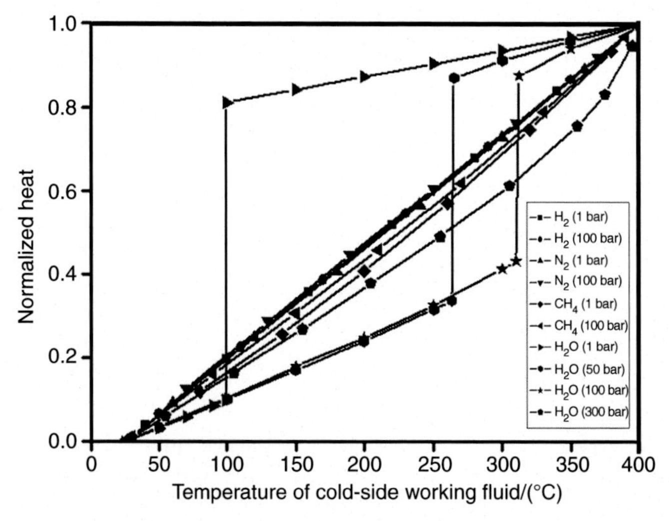

FIGURE 9.4 **Normalized heat as a function of cold-side working fluid temperature.**

4 APPLICATIONS OF LAES THROUGH INTEGRATION

Like other mechanical-based energy storage technologies, issues such as capital cost, roundtrip efficiency, and annual operating hours remain key challenges in the industrial takeup of LAES technology. Integration of LAES with other processes/systems provides a way to address the challenges. Examples are given in the following subsections.

4.1 Integration of LAES with Gas Turbine-Based Peaking Plants

Integration of LAES with a gas turbine-based peaking plant provides an opportunity to use the waste heat from the power generation process leading to high peak-shaving capacity and increased overall efficiency [6]. Integration can also capture CO_2 to give dry ice at no additional efficiency penalty. Fig. 9.5 shows the process diagram of the cycle, which works in the following manner: during off-peak hours, excess electricity generated by the base load is used to power an air separation and liquefaction unit (ASU) to produce oxygen and liquid nitrogen while the rest of the system is powered off. The oxygen and liquid nitrogen produced are stored in a pressurized vessel and a cryogenic tank, respectively,

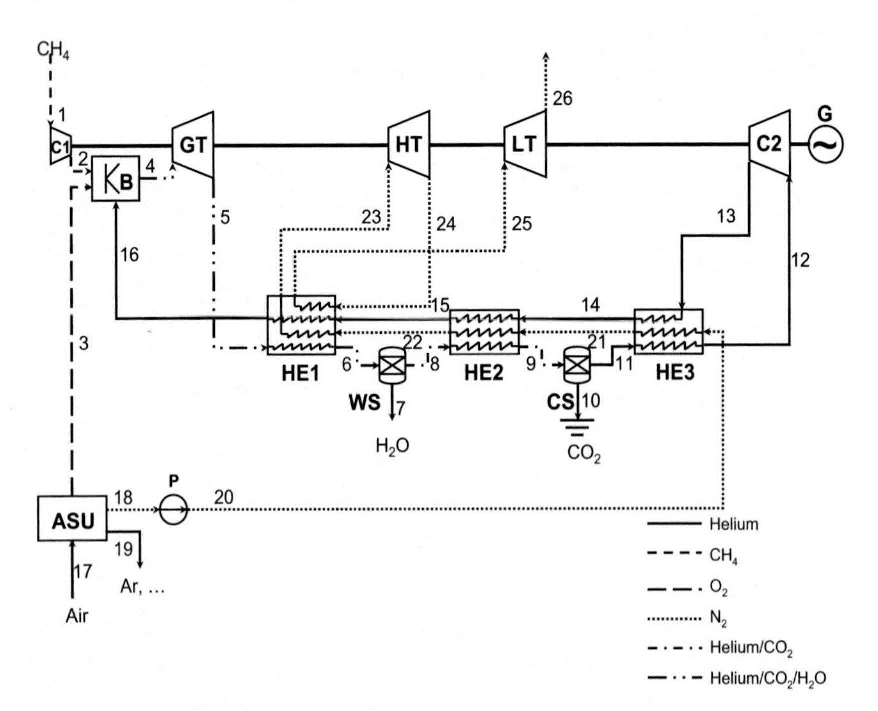

FIGURE 9.5 Process diagram of an integrated LAES and gas turbine-based peaking power system [6]. *ASU*, air separation unit; *B*, combustor; *CS*, CO_2 separator; *G*, generator; *GT*, gas turbine; *HE*, heat exchanger; *HT*, high-pressure turbine; *LT*, low-pressure turbine; *P*, cryogenic pump; *WS*, water separator.

for generating power via the high-pressure turbine (HT) and low-pressure turbine (LT), and assisting combustion in the combustor (B) at peak hours. The liquid nitrogen produced also serves as an energy storage medium.

At peak hours, natural gas is compressed in compressor (C1) to the working pressure. The working fluid then mixes with the oxygen in B where combustion takes place to give high-temperature and high-pressure flue gas consisting of CO_2 and H_2O. Combustion of the natural gas in an oxygen environment can produce a temperature that is too high for the gas turbine (GT). To control such a temperature, an appropriate amount of helium gas is mixed with the flue gas before entering the GT for power generation through a generator (G). Note that the helium gas is not consumed but circulates in the system; see later. The flue gas containing helium from the *GT* then goes through a series of heat exchange processes via heat exchanger 1 (HE1), 2 (HE2), and 3 (HE3) to recover the waste heat by passing the heat to a nitrogen stream from the liquid nitrogen storage tank; see later. During the heat recovery processes, steam in the flue gas is removed via a condenser (WS), whereas CO_2 is removed in the form of dry ice through a solidification process in the CO_2 separator (CS)—the triple point of CO_2 is 571.8 kPa (5.718 bar) and 56.6 °C. As a result the flue gas stream after CO_2 removal contains only helium. The helium stream is then cooled down further in HE3 and compressed in compressor *(C2)* to the working pressure, and finally goes through further heat exchange in HE2 and HE1 before flowing back to the combustor.

The nitrogen stream starts from the cryogenic storage tank where liquid nitrogen is pumped to the working pressure by a cryogenic pump (P). High-pressure nitrogen is then heated in HE3, HE2, and HE1 in series and expands in two stages via, respectively, a high-pressure turbine (HT) and a low-pressure turbine (LT) to generate electricity. HE1 serves as an interheater between the two-stage expansion. After expansion the pure nitrogen can be used to purge the sorbent bed of the ASU dryer.

From the above, it can be seen that the integrated system consists of a closed-loop topping Brayton cycle [6] with $He/CO_2/H_2O$ as the working fluid and an open-loop bottoming nitrogen direct expansion cycle. The topping Brayton cycle can be identified as $4 \rightarrow 5 \rightarrow 6 \rightarrow 8 \rightarrow 9 \rightarrow 11 \rightarrow 12 \rightarrow 13 \rightarrow 14 \rightarrow 15 \rightarrow 16 \rightarrow 4$, whereas the bottoming cycle is $18 \rightarrow 20 \rightarrow 21 \rightarrow 22 \rightarrow 23 \rightarrow 24 \rightarrow 25 \rightarrow 26$. It is the combination of the two cycles that produces electricity at peak hours. The Brayton cycle uses natural gas, which is burned in the pure oxygen produced by the ASU during off-peak hours. Helium is only used to control the turbine inlet temperature (TIT) and is recirculated. The working fluid of the open cycle, nitrogen, is the actual energy carrier of off-peak electricity. As CO_2 is captured, only water and nitrogen are given off by the process.

The optimal energy storage efficiency of such an integrated system is nearly 70% whereas the CO_2 in the flue gas is fully captured. Economic analyses also show that if the integrated system is used for energy arbitrage and peak power generation both the capital and peak electricity costs are comparable with the natural gas combined cycle (NGCC), which are much lower than the oxy-NGCC if the operation period is relatively short [6]. Note that not only helium but also oxygen could be used as the recirculating fluid and similar conclusions could be obtained.

4.2 Integration of LAES with Concentrated Solar Power Plants

Additional heat sources can enhance the roundtrip efficiency of the LAES system. Such heat sources can come from industrial processes and renewable solar radiation. This subsection explores the use of solar heat in large-scale concentrated solar power (CSP) plants. Fig. 9.6 shows an integrated LAES and CSP hybrid power system [15]. As can be seen, no liquefaction process is included in the system so an external supply of liquid air/nitrogen is needed. This is practicable if there is a large-scale centralized liquefaction plant within a reasonable distance. The system shown in Fig. 9.6 consists of a direct expansion (open cycle) of liquid air/nitrogen from an elevated pressure and a closed-loop Brayton cycle operated at medium to low pressure. The use of the Brayton cycle in place of the conventional Rankine cycle gives a more efficient heat transfer process and a much lower working pressure. The expansion occurs sequentially in three stages (high-, medium-, and low-pressure turbines), and solar heat is used to superheat the working fluid. Simulation results show that such a system

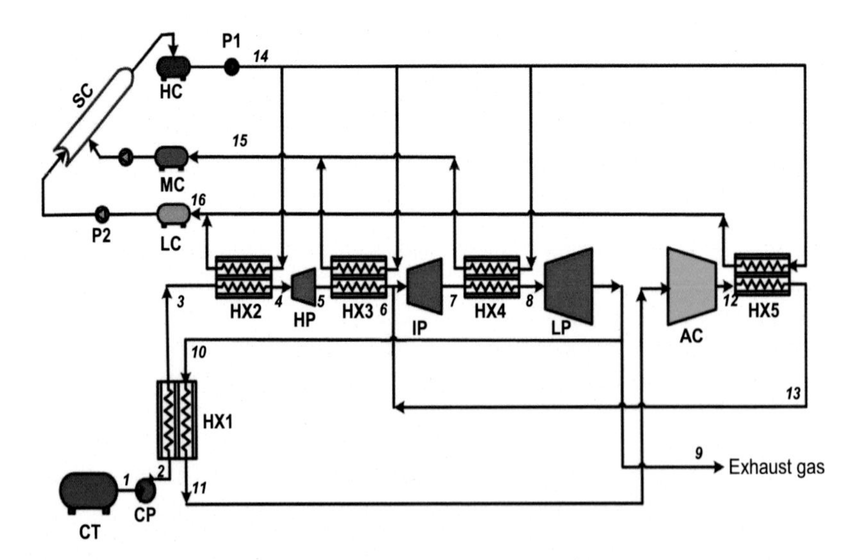

FIGURE 9.6 **Integration of LAES with a solar power system.** *AC*, adiabatic compressor; *CT*, cryogen tank; *CP*, cryogenic pump; *HC*, high-temperature carrier tank; *HP*, high-pressure turbine; *HX*, heat exchanger; *IP*, intermediate-pressure turbine; *LC*, low-temperature carrier tank; *LP*, low-pressure turbine; *MP*, intermediate-temperature carrier tank; *P*, pump; *SC*, solar collector.

provides over 30% more power than the summation of a solar thermal power-only system and an LAES-only powered system.

4.3 Integration of LAES with Nuclear Power Plants

To balance demand and supply at off-peak hours, nuclear power plants often have to be downregulated, particularly when the installations exceed base load requirements. Part load operations not only increase the electricity cost but also impose a detrimental effect on the safety and lifetime of nuclear power plants. Integration of nuclear power generation with LAES provides a promising solution to effective time shift of electrical power output. Fig. 9.7 shows a flow sheet of such integration [16]. The integrated system consists of a nuclear power plant subsystem and an LAES subsystem. The nuclear power plant subsystem in the integrated system is similar to the conventional pressurized water power station. The only difference lies in that there are two three-way valves in the secondary loop, which enables the working fluid to feed into either the steam turbine to produce electricity or heat exchanger 4 to superheat high-pressure air in the LAES subsystem. The LAES subsystem consists of an air liquefaction unit in the left part and an energy extraction unit in the right bottom part of Fig. 9.7. The integrated system could have three operational modes depending on end-user demands as described briefly in the following:

- Energy storage mode: during off-peak hours when demand is much lower than the rated power of the power plant, the power plant operates in a traditional way to drive the steam turbine to produce electricity and excess power is used to drive the air liquefaction unit to produce liquid air.
- Energy release mode: at peak hours when demand is higher than the rated power of the plant, the energy extraction unit is turned on to produce additional power.
- Conventional mode: when supply is approximately balanced by demand, both the air liquefaction unit and energy extraction unit are switched off so that nuclear power operates in a conventional way to drive the steam turbine to produce electricity.

The air liquefaction subsystem works in a similar way to the simplest Linde–Hampson liquefier except for the use of external cold energy in heat exchanger 6 (Fig. 9.7). It should be noted that in the air liquefaction unit a cryoturbine is used to generate a liquid product instead of a throttling device in a conventional setup. The working fluid expands in a near-isentropic manner in the cryoturbine with both temperature and enthalpy decreasing and hence generates more liquid product while producing additional shaft power.

The cold storage and recovery steps act as a bridge between the air liquefaction unit and the cryogenic energy extraction unit. Such an arrangement enables recovery of the cold energy released in the liquid air preheating process. In this process, air is in a supercritical state, and as a result cold energy is produced in

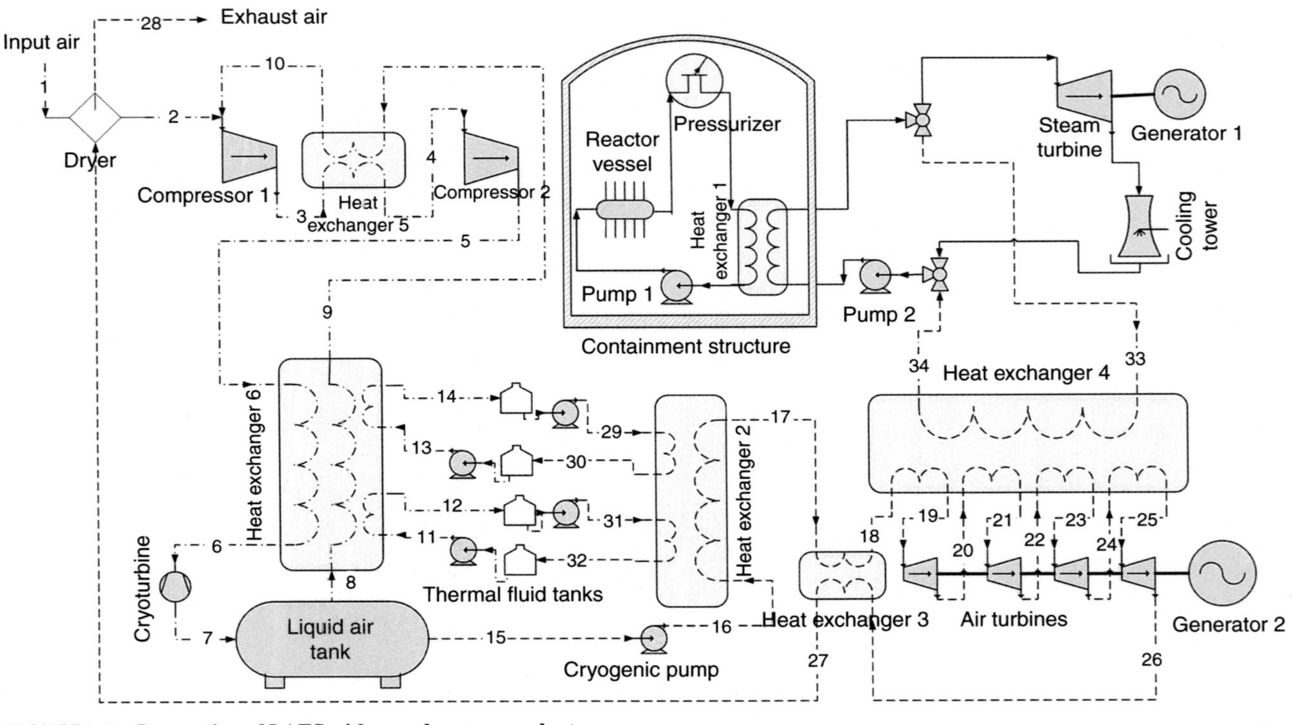

FIGURE 9.7 **Integration of LAES with a nuclear power plant.**

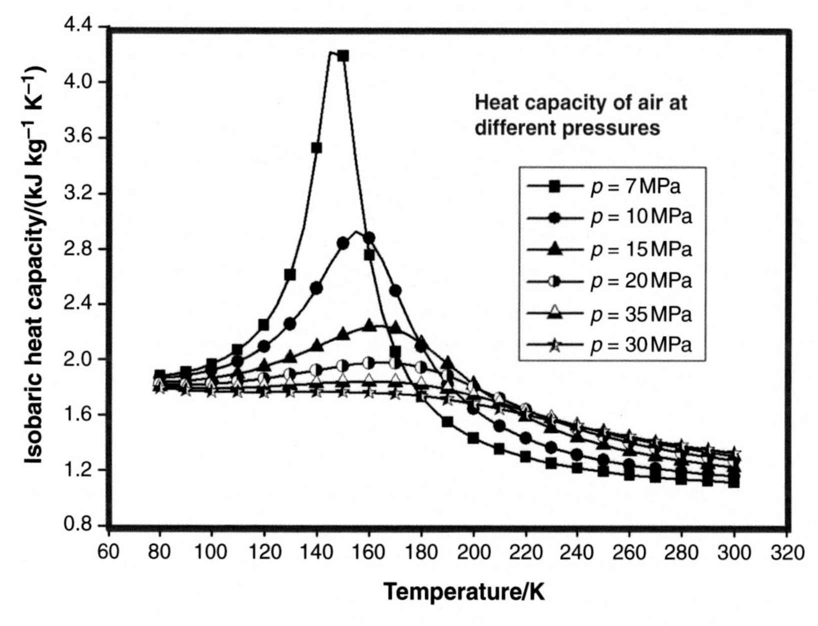

FIGURE 9.8 **Heat capacity of air at different pressures.**

the form of sensible thermal energy. Fig. 9.8 shows the heat capacity of air as a function of temperature at different pressures. One can see that the heat capacity of air changes only slightly in the heating process, particularly at very high pressures. This is similar to the use of liquids as sensible heat storage materials. In fact, cold energy can also be stored in thermal fluids and such fluids can give a good temperature gradient match during heat exchange and, hence, efficient cold recovery. In this process, thermal fluids are used not only as a working fluid but also as a cold storage medium. Fig. 9.9 shows the heat capacities of some commonly used fluids that may be used as storage media. Clearly, no single fluid can fully cover the working temperature region of the liquid air preheating process. However, the combination of propane and methanol could work both as cold storage liquids and working fluids for heat transfer. Such a combination covers the required temperature range and has high heat capacity. For each of the two fluids a two-tank configuration is proposed for cold recovery and storage (Fig. 9.7). The two thermal fluids are pumped from the hot tanks to the cold tanks during the cold storage process (energy storage mode) and flow back during the cold release process (energy release mode). The use of thermal fluids for both transferring and storing thermal energy can greatly simplify the design of the system in that no additional heat exchangers will be needed. Moreover, the operating strategy can be much more straightforward—the amount of cold energy and the objective temperature can easily be adjusted by controlling the flow

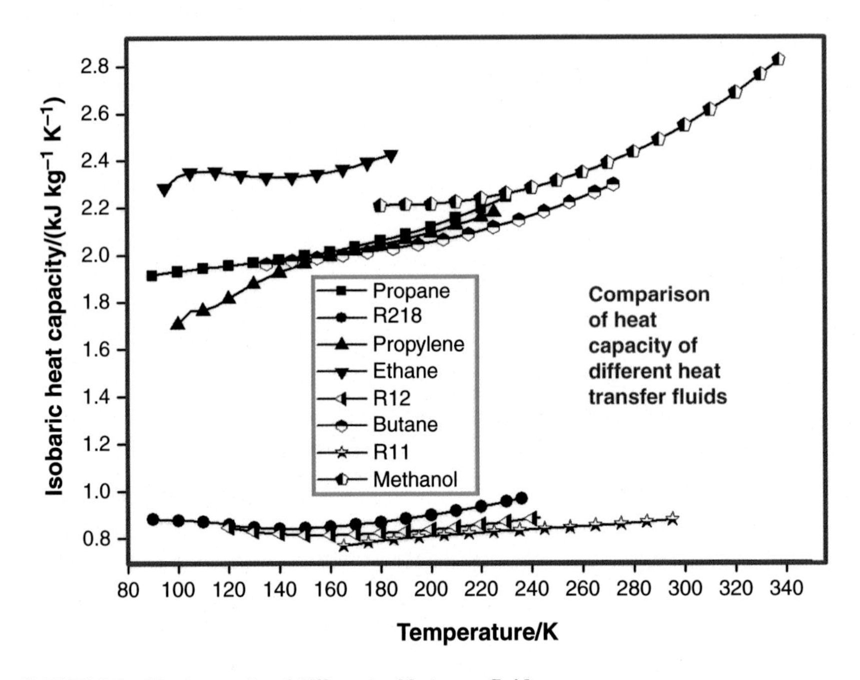

FIGURE 9.9 **Heat capacity of different cold storage fluids.**

rate of the fluids. This is extremely difficult to achieve using the conventional way of storing cold in a packed pebble bed.

The cryogenic energy extraction unit is coupled with the nuclear power plant through the thermal energy utilization process via heat exchanger 4. One can see that hardly any thermal energy is wasted in the cooling process and hence the power output is expected to increase significantly.

By integrating with LAES technology the reactor core and the primary loop of nuclear power plants can operate steadily at full load at all times while net output power is adjusted only by the LAES unit. As the energy extraction process in the LAES subsystem is similar to power generation using a gas turbine a much faster rate of power change could be achieved in comparison with the conventional downregulation of nuclear power plants.

The combination of nuclear power generation and LAES enables the use of cryogen instead of steam as the working fluid in power generation process. As a result this provides an efficient way to use the thermal energy of nuclear power plants, delivering around three times the rated electrical power of the nuclear power plant at peak hours, effectively shaving the peak. Simulations have been carried out on this process, which show that the roundtrip efficiency of LAES is higher than 70% due to the elevated topping temperature in the superheating process.

4.4 Integration of LAES with Liquefied Natural Gas Regasification Process

LAES can also be integrated with liquefied natural gas (LNG) regasification plants to make use of waste cold in the air liquefaction process [16]. The waste cold in LNG import terminals is significant due to large-volume LNG storage. LNG is normally regasified by heating with seawater and burning some natural gas. This leads to wasting of cold contained in the LNG and the burned natural gas. If LAES were colocated at the LNG terminal, and air rather than seawater was used to provide heat for the LNG regasification process, the resulting cold air could then be fed into the air liquefier, potentially reducing its electricity consumption by as much as two-thirds. Currently, there are a number of nitrogen liquefiers in operation at LNG import terminals in Japan and South Korea, which take advantage of this refrigeration to reduce power consumption. The only challenge to be overcome is to reduce the capital cost of such an integrated system.

5 TECHNICAL AND ECONOMIC COMPARISON OF LAES WITH OTHER ENERGY STORAGE TECHNOLOGIES

In this section a brief comparison is made between LAES and other energy storage technologies. The comparison will be from both technical and economic aspects as detailed in the following two subsections.

5.1 Technical Comparison

Only pumped hydro storage can be currently regarded as a mature technology—it has been practiced for over 100 years. Although compressed air energy storage technology has been developed and is commercially available, actual applications are not widespread. LAES—together with flow batteries, hydrogen storage, and a number of other energy storage technologies [10]—is still under development.

Like pumped hydro and compressed air energy storage technologies, LAES offer a long discharge time (hours) compared with coupled energy storage technologies. However, the power discharge rate depends on the scalability of the power-regenerating unit of energy storage technologies. Pumped hydro storage uses hydraulic turbines for power regeneration and as a result offers the largest power discharge rate (up to several gigawatts). Compressed air energy storage uses traditional gas turbines or steam turbines for power regeneration so the power rate is of the order of hundreds of megawatts. An LAES turbine is somewhat like a gas turbine but with a lower expansion temperature so the power rate is expected to be a little lower than compressed air energy storage, but can still reach hundreds of megawatts. However, the scalability of flow batteries and hydrogen storage is a big challenge and hence their power rates are expected to be less than a megawatt.

Flow batteries and pumped hydro storage have a high (system-level) roundtrip efficiency of (65–85)%. The roundtrip efficiency of compressed air energy storage ranges from about 40 (commercialized and realized) to about 70% (still at the theoretical stage). LAES has a low roundtrip efficiency of about (50–60)% mainly due to the low efficiency of the air liquefaction process. However, it should be noted that the roundtrip efficiency of LAES can be significantly enhanced if waste heat is available.

In terms of energy density, hydrogen storage has the highest volumetric energy density of (500–3000) W h L^{-1} depending on the storage methods (e.g., compressed gas, liquid, physical/chemical adsorption etc.). As an extremely flammable gas, however, the technical requirements for hydrogen storage are high. The energy storage density of LAES is an order of magnitude lower at (60–120) W h L^{-1}, but the energy carrier can be stored at ambient pressure. Pumped hydro storage has the lowest energy density of (0.5–1.5) W h L^{-1} while compressed air energy storage and flow batteries are at (3–6) W h L^{-1}.

5.2 Economic Comparison

Economic comparison can be based on the costs per unit amount of power that storage can deliver (dollars per kilowatt) and costs per unit amount of energy (dollars per kilowatt-hour) that is stored in the system. It is difficult to evaluate a specific technology since the costs are influenced by many factors such as system size, location, local labor rate, market variability, local climate, environmental considerations, and transport/access issues. The capital costs provided in this section are intended to provide a high-level understanding of the issues and are not intended as cost inputs into a design.

The capital costs per unit amount of power relate mainly and directly to the cost of charging/discharging devices. Hydrogen storage is characterized by high capital costs for power (>\$10 000 kW^{-1}). Pumped hydro storage, compressed air energy storage and flow batteries, and LAES have more or less the same capital cost for power (about \$400–2000 kW^{-1}). The capital costs per unit amount of energy cannot be used accurately to assess the economic performance of energy storage technologies mainly because of the effect of discharging durations. An alternative measure is the capital cost of storage devices such as a dam for pumped hydro storage and a storage tank for LAES. It is expected that hydrogen storage and compressed air energy storage have the highest storage costs, as the energy carriers are either combustible or at a high pressure. Pumped hydro storage has a low cost due to low energy density. LAES and flow batteries have the lowest cost even though insulation is required.

In terms of lifecycles, mechanical-based technologies including pumped hydro storage, compressed air energy storage, and LAES should last (20–60) years, as these technologies are based on conventional mechanical engineering and the lifecycle is mainly determined by the lifetime of mechanical components. By contrast, the lifetimes of hydrogen storage and flow batteries are currently expected to be about 5–15 years.

REFERENCES

[1] Chen H, Ding Y, Peters T, Berger F. Energy storage and generation. US Patent US20090282840; 2009.

[2] Smith EM. Proceedings of the Institution of Mechanical Engineers 1847–1982. P I Mech Eng 1977;191:289–98.

[3] Kenji K, Keiichi H, Takahisa A. Technical Review 35. Yokohama, Japan: Mitsubishi Heavy Industries Ltd.; 1998. p. 4.

[4] Chino K, Araki H. Evaluation of energy storage method using liquid air. Heat transfer. Asian Res 2000;29:347–57. doi:10.1002/1523-1496(200007)29:5 <347::aid-htj1> 3.0.co;2-a

[5] Wakana H, Chino K, Yokomizo O. Cold heat reused air liquefaction/vaporization and storage gas turbine electric power system. US Patent US20030101728; 2005.

[6] Li Y, Jin Y, Chen H, Tan C, Ding Y. An integrated system for thermal power generation, electrical energy storage and CO_2 capture. Int J Energy Res 2011;35:1158–67. doi:10.1002/er.

[7] Li Y, Wang X, Ding Y. A cryogen-based peak-shaving technology: systematic approach and techno-economic analysis. Int J Energ Res 2013;547–57. doi:10.1002/er.1753 (2011).

[8] Li Y, Chen H, Ding Y. Fundamentals and applications of cryogen as a thermal energy carrier: A critical assessment. Int J Therm Sci 2010;49:941–9. doi:10.1016/j.ijthermalsci.2009.12.012 (2010).

[9] Li Y, Chen H, Zhang X, Tan C, Ding Y. Renewable energy carriers: Hydrogen or liquid air/nitrogen?. Appl Therm Eng 2010;30:1985–90. doi:10.1016/j.applthermaleng.2010.04.033.

[10] Chen H, Cong TN, Wang W, Tan C, Li Y, Ding Y. Progress in electrical energy storage system: a critical review. Prog NatSci 2009;19:291–312. doi:10.1016/j.pnsc.2008.07.014 (2009).

[11] Vaivudh S, Rakwichian W, Chindaruksa S. Heat transfer of high temperature thermal energy storage with heat exchanger for solar trough power plant. Energ Convers Manage 2008;49:3311–7.

[12] Shin JY, Jeon YJ, Maeng DJ, Kim JS, Ro ST. Energy-exergy analysis and modernization suggestions for a combined-cycle power plant. Energy 2002;27:1085–98.

[13] Saleh B, Koglbauer G, Wendland M, Fischer J. Working fluids for low temperature organic Rankine cycles. Energy 2007;32:1210–21.

[14] Chen Y, Lundqvist P, Johansson A, Platell AP. A comparative study of the carbon dioxide transcritical power cycle with an organic rankine cycle with R123 as a working fluid in waste heat recovery. Appl Therm Eng 2006;26:2142–7.

[15] Li Y, Wang X, Jin Y, Ding Y. An integrated solar-cryogen hybrid power system. Renew Energ 2012;37:76–81. doi:10.1016/j.renene.2011.05.038.

[16] Strahan D. Liquid air in the energy and transport systems: opportunities for industry and innovation in the UK. Report No. 021; 2013.

Chapter 10

Flywheels

Donald Bender
System Surety Engineering, Sandia National Laboratories, California, United States of America

1 INTRODUCTION

A flywheel comprises a rotating mass that stores kinetic energy. When charging, a torque applied in the direction of rotation accelerates the rotor, increasing its speed and stored energy. When discharging, a braking torque decelerates the rotor, extracting energy while performing useful work.

Since the invention of the potter's wheel, flywheels have been used as a component in machinery to smooth the flow of energy. In engines or industrial equipment the purpose of the flywheel is to damp out changes in speed due to a pulsed motive source or a pulsed load. Here, the torque may vary significantly between pulses while the speed of the flywheel varies little. Many shapes of flywheel have been used ranging from the "wagon wheel" configuration found in stationary steam engines to the mass-produced, multipurpose disks found in modern automotive engines.

Since the late 20th century a new class of standalone flywheel systems has emerged. The modern flywheel, developed expressly for energy storage, is housed in an evacuated enclosure to reduce aerodynamic drag. The flywheel is charged and discharged electrically, using a dual-function motor/generator connected to the rotor. Flywheel cycle life and calendar life are high in comparison with other energy storage solutions [1].

These modern flywheels are found in a variety of applications ranging from grid-connected energy management to electromagnetic aircraft launch. The prevalent rotor configurations comprise disks, solid cylinders, and thick-walled cylinders made from carbon and glass composite or high-strength steel.

2 PHYSICS

The kinetic energy of a rotating object is given by:

$$E = \frac{1}{2} I \omega^2$$

Storing Energy. http://dx.doi.org/10.1016/B978-0-12-803440-8.00010-5

where E is kinetic energy; I is moment of inertia; and ω is angular velocity. While many rotor shapes have been explored, nearly all flywheels in use are built as solid cylinders or hollow cylinders. Axial extent ranges from short and disk-like to long and drum-like. For a disk or solid cylinder, the moment of inertia is given by:

$$I = \frac{1}{2}mr^2$$

where m is the mass of the rotor; and r is its outer radius. For a thin-walled hollow cylinder, mass is concentrated at the periphery and the moment of inertia is given by:

$$I = mr^2$$

The maximum speed at which a flywheel may operate is limited by the strength of the rotor material. The stress experienced by the rotor must remain below the strength of the rotor material with a suitable safety margin. For a uniform disk or solid cylinder the maximum stress occurs at the center and has a value given by [2]:

$$\sigma_{\max} = \frac{1}{8}\rho r^2\omega^2(3+\nu)$$

where $\sigma_{\max}$ is maximum stress; ρ is the density of the rotor material; and ν is the Poisson ratio of the rotor material. Stress in a rotating thin-walled cylinder is given by [3]:

$$\sigma_\theta = \rho r^2\omega^2$$

where σ_θ is stress in the circumferential direction. The surface speed of a flywheel is given by $V = r\omega$ and the specific energy, or energy per unit mass, of a flywheel rotor can be expressed simply as:

$$\frac{E}{m} = KV^2$$

where K is a shape factor with a value of 0.5 for a thin-walled cylinder and 0.25 for a disk. Flywheel rotors will often be designed to operate at the highest surface speed allowed by the rotor material. High-performance carbon composite rotors have a maximum operating surface speed in the range of (500–1000) m s^{-1} while high-performance steel rotors have a maximum operating surface speed in the range of (200–400) m s^{-1}.

Specific energy may also be expressed in terms of rotor material properties:

$$\frac{E}{m} = K_s\frac{\sigma}{\rho}$$

where K_s is a second shape factor with a value 0.5 for a thin-walled cylinder and 0.606 for a disk with a Poisson ratio of 0.3. This equation reveals that a light strong material, such as a carbon composite, stores considerably more energy per unit mass than a heavy strong material, such as high-strength steel, and that a disk stores more energy per unit mass than a hollow cylinder with the same strength.

3 HISTORY

Flywheels have been used in the manufacture of pottery in China and Mesopotamia since as early as 6000 BC [4]. The inertial disks of the potter's wheel were usually made from wood, stone, or clay [5]. In at least one instance a composite flywheel rotor was constructed using bamboo embedded in clay [6]. Concurrently, thread was made by drawing fiber crop from a holder or distaff onto a hand-held spindle [7]. In a second ancient application the addition of a small stone flywheel to the base of the spindle sped up thread making considerably [8].

The spinning wheel began to displace the hand-held spindle starting around 1200 AD. In this application the operator turns a large drive wheel that functions as a flywheel. The drive wheel is connected to a much smaller bobbin via a drive band. The bobbin spins at a much higher speed than a hand-held spindle, improving on the productivity of the hand-held spindle by an order of magnitude or more [9].

Flywheels remained small and human powered until the steam engines of Watt, Boulton, and Picard in the 1780s. These machines used cranks and flywheels to convert reciprocating force into far more useful uniform rotary motion [10] The "wagon wheel" configuration found in these early engines remained the most common flywheel shape into the 20th century and is still in use today. In the embodiment of this era, flywheels used heavy rims built from cast iron and later steel to damp pulsations in reciprocating engines or reciprocating loads.

Machines using flywheels grew in power and size culminating in the massive stationary steam engines of the late 1800s. The largest engines, such as the Centennial Engine shown in Fig. 10.1 [11], produced 1.04 MW (1400 hp), stood more than 12 m (40 ft.) tall, and employed flywheels 9 m (30 ft.) in diameter.

While modern flywheels operate at a surface speed of 500 m s^{-1} or more, flywheels in stationary steam engines seldom ran at surface speed exceeding 20 m s^{-1} [12]. Consequently, since kinetic energy scales with the square of speed, a 50 t (where t is a metric tonne) flywheel from the industrial age would store just 5 kW h. In comparison, a modern flywheel in use today for stabilizing the electric grid weighs about 1 t [13] and stores more than 25 kW h of usable energy. In the era of the steam engine, flywheel bursts were fairly common, often due to failure of a governor [14]. A particularly large failure could result in the destruction of the building in which it was housed [15].

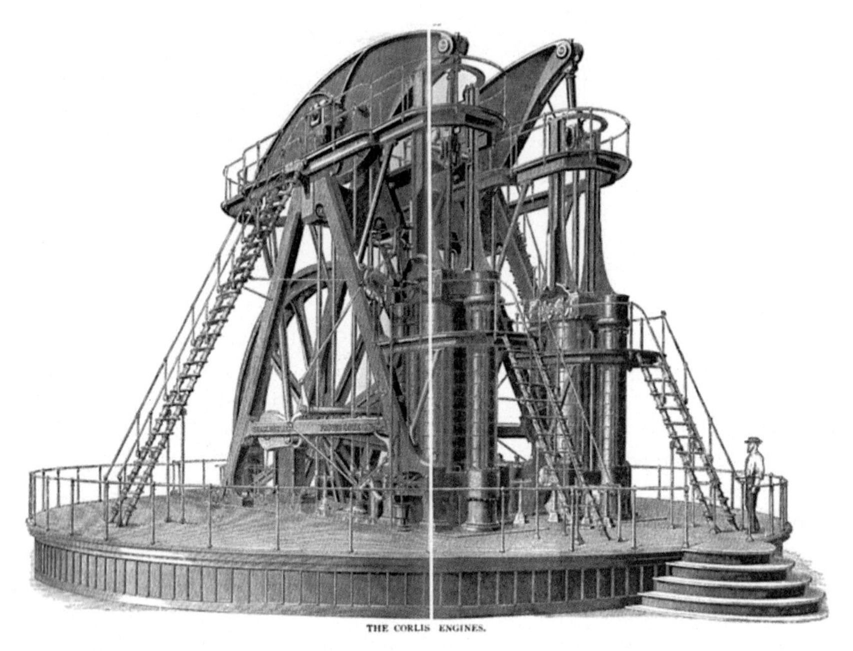

FIGURE 10.1 **Corliss Centennial Engine.**

4 THE DESIGN OF MODERN FLYWHEELS

Standalone flywheels systems are designed expressly for energy storage and power management. A number of attributes differentiate these systems from the flywheels used as engine components. With few exceptions the flywheel power management system is electrically connected to the application that it serves. The flywheel rotor is generally located in its own dedicated housing which is evacuated or held at reduced pressure to minimize aerodynamic drag. The rotor will operate at high speed to make the best use of the rotor material. Charging and discharging events will take place over many revolutions and will usually involve a substantial change in the spin speed of the rotor.

The power delivered by the flywheel and the kinetic energy stored in the flywheel are specified independently. The degree of independence is considerable with flexibility in selecting flywheel discharge time spanning several orders of magnitude.

Since the ratio of energy to power has units of time, it is useful to express the capability of a flywheel in terms of output power that is provided for a specified duration. In one example a flywheel system designed to serve a ridethrough application may provide 1 MW for 3 s. This system provides 0.8 kW h of usable energy. In a second example a flywheel system designed for frequency regulation services may provide 100 kW for 15 min. This system provides 25 kW h of

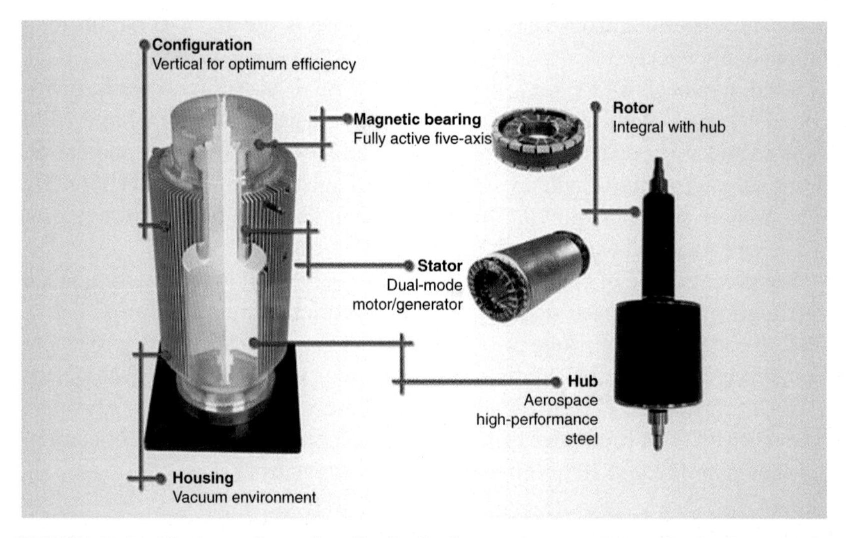

FIGURE 10.2 **Elements of a modern flywheel.** *(Source: Courtesy Calnetix Technologies LLC.)*

usable energy. These two systems will have very different design criteria. The machine of the first example will have a powerful motor and will be optimized to minimize motor cost. The machine of the second example stores much more energy and will be optimized to minimize rotor cost.

Flywheels have inherently long cycle and calendar life. The material properties of the metals and composites used in flywheels are well understood and allow for a design life exceeding 10^6 cycles. The state of charge of a flywheel and its availability are known with high precision and accuracy. Individual modules in use today range in energy capacity from a fraction of a kilowatt-hour to hundreds of kilowatt-hours (Fig. 10.2).

4.1 Rotor Design

Rotors used in flywheel energy storage systems are designed with one of two shapes, depending on the material of construction. Rotors constructed from isotropic materials, such as steel, are in the shape of solid disks or long, solid cylinders. In theory, a tapered disk known as a Stodola hub [16] can store more energy per unit mass than a disk of uniform thickness, but it is impractical in machine design and not used in practice. Rotors built from oriented material, such as carbon and glass fiber, are fabricated in the shape of hollow cylinders.

A tradeoff exists between the performance of a composite rotor and the simplicity of a solid metal rotor. A rim made from high-strength carbon fiber offers much higher specific energy than a solid cylindrical metal rotor and will be much lighter when storing a comparable amount of energy. However, a disk or solid cylinder is much simpler to construct than a rotor assembly using composite materials. Consequently, metal flywheels are more common than composite flywheels.

Solid flywheel rotors may be built to operate at surface speeds up to (200–300) m s^{-1} with readily available grades of steel. However, large solid rotors intended for operation at very high speed (>400 m s^{-1}) must address an additional engineering challenge. To operate at such high speed, exceptionally high–strength steel is required. These steels tend to be brittle and have poorer fatigue and fracture behavior than mild steel.

The potential life of composite rims is extremely long. Presently, composite centrifuge rotors are used on a large scale to enrich uranium. Approximately 500 000 composite centrifuge rotors have been spinning continuously for more than 20 years. These rotors are several meters long and operate at surface speeds in excess of 1000 m s^{-1}. The design life of these rotors is 35 years. Flywheel rotors derived from centrifuge technology are expected to be capable of comparable calendar life and 10 million deep-discharge cycles [17].

4.2 Bearings

Bearings support the flywheel rotor while allowing it to spin freely. Bearing requirements tend to be more severe for flywheels than for other rotating machines and bearings are usually the life-limiting element in a flywheel design. Flywheel rotors tend to be unusually heavy when compared with other rotating machines operating at comparable speed. The need to support the greater weight of the rotor leads to the use of larger bearings which have greater drag losses and inherently poorer life than smaller bearings [18]. Since the flywheel operates in vacuum or reduced pressure, thermal management and lubrication are also difficult.

The two most prevalent types of bearings found in flywheel systems are active magnetic bearings and ball bearings. Active magnetic bearings levitate and actively position the rotor. They are free from contact and therefore free from wear.

Ball bearings represent a simpler, more common, less expensive alternative to active magnetic bearings but are challenged by the life and load requirements of flywheels. For instance, a flywheel designed to operate at 270 Hz (16 000 rpm) will accumulate 1.7×10^{11} revolutions over 20 years. But conventional bearing theory fails to predict reliable life beyond 10^{10} revolutions. Recent advances in ball bearing theory indicate that maintaining peak contact pressure between the ball and the bearing race below 2000 MPa (300 000 psi) can increase rotation life by more than an order of magnitude over conventional theory [19]. In practice, permanent magnets or solenoid coils are often used in conjunction with ball bearings. This reduces load on the ball bearings allowing the use of smaller bearings and lower contact pressure thereby improving bearing life. To simplify design of the levitation system and manage bearing loads, most standalone flywheel systems are built using a vertical spin axis.

4.3 Motor/Generator

The standalone flywheel module is charged and discharged by an integral motor/generator. The motor may be integrated into the steel or composite rotor or may be attached to the rotor by a hub and shaft. A wide variety of motor types have been deployed including homopolar, synchronous reluctance, induction, as well as many types of permanent magnet machine. The selection of a motor type is usually dictated by consideration of thermal management of the flywheel rotor. As the rotor is surrounded by vacuum, removing heat from the rotor occurs through radiation to the housing and is ineffective unless high rotor temperature is allowed. Consequently, a goal of flywheel motor design is to minimize heat dissipated in the rotor. This is not a significant concern for steel rotors used in uninterruptable power supplies as the flywheel motor is operated only occasionally and little energy is deposited in the rotor. For carbon and glass composite rotors that will be cycled frequently, design for low on-rotor loss is critical and permanent magnet machines are usually used.

5 COST AND COMPARISON WITH OTHER TECHNOLOGIES

Cost is the deciding factor in the selection of one energy storage technology over another. Flywheels must compete with batteries and ultracapacitors on the basis of cost where cost is evaluated over the life of a system. For low-cycle applications, such as electric vehicles, battery prices are already nearing the long-sought goal of \$100 $(\mathrm{kW\,h})^{-1}$ [20]. Flywheels are highly unlikely to achieve this incremental energy cost using reasonably foreseeable materials and subsystems.

However, applications requiring 10^6 cycles and a calendar life of decades are well served by flywheels as battery cycle life remains at least two orders of magnitude lower than this. In these applications, flywheels compete with ultracapacitors on the basis of cost per unit energy delivered.

Ultracapacitors have a cycle life as high as 10^6 and an incremental energy cost that has declined to \$20 000 $(\mathrm{kW\,h})^{-1}$ [21]. In theory, ultracapacitors should be cost competitive at any power level for discharge times up to several seconds. However, current applications requiring short-duration discharge (3 s) in excess of 1 MW, such as electromagnetic aircraft launch and ridethrough backup power, are presently served by rotary systems.

The relative cost competitiveness of ultracapacitors, batteries, and flywheels may be presented in terms of power and discharge time. Flywheels are a cost-effective solution for applications requiring power for more than several seconds and up to several or tens of minutes, particularly when high cycle life is required. For applications requiring less than 100 kW, balance-of-system costs make flywheels less cost competitive.

Fig. 10.3 shows regions where flywheels, capacitors, and batteries are most cost effective. Also shown are the ratings of flywheel systems from a number of current manufacturers. The shaded area indicates the region of the parameter space where flywheels are particularly advantageous.

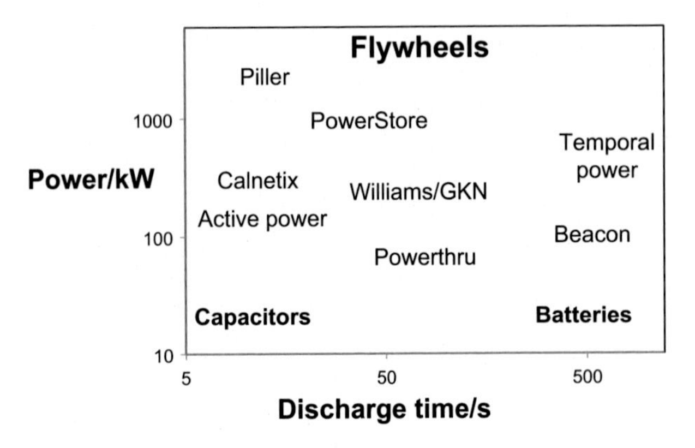

FIGURE 10.3 **Flywheels, capacitors, and batteries.**

Cost drivers for flywheel systems are spread out over a number of subsystems including the rotor, bearings, power electronics, and the balance of system. The total cost of a flywheel system comprises three scalable cost centers:

- *Elements that scale with stored energy*: for a particular geometry and rotor material, rotor weight and cost scale with stored energy. Components and subsystems that scale with rotor weight include the bearings, the housing, and structural hardware.
- *Elements that scale with power*: for a flywheel system with an integral motor/generator, elements that scale with power include the motor itself, the motor drive, and electrical equipment.
- *Balance of system*: balance of system includes the vacuum pump, sensors, telemetry, diagnostics, and controls and other components required for operation of the flywheel that do not scale with energy or power.

The cost for a complete flywheel system may be expressed as follows [1]:

$$C = A \times \text{Power} + B \times \text{Energy} + C_{\text{BOS}}$$

Elements that scale with power A have a cost expressed in dollars per kilowatt, elements that scale with stored energy B have a cost expressed in dollars per kilowatt-hour, and balance-of-system costs C_{BOS} have units of dollars.

Flywheel systems in service today have costs spread across all three cost centers. There appears to be no reported instance of an existing system where the cost of the rotor exceeds 20% of the cost of the system. Consequently, it is not valid to scale flywheel system cost on the basis of dollars per kilowatt-hour absent a consideration of the composition of flywheel system cost.

The incremental cost per unit of stored energy is calculable for rotor materials. To reflect the high-cycling capability of flywheels, it is important to allow for a 50% reduction in strength typical in steel subjected to 10^6 cycles.

TABLE 10.1 Flywheel Rotor Material Cost Per Unit Stored Energy

Material	$/(kW h)$^{-1}$	Mass/(kW h)$^{-1}$
Carbon composite	1 200	1
1 800 MPa (260 000 psi) steel	1 800	7×
1 100 MPa (160 000 psi) steel	2 000	12×
600 MPa (90 000 psi) steel	4 000	24×

Table 10.1 gives an approximation of the incremental cost of rotor material for high-cycle flywheel applications.

The first column refers to the yield strength of various grades of steel when new. Carbon composite values are based on filament-wound construction using 4 800 MPa (700 000 psi) fiber with 65% fiber fraction and proven safety factors for high cycle life [1]. It is important to recognize that this metric applies only to the incremental cost of increasing the mass of a rotor to store more energy. This metric does not reflect other costs such as the motor, bearings, and the housing that are generally greater than the cost of the rotor itself. The third column indicates the mass of a steel flywheel rotor relative to a carbon composite rotor storing the same energy when both are designed for a life of 10^6 cycles. A heavier rotor requires higher capacity bearings and a heavier, more costly housing. Therefore, not only does a carbon composite rotor have lower incremental cost per unit of stored energy, balance-of-systems costs can be reduced as well.

6 APPLICATIONS

Applications for flywheels are viable when two conditions are met. First, the flywheel must represent a more cost-effective solution than competing forms of energy storage. Second, a market must exist so that the deployment of a flywheel system results in an economic return. This section describes and estimates the scale of application areas where flywheels currently represent solutions that are technically effective and cost competitive. These include grid-connected power management, industrial and commercial power management, pulsed power, uninterruptable power supplies, and mobile applications

6.1 Grid-Connected Power Management

Stationary, grid-connected applications exist on the utility side of the electric meter. Here, the sale of services and products is highly regulated and a market for an energy storage solution only exists after being created by a regulatory agency. Flywheels are used in two such applications which are related: frequency regulation and management of ramping due to fluctuating renewable generating resources.

6.1.1 Frequency Regulation

A large electrical grid must operate at a nearly constant frequency for the generators to remain synchronized. When the amount of electricity consumed changes, generator output must be controlled to follow the load. For instance, if the load increases faster than a turbine generator can respond, the generator slows down, momentarily operating at lower frequency. If the load changes are severe enough, or if a large generating asset suddenly drops off-line, other generators may not remain synchronized and a widespread power outage may occur.

Frequency regulation is provided by generators as an ancillary service to improve the stability of the grid. A power plant may sell the frequency regulation service to the grid operator in addition to selling electricity by operating slightly below peak power so that it may regulate up or down. To provide this service effectively the power plant must be able to ramp up and down quickly, responding to a control signal from the grid operator that may change every few seconds or less. Flywheels are ideally suited to this application as they are capable of millisecond response times and nearly constant cycling.

Commercialization of energy storage for frequency regulation is realized through the construction of an energy storage plant. The plant is typically owned by a private entity rather than the utility and is typically installed at an existing substation to facilitate interconnection. Once commissioned the private entity sells frequency regulation services to the grid operator.

Beacon Power LLC pioneered the use of flywheels for frequency regulation with 20 MW plants located in Stephentown, New York and Hazel Township, Pennsylvania. The Stephentown plant (Fig. 10.4) provides approximately 10% of New York's overall frequency regulation needs [22].

FIGURE 10.4 **20 MW flywheel frequency regulation plant.** *(Source: Courtesy Beacon Power LLC.)*

6.1.2 Ramping

For large grids the impact of variations in load and generation is managed through frequency regulation. Islands and isolated grids are even more susceptible to instability but markets for frequency regulation services do not exist in these areas. Here the problem manifests itself as excessive ramping of the output of conventional generators that are used in conjunction with renewable energy sources.

Ramping of conventional generating assets results in inefficient operation and high operating and maintenance cost. This becomes more problematic as renewable energy becomes more available. At (20–30)% renewable penetration the impact on the grid becomes increasing difficult to manage without storage [23]. Some very large islands are targeting wind penetration exceeding 40% on a capacity basis where storage will be an essential element in implementation [24]. Island grid operators are beginning to address this problem. As an example, Puerto Rico is requiring that new solar installations limit ramp rates to 10% per minute [25].

The fluctuations in power produced by wind and solar vary considerably in frequency, severity, and duration. Variations in solar energy are usually gradual and occur over the course of a day. Wind, on the other hand, can have frequent variations of ±20% lasting less than 2 min [26]. Flywheels are particularly well suited for smoothing out the frequent, short-duration variations in electricity produced from wind. There are 50 islands with a combined average power consumption of 53 GW where the potential application of flywheel energy storage would amount to 5% of this value [27,28].

6.2 Industrial and Commercial Power Management

Flywheels can provide high peak power for short duration and retrieve regenerated electricity that would otherwise be lost. Industrial and commercial power management applications are found on the customer side of the electric meter. Here the market for power management products is far less regulated than on the utility side of the meter. Consequently, the process of assessing the economic viability of a flywheel solution is more straightforward and time to market is much shorter than for utility-side solutions.

6.2.1 Transit

Flywheels produced by Calnetix and URENCO have been demonstrated in a number of transit systems for trackside energy recovery [29]. In this application the flywheel is installed at a station or a transit system substation. The flywheel captures energy recovered through regenerative braking and uses this energy to accelerate the train as it leaves its stop. This allows for heavier and longer trains without increasing transmission or distribution line capacity. To mitigate voltage sag or increase transit system capacity in an existing system without using energy storage a new substation has to be installed. Flywheel energy storage installed at a transit station can provide the same mitigation of voltage sag as a new substation but in a small footprint with no new utility feed and at a much lower cost. Given the high rate of charge–discharge cycles, flywheels are particularly well suited for this application.

Globally, 190 metro systems operate 9 477 stations and over 11 800 km of track [30]. Using energy storage to recover energy lost in braking has the potential to reduce metro rail electricity consumption on the order of 10% [31] while achieving energy cost savings of $90 000 per station [32]. When installed in regions where the utility tariff structure includes demand charges, additional savings of up to $250 000 per station per year are attainable [33]. In studies and tests to date, trackside storage sized to provide (1–3) MW of launch power or energy recovery per station was found to be an effective rating for metro rail application [34].

6.2.2 *Mining*

Flywheels have potential application in mining. Open-pit mines around the world use electrically powered draglines to excavate material. The load profile of a dragline is cyclic, highly nonuniform, and produces regenerated electricity which is generally lost. During the lifting phase, peak loads of 6 MW are typical. Lowering the load into a conveyance regenerates as much as 3 MW. This cycle repeats approximately once per minute continuously.

In one instance the Usibelli coal mine in Healy, Alaska operates a 6 MW dragline that is fully electric and is connected to the Golden Valley Electric Association (GVEA) grid. The impact of the fluctuating load was so severe that routine dragline operation caused the lights of other GVEA customers to flicker. Since 1982 the Usibelli mine has operated a flywheel to smooth the load drawn by the dragline. The 40 t flywheel consists of three, 300 mm thick (1 ft) and 2.5 m diameter steel plates and is connected to the GVEA grid in parallel with the dragline, successfully mitigating the problem [35].

6.3 Pulsed Power

Pulsed power is the collection of energy at a steady rate followed by the rapid, high-power discharge of energy into the application. Flywheels are well suited for this use and may be found in military, research, and motive power applications.

6.3.1 *Electromagnetic Aircraft Launch System*

A key military application is the use of flywheels to energize the Electromagnetic Aircraft Launch System (EMALS) [36] on aircraft carriers to replace steam-powered catapults. Steam catapults are large, heavy, and inefficient. Heretofore, each launch consumed 615 kg (1350 lb) of steam produced by the aircraft carrier's nuclear reactor.

The linear synchronous motor of EMALS is designed to launch heavier aircraft using much less energy. The linear motor is powered by alternators with significant inertia. Each alternator comprises an axial field permanent magnet motor with dual stators. The alternator's rotor disk serves as the energy storage component and the field source during power generation. Average power from the ship's electrical system is fed into the alternator between launch events. The system is sized to charge fully in 45 s. During a launch event the energy stored in the rotors is released in a pulse lasting from (2 to 3) s. Peak alternator output is 81.6 MW when discharged into an impedance-matched load. When fully charged, EMALS' rotors store 121 MJ (33.6 kW h) of extractable energy at a maximum speed of 106 Hz (6400 rpm). Total stored energy is much higher as the rotor speed only decreases by about 25% during a launch event.

6.3.2 *Research Facilities*

Flywheels have been used to provide pulsed power to large research facilities such as the Joint European Torus (JET) in Culham (United Kingdom), which is the largest and most powerful tokamak currently in use. JET has been in operation since 1983. A single plasma pulse at JET requires peak power of 1000 MW and occasionally more. On a typical day, 22 tests are conducted. JET draws power from the grid continuously, charging two enormous steel flywheels. The flywheels provide power for each test.

Each of the two JET flywheels has a diameter of about 9 m and weighs 775 t. At full speed the rotors spin at 3.7 Hz (225 rpm) and attain a tip speed of around 100 m s^{-1}. Between shots the wheels are accelerated from half speed to full speed over a period of 540 s (9 min) using 8.8 MW motors. During a 20 s shot each flywheel can discharge 700 kW h of energy at a peak power of 500 MW [37].

6.3.3 *Roller Coaster Launch*

Traditionally, roller coaster launch systems use a chain drive to bring the train to the top of the first hill followed by the familiar plunge. In a new class of coasters, energy is accumulated in a flywheel and then used to rapidly accelerate the train using electromagnetic, hydraulic, and friction wheel propulsion [38]. The Incredible Hulk roller coaster at Universal's Island of Adventure theme park in Orlando, Florida is a noteworthy example.

The Incredible Hulk uses a friction wheel drive system to accelerate the train at 1 G up an incline, reaching a speed at the top of 18 m s^{-1} (40 mph). The launcher comprises 230 motors powering wheels that grip a rail attached to the bottom of the train. The launch event draws 8 MW for 2 s and is repeated every 90 s. In order to avoid disruption to the local utility the constructor installed several 4 500 kg (10 000 lb) flywheels that charge continuously at about 200 kW and then discharge at 8 MW to launch the train [39].

6.4 Uninterruptible Power Supplies

Flywheel systems are in global use providing temporary backup electrical power. The purpose of the flywheel in this application is to support the load of a critical facility or system during a power outage until backup diesel generators can be brought up to speed and synchronized. Flywheels compete directly with batteries and offer the advantages of much longer service life and avoidance of the need to periodically replace and recycle the batteries. Here flywheels are implemented in one of two ways.

When used as a standalone energy storage device the system is referred to as a flywheel uninterruptible power supply (UPS). The flywheel provides electrical power to a DC bus and an inverter converts this into AC electricity to power the load. In this application the flywheel replaces or augments a battery. Discharge times of tens of seconds are typical. Rotors in flywheel UPS systems generally spin about a vertical axis in vacuum or reduced pressure.

Rotary UPS are variously known as ridethrough systems, engine-coupled UPS, or diesel rotary uninterruptible power supplies (DRUPS). A typical rotary system comprises a diesel generator, an inductive coupling with a substantial moment of inertia, and an alternator all mounted coaxially on a common base frame. A clutch may be located between the inductive coupling and the generator. During an outage, kinetic energy stored in the inductive coupling drives the alternator to support the load while the diesel generator starts. Power from the generator may be available in as little as 3 s after an outage begins. Rotary UPS are large, the smallest having a rating around 1 MW.

The global market for UPS systems is on the order of $\$8 \times 10^9$–$\$10 \times 10^9$ (\$8 billion–\$10 billion) per year. Rotary systems account for about 5% of the total UPS market. However, when only large systems (>2 MW) are considered, rotary UPS account for 35% of the market [40]. In Europe, where rotary UPS are well established, half of all new UPS installations that are rated at more than 1 MW are rotary UPS [41].

6.5 Mobile

Mobile applications are those in which the flywheel is installed in a vehicle. Flywheel energy storage has been demonstrated in buses and may now be found in materials handling and motorsport.

6.5.1 Materials Handling

Materials handling involves the intermittent, repeatable application of power to move loads. Often, the peak power required to move a load is much greater than the average power of the process and there is no convenient way to recover energy while lowering a load. Flywheels are well suited for this application as load duration is short and repeated frequently. One application is found in rubber-tired gantry (RTG) cranes (Fig. 10.5). Approximately 8000 RTGs operate

FIGURE 10.5 **Flywheels installed in rubber-tired gantry cranes.** *(Source: Courtesy Calnetix Technologies LLC.)*

in container terminals around the world. While ship-to-shore cranes are grid connected, RTG cranes are free to move about the terminal and are often powered by an onboard diesel genset. Without energy storage the nonuniform load results in inefficient operation and high emissions.

Since 2006, flywheels produced by Calnetix have been deployed in RTGs to move shipping containers. Lifting a container typically draws about 240 kW for around 10 s. The power to perform the lift is provided by the flywheel allowing the genset to operate at more uniform output. During lowering the hoist motors function as generators, returning energy to the electrical system to be stored in the flywheels for reuse. The flywheel system has been demonstrated to reduce fuel consumption by (32–38)% [42], nitrous oxide emissions by 26%, and particulate emissions by 67% [43].

6.5.2 *Motorsport*

Since the late 2000s hybrid propulsion systems have powered the cars in top-tier motorsport beginning with Formula 1 followed by the highest class of WEC (World Endurance Championship) racing: the Le Mans LMP1 series. Hybrid powertrains improve fuel efficiency reducing the number of pit stops required to complete a 24 h race that covers approximately 5000 km (3000 miles). Williams Hybrid Power (WHP) pioneered the use of flywheel energy storage in motorsport. WHP flywheels were used successfully in the Audi R18 e-Tron LMP1 that won at Le Mans in 2012, 2013, and 2014 [44].

6.5.3 Spacecraft

Throughout the history of space flight, flywheels have been used to stabilize and point spacecraft of all types. These flywheels are implemented as control moment gyros (CMGs) or reaction wheels, which are also referred to as momentum wheels. A reaction wheel may have a nominal fixed spin speed or a nominal spin speed of zero. When torque is applied to the wheel the opposing moment rotates the spacecraft. Reaction wheels are useful when the spacecraft must be rotated by a very small amount, such as when pointing at a star or target. Reaction wheels are more common in smaller spacecraft. In contrast, CMGs spin continuously creating gyroscopic moment. Mounted in motorized gimbals, tilting the spin axis of the CMG with respect to the inertial frame of the spacecraft can produce large steering torque with very little power. CMGs are found in spacecraft of all sizes, up to and including the International Space Station.

For either reaction wheels or CMGs, the use of an inertial wheel shifts the burden of attitude control from limited propellant to inexhaustible solar power. Thousands of inertial wheels have flown and a mature industry exists [45].

7 OUTLOOK

As a component of rotating machinery, the continued use of flywheels is both certain and unremarkable. As electrically connected energy storage systems, flywheels must compete with batteries and ultracapacitors on the basis of cost.

For low-cycle applications, such as electric vehicles, flywheels are unlikely to achieve the already low incremental energy cost of batteries [20]. However, applications requiring 10^6 cycles and a calendar life of decades may continue to be well served by flywheels as battery cycle life remains at least two orders of magnitude lower than this. The extent to which the use of flywheels will expand or decline will depend on trends in cost reduction for flywheels and on the various competing technologies.

Flywheels will benefit as other industries drive increasing performance and declining cost in materials and electronics. The materials and processes currently used to produce flywheel rotors are highly mature and large cost reductions are unlikely. However, given the geometric impact of rotor material performance on flywheel cost, the development of potentially transformative materials, perhaps carbon nanotube composite, might not only substantially improve energy per unit mass of the rotor but also lead to much smaller and less costly bearings and housings.

The increasing use of energy storage and electric motors in hybrid and electric vehicles is already impacting the power electronics supply chain used by flywheel developers for motor drive and magnetic bearings. To the benefit of flywheels, motor drive power electronics costs have dropped dramatically over the last decade and are approaching \$5 kW^{-1} [46].

Given the increasing need in areas where flywheels are already in use combined with performance and cost trends in the underlying technology, flywheels should remain a competitive energy storage solution for short-duration, high-cycle applications for the foreseeable future.

ACKNOWLEDGMENTS

The author gratefully acknowledges the support of the US Department of Energy, Office of Electricity, Dr Imre Gyuk, Director, Energy Storage Program. Any errors or omissions in the article are the responsibility of the author alone.

REFERENCES

[1] Bender D. Flywheels. In: Crawly G, editor. The World Scientific handbook of energy. Vol 4. London, UK: World Scientific; 2015.
[2] Young WC, Budynas RG. Roark's formulas for stress and strain. 7th ed. New York: McGraw Hill; 2001. p. 746.
[3] Young WC, Budynas RG. Roark's formulas for stress and strain. 7th ed. New York: McGraw Hill; 2001. p. 745.
[4] Potter's wheel. New World Encyclopedia. http://www.newworldencyclopedia.org/entry/Potter's_wheel
[5] Bryant V. The origins of the potter's wheel. Ceram Today. http://www.ceramicstoday.com/articles/potters_wheel.htm
[6] Chang JB, Christopher DA, Ratner JKH. Flywheel rotor safe-life technology: literature search summary. Darby, PA, USA: Diane Publishing; 2002. ix.
[7] Hodges H. Technology in the ancient world. New York: Barnes & Noble; 1992. 47.
[8] Barber EW. Prehistoric textiles: the development of cloth in the Neolithic and Bronze Ages with special reference to the Agean. Princeton, NJ: Princeton University Press; 1991. p. 41–44.
[9] Pacey A. Technology in world civilization: a thousand-year history. First MIT Press paperback ed Cambridge, MA: MIT Press; 1991.
[10] Dickinson HW. A short history of the steam engine. Cambridge, UK: Cambridge University Press; 2011. p. 79–82.
[11] Walker FA, editor. United States Centennial Commission. International Exhibition, 1876, Reports and Awards Group XV. Philadelphia, PA: Lippincott; 1877.
[12] Industry: a magazine devoted to science, engineering, and the mechanic arts, especially on the Pacific Coast, vol. 5. Detroit, MI: Industrial Publishing Company; 1892. p. 776.
[13] Carbon fibre flywheels. Beacon Power Corporation. http://beaconpower.com/carbon-fibre-flywheels/
[14] Insurance engineering, vol. 10. Insurance Press; 1905. pp. 384 and 579.
[15] http://www.farmcollector.com/steam-traction/100-years-ago-in-american-machinist.aspx#axzz3DIFKslUY

[16] Genta G. Flywheel energy storage. London, UK: Butterworths; 1985.
[17] Gardiner G. Composite flywheels: finally picking up speed? Compos World 2014; March 4. http://www.compositesworld.com/blog/post/composite-flywheels-finally-picking-up-speed
[18] Hertz H. Contact between solid elastic bodies. J Reine Angew Math 1882;92.
[19] Super precision bearings, NSK motion and controls. Life—a new life theory, Part 5: technical guide. file e1245f.pdf, p. 138–144. http://www.jp.nsk.com/app01/en/ctrg/index.cgi?rm=pdfView&pno=e1254f
[20] Shahan Z. Are EV battery prices much lower than we think? Under $200/kWh. Clean Technica. http://cleantechnica.com/2014/01/07/ev-battery-prices-much-lower-think/
[21] Pricing information for Maxwell and Ioxus cells and modules obtained by the author, March 2014.
[22] Beacon power Stephentown advanced energy storage case study. Clean Energy Action Project. http://www.cleanenergyactionproject.com/CleanEnergyActionProject/CS.Beacon_Power_Stephentown_Advanced_Energy_Storage___Energy_Storage_Case_Study.html
[23] IRENA. Electricity storage and renewables for island power. Masdar City, UAE: International Renewable Energy Agency; May 2012.
[24] EirGrid. All-island generating capacity statement 2012–2021. Dublin, Ireland: EirGrid, System Operator for Northern Ireland.
[25] Gevorgian V, Booth S. Review of PREPA technical requirements for interconnecting wind and solar generation. Report NREL/TP-5D00-57089. Golden, CO: National Renewable Energy Laboratory; November 2013.
[26] Gevorgian V, Corbus D. Ramping performance analysis of the Kahuku Wind–Energy Battery Storage System. Report NREL/MP-5D00-59003. Golden, CO: National Renewable Energy Laboratory; November 2013.
[27] http://en.wikipedia.org/wiki/List_of_countries_by_electricity_consumption
[28] http://en.wikipedia.org/wiki/List_of_islands_by_population
[29] Tarrant C. Kinetic energy storage wins acceptance. Railway Gazette 2004; April 1. http://www.railwaygazette.com/news/single-view/view/kinetic-energy-storage-wins-acceptance.html
[30] http://en.wikipedia.org/wiki/List_of_metro_systems
[31] http://www.abb.com/cawp/seitp202/265455d72a797481c1257b59003b8600.aspx
[32] Schroeder MP, Yu J, Teumin D. Guiding the selection & application of wayside energy storage technologies for rail transit and electric utilities. Contractor's Final Report for Transit Cooperative Research Program (TCRP) Project J-6/Task 75. Washington, DC: Transportation Research Board; November 2010.
[33] APTA Whitepaper. SEPTA Recycled Energy Optimization Project with regenerative braking energy storage (Jacques Poulin, Product Manager—Energy Storage). Montreal, Quebec, Canada: ABB Inc.
[34] McMullen P. Green ovations: innovations in green technologies, reducing peak power demand with flywheel technology. Electr Energ Online 2013; Jan/Feb, http://www.electricenergyonline.com/show_article.php?mag=&article=680
[35] Usibelli coal mine flywheel. Alaska Energy wiki. http://energy-alaska.wikidot.com/usibelli-flywheel
[36] GlobalSecurity.Org. http://www.globalsecurity.org/military/systems/ship/systems/emals.htm
[37] Huart M, Sonnerup L. JET flywheel generators. P I Mech Eng A-J Pow 1986; 200:295–300.
[38] Bleck & Bleck Architects LLC. http://bleckarchitects.com/2014/08/flywheel-launched-coaster/
[39] Total immersion: theme park for the 21st century. Orlando, FL: USA Networks. https://www.youtube.com/watch?v=VNDlM-1gvMY

[40] Kinetic energy storage vs. batteries in data centre applications. Hitec Power. http://www.datacentre.me/downloads/Documents/KINETIC%20ENERGY%20STORAGE%20VS%20BATTERIES%20IN%20DATA%20CENTRE%20UPS%20APPLICATIONS.pdf
[41] Gagliano G. Applications Director, S&C Electric Company, private communication; March 2014.
[42] Flywheel energy storage for rubber tired gantry cranes. Green Car Congress; April 9, 2009. http://www.greencarcongress.com/2009/04/flywheel-energy-storage-system-for-rubber-tired-gantry-cranes.html
[43] Flynn M, McMullen P, Solis O. Saving energy using flywheels. IEEE Ind Appl Mag Nov/Dec 2008;69–73.
[44] Cotton A. Audi R18 (2014). Racecar Eng June 1, 2014.
[45] Votel R, Sinclair D. Comparison of control moment gyros and reaction wheels for small Earth-observing satellites. 26th Annual AIAA/USU Conference on Small Satellites.
[46] Rogers S. Advanced Power Electronics and Electric Motors R&D. US DOE Presentation; May 14, 2013. http://energy.gov/sites/prod/files/2014/03/f13/ape00a_rogers_2013_o.pdf

Part C

Electrochemical

Chapter 11

Rechargeable Batteries with Special Reference to Lithium-Ion Batteries

Matthias Vetter, Stephan Lux
Fraunhofer Institute for Solar Energy Systems ISE, Freiburg, Germany

1 INTRODUCTION

There are many different types of rechargeable batteries such as: lead–acid, alkaline, nickel–cadmium, nickel–hydrogen, nickel–metal hydride, nickel–zinc, lithium cobalt oxide, lithium-ion polymer, lithium iron phosphate, lithium sulfur, vanadium redox, lithium nickel manganese cobalt and sodium sulfur. Much has been written about the lead–acid battery and it is still the most popular of the rechargeable batteries. Lithium–ion and vanadium redox batteries offer specific characteristics, which are of special interest for use as stationary storage and possibly grid-level storage. The vanadium battery is discussed in chapter: Vanadium Redox Flow Batteries and the lithium-ion battery is the focus of this chapter.

The main advantages of lithium-ion batteries are: high energy density, high cycle and calendar lifetimes, fast and efficient charging, with little energy wasted, low self-discharge rate, no need to be held upright, fairly maintenance free, and little voltage sag. Its main disadvantages are its relatively high price (at the moment but this is improving) and the possibility of thermal runaway. The latter is being solved by elaborate protective circuits such as battery management circuitry which is discussed in this chapter.

Lithium-ion batteries are a relatively new invention and have only been around commercially since the 1980s. The chemistry and the technology are now reasonably well proven and this battery has edged out older rechargeable batteries such as the nickel–cadmium (NiCd) battery. In the 1990s one heard of lithium–ion batteries bursting into flames. The type of batteries used at this time were lithium-cobalt oxide ($LiCoO_2$) batteries. These have been superseded by the lithium iron phosphate ($LiFePO_4$) battery, and the lithium nickel manganese cobalt oxide (LiNMC) batteries which are very much safer than the $LiCoO_2$ battery.

Storing Energy. http://dx.doi.org/10.1016/B978-0-12-803440-8.00011-7

TABLE 11.1 Comparison of Different Selected Battery Technologies [1,2]

	Lead acid	NiMh	Li NMC/ graphite	$LiFePO_4$/ graphite	Vanadium redox-flow
Energy density/ (Wh kg^{-1})	40	75	160	110	45
Power density/ (W kg^{-1})	350	600	1 300	4 000	120
Cycle lifetime	600	900	2 500	5 000	12 000
Calendar lifetime/a	7	5	7	14	15
Efficiency/%	85	75	93	94	80
Monthly self-discharge/%	8	20	3	3	5
Cost/€ $(kW\ h)^{-1}$	60–300	400–600	200–2 000	200–2 000	150–800

Note: NiMh and LiNMC refer to nickel metal hydride and lithium nickel manganese cobalt batteries, respectively.

2 PHYSICAL FUNDAMENTALS OF BATTERY STORAGE

Stationary battery storage is becoming more important with increasing shares in renewable energies in power supply systems and in grids, both off-grid and on-grid. In principle there exist a variety of different battery technologies suitable for these stationary applications. In Table 11.1 selected technologies are listed with their main parameters.

2.1 Lead–Acid Batteries

The lead–acid battery was invented in 1859 by French physicist Gaston Planté and is the oldest type of rechargeable battery. Although it has a very low energy-to-weight ratio and a low energy-to-volume ratio, it is able to supply high surges of current. Its low cost makes it attractive for many uses including in motor vehicles to provide a high starting current.

The battery consists of lead plates in a solution of sulfuric acid. The basic cell reaction is given at the anode by:

$$\mathrm{Pb(s) + HSO^{-}(aq) \rightarrow PbSO_4(s) + H + (aq) + 2e^{-}}$$

and at the cathode by:

$$\mathrm{PbO_2(s) + HSO^{-}(aq) + 3H^{+}(aq) + 2e^{-} \rightarrow PbSO_4(s) + 2H_2O(l)}$$

The total reaction can be written as:

$$Pb(s) + PbO_2(s) + 2H_2SO_4(aq) \rightarrow 2PbSO_4(s) + 2H_2O(l)$$

In the discharged state both the positive and negative plates become ($PbSO_4$), and the electrolyte loses much of its dissolved sulfuric acid. The discharge process is driven by the conduction of electrons from the negative plate back into the cell at the positive plate in the external circuit.

The liquid electrolyte did limit its application until the introduction of a gel electrolyte in the 1930s. This extended use of the lead–acid battery to be used in different positions without leakage. Further improvements in the 1970s led to the introduction of the valve-regulated lead–acid battery (often called "sealed"); this development made it possible to have the battery in any position.

2.2 Lithium-ion Batteries

The lithium-ion battery is rapidly becoming the most important battery for both portable applications (computers and power tools) and stationary applications (storing grid-level energy). The term lithium-ion is used as there is no elemental lithium in the battery. As the lithium ions move from one host to another the process has been likened to a rocking chair.

The basic half-cell reactions at each electrode (using the $LiCoO_2$ battery as an example) are at the anode:

$$x\text{Li} \rightarrow x\text{Li}^+ + xe^-$$

and at the cathode:

$$x\text{Li}^+ + xe^- + \text{LiCoO}_2 \rightarrow \text{Li}_{1+x}\text{CoO}_2$$

With the overall reaction:

$$x\text{Li} + \text{LiCoO}_2 \rightarrow \text{Li}_{1+x}\text{CoO}_2$$

3 DEVELOPMENT OF LITHIUM-ION BATTERY STORAGE SYSTEMS

Lithium-ion battery systems are assembled from a number of modules interconnected parallel and/or serial, whereas a module typically consists of a number of cells which may be switched in parallel and/or in serial. The principle behind the design of lithium-ion battery systems will be discussed in the following section using as an example the battery used for photovoltaic (PV) applications. This fulfills the task of increasing self-consumption of generated PV electricity. In general, the development steps for a battery system are as follows:

- Cell characterization and selection of appropriate cell technology
- Module and system design

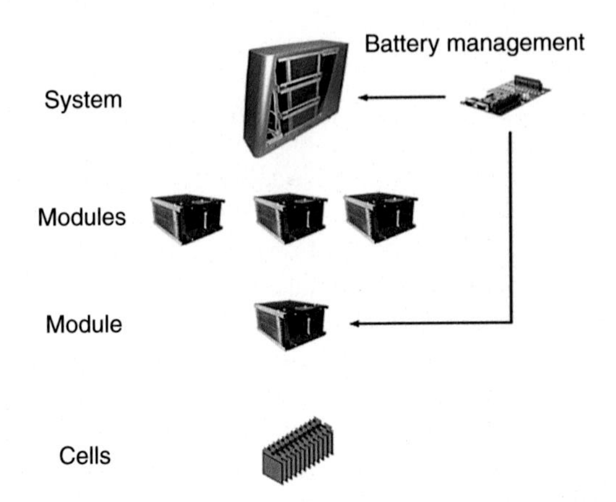

FIGURE 11.1 Concept of a 5.33 kW h lithium-ion battery system for residential photovoltaic applications consisting of three modules switched in parallel to be connected to a 48 V battery inverter. One module consists of 12 cells switched in serial [1]. The list of parameters is given in Table 11.2.

- Cell interconnection
  - Electrical
  - Mechanical
  - Thermal
- Cooling system
- Safety concept
- Battery management
- Interfaces and integration in energy systems.

Justifiable storage sizes for residential PV applications are typically in the range (2–10) kW h. An example of such a storage system is the battery system based on lithium-ion pouch bag cells shown in Fig. 11.1; corresponding parameters are listed in Table 11.2. The system was designed for a battery inverter with a nominal input voltage of 48 V.

3.1 Design of Battery Modules and Systems for Stationary Applications

For the construction of a battery module several aspects have to be considered. Among others, safety, reliability, efficiency, long lifetime, and reduced maintenance efforts play key roles. As temperatures have a huge impact not only on safe operation but also on aging mechanisms, the design of an appropriate cooling system is one of the key aspects. As a result certain temperature levels should not be exceeded, and furthermore a homogeneous temperature distribution within one cell, within a module, and also within the whole battery system is important. In the following section the simulation-based design of the battery module is described.

TABLE 11.2 Parameters of the Lithium-Ion Battery System, Shown in Fig. 11.1

Parameter battery details	Value and unit
Parameter battery module	
Cell chemistry	Graphite/NMC
Number of cells per module	12 (switched in serial)
Rated voltage/V	44.4
$L \times W \times H$/cm	30 × 24 × 15
Gravimetric energy density/(Wh kg^{-1})	105.3
Volumetric energy density/(Wh L^{-1})	164.8 l
Energy content/(kW h)	1.78
Efficiency/%	95 (0.5C), 97 (0.2C)
Parameter battery system	
Number of modules per system	3 (switched in parallel)
$L \times W \times H$/cm	82 × 25.5 × 57.5
Gravimetric energy density/(Wh kg^{-1})	82.2
Volumetric energy density/(Wh L^{-1})	44.4
Energy content/(kW h)	5.34
Cooling	2 fans
Rated current/A	100
Battery management	1 central unit and 3 balancing boards

NMC refers to nickel manganese cobalt (the use of the abbreviation 0.5C and 0.2C will be described in the next section).

For the design and construction of the modules and the battery system, thermal simulations were carried out using the simulation tool Dymola [3] and the model description language Modelica [4]. The approach is based on partition of the cells and cooling plates into segments. Each cell was divided into 49 segments and the cooling plates between the cells were divided into 81 segments (Fig. 11.2). For every segment an electrical and thermal model was developed. The electric model is based on an inner resistance model of a used NMC cell, describing the dependence on state of charge (SOC), temperature, and C rate. The C rate is a measure of the charge and discharge current of the battery and a discharge of 1C draws a current equal to the rated capacity. For example, a battery rated at 1 000 mA h provides 1 000 mA for 1 h if discharged at the 1C rate.

Based on the results of the thermal simulation the cooling area and the thickness of the cooling plates can be calculated. The challenge was to reduce the temperature difference between the single battery cells to a minimum. The low-temperature difference should avoid too much cell balancing and allows homogeneous aging of single cells.

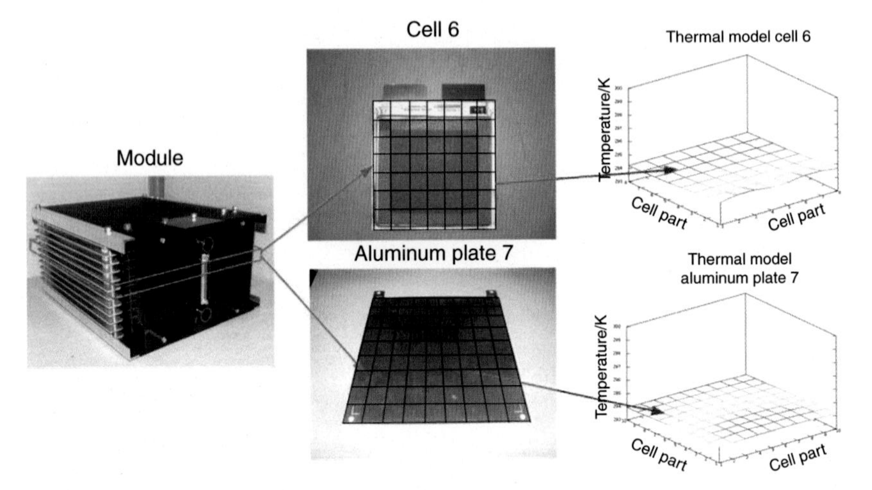

FIGURE 11.2 **Segmentation of lithium-ion cells and aluminum cooling plates for simulation of thermal behavior [5].**

To assure a homogeneous air flow over each cell and the three modules the system was simulated within a computational fluid dynamics (CFD) simulation program. By using this CFD simulation an optimized angle of skewness could be identified (Fig. 11.3). The skewness enables a uniform air flow over every module of the system.

To verify the goal of a maximum temperature difference of 2 K, tests in a climate chamber were carried out with one battery module. Using the flow channel the air distribution over the cells could be investigated. Tests in the climate chamber showed that the temperature difference between the cells of

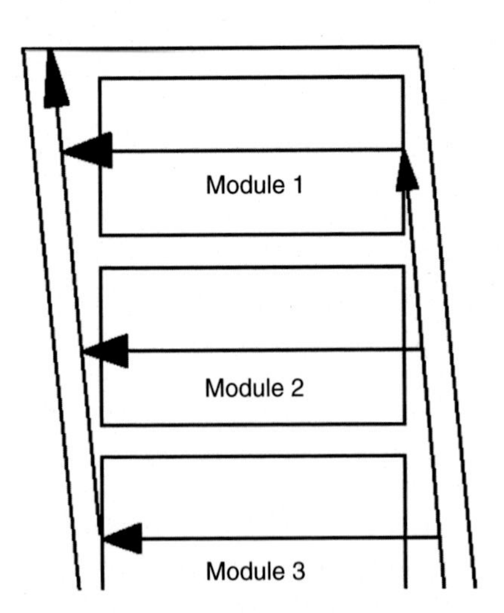

FIGURE 11.3 **Principle of the air cooling system showing built-in skewness [6].**

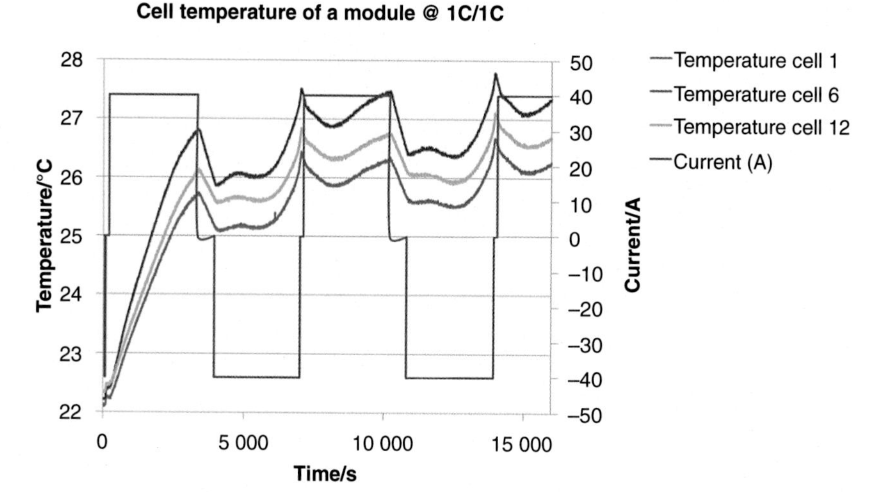

FIGURE 11.4 **Temperature profile of three lithium-ion cells of a battery module for a 1 C charge/discharge test in the laboratory [7].**

one module is nearly constant and almost below 1 K for a charge/discharge C-rate of 1 (Fig. 11.4).

For the analysis of critical temperature segments, module tests were carried out and a thermal imaging camera was used to identify the hotspots. The results (Fig. 11.5) show that the module warmed up mostly at the cells and the cooling plates between the cells. Within these tests no critical sectors could be detected.

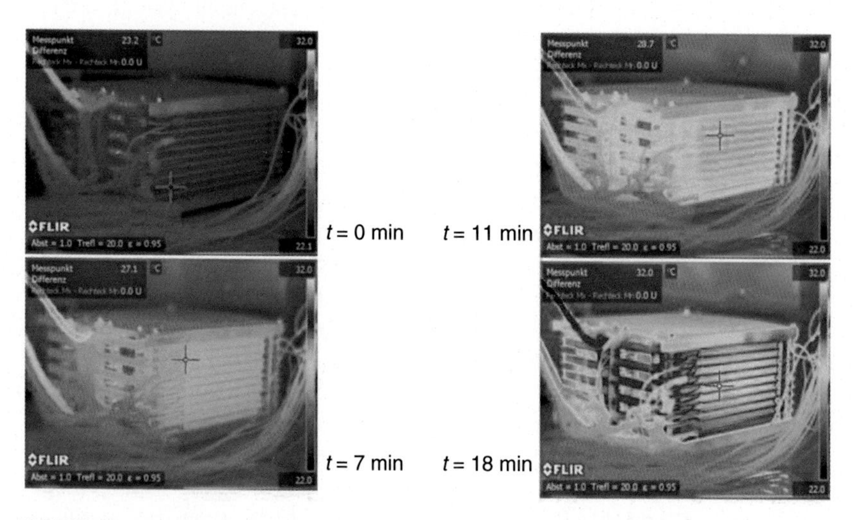

FIGURE 11.5 **Analysis of a thermal camera with 1C in a climate chamber [7].**

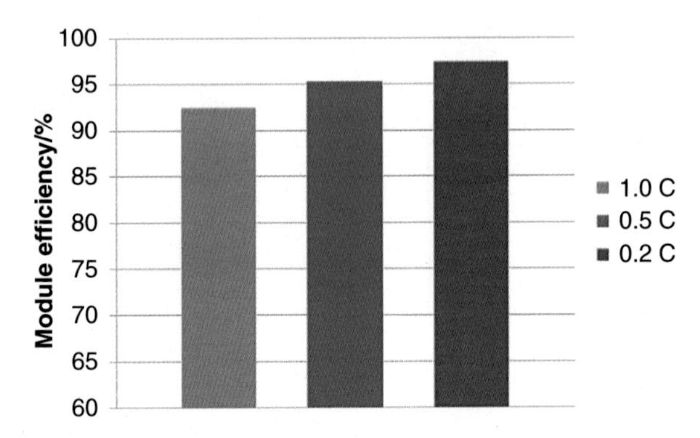

FIGURE 11.6 Efficiencies of a lithium-ion battery module, consisting of 12 cells switched in serial for different C rates [1].

3.1.1 Consideration of Efficiencies

Compared with all other battery technologies, lithium-ion batteries offer very high efficiencies. In Fig. 11.6, achieved efficiencies of the developed battery module (Fig. 11.2) are shown. Due to internal losses, within the cells, and also at the cell interconnectors the efficiencies decrease with higher C-rates. But even at a C-rate of 1, which is not a common operation mode for storages in residential PV applications, the results show values clearly above 90%.

3.2 Battery Management Systems

Battery systems, for example, those using lithium-ion technology, need to be managed. Battery cells have to be monitored and controlled. Challenges in terms of safety, electrical isolation, and energy efficiency have to be considered [11]. Within this section the principles of battery management systems are explained and different concepts introduced.

There are several functions to be handled by a battery management system (BMS). Fig. 11.7 provides an overview of these functions. Every BMS needs a safety layer to prevent the batteries from being overcharged or deep discharged. Furthermore, the temperature of the cells has to be controlled. Therefore, monitoring of system temperature, load current, and cell voltages is required. Functions such as switches and coolers or fluid pumps have to be triggered by the BMS.

Advanced BMS possess sophisticated and precise state estimation algorithms for the state of charge and state of health (SOH) as well as end-of-life predictions and model-based thermal control. To secure high overall system efficiencies, optimization algorithms for smart cell balancing, load, and thermal management are necessary. Furthermore, the BMS needs a communication interface to internal and to external components like a power electronics or supervisory energy management device to secure safe and reliable system integration.

To handle these functions, there exist several types of BMSs with their specific advantages and disadvantages. One may classify the types into modular, central, and single-cell BMS approaches [11].

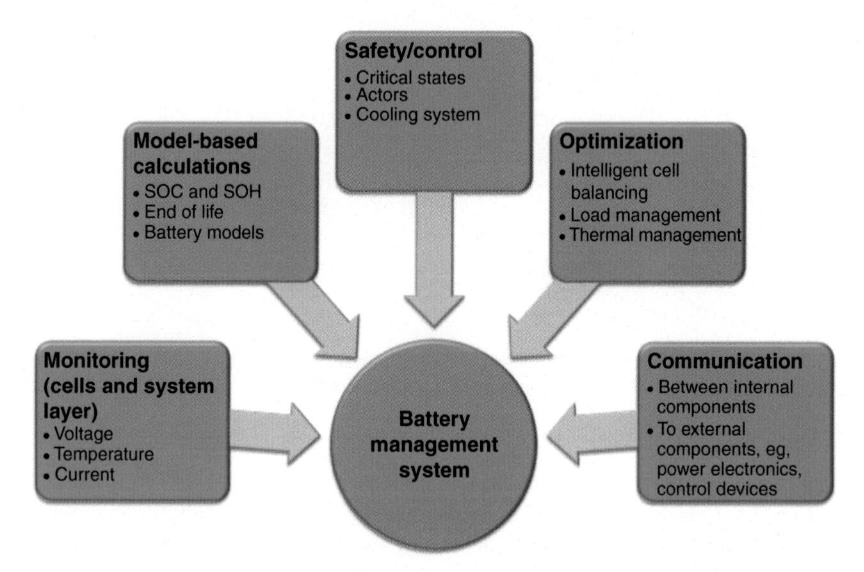

FIGURE 11.7 **Overview of the management functions of a battery management system.**

3.2.1 *Modular Concept*

In a modular approach the BMS contains a central control unit and module management systems (MMSs). The latter are responsible for the measurement and control of each lithium-ion battery module, which contains several battery cells, mostly connected in series to reach higher voltages. The number of cells connected in series strongly depends on the application [10]. Especially for storage applications with higher capacities a higher number of battery cells per module helps to reduce the costs of electronics. Typically, the MMS electronic system carries out the measurement and control on battery cell level and communicates with the central management system (CMS), which collects the measurement and control data of single modules. Furthermore, it controls switches and a cooling system and establishes communication with external components. A schematic representation of the modular approach is shown in Fig. 11.8.

The particular description of an MMS and a CMS is based on a system developed at Fraunhofer ISE. The concept is an example for a modular approach.

The main parts of the MMS are the battery front end unit and a controller unit. The front end unit is responsible for measuring the voltages of the battery cells and for controlling the temperature of the printed circuit board (PCB).

Furthermore, it provides a hardware interface for cell balancing, which is needed to balance the SOC of the cells within a battery module.

The controller unit consists of a low-power reduced instruction set computer (RISC) microcontroller. The microcontroller controls the battery front end unit by communicating via a bus system. In addition, there is a bus interface to send measured data and commands to a CMS or a host PC for further handling. By

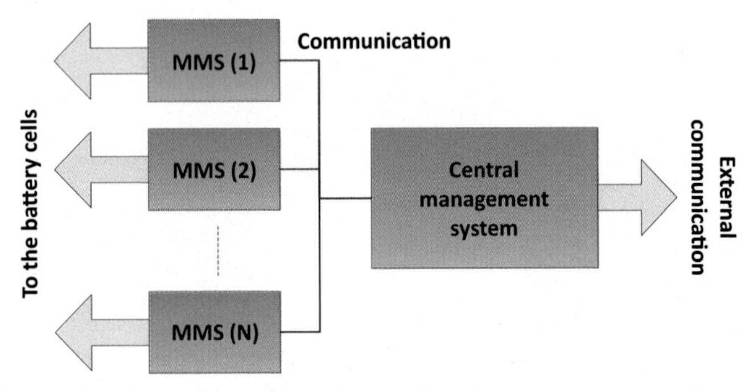

FIGURE 11.8 **Concept of a modular battery management system with a MMS and a CMS system with internal and external communication [11].** *(Source: SMA)*

means of an integrated analog-to-digital (AD) converter, temperature measurement at the cell level [12] is obtained.

Fig. 11.9 shows an photo of such an MMS. It is a prototype developed at Fraunhofer ISE that has been tested in both laboratory and in prototype applications.

The module management system is scalable up to 1000 V of battery voltage. This is insured by integration of a high isolation barrier [11].

Especially in electric vehicle applications, it is important to minimize the self-discharge rate of the battery to prevent deep discharge after extended times without operation. To insure a very low self-discharge rate the MMS system has low current consumption.

The module management control unit contains flexible software modules. Filter algorithms based on Kalman and particle filter approaches for SOC and SOH estimations at the cell level with high accuracy are implemented. Error

FIGURE 11.9 **Photo of a module management system unit with a passive resistive cell balancing, voltage and temperature control and a controller unit [11].**

FIGURE 11.10 **Photo of a central management system with an embedded system as central controller and several interfaces [11].**

management handling of events such as overcharging, deep discharging, or too high temperatures insures reliable operation of the batteries.

The central management unit retrieves all management and measurement data from the modules. Furthermore, it communicates with the modules via a bus system. For communication with external components a second bus interface is implemented. The central management unit controls the battery switches and the integrated cooling system. Signals are integrated for the control of external switches, cooling pumps, and other components. Air cooling, using fluid cooling or a combination of both, is possible.

Additionally, temperature, high-current, and high-voltage measurement are integrated to measure the parameters at the system level via a controller area network (CAN) bus. Important parameters for system safety are monitored. These comprise measurement of the insulation resistance for an electrically isolated system and redundant control of battery switches.

The central controller consists of an embedded system involving thermal management, a central error-handling system, and a charge and discharge management system. With thermal and electric battery models, precise central battery management is needed. Furthermore, with integrated battery models and filter algorithms, lifetime prediction of the battery system can be made available. Fig. 11.10 shows a photo of the central management unit.

3.2.2 Single Central Concept

The single central BMS approach (Fig. 11.11) is based on only one PCB, which controls all the battery cells of a storage system, and which can be switched in serial or parallel. The advantage over the modular system is the lower cost of the electronics. All functions regarding safety, control, and measurement are integrated. Furthermore, energy efficiency is higher because there is only one central controller. Battery systems based on such a

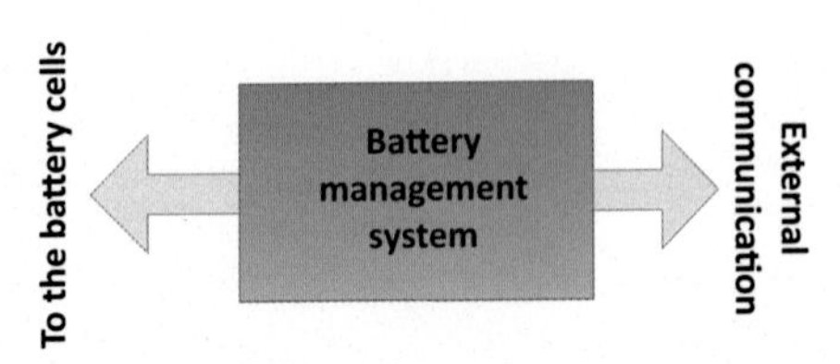

FIGURE 11.11 **Single central BMS concept of a battery management system with one PCB [11].**

BMS are not easily expandable, but this approach enables significant cost reductions in many applications. The decreased system complexity of this solution is often an advantage, too. A disadvantage is the decreased flexibility in comparison with the modular approach. In the modular approach another module is easily integrated on the bus system, which enables easy integration of modules in a spatially distributed system since only a bus cable must be connected and no cable for cell voltage measurement and cell balancing for each single cell are needed.

The battery front end unit and the controller unit are integrated on one PCB (see Section 3.2.1). Several front end units control and measure all cells in the battery pack. The controller unit could contain an advanced RISC machine (ARM) controller with low current consumption.

3.2.3 Single-Cell Concept

A third approach is placing a single-cell control unit (SCU) in each single battery cell. The SCU is a rather simple integrated circuit, measuring cell temperature and voltage. It is also possible to integrate cell balancing. Communication with a CMS is established through a bus interface. Normally, the management functions of the CMS are more complex than for the modular approach. SOC and SOH estimation plus error management have to be executed in this unit. The SCU can only carry out basic safety and measurement functions. Fig. 11.12 shows the architecture of such a system.

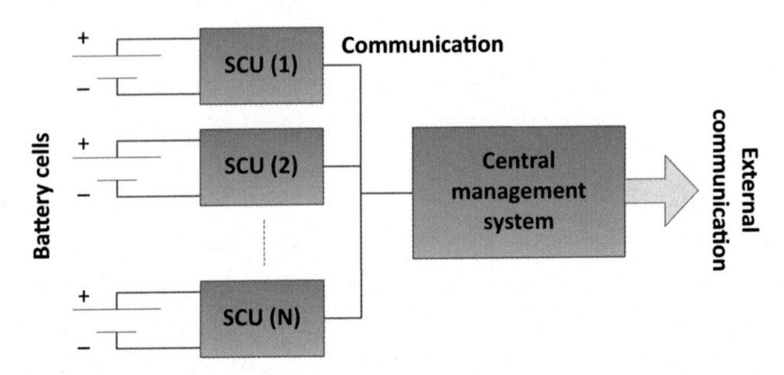

FIGURE 11.12 **Single-cell BMS concept of a BMS with several single-cell control units *(SCU)* and a CMS [11].**

3.2.4 State-of-Charge Estimation

Since accurate knowledge of a battery's SOC is unavoidable for proper usage of a battery system many different approaches have been investigated.

Today, in many sophisticated battery management systems it is state of the art to use Kalman filters to estimate actual SOC. But since the Kalman filter uses some particular assumptions (like Gaussian distributions), its correctness and applicability are limited.

A new approach is the so-called "particle filter" which is derived from the same family (Bayesian filters) as the Kalman filter [13]. By employing Monte Carlo sampling methods the particle filter offers the possibility to deal with any distribution.

As depicted in Fig. 11.13 a Markov chain is assumed for a particle filter:

- u_t, input which will change the system's state over time (it can be measured);
- x_t, state of the system at time t;
- z_t, output is in some correlation with the system's state enabling rough estimation (quantity can be measured).

So one assumes only the quantities of input u_{t-1} and output z_t and not the former values have an effect on the calculated probability of the current state. From Bayes theorem one gets the following equation:

$$P(x_t) = \eta^{-1}(P(z_t|x_t)\int P(x_t|x_{t-1}, u_{t-1})P(x_{t-1})dx_{t-1}$$

The particle filter algorithm (see Fig. 11.14) runs in three steps [14]:

1. State transition—the influence of input u_{t-1} on each sample (particle) s_t^k is calculated. By adding a different random value to the input value for each sample measurement error the uncertainty of this step is taken into account.
2. Weighting—samples are weighted according to the observed measurement z_t and a probability density function $P(z_t|x_t)$. Then the sum of all weights (w_t^k) is normalized to 1.

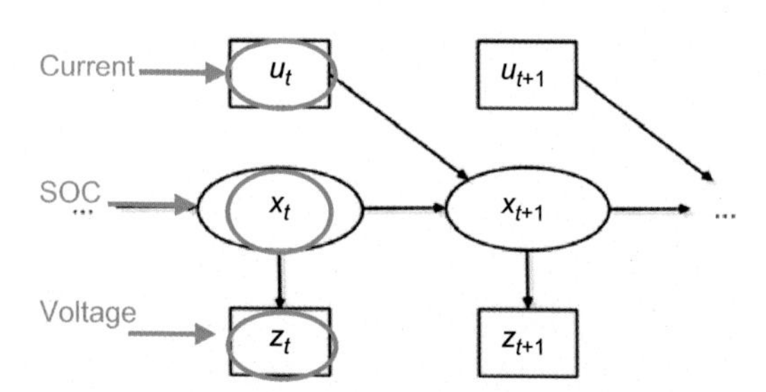

FIGURE 11.13 **Assumption of a Markov chain for SOC determination.**

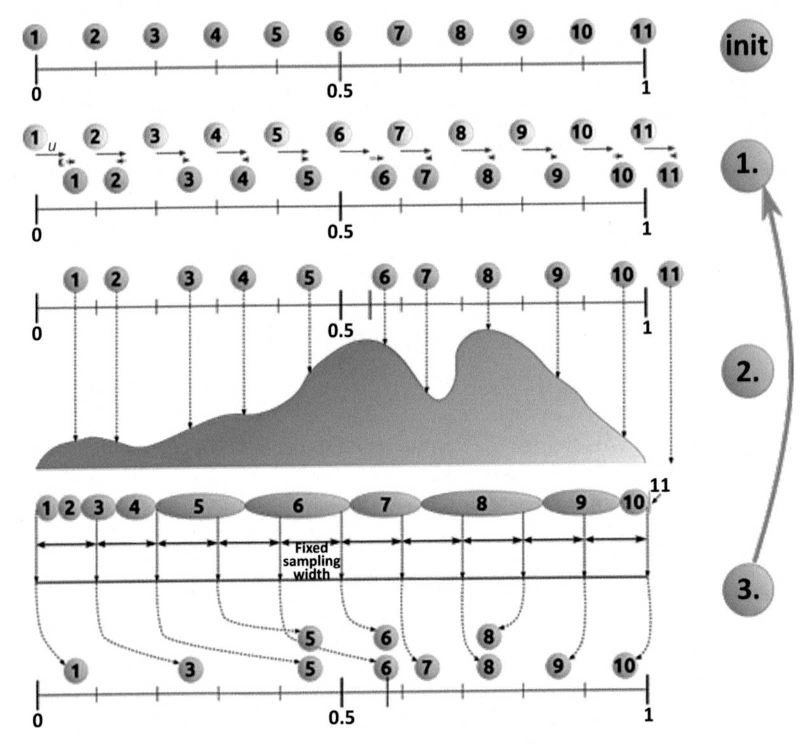

FIGURE 11.14 **Illustration of the particle filter algorithm.** For initialization *(init.)* all samples (or particles) are distributed uniformly over the possible value range of state x. In Step 1 of the algorithm the influence of input u on every sample is calculated. Noise ε, which is taken from a suitable probability distribution, is added to the value of u. Step 2 gives every sample a weight according to probability taken from the measurement value z and the measurement model. Step 3 resamples the weighted particles to gain an unweighted particle set. The low-variance resampling method is depicted [14].

3. Resampling—in this step the samples are resampled according to their weights. After that all samples have the same weight again. This approach uses the low-variance resampling method for low computational afford.

In Step 1 the so-called "process model" is used, which describes the influence of input u_{t-1} on state x_t. For SOC estimation the following model equation is used:

$$s^k_{\mathrm{SOC},t} = s^k_{\mathrm{SOC},t-1} + \frac{(I_{\mathrm{batt}} + \epsilon^k)\,\Delta t}{\mathrm{SOH} \cdot C_n}$$

where ε^k represents the random value which is added to the input value. It can be sampled from any distribution suitable for the application. For this application a Cauchy–Lorentz distribution is used.

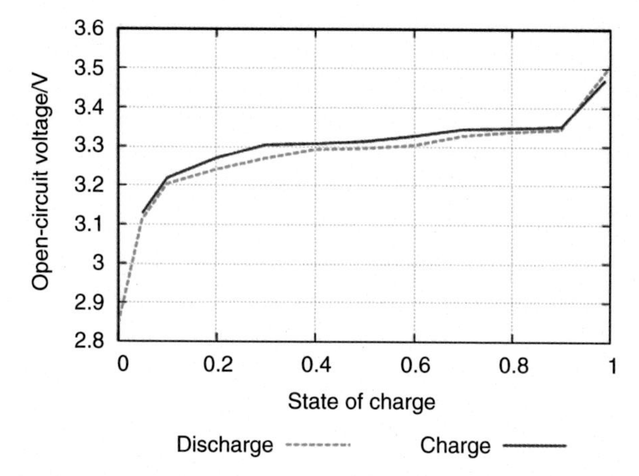

FIGURE 11.15 Open-circuit voltage versus state of charge curve of a lithium iron phosphate battery with a graphite anode. One can see the hysteresis between charging and discharging voltage and the very flat voltage in the medium state-of-charge range [13].

By adding this noise to that of the particles, particle diffusion is increased. This diffusion models—in case of SOC estimation—the increasing uncertainty of ampere-hour counting.

Step 2 decreases this diffusion by weighting particles according to their probability. Therefore, the so-called "measurement model" is used, which calculates estimated terminal voltage at the actual SOC of the particle. For phosphate-based lithium-ion batteries like lithium iron phosphate (LFP) cells the open-circuit voltage (OCV) vs. SOC curve is very flat (see Fig. 11.15) and shows a hysteresis between charging and discharging. For that, two voltages are calculated, one for charging and one for discharging. The following equations show the measurement model for an LFP battery:

$$V^k_{\text{discharge},t} = OCV_{\text{discharge}}\left(s^k_{\text{SOC},t}\right) + R_{\text{i}}\left(s^k_{\text{SOC},t}, T, I_{\text{batt}}\right) \cdot I_{\text{batt}}$$
$$V^k_{\text{charge},t} = OCV_{\text{charge}}\left(s^k_{\text{SOC},t}\right) + R_{\text{i}}\left(s^k_{\text{SOC},t}, T, I_{\text{batt}}\right) \cdot I_{\text{batt}}$$

These two voltages are compared with the measured terminal voltage. So two probabilities are determined with the following formula (using a Cauchy–Lorentz distribution):

$$w^k_{\text{SOC},t} = \frac{1}{\pi} \cdot \left(\frac{\gamma}{\gamma^2 + \left(V^k_{\text{discharge},t} - V_{\text{meas}}\right)^2} + \frac{\gamma}{\gamma^2 + \left(V^k_{\text{charge},t} - V_{\text{meas}}\right)^2} \right)$$

After that, in Step 3 these weighted particles are resampled to regain an unweighted set of particles. In this approach the low-variance resampling method is used as depicted in Fig. 11.15. Low-variance resampling puts all particles in a row, each particle has a length according to its weight. After that, this row is sampled as many times as the number of particles using a fixed sampling width. The sampled particles now represent an unweighted set of particles and the filter goes back to Step 1. The estimated value for the SOC is the mean value of all particle values.

3.2.5 State-of-Health Estimation

The estimation of SOH is also done using a particle filter [13]. Both filters run in parallel and share their results. As the particle filter for SOC estimation needs the SOH as a parameter, the filter for SOH estimation needs the SOC difference between two calculation steps as a parameter. Therefore, the approach is called "parallel particle filter" for SOC and SOH estimation.

SOH in this section is defined as the ratio between actual battery capacity and nominal battery capacity:

$$\mathrm{SOH} = \frac{C_{\mathrm{act}}}{C_n}$$

In general, the particle filter for SOH estimation works similar to that for SOC estimation. as described in the previous section.

Due to the very slow change in SOH, it is assumed in the process model that no input value u influences the state. Only the noise value ε is added to every particle:

$$s^k_{\mathrm{SOH},t} = s^k_{\mathrm{SOH},t-1} + \epsilon^k$$

The weighting step is done by using the SOC change of the SOC filter during the last step ($\Delta\mathrm{SOC}_{\mathrm{meas}}$) as the measurement value z and following equation as the measurement model:

$$\Delta\mathrm{SOC}^{\mathrm{k}}_t = \frac{Q_{\mathrm{step}}}{s^k_{\mathrm{SOH},t} \cdot C_n}$$

Variable Q_{step} represents integrated charge flowing into and out of the battery.

If using a Cauchy–Lorentz distribution the weight is calculated as follows:

$$w^k_{\mathrm{SOH}} = \frac{1}{\pi} \cdot \frac{\gamma}{\gamma^2 + \left(\Delta\mathrm{SOC}^{\mathrm{k}}_t - \Delta\mathrm{SOC}_{\mathrm{meas}}\right)^2}$$

Thereafter, the low-variance resampling step is performed as described earlier. The SOH estimated value is gained by taking the mean value of all particle values.

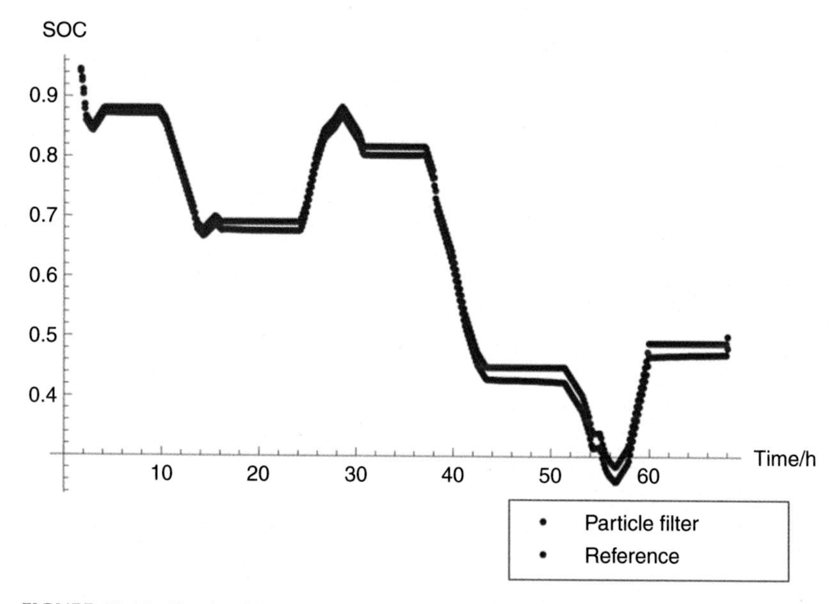

FIGURE 11.16 **Depicted is the state of charge during a PV current profile.** The battery used is a lithium iron phosphate battery with a graphite anode [14].

3.2.6 Validation

The dual-particle filter approach is validated for different types of lithium-ion batteries (LFP and NMC, both with graphite anode) and different types of current profiles including electric vehicle (EV) cycles and PV applications [14]. The EV profile is characterized by large currents, high dynamics, but also relatively long pauses with no current. The PV profile, on the other hand, shows lower currents, but there are virtually no phases without any current.

In Fig. 11.16 a validation sequence for LFP-graphite lithium-ion cells is shown by using a PV profile [14]. Estimation of LFP is much more complicated still, as a result of flat OCV and hysteresis; however, exact SOC is found to be fast and estimated reliably. SOH estimation is very accurate as well, as depicted in Fig. 11.17.

4 SYSTEM INTEGRATION

4.1 Configuration

Battery systems and DC power sources like PV generators can be coupled via power electronics on a DC bus bar or on the AC side. Exemplarily, an AC coupled system is introduced in the following (Fig. 11.18), which allows the integration of lithium-ion battery systems in PV systems by using a market-available battery inverter.

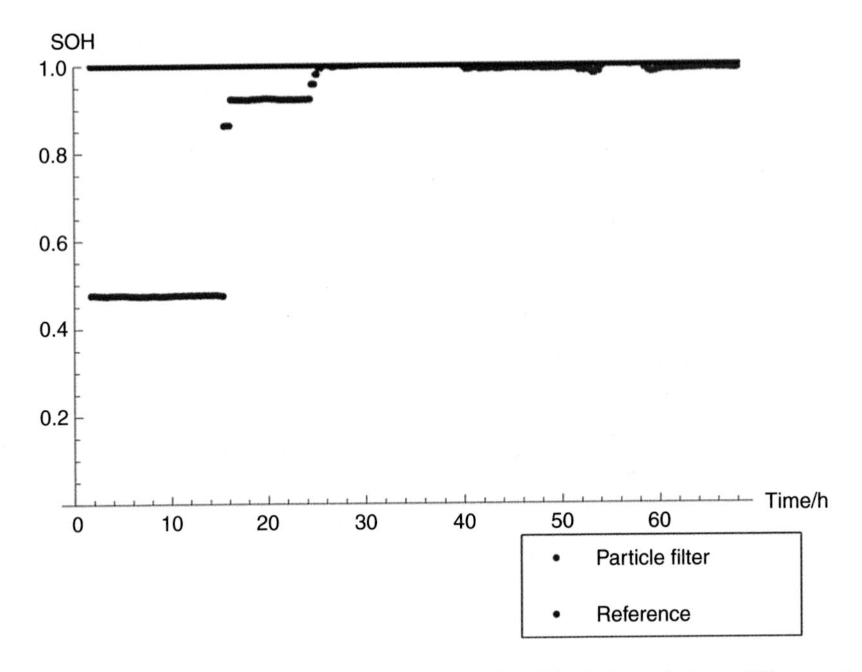

FIGURE 11.17 SOH of a lithium iron phosphate/graphite battery during a PV current profile. The reference value was determined with a standard charge–discharge regime [14].

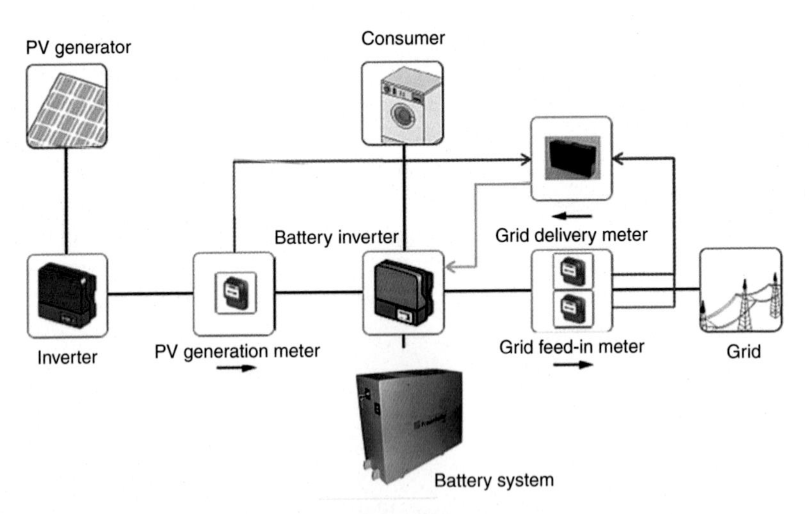

FIGURE 11.18 Integration of the developed lithium-ion battery system (Fig. 11.1) in a residential PV system using a market-available battery inverter [7].

In these AC coupled system configurations the PV generator and the battery system are connected to the AC grid via two separate inverters. The conventional PV system, consisting of PV modules and a PV inverter, is in principle not affected by integration of a battery. Therefore, installed PV systems can easily be complemented with storages later without any adaptation. Due to the modular concept, sizing of the battery system is almost independent of the size of PV system components like the PV inverter.

The disadvantages of this topology are limited cost reduction potential, as two full inverters are needed, as well as the limited voltage levels of market-available battery inverters for residential applications, which are in the range (24–48) V. Therefore, the inverters possess transformers and offer only relatively low efficiencies (~94%) at the nominal operating point and efficiencies much below this value across a wide range of the typical operating window.

4.2 Communication Infrastructure

Lithium-ion batteries are a very promising storage technology especially for decentralized grid-connected PV battery systems. Due to several reasons, for example, safety aspects, battery management is part of the lithium-ion battery system itself and is not integrated into the battery inverter or the charge controller as is usual for lead–acid and nickel-based batteries. This battery management system has to control the battery system itself and the connected power electronics. Furthermore, it has to exchange all relevant data with the supervisory energy management system. Field bus communication is necessary for both tasks (Fig. 11.19), but market-available products offer only proprietary solutions. Therefore, system integrators are not free to choose different system components for specific solutions. Furthermore, it is predefined which battery systems can be operated with which inverters or charge controllers avoiding the need for a huge adaptation effort for the communication system.

To enable higher degrees of freedom in system assembly, a standard for communication at the field bus level among battery systems, power electronics, and energy management systems is necessary. Such an approach is encapsulated currently by the so-called "EnergyBus" [17], which was initially developed for simplified connection of system components of light electric vehicles. This standard defines the communication protocol as well as the power connectors. The CANopen user profile CiA 454 "energy management systems" is used as the communication protocol [16]. This protocol specifies data exchange between single components such as storages, generators, loads, and energy management systems and enables implementation of optimized operating control strategies. Based on this field bus communication standard, components of different manufacturers can be assembled by system integrators. The power connectors are defined especially for light electric vehicles, whereas the communication protocol was extended for stationary applications like PV battery systems. New specification parts for generators and loads have been designed

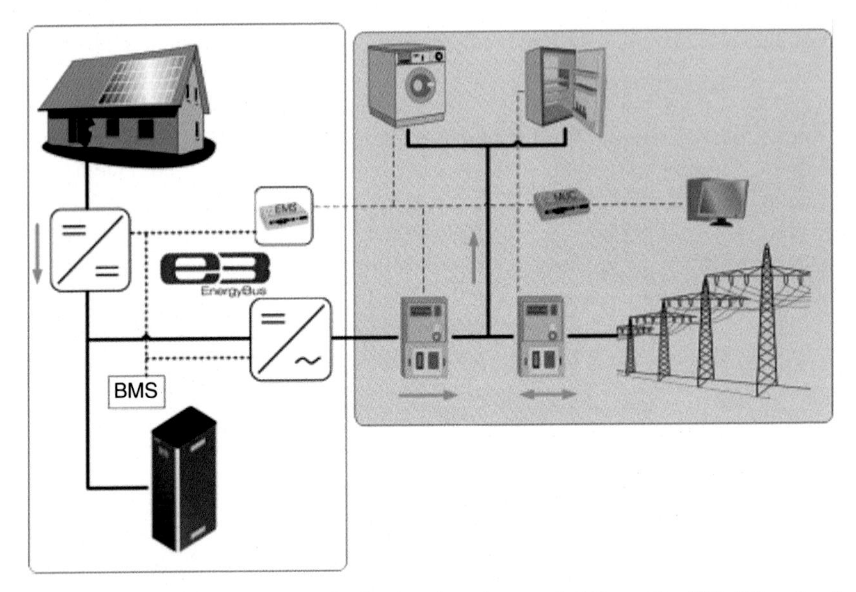

FIGURE 11.19 Field bus communication based on the so-called EnergyBus/CiA 454 protocol [16–18] for setting up a network of different generators, storages, power electronics, and energy management systems.

in such an abstract way that general manageability by a supervisory control system is enabled. For example, a PV generator and a CHP (combined heat and power) unit can be described by the same specification part. Furthermore, smart-metering components can be easily integrated.

5 CONCLUSIONS

In conclusion, in the relatively short time of the development and application of the lithium-ion battery, its main advantages such as high energy density, long lifecycle, fast and efficient charging, little maintenance, and little voltage sag have been undercut by the possibility of thermal runaway. This has resulted in much research and development resulting in elaborate battery management protective circuits and programs which have been the focus of this chapter.

REFERENCES

[1] Vetter M. Energy storage—renewable energy's key "blade" for grid integration. Canada Energy Storage Summit, Toronto; November 12, 2014.

[2] Vetter M. Decentralized PV battery storage systems—system design, integration and optimization. Intersolar Conference North America; 2014.

[3] www.3ds.com/products/catia/portfolio/dymola

[4] www.modelica.org

[5] Lux S, Dennenmoser M, Becker M, Lang N, Jung M, Vetter M. Entwicklung und thermische Modellierung eines innovativen Lithium-Ionen-Speichers für den stationären dezentralen Einsatz. Entwicklerforum 2013, Aschaffenburg; August 25, 2013.
[6] Fraunhofer Institute Project Urban Hybrid Energy Storage; 2014.
[7] Vetter M. Development of an optimized battery system for residential PV applications. Intersolar Conference North America; 2014.
[8] Goodenough JB. Journal of Power Sources 2007;174:996–1000.
[9] Wang J, Liu P, Hicks-Garner J, Sherman E, Soukiazian S, Verbrugge M, Tataria H, Musser J, Finamore P. J Power Sources 2011;196:3942–8.
[10] Stuart TA, F, Wei Zhou. Journal of Power Sources 2011;196:458–64.
[11] Jung M, Schwunk S. High end battery management systems for renewable energy and EV applications. Green 2013;3(1):19–26.
[12] Rao L, Newman J. J Electrochem Soc 1997;144:2697–704.
[13] Schwunk S, Ambruster N, Straub S, Kehl J, Vetter M. J Power Sources 2013;239:705–10.
[14] Armbruster N et al. Particle filter for state of charge and health estimation for both NMC and LFP based batteries. Advanced Automotive Battery Conference, Strasbourg; 24–28, 2013.
[15] http://dx.doi.org/10.1016/j.jpowsour.2012.10.058
[16] www.can-cia.org
[17] www.energybus.info
[18] Vetter M. Dezentrale netzgekoppelte PV-Batteriesysteme. Intersolar—PV Energy World, Munich; June 8, 2011.

Chapter 12

Vanadium Redox Flow Batteries

Christian Doetsch, Jens Burfeind
Fraunhofer Institute for Environmental, Safety, and Energy Technology UMSICHT, Oberhausen, Germany

1 INTRODUCTION AND HISTORIC DEVELOPMENT

The redox flow battery was first developed in 1971 by Ashimura and Miyake in Japan [1]. In 1973 the National Aeronautics and Space Administration (NASA) founded the Lewis Research Center at Cleveland, Ohio (USA) with the object of researching electrically rechargeable redox flow cells. The Exxon Company (USA), Giner Ind. (USA), and Gel Inc. (USA) were awarded contracts to develop a hybrid redox flow battery [2] and in the following 6 years research was done on different redox couples, membrane development, electrodes, etc. [3] Iron chloride ($FeCl_3$) and titanium chloride ($TiCl_2$) were proposed as electrolytes. In 1975 a patent (US Patent 3 996 064) was filed by Lawrence H. Thaller. The description of present-day systems is identical to that described in the original patent which involved a two-tank system and a cell with a separator and two graphite electrodes. With the present-day application of flow batteries to store sustainable electricity, they appear to have foreseen future problems: "Because of the energy crisis... and due to economic factors within the electric utility industry, there is a need for storing bulk quantities of electrical power [...] be produced intermittently [...] by devices such as wind-driven generators, solar cells or the like" (US Patent 3 996 064). Later Thaller [4] replaced titanium with chromium. This electrolyte couple is in the spotlight today with one plant being set up by Enervault (USA) (www.enervault.com) and discussions taking place about further developments. In 1978 an all-vanadium system was proposed for the first time. In this special case both electrolytes consist of the same metal, but at different stages of oxidation. Therefore, there is no problem with cross-contamination through the separating membrane. This technology was further developed in the 1980s by Maria Skyllas-Kazakos at the University of New South Wales (Australia). This vanadium-based redox flow battery is today the most developed and popular flow battery and its sales exceed those of other flow batteries. Also in the 1980s the Japanese company, Sumitomo, was very active in filing patents and developing new membranes and electrolytes. This activity

Storing Energy. http://dx.doi.org/10.1016/B978-0-12-803440-8.00012-9

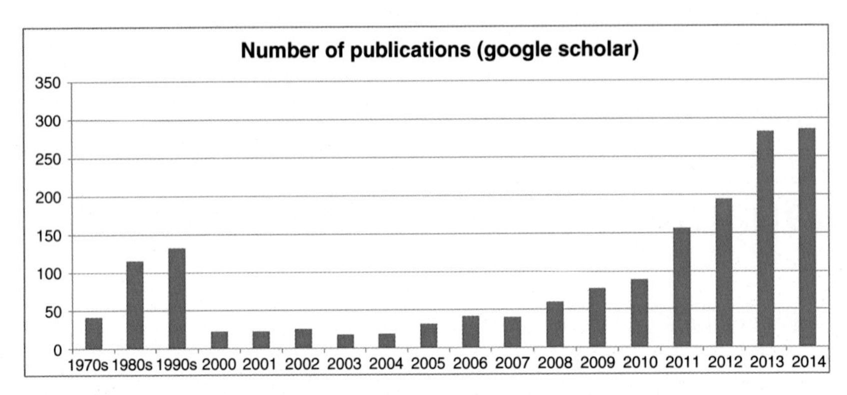

FIGURE 12.1 **Number of publications (research with Google Scholar; allintitle "redox flow" July 2015).**

stopped at the end of the 1990s and was restarted 5 years ago. The Canadian company, VRB Power (CA), was another very active company from 2000 to 2008. At the end of 2008 it filed for bankruptcy and was bought out by Prudent Energy VRB Systems. In Austria, Martha Schreiber developed electrochemical storage systems, with special emphasis on redox flow; she founded the company Cellstrom GmbH (Austria), which today, together with of the German company Gildemeister Energy Solutions, is now part of Deckel-Maho-Gildemeister (DMG) Mori Seiki AG of Japan.

The development of scientific interest in redox flow technology can be seen in Fig. 12.1.

The 1970s witnessed the start of a small number of publications each year (5–12). Since 2004 the number of publications has risen significantly and has almost doubled in number every 3 years. Currently, there are nearly 300 publications per year.

The current strong interest in flow batteries can also be seen in the overview of patents involving redox flow batteries (RFBs): over the past 5 years (2010–14) there have been roughly twice the number of patents filed over the previous 40 years (1970–2009) (Fig. 12.2).

The inventors responsible for these patents come mainly from Japan, China, USA, South Korea, Europe (especially Germany and Austria), Canada, Australia, Taiwan, and India (see Fig. 12.3).

Today, the companies working with RFBs include large companies such as Sumitomo (Japan), DMG Mori Seiki AG (former Gildemeister) (Germany/Japan), and Prudent Energy VRB Systems (USA and Canada); medium-sized enterprises such as UET-Uni Energy Technologies (USA) in cooperation with Dalian Rongke Power (China); and a few startup companies with new ideas such as Enervault (Fe/Cr) (USA) and Volterion (compact welded stacks) (Germany).

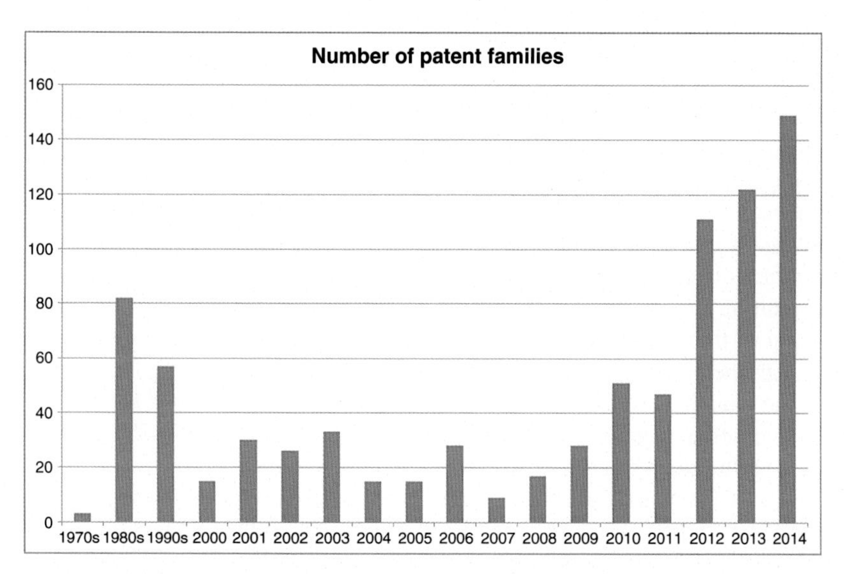

FIGURE 12.2 Development of filed patents regarding redox flow in the world [research with depatisnet; »redox flow« and (»batter*« or »cell« or »cell«) in claims, title, abstract; May 2015].

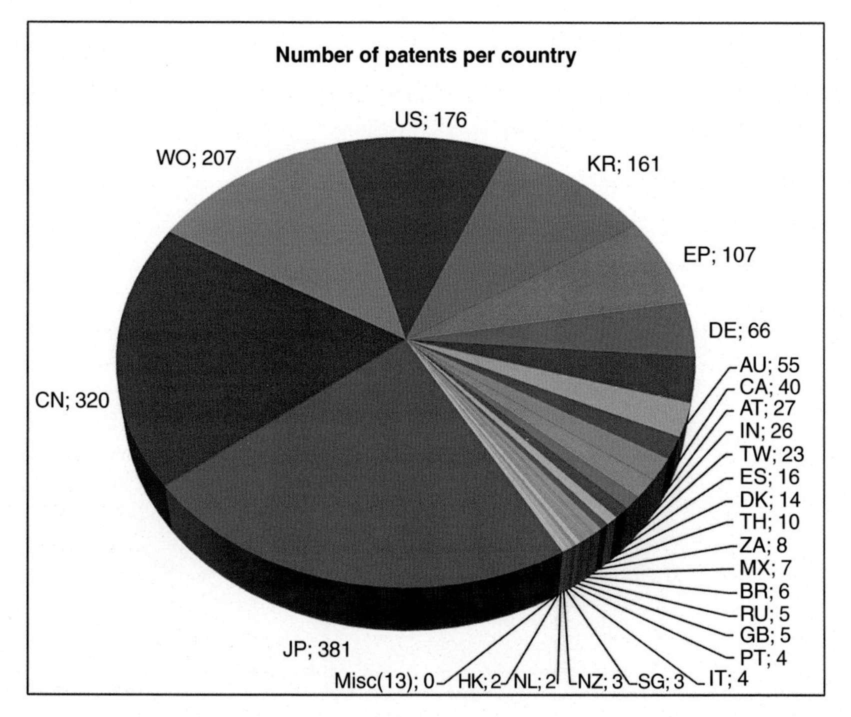

FIGURE 12.3 Filed patents regarding redox flow in different countries [research with depatis net; »redox flow« and (»batter*« or »cell« or »cell«) in claims, title, abstract; May 2015].

2 THE FUNCTION OF THE VRFB

The electrochemical redox flow cell consists of two half-cells which are separated by a separator which can be an anionic exchange membrane, a cationic exchange membrane, or a porous membrane. The liquid electrolyte stores electrical energy in the form of chemical ions which are soluble in liquid aqueous or nonaqueous electrolytes. The electrolytes of the negative half-cell (anolyte) and the positive half-cell (catholyte) are each circulated by a pump in separate circuits. Both electrolyte recirculation circuits are separated by a separator. The function of the separator is to prevent electrical short circuits, prevent cross-mixing of the electrolyte, and to insure the ion exchange across the separator balances the electrical charge of the anolyte and catholyte (Fig. 12.4).

The ions that are exchanged depend on the kind of redox flow battery; the most common types are cationic exchange membranes such as NAFION®. These perfluorinated and sulfonated membranes have been used for decades and are very stable against chemical attack and oxidative corrosion caused by high potentials. These membranes are mostly used in acid electrolyte systems for vanadium redox flow cells or iron chromium cells. Charge balancing is easily done by the transport of hydrated protons (hydronium-ion) through the membrane. These membranes are available worldwide through several commercial suppliers and because they are fluorinated membranes the price is quite high.

A second type of membrane is the anionic exchange membrane. In this case the counter ion of the active species is responsible for the charge balance. These anionic-type membranes are in general cheaper, but chemical stability has to be carefully checked.

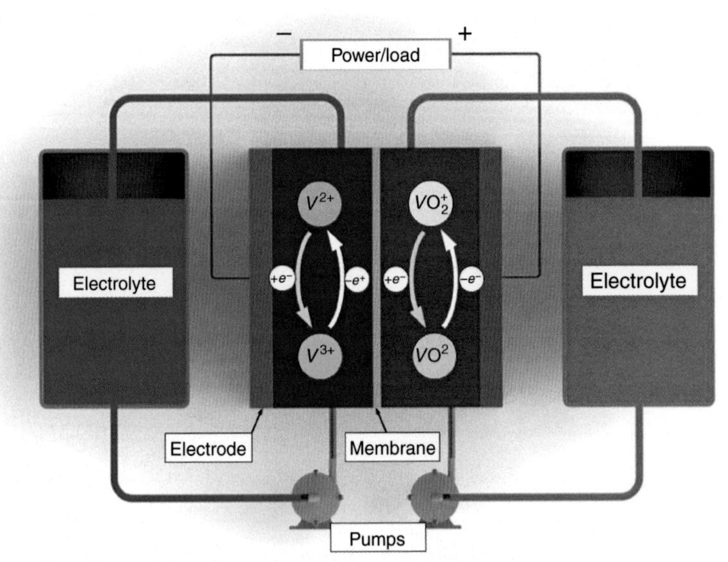

FIGURE 12.4 **Working principle of vanadium redox flow batteries.**

A third class of membranes is the so-called micro- or nanofiltration membranes. The working principle of these membranes is quite different because the function is based on ion exclusion. Small ions such as hydronium (H_3O^+) or sulfate (SO_4^{2-}) ions can freely cross the membrane for charge balancing, while much larger ions such as vanadyl ions are too bulky to cross the membrane so cross-mixing of the electrolyte is prevented.

A number of variations and modifications to such membranes have been made during the past few years. One notable development involves a class of nonfluorinated membranes known as SPEEK membranes—sulfonated poly(tetramethyldiphenyl)etheretherketone—which show promising chemical stability and proton conductivity.

Some attempts have been made to implement inorganic fillers such as SiO_2 and ZrO_2 into the membrane, with the purpose of preventing vanadium ions from crossover by inducing a charge in the membrane which excludes positive charge vanadium ions. The advantage of ionic conducting membranes is that they are effective at preventing the crossover of electrolytes, and hence make for an effective and efficient cell.

Microporous ion exclusion membranes are an order of magnitude cheaper than cationic ion conducting membranes, but because of their relatively poor efficiency they have not been produced commercially.

The membrane separates only the anolyte and catholyte and allows for charge balance. The chemical oxidation and reduction that result from electron transfer take place at the electrode. In most cases the reaction takes place at a graphitic carbon felt. In the VRFB no extra catalyst is necessary, but activation of the graphite felt does help to accelerate the reaction. This is done by adding carboxcylic groups to the graphite surface using either heat treatment or chemical and electrochemical oxidation. Graphite is the material of choice because of its chemical resistance and low cost. The graphite felt must have a high surface area and good electrical conductivity. High surface area is necessary to provide enough reaction sites. Due to its operation in flowthrough mode the graphite felt must have a very open structure (95% void volume) with a thickness of (3–5) mm. The open structure of the felt is necessary to achieve a low-pressure drop between the electrodes. On the other hand, the open structure leads to a relatively high inner resistance of the overall stack in comparison with cells which are constructed in flowby mode as seen in fuel cells and electrolyzers.

As shown in Fig. 12.5 (left and right), respectively, compound-based bipolar plates separate the anolyte and the catholyte from each other. This arrangement gives the opportunity for electrical connection in series. The bipolar plates are often made of a compound-based material, which could be processed by hot pressing or injection molding. Hot pressing is often done with duroplastic resins; injection molding uses thermoplastic material such as polyethylene and polypropylene which are also low-cost materials. To achieve sufficient electrical conductivity a very high amount of graphite and carbon black up to 85%

FIGURE 12.5 **Left, schematic view of a single redox flow cell; right, schematic view of a redox flow stack 60 cell.**

by mass is necessary. This, however, limits the ability to process and limits mechanical stability. In a flowthrough design no flow field is necessary.

The contact resistance between graphite felt and compound-based bipolar plate dominates the overall inner resistance of the stack; as a result much work has been done to lower contact resistance by gluing the graphite felt to the bipolar plates with conducting glue.

The role of the gaskets in a stack is often underestimated and several gaskets are usually needed to properly seal the stack. Gaskets on each side of the separator are used to insure a good seal toward the outside; furthermore, additional gaskets are needed to seal the bipolar half-plates and internal manifolds.

The thickness of the graphite felt, frames, and gaskets must be properly adjusted to insure good compression of the felt and sufficient strength given to the gaskets.

The gaskets are made from elastomeric materials such as ethylene propylene diene monomer (EPDM) rubber or fluoroelastomeric materials.

Single cells are stacked together to achieve higher voltages. The stack itself is compressed by strong end-plates and tension rods.

The vanadium electrolyte consists of vanadium salts which are dissolved in aqueous sulfuric acid. The liquid electrolyte corresponds to the active mass in a conventional battery. The amount of liquid electrolyte which is stored in tanks determines the capacity of the RFB. The big advantage of RFBs is that power and capacity can be scaled independently.

To operate an RFB additional pumps, piping, valves, and storage tanks are necessary. All such equipment must be able to withstand the harsh conditions of sulfuric acid and strong oxidizing power of vanadium ions in the valence state of 5.

To insure an efficient system, each vanadium redox flow system has a simple battery management program which controls the flow rate of pumps with respect to load requirements and state of charge.

The nominal charge voltage of each single cell is usually limited to 1.6 V to avoid the potential at which water is decomposed into oxygen and hydrogen. This is known as the oxygen evolution reaction (OER) which causes carbon corrosion that rapidly destroys the graphite electrodes [5–12]. The end of discharge voltage is kept to (0.9–1.0) V to reach a reasonable efficiency of the cell. It is worth mentioning that an electrochemical cell could be discharged down to 0 V without destroying the cell.

3 ELECTROLYTES OF VRFB

A vanadium-based electrolyte is widely used in flow batteries. This is due to the simplicity and stability of the electrolyte system in the aqueous phase. In an aqueous solution, four different but stable valence states of vanadium exists (V^{2+}, V^{3+}, V^{4+}, and V^{5+}). In anolyte vanadium (+2 and +3) ions exist as V^{2+}, V^{3}, while the +4 and +5 valence states of vanadium exist only as oxo-complexes (VO^{2+}, VO_2^+). By changing the valence states of vanadium species, energy could be stored electrochemically. These basic redox reactions are:

$$V^{2+} \rightleftharpoons V^{3+} + e^- \qquad E^{\ominus} = -0.255\,V \tag{12.1}$$

and

$$VO^{2+} + H_2O \rightleftharpoons VO_2^+ + e^- + 2H^+ \qquad E^{\ominus} = +1.004\,V \tag{12.2}$$

The oxidation of V^{2+} releases one electron and V^{3+} is formed. This creates a standard potential of 0.255 V. The oxidation of V^{4+} to V^{5+} by simultaneous splitting of a water molecule releases a proton and one oxygen atom which form the oxo complex. This delivers a standard potential of +1.004 V [13]. The overall standard potential for the reaction is 1.259 V.

In an aqueous electrolyte the vanadium salts in all the four different valence states—V^{+2}, V^{+3}, V^{+4}, and V^{+5}—must be soluble in concentrations which should be as high as possible. The more vanadium salts held in a stable solution without precipitation the higher the volumetric energy density of the electrolyte. The vanadium salt in valence state 5 has the lowest solubility. The following equation describes the reaction equilibrium between solid vanadium pentoxide and vanadium +5 in solution [14]:

$$\frac{1}{2}V_2O_5(s) + H^+ \rightleftharpoons VO_2^+ + \frac{1}{2}H_2O \tag{12.3}$$

The higher the proton concentration (acid concentration) the more the equilibrium is shifted to the right side (principle of le Chatellier) and the more the V^{+5} vanadium (in the form of VO_2^+) can be kept in the solution. Furthermore, the high proton concentration of the electrolyte results in high electrolyte conductivity which in turn leads to good cell performance. The sulfuric acid is usually at a concentration of between 2 mol L^{-1} and 6 mol L^{-1} [15].

Very recently mixed electrolytes of sulfuric and hydrochloric acids have been used as electrolytes. The addition of hydrochloric acid and therefore chlorine ions allows for very high vanadium concentrations in the solution at even higher temperatures. The higher stability of the electrolyte is caused by the formation of a chloro–oxo complex, and this stable complex prevents the condensation reaction and precipitation of vanadium pentoxide [16]. This mixed electrolyte has the advantage of higher energy density and temperature stability but, on the other hand, can result in possible release of poisonous chlorine gas at the positive electrode during the charge process. Furthermore, the presence of chlorine could compromise material stability in the cell.

4 VRFB VERSUS OTHER BATTERY TYPES

VRFBs like all other flow batteries are in competition with batteries such as lead–acid batteries. This competition is driven by technoeconomic needs for different applications. These storage device needs could be for:

- an uninterruptable power supply (UPS) for which total efficiency and cycle lifetime are not crucial, but capital expenditure (CAPEX), lifetime, and response time are the most important criteria
- home applications coupled with photovoltaics for which efficiency and cycle time are very important issues
- grid operators for which CAPEX, cycle lifetime, and efficiency are important criteria.

In Table 12.1 a list of the main specifications of major electrochemical storage systems are shown.

So far, we have discussed only a few applications and their specifications. To insure the best battery type for a particular application, all specifications must be considered. In the following section the VRFB is compared with other batteries for particular applications.

From an energy density (gravimetrical and volumetrical) point of view, the VRFB is low compared with zinc–air and lead–acid batteries. As a result the VRFB is more suitable for stationary applications. Mobile applications of VRFB could only be possible in niche areas such as on ferries.

Vanadium flow batteries have the highest cycle lifetime of all presently available batteries including lithium-ion batteries.

One big advantage of VRFBs is that they have a long life, because the liquid electrolyte does not degenerate to any great extent and can be used for decades without replacement. Electrodes made of graphite felt are also very stable, and furthermore membrane failure occurrences are extremely rare.

Another advantage of VRFBs is that self-discharging is extremely low due to the fact that self-discharging could only occur in the reaction chamber (cell) and not in separated storage tanks.

TABLE 12.1 Main Specs of VRFB Compared with Other Battery Types [17]

Battery type	Cycle lifetime (cycles)	Energy efficiency/(%)	CAPEX/ [€ (kW h)$^{-1}$]	Status
VRFB	20 000	70–85	500–650	Under development; near to market launch
Lead–acid	300–2 000	75–90	300–600	Commercial
NiCd	1 000	60–65	–	Commercial
NiMh	1 400	70	–	Commercial
Li-Ion	500–15 000	90–95	1 000–1 500	Commercial (consumer size), demonstration (huge, stationary)
NaS (high temperature)	3 000–7 000	70–85	130–230	Commercial
Zn–air	500	60–70	250–300	Under development

The energy efficiency of VRFBs is high and its mean value is better than that of Zn–air and NiCd batteries and is in the same range as that of NaS batteries, but less than that of Li-Ion batteries.

Two unique characteristics of flow batteries are: capacity is dependent on the size of storage tanks; and power is dependent on stack size. As a result any existing flow battery system can be improved and extended by adding additional electrolyte tanks.

These characteristics lead to flow batteries being used for stationary applications (low energy density) with high cycling rates (up to 365 full cycles per year) with a long-lasting lifetime and the capacity for long storage times. In short, flow batteries have high storage capacities in relation to power.

5 APPLICATION OF VRFB

Different researchers have proposed a large number of different possible applications for VRFBs [18]. In Table 12.2 there are nine important properties, with the advantages/disadvantages of RFBs compared with other types of batteries [19].

From Table 12.2 it can be seen that RFBs are promising in all fast and bulk applications (1–6), but less useful in applications with fewer cycles (7—power system startup) and lower capacity (8 and 9), where supercapacitors are very promising. There is however serious competition with other batteries such as lead–acid, NaS and Li-Ion.

TABLE 12.2 Assessment of Redox Flow and Batteries and Other Storage Devices Matched with Different Battery Properties

Application	Redox flow	Lead–acid	Sodium–sulfur	Lithium–ion	Supercapacitor
(1) Time shift	+	+	+	+	–
(2) Renewable integration	+	+	+	+	–
(3) Network investment deferral	+	±	+	+	–
(4) Primary control power	+	+	+	+	–
(5) Secondary control power	+	+	+	+	–
(6) Tertiary control power	+	+	+	+	–
(7) Power system startup	±	+	+	+	–
(8) Voltage support	±	+	+	+	+
(9) Power quality	±	±	±	±	+

5.1 Applications

The most promising applications for redox flow batteries are:

1. Time shift applications
   a. Economics-driven systems which charge the storage plant with inexpensive electric energy purchased from the grid during low-price periods and discharge the electricity back to the grid during periods of high price.
   b. Technology-driven systems which charge the storage plant with surplus energy from the grid during low demand periods and discharge the electricity back to the grid (or island grid) during periods of high demand.
2. Renewable integration systems which assist in wind- and solar-generation integration by reducing output volatility and variability, reducing congestion problems, providing backup for unexpected generation shortfalls, and reducing minimum load violation.
3. Network investment deferral systems which postpone or avoid the need to upgrade the transmission and/or distribution infrastructure.
4. Primary, (5) Secondary, and (6) Tertiary control power: selling positive (discharging) and negative (charging) control power to grid operation in a timeframe of less than 30 s (primary), less than 5 min (secondary), or less than 30 min (tertiary control power).

5.2 Current Large-Scale Applications

In Japan the Tomamae wind farm on Hokkaido Island is linked to a 4 MW VRFB storage system which is used to smooth out wind-generated energy peaks and valleys. In addition to this renewable integration application, there is a showcase at Sumitomo's head office in Osaka, which provides 3 MW peak shaving.

In China the Zhangbei National Wind and Solar Energy Storage and Transmission Demonstration Project is attached to a 2 MW VRFB provided by Prudent Energy.

In the United States, Prudent Energy is also responsible for the current largest VRFB storage system in the United States (0.6 MW) which is at Gills Onions, California. It is used to store energy produced from the methane generated from a biowaste plant [20].

In Germany a 2 MW RFB coupled to a wind turbine is being developed by Fraunhofer at Pfinztal [21]. Fraunhofer [22] is also developing a 0.5 m^2 RFB stack (planned for 25 kW) (Fig. 12.6).

These developments have led to discussions on the future of flow battery systems in terms of "numbering up" versus "scale up". On the one hand, mass production (numbering up) will rapidly reduce the cost of battery modules while, on the other hand, scale up of stack size will also lead to lower cost per kilowatt and less installation demand. Currently, most stacks are at the one-digit kilowatt level and high capacities are realized by linking hundreds of stacks.

FIGURE 12.6 **Redox flow battery with 0.5 m^2 cell area at Fraunhofer UMSICHT [23].**

	Country	Project name	Rated power /kW	Duration at rated power/ (HH:MM)	Energy/ MW h	City
A	Germany	Bosch Braderup ES Facility	325	3:5.00	1.0	Braderup
B	Germany	SmartRegion Pellworm	200	8:0.00	1.6	Island Pellworm
C	Germany	DMG Gildemeister CellCube Industrial Smart Grid	260	2:30.00	0.7	Bielefeld
D	Denmark	RISO Syslab Redox Flow Battery	15	8:0.00	0.1	Kongens Lyngby
E	Italy	Enel Livorno Test Facility	10	10:0.00	0.1	Livorno
F	Netherlands	Fotonenboer 't Spieker Dairy Farm	10	8:0.00	0.1	Vierakker
G	Portugal	PVCROPS Evora	5	12:0.00	0.1	Evora
H	Spain	CENER VRB	50	2:0.00	0.1	Sarriguren
I	United States	Prudent Energy VRB-ESS® - Gills Onions, California	600	6:0.00	3.6	Oxnard
J	United States	NREL American Vanadium CellCube Test Site	20	4:0.00	0.1	Golden
K	United States	1 MW Avista UET Flow Battery	1000	3:12.00	3.2	Pullman
L	China	Zhangbei National Wind and Solar Energy Storage	2000	4:0.00	8.0	Zhangbei
M	China	Prudent Energy Corporation/CEPRI	500	2:0.00	1.0	Zhangbei
N	China	Gold Wind Smart Micro-grid	200	4:0.00	0.8	Beijing
O	China	GuoDian LongYuan Wind Farm	5000	2:0.00	10.0	Shenyang
P	China	Snake Island Independent Power Supply System	10	20:0.00	0.2	Dalian
Q	China	Dalian EV Charging Station	60	10:0.00	0.6	Dalian
R	China	RKP R&D Center Building	60	5:0.00	0.3	Dalian
S	India	Sun-carrier Omega Net Zero Building in Bhopal	45	6:40.00	0.3	Bhopal
T	Indonesia	Sumba Island Microgrid Project	400	1:15.00	0.5	Sumba
U	Japan	Sumitomo Densetsu Office	3000	0:16.00	0.8	Osaka
V	Japan	Yokohama Works	1000	5:0.00	5.0	Yokohama
W	Japan	Tomamae Wind Farm	4000	1:30.00	6.0	Tomamae
X	Korea, South	Samyoung	50	2:0.00	0.1	Gongju
Y	Korea, South	KIER/Juju Island	100	2:0.00	0.2	Juju Island

FIGURE 12.7 **Redox flow battery projects in operation worldwide.** Here HH and MM refer to hours and minutes, respectively. *(Data from Sandia [24].)*

With increasing market share the size of each stack will increase as demand for larger units grows.

Most present applications involve large-scale systems because VRFBs are getting cheaper as a result of upscaling than other systems. Moreover, they have in most cases a high capacity so that the time for charging or discharging takes a few hours (4–10 h). For VRFBs this power-to-capacity ratio is therefore between 1:4 and 1:10. A listing of current worldwide installations is given schematically in Fig. 12.7.

Despite this there are developments aimed at using VRFBs for small-scale applications. In Germany, for example, where the home solar market is becoming increasingly important, small batteries rated at (1–2) kW and (5–10) kW h are being developed for home applications by VOLTERION, for example (Fig. 12.8).

FIGURE 12.8 **Stack for home applications designed by VOLTERION [25].**

6 RECYCLING, ENVIRONMENT, SAFETY, AND AVAILABILITY

One of the most important advantages of RFBs is that the electrolyte could be regenerated while operating and recycled after the lifetime of storage systems. While the stack consists mostly of uncritical material like graphite and plastic, which does not need recycling, the electrolyte (anolyte and catholyte) consists of vanadium and sulfuric acid and does need recycling. Vanadium is a high-priced material that can be almost 100% reclaimed, as can sulfuric acid [26]; from the economic point of view vanadium is the important component. Reclaimed vanadium can be used to produce new electrolyte for RFBs or for other purposes (steel industry).

From the environmental protection point of view, only VRFB electrolyte has to taken into account. This is because sulfuric acid is corrosive and vanadium is a heavy metal. As a result, double-wall storage vessels/catch basins and splash guards have to be provided for the whole system. In this respect the electrolytes of VRFBs can be compared with the electrolytes of lead–acid batteries.

Increasing vulnerability and / or supply risk

VI	Highest criticality	Ge, Re, Sb
V	High criticality	Ag, Bi, Cr, Ga, In, Nb, Pd, Sn, W
IV	Medium criticality	Be, Co, Cu, Li, Mo, Ni, Pt, V, Zn
III / II	Low criticality	Al, Mg, Si, Ti
I	Very low criticality	Fe, Mn, Pb, Ta

FIGURE 12.9 **Criticality and vulnerability of some metals in Germany.** *(Data from Ref. [29].)*

From a safety point of view, VRFBs are safer than many other types of batteries and there is almost no risk of fire because of the larger amount of water present in the system. Furthermore, in case of a short circuit [27] or mixing of anolyte and catholyte (comparable with the "nail test" for lithium-ion batteries) there is only a minor exothermic reaction with less than a 1 °C temperature rise.

In light of being a crucial component the availability of vanadium has often been discussed. Vanadium is an important byproduct of a number of mining operations and is used almost exclusively in ferrous and nonferrous alloys. Vanadium consumption in the iron and steel industry represents about 85% of the vanadium-bearing products produced worldwide [28]. The global supply of vanadium originates from primary sources such as ore feedstock, concentrates, metallurgical slags, and petroleum residues. The main supplier countries are South Africa, China, Russia, but supplies are also exported from Canada, USA, Argentina, etc. From the German/European point of view, vanadium has been assessed as a medium critical metal by Erdmann et al. (2011) [29] (Fig. 12.9).

7 OTHER FLOW BATTERIES

During the past 40 years nearly every chemically possible electrolyte combination has been evaluated as a suitable electrolyte for flow batteries. Due to limitations in chemical stability, energy density, poisoning, or radioactivity only a few electrolyte systems can be considered for practical flow battery application. These applications include hybrid flow batteries; in these batteries the anode is in a fully charged state and is usually a solid metal which dissolves during discharge to form the corresponding salt. Frequently used anode materials are zinc, iron, and possibly copper. In the following section the most important variations are briefly described, and in Table 12.3 the most investigated flow batteries are listed.

7.1 Iron–Chromium Flow Battery

One of the first flow battery electrolyte chemistries studied was the iron–chromium flow battery (ICB). It has been extensively studied by NASA (USA)

TABLE 12.3 Most Investigated Flow Batteries

System type/ active material	Cell voltage/V	Chemistry	Electrolyte
Redox flow		Anode/cathode	Anode/cathode
Vanadium VRB	1.4	V^{2+}/VO_2^+	H_2SO_4/H_2SO_4
Vanadium–bromine	1.3	$V^{2+}/1/2Br_2$	VCl_3/NaBr (HCl)
Polysulfide–bromine PSB	1.5	$2S_2^{2-}/Br_2$	NaS_2/NaBr (NaOH)
Iron–chromium	1.2	Fe^{2+}/Cr^{3+}	HCl/HCl
Hydrogen–bromine	1.1	H_2/Br_2	NaBr (NaOH)
Hybrid flow		Anode/cathode	Anode/cathode
Zinc–bromine	1.8	Zn/Br_2	$Zn/ZnBr_2$ (NaOH)
Zinc–cerium	2.4	$Zn/2Ce^{4+}$	CH_3SO_3H/CH_3SO_3H

and Mitsui (Japan). The iron–chromium battery is a real RFB with energy stored in Fe^{2+}/Fe^{3+} and Cr^{2+}/Cr^{3+} couples which are dissolved in hydrochloric acid. During discharge Fe^{2+} is oxidized to Fe^{3+} and simultaneously Cr^{3+} is reduced to Cr^{2+}. To keep the overall charge in balance a proton is exchanged through the separator which separates the anolyte and catholyte.

In the original iron–chromium system, cross-mixing of the electrolyte was a serious problem. Over time the iron and chromium ions diffuse through the membrane, so an irreversible capacity loss occurs. To avoid this cross-mixing effect, expensive ion exchange membranes were used.

Modern iron–chromium batteries work with a mixed electrolyte which uses iron and chromium on both sides. This allows the use of inexpensive porous separators. The optimal working temperature of the iron–chromium flow battery is (40–60) °C, which is quite high for a battery and thus makes this battery suitable for hot climates. The electrolyte is cheap and nonflammable. One disadvantage is the possibility of hydrogen evolution, which causes a loss in efficiency.

7.2 Polysulfide Bromine Flow Battery

The polysulfide bromine (PSB) redox flow battery is a well-investigated battery type. The great advantages of this type of battery is the very cheap and abundant electrolyte and the high voltage of 1.5 V. The electrolyte is an alkaline solution of sodium polysulfide and sodium bromide as anolyte and catholyte, respectively. During charging and discharging, sodium ions exchange through an ionic exchange membrane to keep the charge in balance. The efficiency of the battery is about 75%. As in every bromine-based electrolyte the bromine must be dissolved in the electrolyte with the aid of a complexing agent. The reactions are:

$$\text{Positive electrode}: NaBr_3 + 2Na^+ + 2e^- \rightarrow 3NaBr\,(\text{discharge})$$
$$\text{Negative electrode}: 2Na_2S_2 \rightarrow Na_2S_4 + 2Na^+ + 2e^-\,(\text{discharge})$$

The active species are highly soluble in aqueous electrolyte and therefore the electrolyte has a relative high energy density at low cost.

In 2002 Regenesys built a 15 MW, 120 MW h PSB flow battery system at Little Barford in the United Kingdom, but the project was never fully commissioned. The business was owned by RWE Power and it left the project before final commissioning.

7.3 All-Organic Redox Flow Battery

Very recently a new type of flow battery has been under development which involves organic molecules that are soluble in aqueous phase and could easily be oxidized and reduced.

A benefit of these organic flow batteries is that the electrolyte can be very cheap and not based on limited resources like vanadium. A promising candidate is the sulfonated anthraquinone redox couple. In 2013 researchers suggested the use of 9,10-anthraquinone-2,7-disulfonic acid (AQDS), a quinone, as a organic redox molecule in metal-free flow batteries [30]. AQDS easily undergoes rapid and reversible two-electron two-proton reduction at a carbon electrode in sulfuric acid. Each of the carbon-based molecules holds two functional groups which can be oxidized and reduced. This is a promising research area as it has the potential to offer a low-cost flow battery electrolyte. By modifying the chemical structure of the basic anthraquinone molecule, solubility in the aqueous phase could be increased. Moreover, the potential could be shifted even higher.

7.4 Hybrid Flow Batteries

7.4.1 Zinc–Bromine Flow Battery

The zinc–bromine flow battery is a so-called hybrid flow battery because only the catholyte is a liquid and the anode is plated zinc. The zinc–bromine flow battery was developed by Exxon in the early 1970s. The zinc is plated during the charge process. The electrochemical cell is also constructed as a stack. Storage capacity is determined by the size and thickness of the plated zinc plate and of the catholyte storage reservoir, and as a result the power rating and capacity correspond to each other.

The catholyte contains an organic complexing agent to keep the generated bromine in solution during the charging process. A microporous separator is used in most cases. During charging, zinc is plated on a carbon composite plate. The morphology of the plated zinc is strongly related to current density, temperature, and flow velocity. At high current densities, zinc tends to dendritic growth which might cause short circuits through the separator. During charging, bromine is generated. Bromine is highly oxidative and is a poison. Its solubility in water is limited, so to increase the solubility an organic complexing amine is added; this interacts with the bromine to keep it in solution. The organic dense phase behaves like oil and forms a separate phase; this has to be considered for a system layout. An important issue is the toxicity of the bromine. Its high oxidative power necessitates the use of chemically resistant parts for the flow battery, which are expensive. Temperature stability of the complexed bromine is also an issue, since temperature must be kept below 50 °C.

7.4.2 Zinc–Cerium Flow Battery

The zinc–cerium battery is a nonaqueous battery. It is an important battery because of its high potential. The electrolyte used is methanesulfonic acid. The high potential of the catholyte cerium requires the use of very expensive electrode materials (titan electrodes and precious metal coatings) for the cathode. Graphite felt cannot be used for the reaction as the cathode side. It would be oxidized because of the high potential. At the anode, zinc is electroplated on and stripped off the carbon polymer electrode during charge and discharge, respectively [31–33]:

$$Zn^{2+}_{(aq)} + 2e^- \leftrightarrow Zn_{(s)} \ (-0.76 \text{ V } \textit{vs}. \text{ SHE})$$

At the positive electrode (cathode), Ce(III) oxidation and Ce(IV) reduction take place during charge and discharge, respectively:

$$Ce^{3+}_{(aq)} - e^- \leftrightarrow Ce^{4+}_{(aq)} \ (\text{ca.} +1.44 \text{ V } \textit{vs}. \text{ SHE})$$

Because of the large cell voltage, hydrogen (0 V vs. SHE) and oxygen (+1.23 V vs. SHE) could evolve theoretically as side reactions during battery operation (especially on charging). The positive electrolyte is a solution of cerium(III) methanesulfonate.

Due to the high standard electrode potentials of both zinc and cerium redox reactions the open-circuit cell voltage is as high as 2.43 V. Methanesulfonic acid is used as electrolyte, as it allows high concentrations of both zinc and cerium; the solubility of the corresponding methanesulfonates is 2.1 mol for Zn, 2.4 mol for Ce(III), and up to 1.0 mol for Ce(IV). Methanesulfonic acid is particularly well suited for industrial electrochemical applications and is considered to be a green alternative to other support electrolytes.

7.4.3 Iron/Iron Flow Battery

One simple approach is the all-iron hybrid flow battery which uses a very cheap electrolyte: \$7 (kW h)$^{-1}$. Iron is plated at the anode and the Fe^{2+}/ Fe^{3+} is in the form of a complex in alkaline solution. The iron electrode is well known and has been used for decades in the nickel–iron battery; the reaction is highly reversible and very stable. As the catholyte, ferro/ferricyanide can be used; this is also a well-investigated reaction and known as a stable redox system. The kinetics of the reaction are fast, so high current densities up to 200 mA cm^{-2} could be achieved. The cost should be about \$150 kW^{-1}. The reactions involved are:

$$Fe^{2+} \rightarrow Fe^{3+} + e^{-} + 0.77\ V$$
$$Fe^{2+} + 2e^{-} \rightarrow Fe^{0} - 0.41\,V$$

A challenge is hydrogen evolution as a side-reaction; this reduces the efficiency of the system [34].

7.4.4 Copper/Copper Flow Battery

A relatively simple approach is the use of an all-copper hybrid flow battery. The idea is to stabilize Cu (I) as a chloro complex in solution using suitable anions such as chloride or amine. There are three different valence states of copper available for use with this battery.

One interesting effect is that this battery could in principle be recharged by applying higher temperatures in which case the Cu(I) complex becomes unstable and disproportionate to metallic copper and Cu(II), which is the starting material of the charged battery.

The energy density (20 W h L^{-1}) achieved is comparable with traditional VRFBs. This is due to the high solubility of copper (3 M), which offsets the relatively low cell potential (0.6 V). The electrolyte is cheap, simple to prepare, and easy to recycle since no additives or catalysts are used. The system can be operated at 60 °C eliminating the need for a heat exchanger and delivers energy efficiencies of 93, 86 and 74% at 5, 10, and 20 mA cm^{-2}, respectively [35].

7.4.5 Hydrogen–Bromine Battery

The hydrogen–bromine battery works with sodium–bromine in alkaline solution, which is a low-cost and well-known electrolyte. The combination with the hydrogen evolution reaction (HER) and hydrogen reduction reaction (HRR) has the advantage of being very fast with low overpotential in combination with the high oxidative power of bromine. These advantages are tempered by disadvantages such as on both sides catalysts are needed to enhance the reaction. New developments include working with non-noble catalyst systems, but the state of the art does involve precious metal catalysts at the anode as well as at the cathode. To insure reasonable storage capacity the hydrogen must be stored under pressure. This could partly be achieved by hydrogen evolution in the stack of up to 1 MPa (10 bar) to 2 MPa (20 bar). For higher pressures a compressor is needed, making the overall system both complex and expensive [36].

Very recently, the Israeli company EnStorage has developed such a system with a target capacity of 150 kW and 900 kW h.

REFERENCES

[1] Ashimura S, Miyake Y. Denki Kagaku 1971;39:977.
[2] Bartolozzi M. J Power Sources 1989;27:219–34.
[3] Thaller LH. Ninth Intersociety Energy Conversion Engineering Conference, San Francisco, CA; August 26–30, 1974. p. 924–928. NASA TM X-71540.
[4] Thaller LH. Electrically rechargeable redox flow cell. US Patent 3996064; 1976.
[5] Chen CL, Yeoh HK, Chakrabarti MH. Electrochim Acta 2014;120:167.
[6] Oh K, Yoo H, Ko J, Won S, Ju H. Three-dimensional, transient, non isothermal model of all-vanadium redox flow batteries. Energy 2015;81:3–14.
[7] Xiong B, Zhao J, Jinbin L. Modeling of an all-vanadium redox flow battery and optimization of flow rates. IEEE Power and Energy Society Meeting, Vancouver, Canada; 2013.
[8] Xu Q, Zhao T, Leung P. Appl Energ 2013;105:47.
[9] Qiu G, Dennison CR, Knehr KW, Kumbur EC, Sun Y. J Power Sources 2012;219:223–34.
[10] Qiu G, Joshi AS, Dennison CR, Knehr KW, Kumbur EC, Sun Y. Electrochim Acta 2012;64:46–64.
[11] Ma X, Zhang H, Xing F. Electrochim Acta 2011;58:238.
[12] Shah A, Watt-Smith M, Walsh F. Electrochim Acta 2008;53:8087.
[13] Fabjan C, et al. . Electrochim Acta 2001;47:825.
[14] Hashimoto M, Kubata M, Yagasaki A. Crystal structure communications. Acta Crystallogr C Cryst Struct Commun 2000;56:1411.
[15] Kazacos M, Cheng M, Skyllas-Kazacos M. J Appl Electrochem 1990;20:463.
[16] Kim S, Thomsen E, Xia G, Nie Z. 1 kW/1 kW h advanced vanadium redox flow battery utilizing mixed acid electrolytes. J Power Sources 2013;237:300–3009.
[17] Ausfelder et al. Energiespeicherung als Element einer sicheren Energieversorgung. Chem Ing Tech. 2015; 87(1–2):17–89. DOE: 10.1002/cite.20100183.
[18] Sandia National Laboratories on behalf of the US Department of Energy's (DOE) Office of Electricity Delivery and Energy Reliability and the DOE's Office of Energy Efficiency and Renewable Energy Solar Technologies Program Electric Power Industry Needs for Grid Scale Storage Applications. http://energy.tms.org/docs/pdfs/Electric_Power_Industry_Needs_2010.pdf
[19] European Association for Storage of Energy (EASE). Energy storage roadmap. p. 40. http://www.ease-storage.eu/tl_files/ease-documents/Stakeholders/ES%20Roadmap%202030/EASE-EERA%20ES%20Tech%20Dev%20Roadmap%202030%20Final%202013.03.11.pdf
[20] http://energystoragereport.info/redox-flow-batteries-for-energy-storage/
[21] http://www.ict.fraunhofer.de/en/comp/ae/rfb/redoxwind.html
[22] http://www.umsicht.fraunhofer.de/content/dam/umsicht/en/documents/energy/redox-flow-battery-lab.pdf

[23] http://www.umsicht.fraunhofer.de/en/press-media/2013/scale-up-redox-flow.html

[24] Sandia Corporation. Global energy storage data base. http://www.energystorageexchange.org/projects?utf8=%E2%9C%93&technology_type_sort_eqs=Vanadium+Redox+Flow+Battery&technology_type_sort_eqs_category=Electro-chemical&technology_type_sort_eqs_subcategory=Electro-chemical%3AFlow+Battery&technology_type_sort_eqs_child=Electro-chemical%3AFlow+Battery%3AVanadium+Redox+Flow+Battery&country_sort_eq=&state_sort_eq=&kW=&kWh=&service_use_case_inf=&ownership_model_eq=&status_eq=Operational&siting_eq=&order_by=&sort_order=&search_page=1&size_kw_ll=&size_kw_ul=&size_kwh_ll=&size_kwh_ul=&show_unapproved=%7B%7D

[25] http://www.volterion.com

[26] Sterner M, Stadler I. Energiespeicher. Berlin: Springer Vieweg Verlag; 2014. p. 292.

[27] Whitehead A, Trampert N, Pokorny P, Binder P, Rabbow T. Critical safety features of the vanadium redox flow battery. International Flow Battery Forum, Glasgow; Jun 2015. p. 62.

[28] Moskalyk RR, Alfantazi AM. Miner Eng 2003;16:793–805.

[29] Erdmann L, Behrend S, Feil M. Kritische Rohstoffe für Deutschland. Berlin, Germany: Institut für Zukunftsstudien und Technologiebewertung (IZT) and Adelphi; 2011.

[30] Huskinson B, Marshak MP, Suh C, Er S, Gerhardt MR, Galvin CJ, Chen X. Nature 2014;505:195–8.

[31] Nikiforidis G, Berlouis L, Hall D, Hodgson D. Evaluation of carbon composite materials for the negative electrode in the zinc–cerium redox flow cell. J Power Sources 2012;206:497–503.

[32] Nikiforidis G, Berlouis L, Hall D, Hodgson D. A study of different carbon composite materials for the negative half-cell reaction of the zinc cerium hybrid redox flow cell. Electrochim Acta 2013;113:412–23.

[33] Leung PK, Ponce de León C, Low CTJ, Walsh FC. Zinc deposition and dissolution in methanesulfonic acid onto a carbon composite electrode as the negative electrode reactions in a hybrid redox flow battery. Electrochim Acta 2011;56:6536–46.

[34] Sassen J, Goldstein J, Eliad L, Baram N. A novel/iron/iron flow battery for grid storage. International Flow Battery Forum, Glasgow; June 2015. p. 48.

[35] Leung P, Garcia-Quismondo E, Sanz L, Palma J, Anderson M. Evaluation of electrode materials towards extended cycle-life of all copper redox flow batteries. International Flow Battery Forum, Glasgow; June 2015. p. 36.

[36] Tucker M, Weber A, Wycisk R, Pintauro P. Improving the durability, performance, and cost of the Br_2–H_2 redox flow cell. International Flow Battery Forum, Glasgow; June 2015. p. 56.

Part D

Thermal

Chapter 13

Phase Change Materials

John A. Noël, Samer Kahwaji, Louis Desgrosseilliers, Dominic Groulx, Mary Anne White
Dalhousie University, Halifax, Nova Scotia, Canada

1 INTRODUCTION

1.1 Thermal Energy Storage

To reduce our dependence on fossil fuel energy sources, renewable energy sources must be developed to meet our power needs and to reliably provide energy. The difficulty with some renewable energy sources, such as the Sun or wind, is their intermittency. Wind turbines cannot operate when the wind is not strong enough or is too strong. Photovoltaic (PV) panels cannot generate electricity at night or on days with thick cloud cover. For direct electricity–generating technologies, the intermittency of the power source can be overcome by charging charge-storage devices, for example, batteries and fuel cells, during peak hours to store energy for later use. For applications using the Sun's thermal energy, thermal energy storage (TES) materials can be used to store heat. TES can be broadly categorized into two classes: chemical storage and physical storage.

Chemical storage involves the breaking and formation of chemical bonds [1,2], employing a reversible chemical reaction with a large enthalpy change. Heat from the Sun, or other thermal source, drives the reaction to form high-energy products, charging the system. During discharge, the reaction proceeds in the opposite direction, releasing heat. Physical storage utilizes the thermal properties of the TES material to store heat and is the focus of the chapter.

Physical TES can be considered in two categories: sensible heat storage and latent heat storage [1–4]. Sensible heat storage materials store heat via their heat capacities. For a given material the amount of heat that can be stored, Q, depends on both the temperature range over which the material is heated and the mass of the material, m, such that [5]:

$$Q = \int_{T_1}^{T_2} mc_{\mathrm{p}}\, dT, \qquad (13.1)$$

Storing Energy. http://dx.doi.org/10.1016/B978-0-12-803440-8.00013-0

where c_p is the specific heat capacity at constant pressure; T_1 is the initial temperature; and T_2 is the final temperature. The best sensible heat storage materials are those with high heat capacity. In this regard, water is one of the best sensible heat storage materials, with a specific heat capacity of 4.2 J K^{-1} g^{-1} [5]. TES by water is exemplified by differences in climate: consider the moderated temperatures of a maritime region compared with the large temperature swings of dry continental climates. The range of temperatures encountered by the material is also important. The quantity of heat stored is related to the integral over the temperature range, so a wider temperature range leads to more heat stored. The other central term here is the mass. Dense materials such as concrete or stone present a large thermal mass per unit volume and can store a significant quantity of heat.

Over a small temperature range, latent heat storage materials can outperform even the best sensible heat storage materials. These materials store heat through the latent heat of a phase transition and are referred to as phase change materials (PCMs) [1–4,6–8]. When a PCM is heated to its transition temperature, it is converted from one phase to another. If the transition of a pure substance is first order, it occurs isothermally and requires an input of energy at the transition temperature [9]. This input of energy is the transition enthalpy change, also called latent heat. If the phase change is reversible the energy can be recovered through cooling. Furthermore, any input of energy required to heat the material to its transition temperature, and any energy input to raise the temperature of the material after the phase transition, is stored as sensible heat [4]. Therefore, the total quantity of heat that can be stored by a PCM over the temperature interval T_1 to T_2 is [5]:

$$Q = \int_{T_1}^{T_{trs}} mc_{p,1}\,dT + m\Delta_{trs}H + \int_{T_{trs}}^{T_2} mc_{p,2}\,dT, \tag{13.2}$$

where T_{trs} is the transition temperature; $\Delta_{trs}H$ is the transition enthalpy change; $c_{p,1}$ is the specific heat capacity of the low-temperature phase; and $c_{p,2}$ is the specific heat capacity of the high-temperature phase. Fig. 13.1 compares the sensible heat storage of liquid water with heat storage in octadecane PCM.

In principle, PCMs can utilize any phase change transition, but solid–liquid transitions are most prevalent [1–4,6]. While solid–gas and liquid–gas phase transitions often have large transition enthalpy changes (>300 J g^{-1}), they also involve a very large change in volume, making them impractical in most cases. Solid–solid PCMs are used in some applications [10–12]; these transitions usually have only a small volume change, and both phases are immobile and therefore easier to contain. However, solid–solid transitions generally have low transition enthalpy changes (typically <100 J g^{-1}) [10,12]. Therefore, PCMs with solid–liquid transitions are by far the most applied and studied. These transitions have high enthalpy changes (~200 J g^{-1}), and the transition from solid to liquid phase only has a small volume increase (<10%). The liquid phase requires containment, but this is more easily achieved than the containment of a gaseous phase.

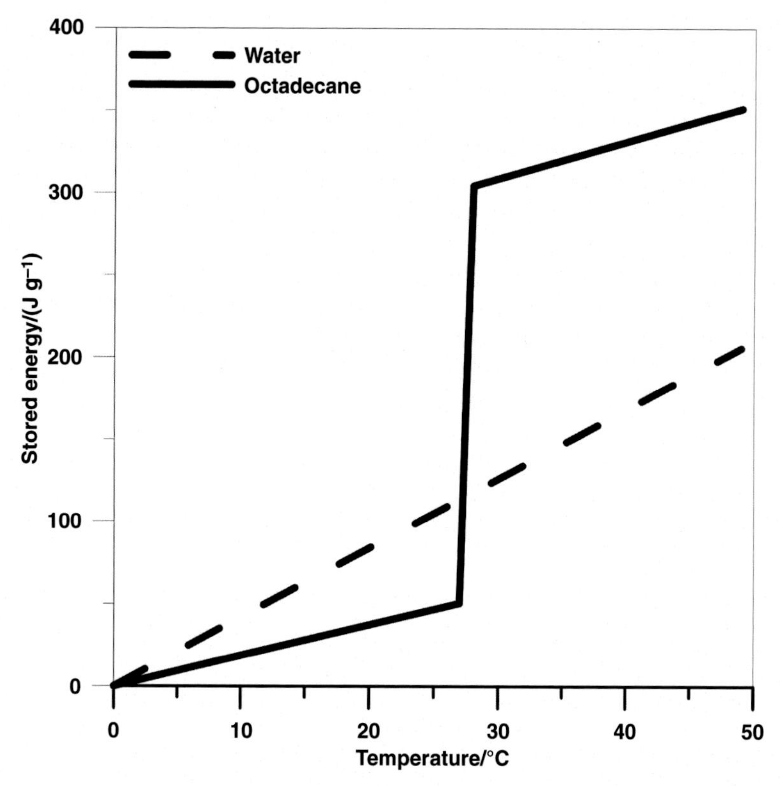

FIGURE 13.1 Comparison of energy stored by liquid water (sensible heat storage) with energy stored (sensible + latent) by octadecane (PCM with melting point 28 °C) over the temperature range (0–50) °C.

1.2 Properties of Phase Change Materials

Many variables must be taken into account when selecting a PCM. Transition temperature is of primary concern. If the transition temperature is not within the temperature range of the application, no phase change will occur, and the material will only store sensible heat. The predicted temperature difference between transition temperature and the charging (hotter) and the discharging (cooler) temperatures also plays a crucial role since it directly influences the overall energy transfer rate to the storage system. In the chapter, materials used in sub-ambient, ambient, moderate, and high-temperature applications are described.

The next major consideration for a PCM is the latent heat (enthalpy of phase change) of the phase transition. To achieve high energy storage density, it is necessary to have as high a latent heat as possible to maximize the energy stored in the phase transition. If space is of concern, the volumetric transition enthalpy change is of more relevance than the gravimetric value and dense materials would be favored in this case.

There are many other important properties to consider for a viable PCM, including thermal conductivity, stability, cyclability, phase segregation, supercooling, containment, cost, and safety [1–4,6–8]. Note that care should be taken when searching the literature for PCM properties: for some compounds, widely varying values of physical properties have been reported for the same material, and some errors have been carried through multiple reviews. For this reason, it is best to consult the original literature sources for the physical properties of PCMs.

Thermal conductivity quantifies the conduction of heat in the material. If thermal conductivity is too low, it will be difficult to charge the PCM, or extract energy from it, in a reasonable amount of time.

For any long-term application the PCM must be stable. The material should not degrade over time, should not react with the ambient air or moisture, and should not degrade its containment vessel. The PCM and container must be stable over thousands of thermal cycles. Over its lifetime, it should not exhibit a significant decrease in latent heat or change in transition temperature. In addition, repeated melting and crystallization should not degrade or otherwise alter the material.

One of the biggest limits to the cyclability of PCMs is separation into different phases [1,2,6]. Such degradation is most common in multicomponent PCMs in which the components differ significantly in density. One component could separate from the other(s) by gravity, altering the melting point of the system. An example is incongruently melting salt hydrates, as discussed in Section 4.1. The segregation of phases can be enhanced over many cycles and lead to a gradual but significant decrease in performance [3,6,7].

A drawback to some otherwise promising PCMs is supercooling. Supercooling, sometimes termed "subcooling," "undercooling," or "supersaturation", refers to persistence of the high-temperature phase below its transition temperature. Below the transition temperature, the high-temperature phase is metastable, but could supercool by 100 K or even more before finally undergoing transition to the stable phase. If the system supercools below the minimum temperature of the application, the latent heat that was stored will not be regained, and after the first heating the PCM would act only as a sensible heat storage material. Fig. 13.2 shows supercooling in the differential scanning calorimetry (DSC) thermogram of a material. In some applications, supercooling can be desirable as it provides a means for long-term storage of energy at ambient temperatures. In that case, nucleation of the solid phase can be artificially initiated when required.

The PCM also needs to be compatible with its container. The PCM should not corrode, degrade, or soften/dissolve the material containing it. For melting phase changes the containment vessel must be able to hold the liquid phase without leaking its contents, and in all cases the containment vessel needs to accommodate any volume change associated with phase change and thermal expansion of the PCM over the application temperature range.

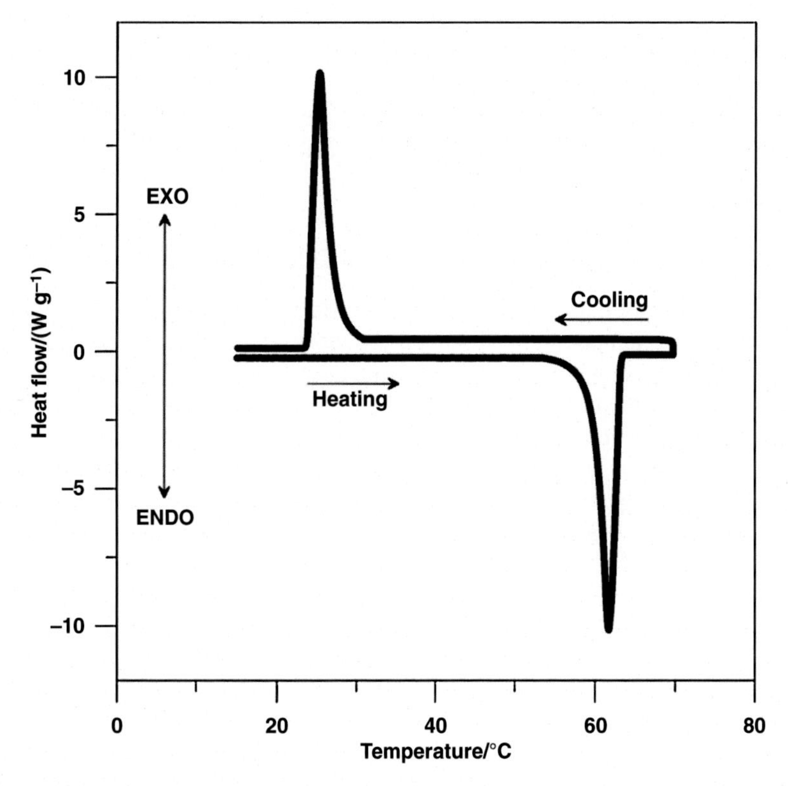

FIGURE 13.2 **Schematic representation of supercooling in a material in terms of its DSC thermogram.** The peaks indicate phase transitions and in this case crystallization does not occur until well below the melting point.

If latent heat storage is to achieve large-scale use in an economical way the PCM must be readily available and at low cost. A material with excellent thermal properties and stability might be unsuitable as a PCM if its cost is excessive. In most circumstances a desirable PCM should be safe to use on a domestic scale. This means that it should have low toxicity, should not be violently reactive, and should not pose an elevated fire risk.

In reality, it is difficult or impossible to find a material that is ideal for all criteria. Material selection must be carried out such that the needs of the specific application are met in an optimal way [3].

1.3 Sustainability

In many circumstances the environmental cost of the PCM also should be considered. For example, if the payback time for the embodied energy of the material is high, it might not be recouped over the lifetime of its use. In this regard, production of PCMs from biological, renewable sources is very attractive [13].

Many oily compounds produced by algae and by plants, including oil palm and rape seed, can be used for TES [13]. These sources reproduce quickly and use solar energy as the primary energy input; thus, there exists the potential to extract organic PCMs in a sustainable manner, at little energetic cost. Lifecycle analysis is a valuable tool to quantify the sustainability aspects of a PCM choice [13–15].

2 HEAT STORAGE AT SUBAMBIENT TEMPERATURES

Several materials that qualify as good PCMs have transition temperatures below ambient temperature [2,16] and are most commonly used for cooling applications. Among these PCMs, ice is the most familiar and least expensive. With its high latent heat of fusion (334 J g^{-1} [5]), it takes about 11 kg of ice to store the equivalent of 1 kW h of energy at 0 °C. Although the ~10% volume change that occurs when ice freezes puts high demands on its containment, ice has been integrated as a PCM in large cooling projects. For eons, natural ice and snow have been collected in a number of countries during the winter season and stored in underground pits or thermally insulated constructions, to be used to cool goods or buildings during the hot summer months. The heat from the surroundings is normally transferred to the ice by air, either from a fan or by natural convection, or by water circulating in pipes in direct contact with the ice. Projects based on cooling by natural ice and snow storage are discussed in Refs. [17] and [18].

In a similar concept, district cooling uses a central facility to produce and store ice for the purpose of cooling residential and commercial buildings. An example is the Minato Mirai 21 Central District in Yokohama (Japan) [19], which uses an ice storage system to transfer chilled water to cool buildings within the district (Fig. 13.3). In addition to ice, encapsulated PCMs based on fatty acids, paraffins, and salt hydrates with fusion temperatures between –33 and 27 °C are also used in many of the district cooling storage tanks installed by Cristopia Energy Systems around the world [20].

On a smaller scale, systems such as the IceBank designed by CALMAC [21] are installed to individually cool large commercial buildings. This system operates a chiller and an antifreeze solution (i.e., water-glycol solution) as a heat transfer fluid to freeze water stored in large tanks at night, when off-peak electricity rates are the lowest. As with the Minato Mirai 21 district cooling system the ice stored in the tanks is then used the next day to cool the water–glycol solution circulating in heat exchange coils. A fan blows air on the coils to deliver cold air to occupants throughout the building. This system has a smaller footprint on the building infrastructure than the traditional air-conditioning system, since up to 80% less piping [21] and smaller size chillers are required for IceBank installation.

The concept of cooling based on thermal energy storage is becoming increasingly popular worldwide and many energy-conscious corporations have

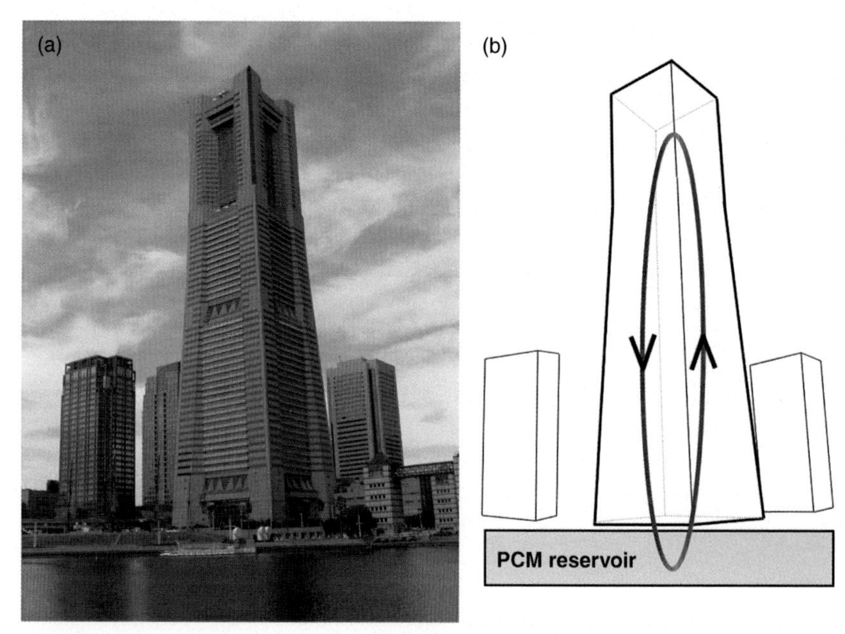

FIGURE 13.3 (a) Yokohama Landmark Tower at Minato Mirai 21, and (b) a simplified schematic of the PCM-based cooling system at Minato Mirai 21 Central District showing the circulation of fluid that removes heat from the buildings. The PCM is frozen at night using off-peak power, thereby shifting the cooling energy from the usual daytime to the night and leveling overall energy demand.

already adopted it. Buildings cooled by ice storage systems include the tallest building in the world, Burj Khalifa in the United Arab Emirates, where the outside temperature reaches 50 °C [22], the Google Data Center in Taiwan, and several buildings in New York City, including the Rockefeller Center, Bank of America Tower, and Goldman Sachs headquarters. The shift of cold production to off-peak hours not only lowers the cooling cost, but also has a positive environmental impact. By easing the load on the power grid during peak hours, CO_2 gas emissions from power plants operated by fossil fuels are significantly reduced. It is also good business: it is estimated that the ice storage cooling system saves Goldman Sachs $50 000 per month on their utility bill during the summer [23].

PCMs with a subambient transition temperature also can be found in many commercially available products. Paraffins, glycols, and salt hydrates have been integrated as PCMs in storage containers to maintain the temperature of perishable goods near the cold phase transition temperature of the PCM. For example, containers designed with built-in PCM packs also offer a simple approach to keep temperature-sensitive medical and pharmaceutical products at the desired temperature during transport [18].

3 HEAT STORAGE AT AMBIENT TEMPERATURE

PCMs with a phase transition temperature falling within the ambient temperature range (21–28) °C are predominantly used to improve thermal comfort in a living space by reducing temperature fluctuations during the day [e.g., reducing the daily temperature range (20–27) °C, to a smaller range such as (22–25) °C)], or to provide thermally regulated clothing for use in hot environments [24]. Fig. 13.4 illustrates that PCMs reduce temperature fluctuations. The addition of a PCM achieves this leveling effect through the absorption of excess heat generated within the space and the storage of thermal energy at the temperature of the PCM's melting point. The stored energy is later released during solidification when the freezing point of the PCM is reached, usually at night when the ambient temperature is lower. As in the case of ice storage systems discussed earlier, PCMs integrated in building materials can ease the load on the air-conditioning system during the peak hours of the day when the power rates are more expensive. A list of organic and inorganic PCMs used in building comfort applications along with some of their thermophysical properties has been compiled by Cabeza et al. [25]. These PCMs can be incorporated passively in several different parts of the building, including the wallboards, concrete, insulation, and the ventilation system [26]. Different methods for the incorporation of PCMs into construction materials have been investigated including [27]: direct incorporation, where the PCMs are directly mixed with the construction materials (e.g., gypsum, concrete, mortar); immersion, where the building materials are impregnated with liquid PCM; encapsulation, where either small particles (1–1000) μm of PCMs are enclosed in thin shells (microencapsulation) or

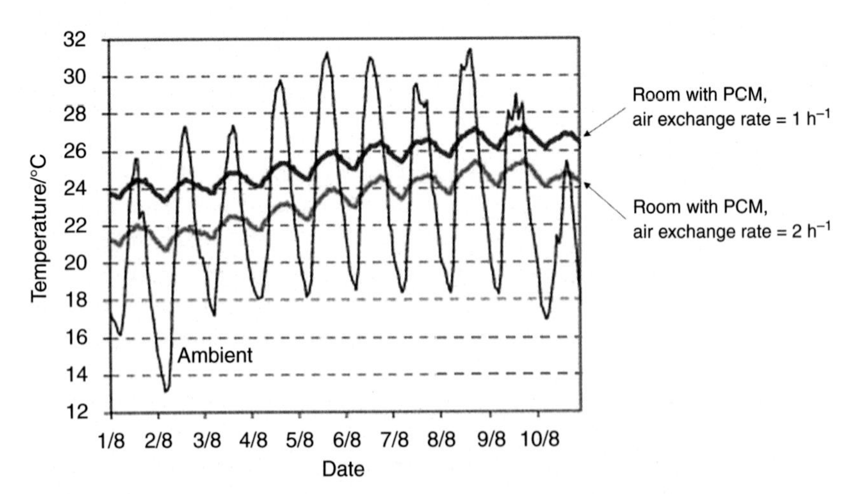

FIGURE 13.4 Temperature-moderating effect of PCMs in a room over 8 days in August in a building in Ljubljana (Slovenia). Without PCM, the ambient temperature underwent large fluctuations, whereas a room with PCM only underwent small fluctuations, and achieved a lower maximum temperature. *(Reproduced from Ref. [28] with permission from Elsevier.)*

several liters of PCMs are packed in containers (macroencapsulation) such as tubes, spheres, and panels before being introduced into building materials.

The behavior of PCMs incorporated in building materials has been widely studied in different climates, and several commercial products are readily available. Octadecane wax is one example of paraffin PCM used to impregnate wallboards and increase their thermal storage capacity [7]. Paraffin waxes are commonly found in commercial products such as Micronal (R) [29], which is used in wallboards and metal ceiling tiles, and Energain (R) thermal mass panels [30]. Mixtures of fatty acids such as decanoic (aka capric acid) and dodecanoic (aka lauric acid) acids, and esters such as butyl stearate and propyl palmitate, have also been intensively studied for integration in gypsum and concrete [27,31]. These nonparaffin organics are particularly attractive for building applications because they are derived from renewable sources and because they are nontoxic, biodegradable, and can be easily recycled. Impregnation of building materials with salt hydrate PCMs has also been shown to increase their thermal storage capacity. For example, encapsulated calcium chloride hexahydrate ($CaCl_2 \cdot 6H_2O$) has been embedded in concrete slabs to develop a floor-heating system with improved heat storage [32].

4 HEAT STORAGE AT MODERATE TEMPERATURES

4.1 Moderate-Temperature PCMs

Many PCMs, including many salt hydrates, have their phase transitions in the moderate-temperature range, 40 °C to just over 100 °C. Salt hydrate materials consist of an inorganic or organic salt with one or more waters of hydration. Salt hydrates were the first materials to be investigated for use as PCMs in the ground-breaking work of Telkes [33,34]. Salt hydrates are generally high in density and tend to have greater volumetric heats of fusion than other materials in the moderate-temperature regime. Sharma et al. [1], Pielichkowska and Pielichowska [4], and Farid et al. [6] all reviewed a number of salt hydrate phase change materials. Most salt hydrate PCMs have moderate to high latent heats of fusion, typically (100–300) J g^{-1} [1,2,4].

While salt hydrates are very attractive PCMs from the standpoint of energy storage density, they do have some shortcomings [1,2,6]. The first challenge is that some salt hydrates melt incongruently: upon heating, a portion of the salt hydrate dehydrates to a less hydrated phase before melting. This phase transition is marked by a peritectic point in the phase diagram. The less hydrated phase typically will not melt in the operational temperature range and will be denser than the solution formed with the water of dehydration. Thus, the less hydrated salt will settle to the bottom of the containment vessel and might not be available to rehydrate upon cooling [3,6]. The resulting phase segregation leads to a gradual irreversible loss in performance. However, there are ways by which this can be overcome. Mechanical agitation and mixing can keep the solid particles suspended, allowing them to be in contact with the fluid phase to recombine on cooling. Thickening agents can be added to increase the viscosity and thereby prevent the solid phase from segregating to the bottom of the containment vessel [35]. In addition, the size of the containment vessel can be modified to reduce the tendency for a salt hydrate to undergo irreparable phase segregation. A smaller container reduces the propensity for the dehydrated solid to settle far from the liquid, increasing the probability of rehydration upon cooling [3].

A second difficulty with salt hydrate PCMs is their tendency to supercool (Fig. 13.2). Even if many crystal nuclei form on cooling, crystallization does not proceed until a crystal nucleus of critical radius is formed. Formation of critical-sized nuclei is a balance between the energy benefit in forming the thermodynamically stable solid phase and the energy cost to create high-energy surfaces at the interface between the solid and liquid. The addition of nucleating agents can help remove the effects of supercooling. Nucleating agents provide surfaces on which crystal nuclei can form, reducing the requirement to form new solid surfaces [33,36]. The nucleating agent must remain solid over the entire temperature range of the application, and the most effective nucleating agents often have crystals similar to the PCM that they nucleate. Another technique used to overcome supercooling is the application of a "cold finger," which is a piece of thermally conductive material extending into the PCM, attached to a cold source. The cold finger provides a locally cooled area in which critical nuclei can form more readily due to the larger temperature drop, and thereby initiate crystallization throughout the bulk of the material.

A third material property concern for salt hydrates is that in their molten state they are aqueous salt solutions and thus very corrosive to many metals. Care must be taken when selecting materials for their containment to avoid corrosion limiting the system's lifetime.

Numerous organic materials also undergo phase transitions in the moderate-temperature range including: paraffins ($H_3C(CH_2)_nCH_3$) [37], fatty acids ($H_3C(CH_2)_nCOOH$) [38–40], and sugar alcohols [41–43]. Fatty acids and sugar alcohols have the attractive feature of availability by extraction from renewable, plant-based sources [13]. Fatty acids have been shown to cycle well, maintaining their thermodynamic properties over hundreds of cycles as illustrated in Fig. 13.5 [38,39,41,42].

Organic paraffinic and fatty acid PCMs do not undergo the supercooling or phase segregation processes that plague salt hydrates. The melting points of paraffins and fatty acids are correlated with their alkyl chain length, with longer chain compounds melting at higher temperatures. Therefore, by selection of the appropriate chain length, these PCMs can be used in a variety of applications which require phase transition temperatures in the range −56 °C to 80 °C [1,2,4,16,37]. While the gravimetric latent heat of these materials is typically moderate to high (100–250) J g^{-1} [1,2,4], they tend to have low densities (<1 g cm^{-3}) and thereby lower volumetric energy density than salt hydrates. The thermal conductivities of these organic materials are also quite

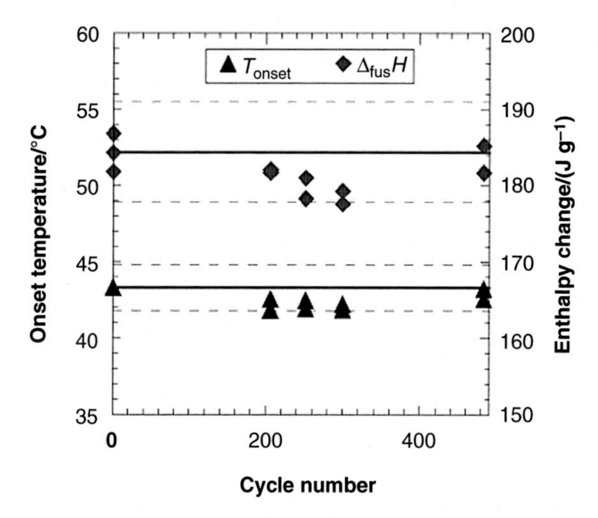

FIGURE 13.5 **Onset temperature for melting and latent heat of fusion for dodecanoic acid PCM over several hundred melt–freeze cycles, as determined by DSC.** The dashed lines indicate standard deviation in the values of onset temperature (i.e., melting point) and enthalpy change for the initial sample. Within 500 cycles, there was no significant change in either property. *(Reproduced from Ref. [39] with permission from Elsevier.)*

low (~0.2 W m^{-1} K^{-1} [2]), resulting in low inherent rates of charge/discharge. There are ways by which the effective thermal conductivity can be increased, such as the introduction of metallic particles or nanoparticles [44–46], insertion of metallic structures such as fins or rods [47–49], or impregnation of the PCM in graphite [50,51].

Sugar alcohols differ from paraffins and fatty acids in that they have higher densities, leading to higher volumetric latent heat, in some instances exceeding 400 J cm^{-3} [1,2,4]. However, sugar alcohols, unlike other moderate-temperature organic PCMs, undergo significant supercooling, as much as ca. 100 K below their melting point. This extreme supercooling is likely due to kinetic limitations in the formation of crystallites, a result of the complex hydrogen-bonding networks in sugar alcohol crystal structures [52,53].

4.2 Applications of Moderate-Temperature Phase Change Materials

4.2.1 Solar Thermal Hot Water

A major application of PCMs in the moderate-temperature range is in domestic solar hot-water applications. A heat transfer fluid, such as water or glycol, circulated through a flat plate or an evacuated tube solar collector, collects solar radiation as heat to be used to heat water for household use. This type of system can replace or supplement electric- or gas-powered hot-water heaters, for substantial energy savings and reduction in CO_2 emissions. Solar hot-water systems require energy storage, sensible or latent. Water could be heated and stored in a large tank to be used directly or by heat exchange with cold water, when there is insufficient solar gain to heat cool water. This type of sensible storage is, however, necessarily quite large and massive, possibly ruling out its use in some domestic or commercial locations, especially for retrofitting where space is problematic. If, instead, latent heat storage with a PCM is used, the volume and mass requirements are much lower [54], in some instances allowing use of solar thermal hot water when otherwise only nonrenewable sources would be feasible. In this case, solar energy can be used to charge a much smaller volume of a PCM, and cool water can then be heated by circulating through a heat exchanger in the PCM tank.

Choice of PCM is important for this application. It is desirable to have a high melting point PCM to provide water at a higher temperature. However, if the melting point of the material is chosen based on the maximum achievable temperature in the heat transfer fluid from the collector on a day with high solar gain, on days with less ideal conditions the PCM would only melt partially, if at all. In the latter case the bulk of the PCM would only provide sensible heat storage. Transfer of heat to the PCM also is important. The time during the day when there would be sufficient solar gain on the collector to melt the PCM is limited, so heat must be transferred efficiently. Most moderate-temperature PCMs have low thermal conductivity, so the heat exchange system must be designed to promote melting across the entire PCM. Numerous solar water heater designs employing PCMs have been explored [55–59].

4.2.2 Seasonal Heat Storage

Long-term "seasonal" heat storage is characterized by multi-month to multi-annual heat storage retention. Two principal mechanisms contribute to the ability of a PCM heat storage system to accomplish seasonal heat storage: insulated thermal mass and stable supercooling. The first mechanism is common to all heat storage media and not unique to PCMs, whereas stable supercooling is a unique property of PCM storage [60].

Insulation minimizes the rate of heat loss from a heat storage mass relative to its storage capacity. Improved insulation around smaller heat storage vessels can be practical but limited by cost and maximum feasible volume, while the inherently low surface-area-to-volume ratio of larger heat stores requires less effort to insulate adequately [60]. Buried and bermed (semiburied) heat storage vessels (Fig. 13.6) are especially effective to insulate large heat stores when the cost of above-ground insulation is prohibitive [60]. Higher temperature PCM storage requires greater efforts to combat self-discharge by heat loss than lower temperature storage.

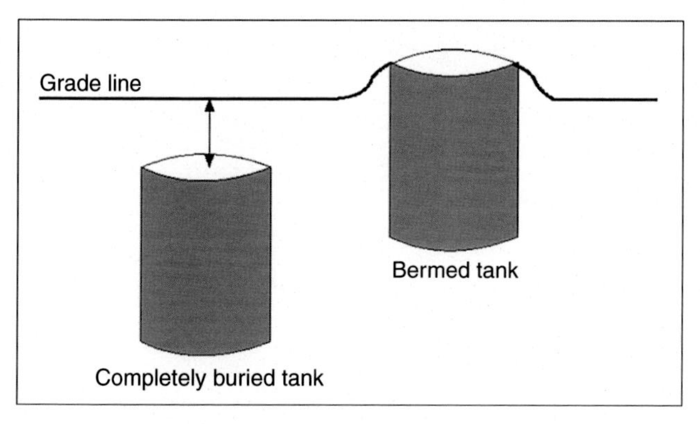

FIGURE 13.6 **Buried and bermed tanks [60].** *(Reproduced with permission from Elsevier.)*

Stable supercooling PCMs are materials that readily supercool and can remain supercooled at ambient temperatures for seasonal durations (e.g., some salt hydrates and sugar alcohols) [61–65]. Supercooling of the PCM to ambient temperature has the advantage of long-term storage without heat loss (i.e., no self-discharge).

Practical supercooling is limited by the volume of the PCM and the degree of supercooling (difference between transition temperature and storage temperature), where larger volumes and greater supercooling increase the probability of autonucleation [3]. There is no available theory to predict the maximum practical contiguous size of a supercooling PCM [3], and limits need to be defined empirically by incremental scaleup.

As a consequence of stable supercooling, these PCMs require a solidification triggering mechanism to initiate heat discharge on demand. Compression springs and other prestressed materials that generate hard surface contacts are useful internal mechanical triggering devices [66,67], while the addition of seed crystals and cascading solidification through a capillary channel require external material input [3,68]. Although PCMs can be chilled to their autonucleation point [69,70], it is rarely practical to use this mechanism to initiate crystallization. Ultrasonic vibration and electric fields do not appear to reliably initiate solidification in most PCMs [68,71].

Heat released during the initial solidification of a supercooled PCM elevates the temperature of the PCM to its corresponding two-phase equilibrium state (isenthalpic process, Fig. 13.7 [61,63,70]). The degree of supercooling influences the quantity of retained thermal energy available for discharge and thus the storage efficiency and, for incongruent PCMs, the maximum discharge temperature achieved at the onset of solidification (Fig. 13.7 [61,63]). These factors limit the effectiveness of higher transition temperature supercooling PCMs when supercooled to room temperature.

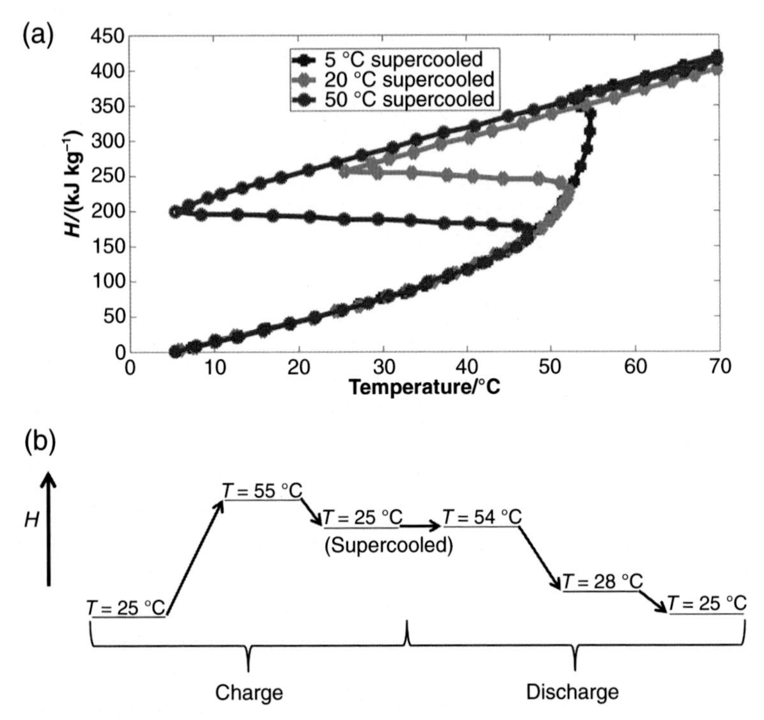

FIGURE 13.7 (a) Enthalpy–temperature profile of diluted $NaCH_3COO{\cdot}3H_2O$ including supercooling and solidification [72], and (b) relative enthalpy profile for a complete charge–discharge cycle for diluted $NaCH_3COO{\cdot}3H_2O$ [63]. *(Part a: reproduced with permission from Elsevier.)*

5 HEAT STORAGE AT HIGH TEMPERATURES

5.1 High-Temperature PCMs

For high-temperature heat storage, in the range of several hundred to well over 1000 °C, the PCMs used are metals and anhydrous salts. For the latter, various carbonates, chlorides, sulfates, nitrates, and nitrites are commonly used, as well as their eutectic mixtures [73]. Nitrates and nitrites typically melt in the range (300–550) °C, but have relatively low latent heat, ~(100–175) J g^{-1} [74]. Carbonates and chlorides do not melt until temperatures above 700 °C, but have higher latent heat, generally above 200 J g^{-1} [74]. Fluoride salts provide very high latent heat, 790 J g^{-1} for LiF/CaF_2 eutectic, but are typically avoided due to cost and material compatibility [74]. Containment of molten salts at high temperatures introduces material-related challenges, and molten salts also have high safety risks. Molten salts are corrosive to many steels and their vapors are often reactive, and therefore expensive alloys and coatings are required to achieve an acceptable lifetime for the system [73]. Salts also have low thermal conductivity, so metal or graphite rods or fins are required to transfer heat within the bulk of the PCM [74]. Although metals and metal eutectics are more expensive than most salts and typically have lower latent heat, molten metals are less corrosive than molten salts, and they have high thermal conductivity, such that in some cases the use of metals is competitive with salts. Liu et al. [74] have reviewed many salt and metal PCMs and their eutectics.

5.2 High-Temperature Applications

5.2.1 *Concentrated Solar Power: Andasol*

Solar thermal power stations use large volumes of molten salt for thermal energy storage. The use of thermal storage in concentrated solar power (CSP) plants decouples power production from solar radiation availability [74,75]. Other solar power systems, such as PVs, require direct solar radiation to produce electricity, but thermal energy storage in CSP allows for night-time power production and mitigates the drop in production on days with nonideal solar conditions. In most thermal storage systems for solar thermal power plants, only the sensible heat of the molten salt is used. The Andasol Solar Power Station in Andalusia (Spain) is one such example. Andasol consists of three 50 MW projects, each producing about 165 GWh of energy annually [76]. Andasol is a parabolic trough-type solar thermal station and operates just under 400 °C [75,76]. In this configuration, parabolic mirrors focus solar radiation onto a system of pipes containing biphenyl–diphenyl oxide eutectic as a heat transfer fluid [77], which carries heat to a steam generator and turbine, and to tanks of molten salt for storage. Andasol 1 has two tanks for thermal energy storage, holding 28 500 t of salt, a mixture of 60% $NaNO_3$ and 40% KNO_3 [75,76]. The 28 500 t of salt used in Andasol 1 gives 1000 MW h of stored energy when heated from 300 to 400 °C and can power the station for 7.5 h [75,76], allowing the station to produce electricity through the night or on cloudy/rainy days.

5.2.2 *Industrial Heat Scavenging*

Excess process heat in continuous processes is routinely and economically captured in heat exchangers called economizers and using combined or cogeneration cycles [78,79]. Continuous capture is not feasible in batch and semibatch processes due to their inherent intermittency, but they present opportunities for heat recovery via heat storage.

High-temperature (>250 °C) batch and semibatch processes are common in metal foundries, pulp mills, and the cement industry [78,80]. Sensible heat storage masses (e.g., bricks) are presently used to store excess heat from exhaust gases in large furnaces, then used to preheat inlet gases to active furnaces [78]. Suitably encapsulated PCMs could be readily deployed to substitute bricks for thermal mass and improve the heat storage densities in these processes since operating temperatures are well defined and storage durations are short.

Similar opportunities could also exist with the secondary heat exchange fluids used in batch or semibatch processes (e.g., heat transfer oil or process steam). PCM heat storage could be implemented in a reservoir serving a closed loop for added thermal mass, and used either to preheat the heat exchange fluid at the beginning of the successive batch operation, or to preheat the equivalent fluid in the alternating batch. The former scenario is analogous to using a PCM to store excess automotive engine heat accumulated by the engine oil; this stored heat is used to mitigate engine cold-start inefficiencies by preheating the engine oil/engine block [81].

6 HEAT TRANSFER IN PCM-BASED THERMAL STORAGE SYSTEMS

Heat transfer is a major issue for PCMs and their storage systems, for both charging (PCM melting) and discharging (PCM solidifying) modes.

When energy is added to a fully solid PCM, either through the flow of a heat transfer fluid (HTF) on the outside wall of the PCM encapsulation or from heat generated internally (e.g., Joule heating from an electrical element), heat is carried to the PCM first by conduction, resulting in the phase transition (melting) of the first layers of PCM. Most of the energy added is stored as latent heat in the PCM with the remaining energy increasing the temperature of the PCM. Once enough PCM has melted, the main heat transfer mode changes from conduction to natural convection resulting in higher temperature liquid PCM moving upward in the system, melting the upper portion of the PCM faster (Fig. 13.8). From this point, natural convection dominates the melting process until the entire PCM is melted, providing higher heat transfer rates than for conduction alone. Higher temperature differences between the heat source and the PCM result in faster energy storage and melting. To a lesser extent the increased flow rate of the HTF, which increases the strength of forced convection on the outside surface of the PCM, also results in faster storage rates [82].

When the PCM is fully liquid and energy is removed from the system through a cold-temperature source (e.g., circulation of a cold HTF), the first layers of PCM contacting the outside surface cool and start solidifying on the surface. (Note that the solidification temperature might be different from the melting temperature due to supercooling.) The solidification process is completely controlled by conduction and is therefore typically slower than melting for the same system geometry [84]. The PCM solidifies in successive layers, increasing in thickness and thereby imparting an ever increasing additional thermal resistance between the cold surface and the remaining liquid PCM. Changing the HTF flow rate has little effect on the energy extraction rate, although reducing the HTF cold-temperature results in an increase in the discharge rate.

The major thermal problem in PCM storage systems could be quantified as a "rate problem," that is, the theoretical quantity of energy that can be stored can be directly calculated based on system volume and PCM properties, but

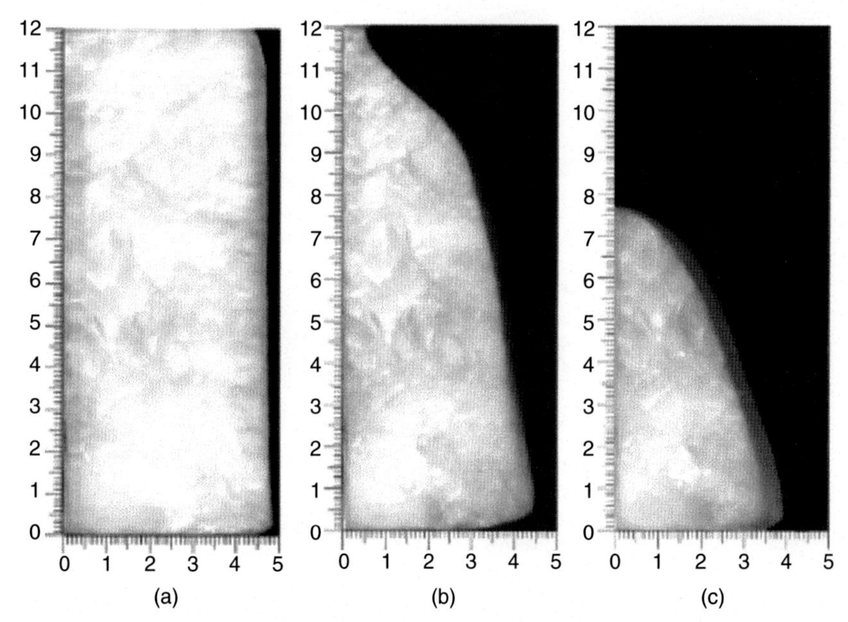

FIGURE 13.8 Melting of dodecanoic acid in a rectangular cavity with the right wall at 60 °C. (a) After 10 min of heating, conduction is the dominant heat transfer mode. However, the influence of convection is clearly apparent at (b) 90 min and (c) 170 min. *(Reproduced from Ref. [83] with permission from Elsevier.)*

the heat transfer rates for energy storage or discharge are inherently small for most PCMs due to their usual small thermal conductivities (~0.2 W m^{-1} K^{-1} [2]). This rate problem requires significant research and design to have a system store the right amount of energy in *the right amount of time*. Various methods are used to increase the overall transfer rates during both charging and discharging, although discharging is often the more limiting factor. Most solutions are geometrical: adding fins to the PCM-side of a system, and tailoring the encapsulation shape to increase the available surface area for heat transfer [85], or using a specifically designed heat exchanger to enhance heat transfer rates [86]. Fig. 13.9 shows several possibilities for increasing heat transfer in a cylindrical system.

A third mode of operation is theoretically possible for PCM storage systems: allowing the heat source and cold source to operate at the same time, leading to simultaneous charging and discharging of the system. Very little research has been reported for this mode of operation in which the system is designed for heat transfer both through the PCM and the enhancement features (geometry, fins) of the system [87].

An additional mode of heat transfer is sometimes present during melting of a PCM: close contact melting (CCM). CCM is present when a solid PCM is in close contact with a warmer solid surface below it [88], such that heat

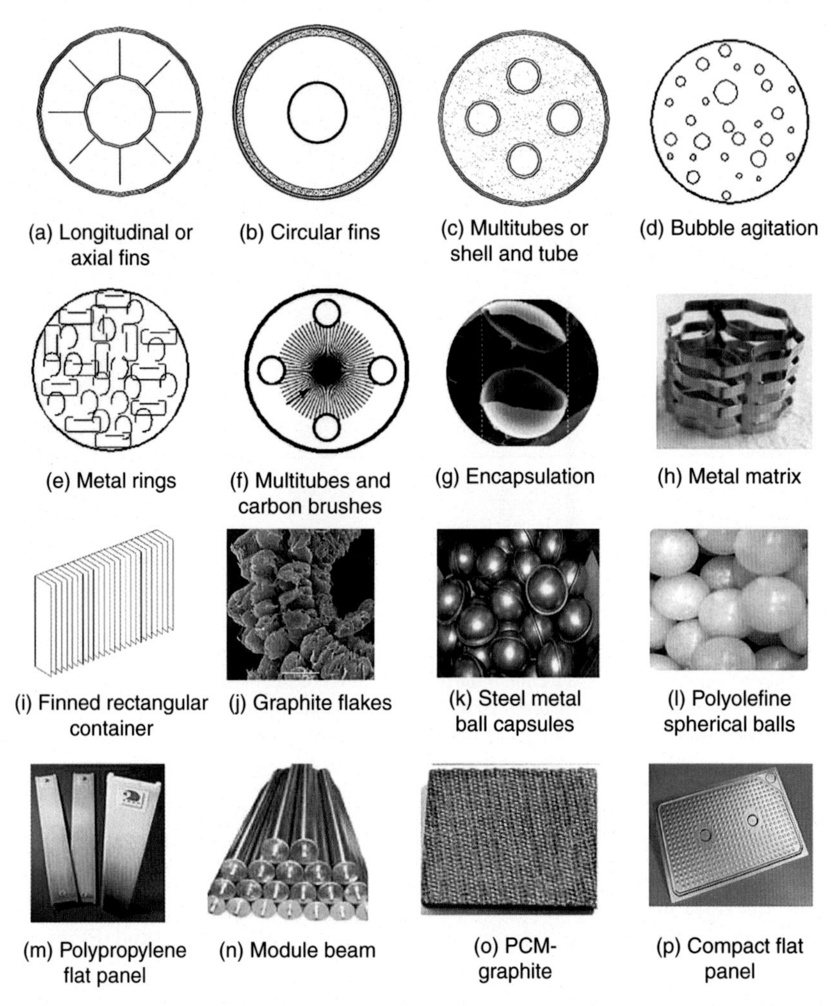

FIGURE 13.9 Various examples of heat transfer enhancement methods used for PCM energy storage and recovery (a–p). *(Reproduced from reference [49] with permission from Elsevier.)*

is transferred from the surface to the PCM through a thin melted PCM layer, providing higher heat transfer rates since the heat conduction resistance of this thin layer is extremely small. Spherical encapsulation in a packed bed storage system is an example of a system using CCM [89].

7 GAPS IN KNOWLEDGE

In both the materials science and applied science aspects of PCM science, there is much yet to be learned to have widespread implementation of PCMs in active and passive energy systems. Disparate efforts by researchers have indeed advanced knowledge of PCMs, but a comprehensive and unified effort in key areas could reduce barriers, uncertainty, and risk for future deployment of PCM heat storage technologies.

In fundamental evaluations of PCM thermophysical properties, liquid-phase properties are known much better than solid-phase properties [39]. Even then, liquid PCM viscosity, essential for proper modeling, and understanding of natural convection during melting, is not always determined or reported. Thorough investigations of both liquid and solid phases of potential PCMs would better equip scientists and engineers to determine the suitability of candidate materials, as well as help identify errors in property determinations that are committed in the absence of meaningful comparisons; for an example, see Ref. [39]. A reliable database of PCM properties would be very useful in this regard.

Conventional PCM groupings oversimplify the nature of phase change itself and provide no additional insight for new material discoveries with respect to T_{trs} and $\Delta_{trs}H$ [90–92]. Details of the atomic and molecular bonding in each of the phases reveals commonalities in the entropy changes associated with the transitions, thus providing an atomic-level basis for PCM classification [90–92] and a rational way to search for candidate PCMs.

The fundamental limitations to practical supercooling of PCMs remain undiscovered. Simple models have been used to approximate intensive limits on the maximum degree of supercooling [3], but the degree and duration of supercooling is fundamentally an extensive property [3], depending largely on the amount of PCM as the determinant of the probability of autonucleation. In the absence of accurate models, supercooling PCM heat storage devices are designed conservatively to insure the desired supercooling behavior and considerable experimental effort would be required to relax these conservative limits.

Validated methodologies to predict PCM heat exchanger sizing and performance, analogous to conventional heat exchanger design, are another gap [93]. The absence of standard heat exchange methodologies for PCM energy storage systems leads engineers to rely heavily on capital-intensive incremental scaleup and computationally expensive computer simulations to predict PCM heat exchange and heat storage performance. A normalized understanding of heat exchanger performance for PCM heat storage would bridge these gaps in scaleup more easily [93], ultimately allowing greater effort to be allocated to optimization.

It has been noted recently that the results of numerical modeling of phase change heat transfer, including melting and solidification, are increasingly published without any experimental validation [94]. Much of this numerical work relies on built-in features in commercial software, for which some parameters, such as the Carman–Koseny or mushy zone constant, are not properly validated through experimental studies [95]. For such studies the limit of applicability of the model is difficult to determine, and in some cases it is apparent that the results are unphysical. With the ever increasing computing power available for numerical studies, it is time to gain a better understanding of the models through validation with well-designed experiments.

8 OUTLOOK

PCM-based thermal storage might be the oldest form of energy storage known to humanity, as our ancestors valued ice for exactly that purpose. Today, what could be termed "cold storage" is well developed and finds applications in food preservation and peak shifting of energy loads from day to night. All this is possible through the use of the best and most abundant PCM: water.

The use of other PCMs, especially for storage at higher temperatures, has seen slower development. With the worldwide need for renewable energy sources and their intermittent nature, along with improved PCM characterization, PCM thermal storage is now becoming an important player in energy technologies. Numerous pilot projects are under way in which PCMs are incorporated in buildings as part of the building structure. Research, backed by various industries, is also under way looking at incorporation of PCMs in electronic components, PV panels, clothing, and packaging for temperature control, as well as in applications for waste heat harvesting.

The largest challenges in the further development of PCM thermal storage are design and integration of PCMs into particular applications, with the rate problem at the forefront of those considerations. From an engineering point of view, most experimental and modeling research concerning PCMs concentrates on accurate understanding of the overall thermal and energy behavior of PCMs and PCM systems, and mechanisms to address the rate problem. Industry is adding to this mix by looking at economical ways to achieve the desired storage.

Encapsulation of PCMs is still an important issue to be considered, with solutions optimized for heat transfer, cost, and ease of manufacture yet to come. Breakthroughs in this area will increase the use of PCM thermal storage. Finally, some chemists, materials scientists, and engineers in the field are investigating another critical matter: development of novel PCMs with enhanced physical properties. Such materials could help solve the rate problem, reducing the required size of storage systems, while facilitating manufacturing and encapsulation processes.

REFERENCES

[1] Sharma A, Tyagi VV, Chen CR, Buddhi D. Review on thermal energy storage with phase change materials and applications. Renew Sust Energy Rev 2009;13:318–45.

[2] Zalba B, Marin JM, Cabeza LF, Mehling H. Review on thermal energy storage with phase change: materials, heat transfer analysis and applications. Appl Therm Eng 2003;23:251–83.

[3] Lane GA. Solar heat storage: background and scientific principles. Boca Raton, FL: CRC Press; 1983.

[4] Pielichkowska K, Pielichowski K. Phase change materials for thermal energy storage. Prog Mat Sci 2014;65:67–123.

[5] Atkins P, de Paula J. Physical chemistry. 9th ed. New York: W.H. Freeman & Company; 2010.

[6] Farid MM, Khudhair AM, Razack SAK, Al-Hallaj S. A review on phase change energy storage: materials and applications. Energy Convers Manage 2004;45:1597–615.

[7] Khudhair AM, Farid MM. A review on energy conservation in building applications with thermal storage by latent heat using phase change materials. Energy Convers Manage 2004;45: 263–75.
[8] Zhou D, Zhao CY, Tian Y. Review on thermal energy storage with phase change materials (PCMs) in building applications. Appl Energy 2012;92:593–605.
[9] White MA. Physical properties of materials. Boca Raton, FL: CRC Press; 2012.
[10] Whitman CA, Johnson MB, White MA. Characterization of thermal performance of a solid-solid phase change material, di-*n*-hexylammonium bromide, for potential integration in building materials. Thermochim Acta 2012;531:54–9.
[11] Alkan C, Gunther E, Hiebler S, Ensari OF, Kahraman D. Polyurethanes as solid-solid phase change materials for thermal energy storage. Sol Energy 2012;86:1761–9.
[12] Xi P, Gu X, Cheng B, Wang Y. Preparation and characterization of a novel polymeric based solid-solid phase change heat storage material. Energy Convers Manage 2009;50:1522–8.
[13] Noël JA, Allred PM, White MA. Life cycle assessment of two biologically produced phase change materials and their related products. Int J Life Cycle Assess 2015;20:367–76.
[14] de Gracia A, Rincon L, Castell A, Jiminez M, Boer D, Medrano M, Cabeza LF. Life cycle assessment of the inclusion of phase change materials (PCMs) in experimental buildings. Energy Build 2010;42:1517–23.
[15] Menoufi K, Castell A, Farid MM, Boer D, Cabeza LF. Life cycle assessment of experimental cubicles including PCM manufactured from natural resources (esters): a theoretical study. Renew Energy 2013;51:398–403.
[16] Kenisarin MM. Thermophysical properties of some organic phase change materials for latent heat storage: a review. Sol Energy 2014;107:553–75.
[17] Cabeza LF. Advances in thermal energy storage systems: methods and applications. Toronto: Elsevier Science; 2014.
[18] Mehling H, Cabeza LF. Heat and cold storage with PCM. Berlin: Springer; 2008.
[19] http://www.mm21dhc.co.jp/english/index.html
[20] http://www.cristopia.com
[21] http://www.calmac.com/
[22] http://www.alfalaval.com/media/stories/industries/worlds-tallest-building-stays-cool-with-innovative-ice-storage-system/?id=10172
[23] http://www.bloomberg.com/news/articles/2014-08-01/goldman-s-icy-arbitrage-draws-interest-to-meet-epa-rule
[24] http://www.glaciertek.com/
[25] Cabeza LF, Castell A, Barreneche C, de Gracia A, Fernández AI. Materials used as PCM in thermal energy storage in buildings: a review. Renew Sust Energy Rev 2011;15:1675–95.
[26] Baetens R, Jelle BP, Gustavsen A. Phase change materials for building applications: a state-of-the-art review. Energy Build 2010;42:1361–8.
[27] Memon SA. Phase change materials integrated in building walls: a state of the art review. Renew Sust Energy Rev 2014;31:870–906.
[28] Arkar C, Medved S. Free cooling of a building using PCM heat storage integrated into the ventilation system. Sol Energy 2007;81:1078–87.
[29] http://www.sustainableinsteel.eu/p/556/pcm_products.html
[30] http://energain.co.uk/Energain/en_GB/products/thermal_mass_panel.html
[31] Feldman D, Banu D, Hawes DW. Low chain esters of stearic acid as phase change materials for thermal energy storage in buildings. Sol Energy Mat Sol Cells 1995;36:147–57.
[32] Farid MM, Kong WJ. Underfloor heating with latent heat storage. P I Mech Eng A-J Power Energy 2001;215:601–9.

[33] Telkes M. Nucleation of supersaturated inorganic salt solutions. Ind Eng Chem 1952;44:1308–10.

[34] Telkes M. Thermal energy storage in salt hydrates. Sol Energy Mat 1980;2:381–93.

[35] Ryu HW, Woo SW, Shin BC, Kim SD. Prevention of supercooling and stabilization of inorganic salt hydrates as latent heat storage materials. Sol Energy Mat Sol Cells 1992;27:161–72.

[36] Lane GA. Phase change materials for energy storage nucleation to prevent supercooling. Sol Energy Mat Sol Cells 1991;27:135–60.

[37] Himram S, Suwono A, Mansoori GA. Characterization of alkanes and paraffin waxes for application as phase change energy storage medium. Energy Sources 1994;16:117–28.

[38] Yuan Y, Zhang N, Tao W, Cao X, He Y. Fatty acids as phase change materials: a review. Renew Sust Energy Rev 2014;29:482–98.

[39] Desgrosseilliers L, Whitman CA, Groulx D, White MA. Dodecanoic acid as a promising phase-change material for thermal energy storage. Appl Therm Energy 2013;53:37–41.

[40] Rozanna D, Chuah TG, Salmiah A, Choong TSY, Sa'ari M. Fatty acids as phase change materials (PCMs) for thermal energy storage: a review. Int J Green Energy 2004;1:495–513.

[41] Sole A, Neumann H, Niedermaier S, Martorell I, Schossig P, Cabeza LF. Stability of sugar alcohols as PCM for thermal energy storage. Sol Energy Mat Sol Cells 2014;126:125–34.

[42] Nomura T, Zhu C, Sagara A, Okinaka N, Akiyama T. Estimation of thermal endurance of multicomponent sugar alcohols as phase change materials. Appl Therm Eng 2015;75:481–6.

[43] Diarce G, Gandarias I, Campos-Celador A, Garcia-Romero A, Greisser UJ. Eutectic mixtures of sugar alcohols for thermal energy storage in the 50–90 degrees C temperature range. Sol Energy Mat Sol Cells 2015;134:215–26.

[44] Khodadadi JM, Hosseinizadeh SF. Nanoparticle-enhanced phase change materials (NEPCM) with great potential for improved thermal energy storage. Int Commun Heat Mass 2007;34: 534–43.

[45] Ho CJ, Gao JY. Preparation and thermophysical properties of nanoparticle-in-paraffin emulsion as phase change material. Int Commun Heat Mass 2009;36:467–70.

[46] Jesumathy S, Udayakumar M, Suresh S. Experimental study of enhanced heat transfer by addition of CuO nanoparticle. Heat Mass Transf 2012;48:965–78.

[47] Nayak KC, Saha SK, Srinivasan K, Dutta P. A numerical model for heat sinks with phase change materials and thermal conductivity enhancers. Int J Heat Mass Transf 2006;49:1833–44.

[48] Strith U. An experimental study of enhanced heat transfer in rectangular PCM thermal storage. Int J Heat Mass Transf 2004;47:2841–7.

[49] Agyenim F, Hewitt N, Eames P, Smyth M. A review of materials, heat transfer and phase change problem formulation for latent heat thermal energy storage systems (LHTESS). Renew Sust Energy Rev 2010;14:615–28.

[50] Mills A, Farid M, Selman JR, Al-Hallaj S. Thermal conductivity enhancement of phase change materials using a graphite matrix. Appl Therm Eng 2006;26:1652–61.

[51] Karaipekli A, Sari A, Kaygusuz K. Thermal conductivity improvement of stearic acid using expanded graphite and carbon fiber for energy storage applications. Renew Energy 2007;32: 2201–10.

[52] Yu L. Nucleation of one polymorph by another. J Am Chem Soc 2003;125:6380–1.

[53] Yu L. Growth rings in D-sorbitol spherulites: connection to concomitant polymorphs and growth kinetics. Cryst Grow Design 2003;3:967–71.

[54] Joseph A, Kabbara M, Groulx D, Allred P, White MA. Characterization and real-time testing of phase change materials for solar thermal energy storage. Int J Energ Res 2016;40:61–70.

[55] Mazman M, Cabeza LF, Mehling H, Nogues M, Evliya H, Paksoy HO. Utilization of phase change materials in solar domestic hot water systems. Renew Energy 2009;34:1639–43.

[56] Prakash J. A solar water heater with a built-in latent heat storage. Energy Convers Manage 1985;25:51–6.

[57] Mehling H, Cabeza LF, Hippeli S, Hiebler S. PCM-module to improve hot water heat stores with stratification. Renew Energy 2003;28:699–711.

[58] de Gracia A, Oro E, Farid MM, Cabeza LF. Thermal analysis of including phase change material in a domestic hot water cylinder. Appl Therm Eng 2011;31:3938–45.

[59] Shukla A, Buddhi D, Sawhney RL. Solar water heaters with phase change material thermal energy storage medium: a review. Renew Sust Energy Rev 2009;13:2119–25.

[60] Pinel P, Cruickshank CA, Beausoleil-Morrison I, Wills A. A review of available methods for seasonal storage of solar thermal energy in residential applications. Renew Sust Energy Rev 2011;15:3341–59.

[61] Sandnes B, Rekstad J. Supercooling salt hydrates: stored enthalpy as a function of temperature. Sol Energy 2006;80:616–25.

[62] Araki N, Futamura M, Makino A, Shibata H. Measurements of thermophysical properties of sodium acetate hydrate. Int J Thermophys 1995;16:1455–66.

[63] Desgrosseilliers L, Groulx D, White MA, Swan L. Thermodynamic evaluation of supercooled seasonal heat storage at the Drake Landing solar community. Eurotherm Seminar 99, Lleida, Spain; 2014.

[64] Wei LL, Ohsasa K. Supercooling and solidification behavior of phase change material. ISIJ Int 2010;50:1265–9.

[65] Zhang H, van Wissen RMJ, Nedea SV, Rindt CCM. Characterization of sugar alcohols as seasonal heat storage media - experimental and theoretical investigations. Eurotherm Seminar 99, Lleida, Spain; 2014.

[66] Sandnes B. The physics and the chemistry of the heat pad. Am J Phys 2008;76:546–50.

[67] Anthony AEM, Barrett PF, Dunning PK. Verification of a mechanism for nucleating crystallization of supercooled liquids. Mat Chem Phys 1990;25:199–205.

[68] Sandnes B. Exergy efficient production, storage and distribution of solar energy. PhD Thesis. Department of Physics, University of Oslo, Oslo, Norway; 2003.

[69] Hirano S, Saitoh TS. Growth rate of crystallization in disodium hydrogenphosphate dodecahydrate. J Thermophys Heat Transf 2002;16:135–40.

[70] Hirano S, Saitoh TS. Long-term performance of latent heat thermal energy storage using supercooling. ISES Solar World Congress 2007 Solar Energy and Human Settlement; 2009.

[71] Wei J, Kawaguchi Y, Hirano S, Takeuchi H. Study on a PCM heat storage system for rapid heat supply. Appl Therm Eng 2005;25:2903–20.

[72] Desgrosseilliers L, Allred P, Groulx D, White MA. Determination of enthalpy-temperature-composition relations in incongruent-melting phase change materials. Appl Therm Eng 2013;61:193–7.

[73] Guillot S, Faik A, Rakhmatullin A, Lambert J, Veron E, et al. Corrosion effects between molten salts and thermal storage material for concentrated solar power plants. Appl Energy 2012;94:174–81.

[74] Liu M, Saman W, Bruno F. Review on storage materials and thermal performance enhancement techniques for high temperature phase change thermal storage systems. Renew Sust Energy Rev 2012;16:2118–32.

[75] Dinter F, Gonzalez DM. Operability, reliability and economic benefits of CSP with thermal energy storage: first year of operation of ANDASOL 3. Energy Proc 2014;49:2472–81.

[76] http://www.nrel.gov/csp/solarpaces/project_detail.cfm/projectID=3

[77] DOW Corning Corporation. http://www.dow.com/heattrans/products/synthetic/dowtherm.htm

[78] Green DW, Perry RH. Perry's chemical engineers' handbook. 8th ed New York: McGraw-Hill; 2008.
[79] Philip K. Power generation handbook: selection, applications, operation, and maintenance. New York: McGraw-Hill Professional; 2003.
[80] Wettermark G, Carlsson B, Stymne H. Storage of heat: a survey of efforts and possibilities. Stockholm: Swedish Council for Building Research; 1979.
[81] Maurer MJ, Bank DH, Soukhojak AN, Sehanobish K, Khopkar A, Sharma S, Shembekar P. Computational fluid dynamic modeling of latent heat discharge in a macro-encapsulated phase change material device. 7th International Energy Conversion Engineering Conference, Denver, CO; 2009.
[82] Murray RE, Groulx D. Experimental study of the phase change and energy characteristics inside a cylindrical latent heat energy storage system: part 1 consecutive charging and discharging. Renew Energy 2014;62:571–81.
[83] Shokouhmand H, Kamkari B. Experimental investigation on melting heat transfer characteristics of lauric acid in rectangular thermal storage unit. Exp Therm Fluid Sci 2013;50:201–12.
[84] Liu C, Groulx D. Experimental study of the phase change heat transfer inside a horizontal cylindrical latent heat energy storage system. Int J Therm Sci 2014;82:100–10.
[85] Fan L, Khodadadi JM. Thermal conductivity enhancement of phase change materials for thermal energy storage: a review. Renew Sust Energy Rev 2011;15:24–46.
[86] Medrano M, Yilmaz MO, Nogues M, Martorell I, Roca J, Cabeza LF. Experimental evaluation of commercial heat exchangers for use as PCM thermal storage systems. Appl Energy 2009;86: 2047–55.
[87] Murray RE, Groulx D. Experimental study of the phase change and energy characteristics inside a cylindrical latent heat energy storage system: part 2 simultaneous charging and discharging. Renew Energy 2014;63:724–34.
[88] Groulx D, Lacroix M. Study of the effect of convection on close contact melting of high Prandtl number substances. Int J Therm Sci 2007;46:213–20.
[89] Fomin SA, Saitoh TS. Melting of unfixed material in spherical capsule with non-isothermal wall. Int J Heat Mass Transf. 1999;42:4197–205.
[90] Mehling H. Enthalpy and temperature of the phase change solid-liquid — an analysis of data of compounds employing entropy. Sol Energy 2013;95:290–9.
[91] Mehling H. Enthalpy and temperature of the phase change solid-liquid — an analysis of data of the elements using information on their structure. Sol Energy 2013;88:71–9.
[92] Mehling H. Analysis of the phase change solid–liquid—new insights on the processes at the atomic and molecular level. Eurotherm Seminar 99, Lleida, Spain; 2014.
[93] Kabbara M. Real time solar and controlled experimental investigation of a latent heat energy storage system. M.A.Sc. thesis, Department of Mechanical Engineering, Dalhousie University, Halifax, NS, Canada; 2015.
[94] Dutil Y, Rousse DR, Salah NB, Lassue S, Zalewski L. A review on phase-change materials: mathematical modeling and simulations. Renew Sust Energy Rev 2011;15:112–30.
[95] Kheirabadi AC, Groulx D. The effect of the mushy-zone constant on simulated phase change heat transfer. Proceedings of CHT-15: ICHMT International Symposium on Advances in Computational Heat Transfer, 2015. p. 22.

Chapter 14

Solar Ponds

César Valderrama*, José Luis Cortina*,, Aliakbar Akbarzadeh†**
**Departament d'Enginyeria Química, Universitat Politècnica de Catalunya, Spain; **Water Technology Center CETaqua, Barcelona, Spain; †School of Aerospace, Mechanical and Manufacturing Engineering, RMIT University, Australia*

1 INTRODUCTION

Climate change and depletion of fossil fuels are of major concern. The world community has prompted a broad discussion of technical and financial resources for promoting increased energy efficiency in the use of renewable energy resources. Solar energy is regarded as one of the most promising substitutes for traditional energy resources; however, its intermittent and unstable nature is a major drawback, which leads to a disparity between supply and demand [1]. To enhance the fraction of energy utilization and make solar energy products more practical and attractive, thermal storage systems today are perceived as crucial components in solar energy applications. Thermal energy storage systems utilize either thermochemical reactions or the sensible or latent heat capacity of materials to provide a heating or cooling resource, which can be replenished as required [2]. In sensible thermal storage, energy is stored by changing the temperature of the storage material. The amount of heat stored is proportional to the density, specific heat, volume, and variation of temperature of the storage material. The performance of a storage system depends mainly on the density and specific heat of the substance used, which determine the necessary storage volume [3]. Water is one of the best storage liquids for low-temperature heat storage [4]. It has a higher specific heat than any other material and is cheap and widely available. Surface water bodies (ponds or lakes) can be used to collect and store solar heat [5]. In a pond the creation of a salinity gradient results in a higher salt concentration and density at the bottom, and the heat absorbed remains trapped there because the salinity gradient inhibits natural convection and the cooler water at the surface acts as an insulator as it does not mix with the saline water. Darkening the bottom surface of the pond also results in more solar radiation being absorbed [2]. There are many factors which affect the economical and operational size of the heat storage solar pond for a particular application. These factors include (1) the purpose of the solar energy system

Storing Energy. http://dx.doi.org/10.1016/B978-0-12-803440-8.00014-2

(load), (2) the area of the collector, (3) the meteorological conditions at the location, and (4) the operational characteristics of the system [6].

Solar ponds are an old natural phenomenon that was first documented by Von Kalecsinsky (1902) for Medve Lake in Transylvania (Hungary) where temperatures up to 70 °C at a depth of 1.32 m were recorded at the end of the summer. Similar observations were reported by Anderson (1958) and Wilson and Wellman (1962) for several other lakes as well as by other authors [7–9].

The concept of an artificial solar pond as a possible means of collection and storage of solar energy was proposed in the middle of the last century [7]. The convection currents that normally develop due to the presence of hot water at the bottom and cold water at the top are diminished or minimized by the presence of a strong density gradient from bottom to top. Thus, the water in the lower zones can be warmer than the water above without, simultaneously decreasing density and causing convection to the surface [9,10]. The most attractive characteristics of solar ponds are, first, the capacity of long-term storage which can supply sufficient heat for the entire year and, second, the annual collection efficiency in the range (15–25)% for all locations and the capacity to supply adequate heat even at higher latitudes. It has been found that solar ponds of areas of the order of 1000 m^2 or higher are more cost effective than flat plate collectors with higher efficiency as their cost per square meter is much lower than those for flat plate collectors [11].

2 TYPES OF SOLAR PONDS

There are several types of solar ponds and most are based on the use of a salinity gradient. Their designs attempt to overcome some of the disadvantages observed in these types of ponds, for example, the need to establish a salinity gradient and the intensive level of maintenance and operation procedures that is required. Several authors have proposed different classifications of solar ponds; among them are the formulations by Kaushika [12], El-Sebaii et al. [13], and Ranjan [11]. Solar ponds can be classified according to four basic factors: (a) convecting or nonconvecting, (b) partitioned (multilayered) or nonpartitioned, (c) gelled or nongelled, and (d) separate collector and storage or in-pond storage [2]. However, most research efforts are presently focused on the nonconvecting salinity gradient solar pond [14] which will be discussed in detail. They are easier to operate and cheaper to construct. Other types will also be briefly discussed based on the most representative studies reported in the literature.

It is observed that salinity gradient solar ponds (see Section 2.1) have the advantage of long-term energy storage over nonsalinity solar ponds such as membrane-stratified ponds and shallow solar ponds, which are more suitable for short-term energy storage because of the higher rate of increase of the temperature of the pond water.

2.1 Salinity Gradient Solar Pond

A solar pond consists of three distinct zones as can be seen in Fig. 14.1 [8,15]. The first zone, which is located at the top of the pond and contains the less dense salt/water mixture, is the absorption and transmission region, also known as the upper convective zone (UCZ), which has the function of protecting the salinity gradient layer. Its stability is controlled by the addition of water onto the solar pond surface and the prevention of wind agitation. The second zone, which contains a variation of salt/water density, increasing with depth, is the gradient zone or nonconvective zone (NCZ), also called the salinity gradient layer. The main purpose of this zone is to act as an insulator to prevent heat from escaping to the UCZ, thus maintaining higher temperatures at the deeper zones. The last zone is the lower convective zone (LCZ) also called the energy storage zone, which consists of saturated brine with almost homogeneous salinity and density. The salts used include sodium chloride, magnesium chloride, or sodium nitrate which are dissolved in water with concentrations varying from (20 to 30)% at the bottom (LCZ) to almost zero at the top (UCZ) [11].

When solar radiation is incident on the solar pond, part of the radiation is reflected away from the top surface while most of the incident sunlight is transmitted down through the top surface of the UCZ. A fraction of the transmitted radiation is rapidly absorbed in the surface layer. However, this absorbed heat is lost to the atmosphere by convection and radiation heat transfer. Some of the remaining radiation is absorbed in the middle NCZ before the rest reaches the bottom of the pond. In the LCZ the absorbed solar energy is converted to heat and stored as sensible heat in the high-concentration brine. Since there are no heat losses by convection from the bottom layer the temperature of this layer can rise substantially. A double-diffusion process occurs where the temperature and salinity fields make opposing contributions to the fluid density [16]. The temperature difference between the top and the bottom can be as high as 60 °C [17]. The thermal energy stored can be used for the heating of buildings, power production, and industrial processing [18]. It has been studied around the world

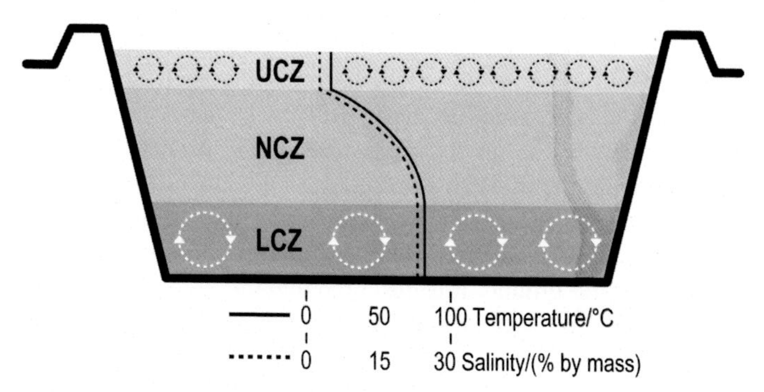

FIGURE 14.1 **Salinity gradient solar pond scheme.**

for more than half a century and successful case studies have been reported in Israel, USA, India, China, Australia, and Spain [19–24].

2.1.1 Design and Construction

The design objective is to meet the energy requirements of the application which integrates the solar pond as a heat source at low temperature. The medium to be heated (water, air, other fluids) and the temperature to be supplied are critical to the design specifications. Economic viability is critically affected both by the location of the facility and the maximum use of local resources. The availability of a salt source (salt or brine), freshwater, and enough land area (flat terrain) are the key requirements in the assessment of potential locations [23]. The next step is sizing the solar pond to meet energy requirements. The most important parameter is solar radiation; the higher the radiation on the site selected the higher the energy efficiency and operating temperature of the pond. A typical solar pond of depth 3 m and a storage zone of 1 m thickness would receive approximately (20–25)% of the radiation incident upon the pond surface. Heat losses to the ground must also be taken into account and, in practice, approximately (15–20)% of the incoming radiation is available for extraction to an application, resulting in a rise of (40–50) °C above the daily average temperature of the location. A simple way to estimate the surface area needed to meet any particular average thermal load was proposed by Akbarzadeh et al. [23]. First, the annual solar energy incident on a square meter of horizontal surface at the location of the pond (e.g., 7 GJ m^{-2} a^{-1}, where "a" refers to annum) is calculated. Then, an estimate of the horizontal surface area on which the incoming solar radiation over a year is required to meet the annual load. If, for instance, the annual load is 2800 GJ, this area would be 400 m^2. Finally, multiply this surface area by a factor of 5–10 to estimate the surface area of a solar pond to meet this annual load. In this example the solar pond would need to have an area between (2000 and 4000) m^2 to supply an annual thermal load of 2800 GJ. The thermal performance of a solar pond is also affected by the pond depth; the pond is usually contained in earth excavated to a depth of (3 or 4) m. The deeper the storage zone the larger the storage capacity and the larger the quantity of heat available for an extended time as a result of the reduction in heat losses and and increased thermal efficiency of the pond.

Solar pond construction involves consideration of several strategies related to the location of the pond, earth excavation (land chosen should be as flat as possible), lining, insulation, and pond shape. Local geology is critical in site selection and care must be taken to avoid connections with aquifers [12]. To minimize heat losses to the ground, it is desirable that the groundwater table should be at least 5 m below the ground surface. For large ponds construction the best option involves the use of soil excavated from the periphery of the pond for the pond walls. The bottom of the pond will thus be below the surrounding ground level. This approach can represent a significant reduction in the construction costs of the pond. A lining is necessary, for both environmental as well as performance reasons. The liner material should be able to withstand the anticipated maximum pond temperature, be resistant to UV radiation, and should not react with the salt. Above all, it should be mechanically strong. Thermal insulation at the base of the pond and the sides may be used to reduce heat losses; this may enhance pond heating but may represent excessive cost, especially for larger ponds.

2.1.2 Settling the Salinity Gradient

The stratification of layers is artificially done. Before the pond is heated the profile set when filling the pond may remain constant provided there are no external disturbances such as wind. It is crucial that when the bottom is heated the density at the bottom is greater than that of the cooler brine on top, otherwise mixing takes place. Three methods have been adopted for the establishment of the initial salt density gradient: natural diffusion, stacking, and redistribution. The first method relies on the natural diffusion between a freshwater layer and the layer of saturated salt solution [12]. In the natural diffusion method the upper half is filled with water; top and bottom concentrations are maintained constant by regularly washing the surface and by adding salt in the bottom. Due to the upward diffusion of salt a salinity gradient will be established [13]. Stacking involves filling the pond with a storage layer of high-concentrated salt solution and several other layers of salt solutions at different concentrations. The concentration of salt in successive layers is changed in steps from near saturation at the bottom to freshwater at the top. The practical approach for stacking used in most solar ponds is that the bottom layer is filled first and successively lighter layers are floated upon the lower denser layers [25]. The redistribution method is considered to be most convenient and recommended for large-area solar ponds. The first step is to fill the pond up to half the depth of the planned gradient zone with high-concentration brine. Freshwater or low-salinity water is then injected horizontally into the brine through a diffuser. The water will stir and uniformly dilute the brine above the diffuser. Injection starts from the desired level of the boundary between the NCZ and the LCZ. As injection proceeds and the pond surface level of the water rises the diffuser is simultaneously raised in increments from its position within the brine solution toward the surface of the pond. The speed of raising the diffuser is twice the speed of the rise of the pond water; that is, after each 50 mm rise of water the diffuser is raised by 100 mm. The timing of diffuser movement is adjusted so that the diffuser and the water surface reach the final level at the same line, which is the boundary between the UCZ and NCZ. Finally, the UCZ is formed by adding freshwater above the current water surface, and at the end of this process the pond is full and the desired salinity gradient created as can be seen in Fig. 14.2 [12,13,23]).

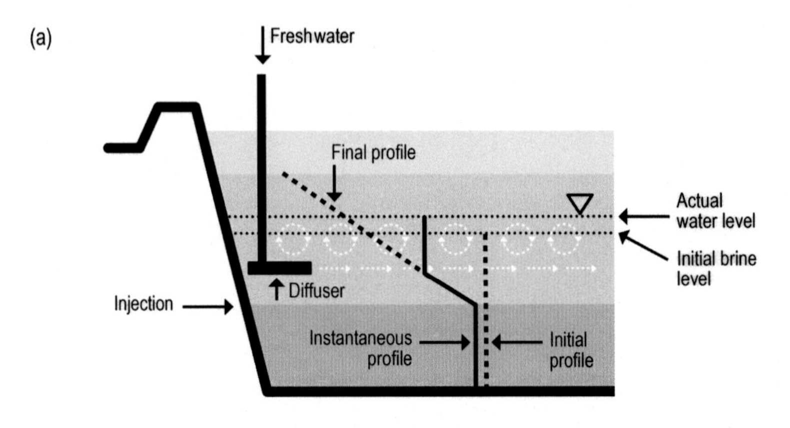

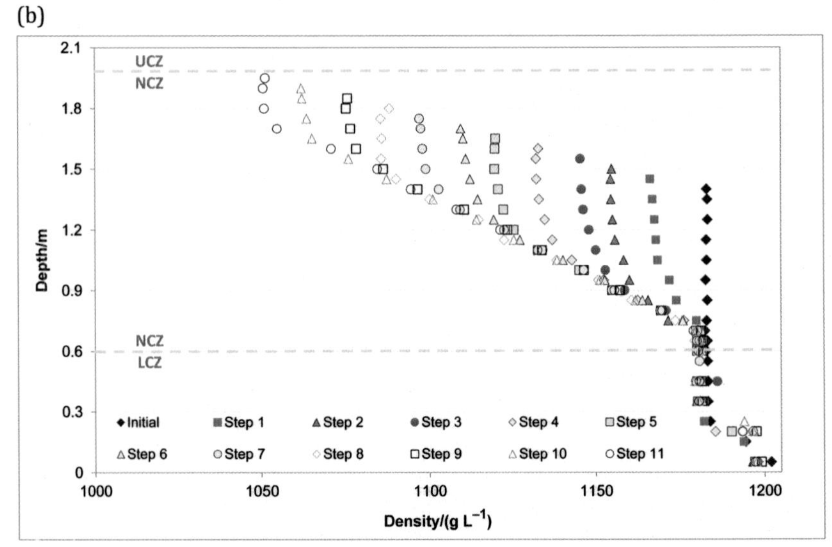

FIGURE 14.2 (a) Scheme of the redistribution method for establishing the salinity gradient and (b) the density profile obtained during salinity gradient settling in the 500 m^2 Escuzar Solar Pond, Granada (Spain) in June 2014.

2.1.3 Control and Maintenance

A solar pond should be maintained by periodical addition of a saturated salt solution at the bottom, washing the surface with freshwater, and monitoring and controlling disturbances such as treatments of algae blooms. Salt chargers and flushing systems have been designed and used recently for salt addition and to compensate the losses caused by evaporation and to renovate the surface water, respectively [24]. Moreover, the addition of freshwater (or even seawater) to keep both the pond depth and the surface concentration constant is a critical parameter in water scarcity areas. Indeed, it is important to point out that salt concentration in the UCZ can be fixed at approximately 4% and water only added when the concentration is higher than this value, thus avoiding high consumption of freshwater. A more complex situation is the case of inland solar pond applications where water is scarce. In such cases the availability of makeup water is a major issue and may limit the construction of a pond [26].

Maintaining a high transparency of water is one of the most important elements in achieving high efficiency and delivering the collected heat at high temperature. The growth of algae and deposition of tree leaves, flying debris such as dust or larger agglomerates that can contribute to an increase of pond water turbidity, and increased resistance and absorption of radiation above the storage zone are the main concerns. Acidification of the pond provides a simple and reliable maintenance procedure for preventing or inhibiting algal blooms and maintaining high transparency. In this case, the pH in the salt gradient zone should be maintained by the addition of hydrochloric acid to keep it below 4 [23,24,26].

Finally, a parameter to be considered is surface mixing, especially in locations with strong winds and for larger solar ponds. Surface turbulence depletes the salt concentration gradient and can promote complete convection in the surface region, causing a reduction in thermal performance. Windbreakers are used to reduce the possibility of wind-induced surface turbulence and most of the windbreaks used in pilot and larger solar ponds are floating rings [23,24].

2.1.4 Heat Extraction

There are two methods of extracting heat from the lower convective zone of a solar pond. The first, which is the most commonly used method, is pumping hot brine from the LCZ through a diffuser to prevent excessive velocity and motion within the pond and thereby minimize erosion of the gradient zone. An external heat exchanger is used to extract heat from brine and the cooled brine is returned to the bottom of the pond (Fig. 14.3a). This method was successfully used at the 5 MW solar pond power plant in Beit Ha'arava (Israel) near the Dead Sea; heat was extracted from the top of the storage zone and the cold and heavier brine was returned to the bottom of the pond at the same side of the pond. Being cooler than the stored hot brine, it remained on the bottom of the pond and spread over the entire area before rising up when filling the volume and/or being heated during the next charging period [26,27].

The second method involves a heat exchanger that is placed in the lower convective zone of the pond. Its most appropriate position is just below the gradient zone, so that the heat removal can stimulate convection throughout the lower convective zone and remove heat from its entire volume (Fig. 14.3b). Since the working fluid is typically freshwater the heat exchanger can be constructed from low-cost materials such as plastics. The low thermal conductivity of the plastic pipes is compensated by increasing the heat transfer area on the in-pond pipes; that is, by installing more pipes and/or increasing the diameter

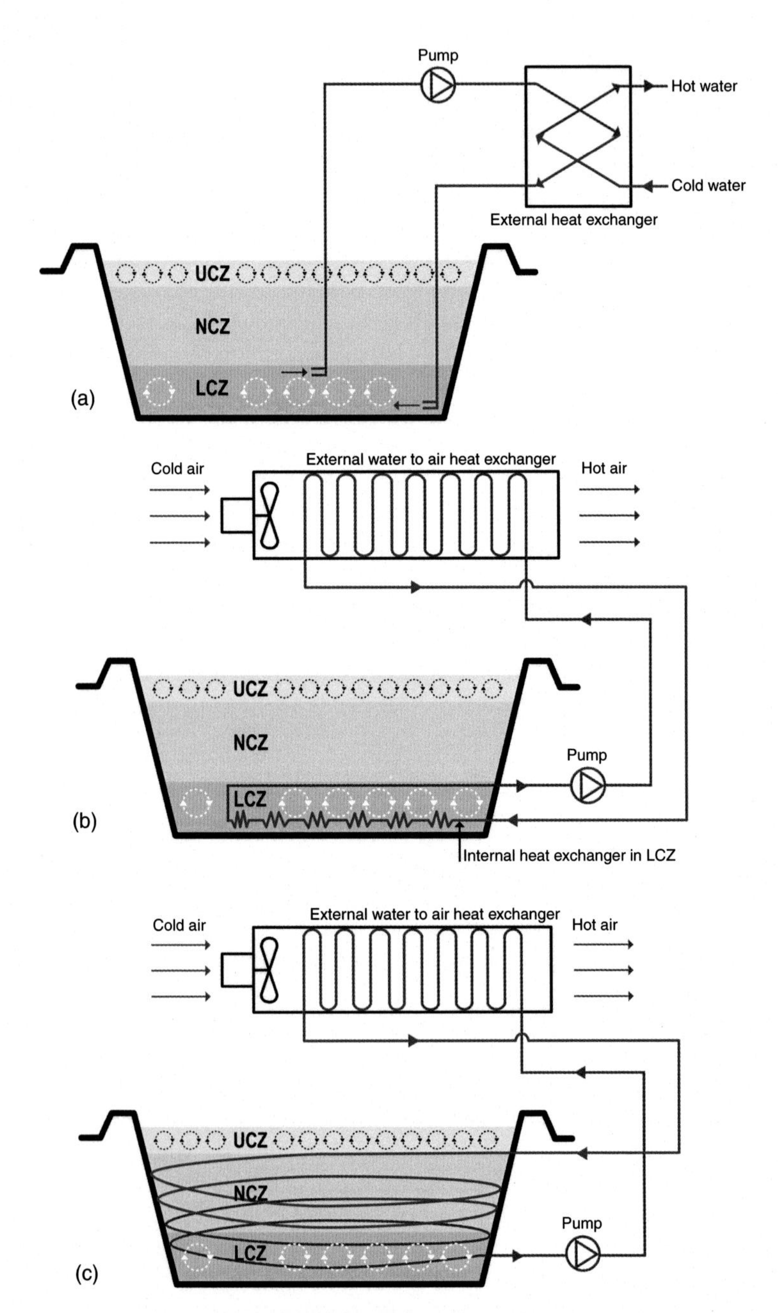

FIGURE 14.3 **Heat extraction methods in solar pond technology.** (a) External heat exchanger, (b) internal heat exchangers, and (c) internal heat exchanger at the gradient zone.

of the pipes. The disadvantages of this method are related to the large quantity of tubes required, difficulties in locating the heat exchanger, and the difficulty in effecting repairs. This method was used at a 3000 m^2 demonstration solar pond in Pyramid Hill (Australia) by using plastic pipes connected to weights on the pond bottom by ropes to overcome buoyancy forces and has proved to be a simple and reliable method of heat extraction [13,23]. These two heat extraction methods are shown in Figs. 14.3(a,b). Recently, a novel system of heat extraction has been proposed and assessed both theoretically and experimentally with the aim of improving overall energy efficiency. In this method, heat is extracted from the NCZ as well as, or instead of, the LCZ (Fig. 14.3c). Theoretical and experimental investigation showed that heat extraction from the NCZ increases overall thermal efficiency by up to 55% compared with the conventional method of heat extraction solely from the LCZ [28,29].

2.2 Saturated Solar Ponds

This type of pond is designed to improve or reduce the level of maintenance of the salinity gradient by making the pond saturated at all levels, with a salt whose solubility increases with temperature. A number of salts besides KNO_3 have been used and are found to be appropriate; these include $Na_2B_4O_7$ (borax), $KAl(SO_4)_2$, $CaCl_2$, $MgCl_2$, and NH_4NO_3 [11]. Such saturated ponds have no apparent diffusion problems and the gradients are self-sustaining depending on local temperature. This gives these ponds the advantage of inherent stability [13]. Harel et al. [30] developed the *equilibrium solar pond* concept as a generalization of the saturated solar pond and the anticipated advantages over the saturated solar pond, which relies on the fact that the fluid in the *equilibrium solar pond* is unsaturated. This advantage has two important consequences: crystallization is only possible under significant cooling of the brine; and the absence of solid salt at the bottom of the pond. Both consequences lead to an increase in the absorption of energy in the LCZ and thus higher thermal performance of the pond.

Two main advantages can be identified in comparison to the conventional solar pond: first, the zero salt flux throughout the pond eliminates the need for addition of salt after the pond is set up and the need for disposal of water from the pond. Second, due to the high concentration in the bottom region a higher bottom temperature can be achieved before the onset of boiling of the salt solution, thus increasing the thermal efficiency of the pond [11].

2.3 Solar Gel and Membrane Ponds

A solar gel or viscosity-stabilized pond is a nonconvective and nonsalt solar pond and was proposed to minimize or eliminate evaporation losses from the surface by reduction of heat losses. These ponds use a transparent polymer gel as a nonconvecting layer. The polymer gel has low thermal conductivity and is used at a near solid state, so that it will not convect [31]. The polymer gel has good optical and thermal insulating properties but the cost of the gel is high.

Materials suitable for viscosity-stabilized solar ponds should have high transmittance for solar radiation, high efficiency of the chosen thickness, and should be capable of performing at temperatures up to 60 °C. Polymers such as gum arabic, locust bean gum, starch, and gelatin are all potentially useful materials for this configuration [13]. Wilkins [32] reported the design and construction of two solar gel ponds of (110 and 400) m^2 (5 m deep with a gel thickness of 0.60 m) in New Mexico to provide process heat for a food company. It was observed that solar gel ponds are superior to salt gradient solar ponds from the point of view of efficiency, ease of operation, and economics [33]. The potential and economic feasibility of gel ponds as a source of hot water (45 °C) for domestic use in the United States was demonstrated. Industrial applicability of gel ponds as a source of hot water (65 °C) for a textile mill in Cairo (Egypt) has also been shown [11].

Membrane-stratified solar ponds belong to the group of nonconvective solar ponds, having a body of liquid between closely spaced transparent membranes to minimize convective heat transfer. The disadvantage of adding a physical layer to a system where solar radiation is the only input is reduction of total transmission of sunlight to the bottom of the pond. Thus, the membrane space for suppressing convection should be very small and a large number of highly transparent films is required [31]. Three types of membranes are suggested for the membrane-stratified solar pond, which are horizontal sheets, vertical tubes, and vertical sheets [13].

2.4 Shallow Solar Pond

The term shallow solar pond (SSP) has been derived from that of the solar still. The name implies that the depth of water in the SSP is relatively small, typically (4–15) cm, which is like a conventional solar still consisting of a blackened tray holding some water [34]. An SSP is essentially a large water bag or pillow placed within an enclosure with a clear upper glazing. Water is placed within the bag, which is generally constructed from clear upper plastic film and a black lower plastic film, in such a way that the film is in contact with the top surface of the water, and thus prevents the cooling effect due to evaporation [35]. The black bottom of the pond absorbs solar radiation; as a result, the water gets heated.

Total solar energy absorbed by system cannot all be used as useful energy. There are losses due to conduction, convection, and radiation. Using a suitable insulation material, conduction losses can be reduced. To reduce the thermal losses by convection and radiation, one or two transparent sheets are used over the pond [36]. Solar energy collection efficiency is directly proportional to water depth, whereas water temperature is inversely proportional to water depth. Solar energy is converted to thermal energy by heating the water during the day.

The water is withdrawn from the SSP before sunset (or more precisely when the collection efficiency approaches zero) for utilization or storage. Different studies have been carried out with the aim of assessing their performance. The Solar Energy Group at Lawrence Livermore Laboratory (USA) developed an SSP to supply heated water (25–60) °C for industrial and commercial uses around 1973 [37]. Later, in the late 1980s, the Erel Company developed a shallow water pond on top of which float thermal diodes composed of an array of translucent honeycombs made of plastic material. The water heated in the pond could reach temperatures of about 85 °C. This type of pond is suitable for supplying warm water for household use and other low-temperature applications, for example, laundries, textile factories, canned food factories, greenhouses. More recently, a pilot system was installed in the Kibbutz Maoz-Haim (Israel) to supply hot water for a housing project of 42 units [38]. El-Sebaii [39] studied theoretically and experimentally an SSP integrated with a baffle plate to suit prevailing weather conditions at Tanta (Egypt). The system could provide 88 L of hot water at a maximum temperature of 71 °C at 3:00 pm with a daily efficiency of 64% when the baffle plate was used without vents. The pond could retain hot water until 7:00 am the next day at a temperature of 43 °C, which represents a benefit for most domestic applications.

3 INVESTMENT AND OPERATIONAL COST

The costs of a solar pond vary widely according to location and application. The major cost factors are the earthwork (excavation, leveling, and compaction), the lining and isolation, the salt source, the freshwater or low-density water, the land, and monitoring and control equipment. Furthermore, economic feasibility analysis of the specific design, the site, and integration of the solar pond with the application is essential. Indeed, economy of scale affects solar pond technology, thus small ponds are considerably more expensive than large ponds, and there is wide variability in per unit area cost estimates of solar ponds operated at different locations throughout the world. This variability was reported by Kaushika [12] for two solar ponds of approximately 2,000 m^2 constructed between 1981 and 1982 in Alice Springs (Australia) and Miamisburg, Ohio (USA). The total costs reported were (15 and 32) $ m^{-2}, respectively. The relative 2015 cost of these both ponds taking into account inflation amounts to c. (39 and 83) $ m^{-2}, respectively. More recently, in 2014 a 500 m^2 solar pond was constructed in Granada (Spain) with the purpose of delivering heated water (>60 °C) to replace or minimize fuel oil consumption at a mineral flotation processing facility. The total cost of construction was 190 $ m^{-2} with 50% accounting for earth and civil engineering work. The site location was difficult to access, penalizing the cost of excavation, levelling, and civil works. Cost analysis was performed for this solar pond by increasing the size to 5 000 m^2. The cost of the expanded pond was estimated to be 90 $ m^{-2}, which was in line with the figure quoted by Kaushika [12] and confirming that economy of scale has a huge impact on solar pond viability.

Another cost factor with wide variability is the lining material. The cost of the lining of a solar pond in El Paso, Texas (USA) in 1991 was reported to be 4 $ m^{-2} [40], while liners for solar ponds in Alice Springs, Miamisburg, and Granada were quoted as (3, 11, and 9) $ m^{-2}, respectively. The lower cost was for cases in which one layer of plastic lining was sufficient, while the higher cost applies to intensive earth moving, sophisticated linings, underground leak detection, and monitoring facilities. Furthermore, the cost of manpower involved in the construction also impacts total cost [22].

Annual operation and maintenance costs for large ponds are approximately (3–5)% of the investment. This includes all previously mentioned tasks, including gradient control, monitoring, clarity control, and makeup water for evaporation compensation. A 5% figure can be used for small ponds, while 3% is more appropriate for large ponds [22]. The annual operation and maintenance cost for the Granada solar pond was estimated to be around 3%.

4 APPLICATIONS OF SOLAR PONDS

Solar ponds are ideal for storing energy for applications needing low-grade thermal energy.

4.1 Industrial Process Heating

Heating applications particularly suited to solar ponds include provision of warm air for commercial salt production [23]; grain, fruit, and wood drying; hot water for the dairy industry in rural locations [21]; remote mining operations [41]; and any industrial process in a rural environment requiring low-grade heat (at temperatures up to 80 °C). This form of heating is particularly important in saving fossil fuel consumption and thus reducing the emissions of greenhouse gases [13].

4.2 Desalination

Solar pond–powered desalination is a promising renewable energy system for producing significant quantities of freshwater. Research undertaken during the El Paso Solar Pond Project from 1987 to 1992 mainly focused on the technical feasibility of thermal desalination coupled with solar ponds [42]. Thermal desalination processes such as multistage flash and multiple effect evaporation may use solar ponds to heat incoming salty water with zero greenhouse emissions [29,43]. Saleh et al. [44] reported that a 3000 m^2 solar pond installed near the Dead Sea is able to provide an annual average production rate of 4.3 L min^{-1} of distilled water pointing out that the solar pond appears to be a feasible and appropriate technology for water desalination. Suarez et al. [45] evaluates the utilization of direct contact membrane distillation (DCMD) coupled to a salt gradient solar pond for sustainable freshwater production at terminal lakes.

Terminal lakes are water bodies that are located in closed watersheds and, therefore, the only output of water occurs through evaporation and infiltration. The majority of these lakes, which are commonly located in the desert and influenced by human activities, are increasing in salinity. Water production of the order of 2.7 L d^{-1} m^{-2} from a solar pond was reported to be possible if the pond is constructed inside a terminal lake.

Economic and technical assessment for solar ponds combined with a multistage flash (MSF) desalination system [43] indicating that large land areas of c. (73–185) m^2 are required to produce desalinated water at the rate of 1000 L d^{-1} assuming that the storage zone temperature ranged between (70–90) °C. Recently, Salata et al. [46] studied the feasibility of integrating a solar pond with an absorption heat transformer, the latter being a thermal machine that extracts heat from a source (at an available temperature) and enables a portion of the heat collected/obtained, to be available at higher temperatures. To produce 1 m^3 d^{-1} of desalinated water a solar pond area ranging from (1000 to 4000) m^2 is needed, together with a thermal flux drawn of between (40 and 20) W m^{-2}, respectively. Thus allowing the absorption heat transformer to increase the temperature of part of the stored energy (about 50%) to reach typical temperatures of up to 130 °C needed for the traditional desalination of seawater by distillation. A further benefit is that the more concentrated brine issuing from the desalination process may be used either to maintain solar ponds or in an integrated salt production system. This approach, called zero discharge desalination, proposes concentrating the rejected streams of brine solutions to near saturation point and using NaCI solutions to fill additional solar ponds. This system will be suitable at places where potable water is in short supply, but brackish water is available.

4.3 Electrical Power Production

Extensive research has also been carried out to utilize the thermal energy produced by solar ponds to produce electrical power. The best showcase for such power generation was a project near the Dead Sea (Israel) involving a large solar pond of 210 000 m^2 having a depth of 4.5 m linked to a Rankine cycle heat engine which produced 5 MW electrical power. In El Paso a 100 kW Rankine cycle turbine was used to generate electrical power from a 3700 m^2 solar pond and the power was fed to the local electricity grid. In Alice Springs (Australia), a 1600 m^2 solar pond supplied heat to run a 20 kW organic vapor screw expander Rankine cycle engine and generator [47,48]. Organic Rankine cycle (ORC) engines developed specifically to produce electric power from lower temperature heat sources (80–90) °C have been used in these applications. An ORC was developed by Ormat's solar pond power plants in Ein-Boqek and in Beit Ha'arava (Israel) [26]. The organic liquid, which absorbs the heat from the hot brine, vaporizes under relatively high–pressure conditions, expands through a special vapor turbine, then condenses at near atmospheric pressure and is pumped back into the vaporizer. Because of low temperature the solar pond power plant requires organic working fluids that have low boiling points such as halocarbons, e.g., freon, or hydrocarbons (such as propane) [13]. The characteristics of the low boiling point organic fluid simplifies the design of the turbine and the overall heat exchange system. However, conversion efficiency is limited due to the low operating temperature of (70–100) °C and the low thermodynamic performance, resulting in low net thermal-to-electric energy conversion efficiencies which are of the order of 7%. The overall efficiency for electric energy production was measured to be in the range (0.8–2) %, and this is due to low Carnot efficiency and low turbine efficiency which adversely affect their economic viability.

In recent times the concept of combining a chimney with a salt gradient solar pond for generation of power has been assessed through several demonstration projects. Incorporation of an air turbine (carrying out a trilateral flash cycle) into a solar chimney for desalination and power production has also been examined for salt-affected areas [49]. The results indicated that the system was able to produce power with the potential benefit of being able to generate power intermittently at any time (day or night) and at times of peak demand (or high cost for electricity) [50]. It is shown that for conditions in northern Victoria (Australia) (with daily annual solar radiation of 19 MJ m^{-2}) 60 kW of power can be generated for the case where air in the chimney has been heated from an initial (20–50) °C [11].

New trends in power generation by solar pond technology involve the application of thermoelectric concepts avoiding low conversion of thermal energy to power and offer remarkable influence on medium- and small-size solar ponds. Recent studies have been carried out by coupling solar pond with thermosyphon and thermoelectric modules for electric power generation at the laboratory scale [51]. A thermoelectric generator is a device which converts heat directly into electrical energy. The process is based on the Seebeck effect [52,53]. The thermoelectric generator system is designed to be powered by the hot and cold water from a salinity gradient solar pond with a temperature difference in the range (40–60) °C between the LCZ and UCZ. The system is capable of producing electricity even on cloudy days or at night as the salinity gradient solar pond acts as a thermal storage system. Preliminary results indicated that these systems have promising potential to produce electricity from low-grade heat for power supply in remote areas [11].

4.4 Salinity Mitigation

The integration of solar ponds with salinity mitigation or interception schemes is a particularly attractive potential application. Many areas of formerly productive land are suffering from rising salinity levels around the world as a result mainly of tree clearing and irrigation. Many salinity mitigation interception schemes involve the use of evaporation basins, into which saline groundwater is pumped. Solar radiation vaporizes the water, leaving the salt behind [18].

Solar ponds could be incorporated in such evaporation basins to produce heat and/or electricity from otherwise unproductive land. If evaporation ponds are established in a chain the first few ponds in the chain provide ideal opportunities for creating salt gradient solar ponds. While the surface of a solar pond acts as an evaporation surface, heat can be withdrawn from the bottom of the pond for industrial process heating or other applications.

4.5 Production of Chemicals

Solar ponds can be used to produce sodium sulfate, chloride salts, fertilizer, and other common industrial chemicals, either in situ or simply by making use of the heat provided by the pond. The most closely related commercial technologies are the evaporation ponds at salt production facilities [23]. This application is undoubtedly, at present, one of the most important uses of solar energy for processing heat [18]. A good example is the 2500 m^2 solar pond constructed and successfully operated in the production of lithium carbonate from the Zabuye salt lake in the Tibet plateau [54–56].

4.6 Aquaculture and Biotechnology

Applications requiring relative low-temperature water heating such as aquaculture and biotechnology farming, including fish, brine shrimps, and various algae, are ideal candidates for solar pond operations. In many cases, the solar pond can be used to control the environment for growth as well as providing desired thermal energy upon demand. This dual benefit may help to make solar ponds an even more attractive economic proposition [18]. China has been very successful in studying and applying solar pond technology in different applications. Such technology has been widely applied in the production of Glauber's salt and in aquaculture during winter periods [57,58].

4.7 Buildings and Domestic Heating

Because of the large heat storage capability in the LCZ of a solar pond, it has ideal use for house heating even over several cloudy days [13]. A 2000 m^2 solar pond at Miamisburg in 1978 was used to supply heat to a municipal swimming pool. It supplied all of the heat needed for the pool during the swimming season and also supplied heat to a bath house during other parts of the year [59]. Styris et al. [60] proposed a formula to determine pond dimensions for the heating requirements of projects involving house heating, winter crop drying, and paper processing in the Richland area, Washington (USA).

More recently, the integration of solar ponds with heat pumps had been proposed. The heat pump could serve as an air conditioner in summer and the freshwater layer above the top partition (in a partitioned solar pond) could be designed to serve as a heat sink to increase the coefficient of performance of the air conditioner [61]. Indeed, gas engine powered heat pumps can greatly increase the effectiveness of a solar pond that is attached to a heating load requiring temperatures above 40 °C such as greenhouse heating during the winter season [62]. At the Ohio Agricultural Research and Development Center (Ohio, USA), a solar pond, (18.5 × 8.5 × 3.0) m, was constructed to supply the winter heat requirements of a single-family residence and partial needs of an adjacent greenhouse. In this system, hot brine was extracted from the pond and passed through shell and tube heat exchangers. When the pond temperature was higher than 40 °C the heat extracted was supplied to a water-to-air discharge heat exchanger in the greenhouse; when the pond temperature was less than 40 °C the heat extracted was first upgraded by a heat pump and then used for heating a greenhouse [12].

REFERENCES

[1] Yu N, Wang RZ, Wang LW. Prog Energy Combust Sci 2013;39:489–514.

[2] Kousksou T, Bruel P, Jamil AT, ElRhafiki , Zeraouli Y. Sol Energy Mater Sol Cells 2014;120:59–80.

[3] Fernandez AI, Martinez M, Segarra M, Martorell I, Cabeza LF. Sol Energy Mater Sol Cells 2010;94:1723–9.

[4] Kousksou T, Bedecarrats J-P, Strub F, Castaing-Lasvignottes J. Int J Energy Technol Pol 2008;6:143–58.

[5] Kurt H, Halici F, Binark AK. Energy Convers Manage 2000;41:939–51.

[6] Dincer I, Rosen MA. Thermal energy storage. Systems and applications. New York: Wiley; 2002.

[7] Tabor H, Matz R. Sol Energy 1965;9:177–82.

[8] Tabor H, Weinberger Z. Non-convecting solar ponds. In: Kreider JF, Kreith F, editors. Solar energy handbook. New York: McGraw Hill; 1981. Chapter 10.

[9] Weinberger H. Sol Energy 1964;8:45–56.

[10] Bansal PK, Kaushika ND. Energy Convers Manage 1981;21:81–95.

[11] Ranjan KR, Kaushik SC. Renew Sust Energy Rev 2014;32:123–39.

[12] Kaushika ND. Solar ponds, reference module in Earth systems and environmental sciences. Encyclopedia of Energy. ; 2004. p. 651–9.

[13] El-Sebaii AA, Ramadan MRI, Aboul-Enein S, Khallaf AM. Renew Sust Energy Rev 2011;15:3319–25.

[14] Angeli C, Leonardi E, Maciocco L. Sol Energy 2006;80:1498–508.

[15] Zangrando F. Sol Energy 1980;25:467–70.

[16] Giestas MC, Pina HL, Milhazes JP, Tavares C. Int J Heat Mass Transf 2009;52:2849–57.

[17] Tundee S, Terdtoon P, Sakulchangsatjatai P, Singh RA, Akbarzadeh A. Sol Energy 2010;84:1706–16.

[18] Akbarzadeh A, Andrews J, Golding P. Adv Sol Energy 2005;16:233–94.

[19] Newell TA, Cowie GR, Upper JM, Smith MK, Cler GL. Sol Energy 1990;45:231–9.

[20] Tabor HZ, Doron B. The Beith Ha'Arava 5 MW(e) solar pond power plant (SPPP). Sol Energy 1990;45:247–53.

[21] Kumar A, Kishore VVN. Sol Energy 1999;65:237–49.

[22] Xu H, Li SS. Acta Energiae Solaris Sinica 1983;4:74–86.

[23] Akbarzadeh A, Andrews J, Golding P. Solar ponds. Encyclopedia of life support systems (EOLSS). Developed under the auspices of UNESCO. Oxford, UK: EOLSS Publishers; 2008. http://www.eolss.net

[24] Valderrama C, GIbert O, Arcal J, Solano P, Akbarzadeh A, Larrotcha E, Cortina JL. Desalination 2011;279:445–50.
[25] Chepurniy N, Savage SB. Sol Energy 1975;17:203–5.
[26] Bronicki Y. Solar ponds. Encyclopedia of Physical Science and Technology, 3rd ed. 2002. p. 149–166.
[27] Tabor H, Doron B. The Beit Ha'arava 5 MW solar pond power plant. Presented at the International Conference on Solar Ponds, Cuernavaca, Morelos, Mexico, 1987; March–April.
[28] Andrews J, Akbarzadeh A. Sol Energy 2005;78:704–16.
[29] Leblanc J, Akbarzadeh A, Andrews J, Lu H, Golding P. Sol Energy 2011;85:3103–42.
[30] Harel Z, Tanny J, Tsinober A. J Sol Energy Eng 1993;115:32–6.
[31] Taga M, Matsumoto T, Ochi T. Sol Energy 1990;45:315–24.
[32] Wilkins E. Sol Energy 1991;46:383–8.
[33] El-Housayni K, Wilkins E. Energy Convers Manage 1987;27:219–36.
[34] El-Sebaii AA, Aboul-Enein S, Ramadan MRI, Khallaf AM. Sol Energy 2013;95:30–41.
[35] Garg HP, Bandyopadhyay B, Rani U, Hrishikesan DS. Energy Convers Manage 1982;22:117–31.
[36] Ramadan MRI, El-Sebaii AA, Aboul-Enein S, Khallaf AM. Energy Buildings 2004;36:955–64.
[37] Casamajor AB, Parsons RE. Design guide for shallow solar ponds. Report UCRL-52385 Rev.1. Livermore, CA: Lawrence Livermore Laboratory, University of California; 1979.
[38] Einav A. Sol Energy Eng 2004;126:921–8.
[39] El-Sebaii AA. Appl Energy 2005;81:33–53.
[40] Lu H, Swift AHP, Hein HD, Walton JC. J Sol Energy Eng 2004;126:759–67.
[41] Bernad F, Casas S, Gibert O, Akbarzadeh A, Cortina JL, Valderrama C. Sol Energy 2013;98:366–74.
[42] Lu H, Walton JC, Swift AHP. Desalination 2001;136:13–23.
[43] Agha KR. Sol Energy 2009;83:501–10.
[44] Saleh A, Qudeiri JA, Al-Nimr MA. Energy 2011;36:922–31.
[45] Suarez F, Tyler SW, Childress AE. Water Res 2010;44:4601–15.
[46] Salata F, Coppi M. Appl Energy 2014;136:611–8.
[47] Hull JR. Sol Energy 1980;25:33–40.
[48] Hull JR, Nielsen CE, Golding P. Salinity gradient solar ponds. Boca Raton, FL: CRC Press; 1989.
[49] Date A, Alam F, Khaghani A, Akbarzadeh A. Procedia Eng 2012;49:42–9.
[50] Akbarzadeh A, Johnson P, Singh R. Sol Energy 2009;83:1345–59.
[51] Tundeea S, Srihajong NS, Charmongkolpradit S. Energy Procedia 2014;48:453–63.
[52] Chen L, Gong J, Sun F, Wu C. Int J Therm Sci 2002;41:95–9.
[53] Singh B, Gomes J, Tan L, Date A, Akbarzadeh A. Procedia Eng 2012;49:50–6.
[54] Cao WH, Wu C. Brine resources and the technology of their comprehensive utilization. Beijing: Geology Publishing; 2004.
[55] Huang WH, Sun ZN, Wang XK, Nie Z, Bu LZ. Mod Chem Ind 2008;28:14–9.
[56] Nie Z, Bu L, Zheng M, Huang W. Sol Energy 2011;85:1537–42.
[57] Ding CL, Ma FY, Yang RL. J Salt Chem Ind 1997;27:17–20.
[58] Ma FY, Ding CL, Yang RL. Chem Ind Prog 1998;4:35–8.
[59] Duffie JA, Beckman WA. Solar engineering of thermal processes. 3rd ed. Hoboken, NJ: John Wiley & Sons; 2006.
[60] Styris DL, Harling OK, Zarworski RJ, Leshuk J. Sol Energy 1976;18:245–51.
[61] Al-Jamal K, Khashan S. Energy Convers Manag 1998;39:559–66.
[62] Taga M, Fujimoto K, Ochi T. Sol Energy 1996;56:267–77.

Chapter 15

Sensible Thermal Energy Storage: Diurnal and Seasonal

Cynthia Ann Cruickshank, Christopher Baldwin
Department of Mechanical and Aerospace Engineering, Carleton University, Ottawa, ON, Canada

1 INTRODUCTION: STORING THERMAL ENERGY

There are a variety of ways to store thermal energy. The most common method is to store the heat as internal energy within the material, increasing its temperature. The process is considered to be one of sensible heat storage when there is no change in chemical composition or phase associated with the heating process. The amount of heat that can be stored in sensible heat storage is directly proportional to the specific heat and mass of the material and the temperature change associated with the process. For this reason, solids (e.g., rock, concrete) and liquids (e.g., water, glycol) that have a high mass and specific heat are often used to increase the amount of heat that can be stored. This increase in energy density allows the storage to be compact, which reduces the cost of the storage unit and its installation. In addition, the smaller surface area associated with compact storage lowers standby thermal losses from the storage to the surroundings. The choice of storage medium is often influenced by the working fluid in the system, for example, if the primary working fluid is a liquid, then the storage medium will often also be a liquid. The most commonly used liquid storage medium for low- to medium-temperature applications is water due to its high volumetric heat capacity, widespread availability, and low cost. As a result, storage tanks filled with water are widely used for thermal storage.

Thermal energy can also be stored using latent heat storage. Latent heat storage uses a phase change material as the storage medium, that is, energy is absorbed or released during a change of phase (e.g., water and ice, salt hydrates, wax) at a particular temperature. Phase change materials (PCM) are of particular interest as they can offer an order of magnitude increase in heat capacity (when compared with conventional materials) with very small or negligible temperature change. For this reason, PCMS have been introduced in the building construction sector and industrial processes. Some examples include PCM-treated wallboards and phase change slurries (used as a cold carrier liquid or for heat storage).

Storing Energy. http://dx.doi.org/10.1016/B978-0-12-803440-8.00015-4

For each storage medium, there is a wide variety of choices depending on the temperature range and application. When utilizing a material that undergoes a change of phase, the total energy stored over a particular temperature range is related to the specific heat of the material in the various phases and latent heat associated with the phase change.

Thermal energy can also be stored using a thermochemical reaction. That is, a reaction for an energy storage would consist of an endothermic reaction resulting in products that separate and do not undergo any future reactions. To permit recovery of the stored energy, the products of the chemical reactions would be recombined and the reaction would be reversed. This form of energy storage has found application in many areas of electrical generation and energy transportation.

2 DESIGN OF THE THERMAL STORAGE AND THERMAL STRATIFICATION

There are a number of aspects of importance in the design of thermal energy storage (TES). These include total capacity; energy density; size, shape, and volume; heat loss; and charge and discharge efficiency [1,2]. Therefore, the optimum choice and sizing of the thermal storage will depend on many factors including the distribution and temperature of the energy supply; the temperature requirements, magnitude, and distribution of the load throughout the day or the season; the required charge and discharge rates; and the spatial limitations related to the installation and placement of the storage [3].

The capability of the storage to deliver its stored energy as high-quality energy is also an important aspect in the design of TES. This concept is referred to as exergy and attempts to quantify the quality or work potential of energy. Maintaining a high exergy level in the thermal storage tends to equate to maintaining as high a temperature as possible in the storage. This is most often and simply accomplished by ensuring that the storage remains thermally stratified with the hot-charge fluid stored with as little mixing as possible. Thermal stratification corresponds to the existence of a temperature gradient in the storage that allows the separation of fluid at different temperatures. When observing temperature distribution in a real tank, one concept used to characterize the level of stratification within TES is to quantify the temperature gradient (dT/dx) and thickness of the thermocline (intermediate region) that separates hot and cold regions within the storage [4].

The generation and maintenance of highly stratified storage depends on many factors, including those related to fluid dynamics and heat transfer. For example, inlet fluid streams to the storage and storage construction, including its geometry and material properties, may significantly affect the performance of TES. Heat losses through the walls of the storage to the surrounding environment may also degrade storage performance and lead to thermal destratification. A study conducted by Lavan and Thompson [1] indicated that an increase in flow rate had an adverse effect on stratification, and that stratification did improve with increasing tank aspect ratio (height to diameter ratio); however, a limit existed where the heat loss to the surroundings became greater as taller tanks have a larger surface area. Lavan and Thompson [1] recommended an aspect ratio between 3 and 4 for a reasonable compromise between performance and cost.

Regarding thermal stratification, there are two approaches that are typically followed. The first and most common is referred to as "natural stratification" where the charge fluid is circulated into a storage vessel of simple geometry and the fluid in the storage does not mix appreciably. This is most commonly achieved in vertical storages, where the inlet velocities are very low, thereby not mixing the existing fluid in the storage. As the fluid velocity is increased the momentum of the inlet fluid stream is seen to "entrain" or mix with the fluid in the tank, increasing entropy and reducing exergy [5,6]. It should be noted, as well, that mixing can occur from the inflow of the charging fluid, as well as from the inflow associated with discharging of TES.

Numerous designers (especially in Europe) have attempted to increase thermal stratification in storage tanks by modifying the design and geometry of storage vessels. In particular, they have added various diffusers and baffles to the interior of the storage tank to reduce the velocity and momentum associated with the flow of fluids into and out of the storage. These devices can be successfully deployed, but tend to increase the cost of the storage. To further encourage stratification, studies have also been conducted on inlet stratifiers in tanks [7–9]. Typically built of rigid material or fabric, stratifiers reduce the momentum of water entering the tank thus allowing buoyancy forces to direct the incoming fluid to the location in the tank where the temperature of the two fluids is equal.

2.1 Heat Exchangers

A variety of heat exchanger styles are available for energy storage systems, but can generally be classified into three categories: immersed coils, external sidearm heat exchangers, and mantle types. Immersed coil heat exchangers are generally located at the bottom of thermal storage tanks to take advantage of the greatest temperature differences between the heated fluid and incoming potable water. It is commonly accepted that if an immersed coil is placed at the bottom of a fluid-filled thermal storage, it will promote mixing of the portion of the storage above the heat exchanger, that is, promoting destratification (uniform temperature distribution) and entropy production. External heat exchangers represent flexible options for TES as they allow standard (i.e., low cost) storage tanks to be used. If the flow through the tank side of an external exchanger is pumped, this can also lead to mixing and destratification. Successful design of this configuration includes a careful balance between buoyancy-induced flow velocity and heat exchanger effectiveness. The use of external natural convection heat exchangers is increasingly being used as a simple means to promote stratification. Mantle heat exchangers consist of a double-walled storage tank that allows the heat transfer fluid to be circulated through the storage mantle (i.e., a cavity formed by the two walls) transferring heat to the stored water. Mantle tank storage systems tend to have a large heat transfer surface area that increases performance but requires specialized tanks that may be more costly.

2.2 Destratification in Storage Tanks

Stratification of TES may be destroyed by different physical processes such as mixing caused by "plume entrainment" of the incoming liquid during charging and discharging [10], high conductivity within the working fluid that will tend to promote mixing by transferring heat through the storage medium, and heat loss and conduction in and through the storage vessel walls [11]. These processes are caused by several factors: the kinetic energy of the fluid jet entering the tank, heat conduction in tank components, and inverse temperature gradients that lead to buoyancy-induced flow. Mixing introduced during charge and discharge cycles is generally the major cause of destratification [12].

The magnitude and distribution of daily hot-water draw profiles also affect the temperature profile within a storage tank and consequently have been the subject of a number of studies. As a result of these previous works, it is common practice to use standard draw volumes and load profiles for comparing the performance of energy storage systems. As such, standard draw profiles have been established for the evaluation of domestic hot-water storages and subsequently employed to investigate the effects of various load profiles on storage stratification.

3 MODELING OF SENSIBLE HEAT STORAGE

Computer models of storage operation have been developed and implemented within various simulation environments, including the widely used TRNSYS simulation package [13]. As well, it is now possible to model water-based thermal storage with considerable accuracy through detailed multidimensional computational fluid dynamics modeling [14]. In the case of annual performance evaluations, however, it is standard practice to use simplified computer algorithms to reduce computational overhead and computing times. The complexity of both component and system models is often weighed against "user convenience," computing time and resources, and desired accuracy. In many instances, due to the lack of detailed information, a number of simplifying assumptions are usually made in the model. The success of this process relies on accurate specification of the system's physical and thermal characteristics and the complexity and underlying assumptions of the computer algorithm.

Current storage algorithms are often based on one-dimensional (1-D) finite volume assumptions which incorporate basic models of tank heat loss, thermal diffusion, flow, and buoyancy-induced mixing [6]. These approaches have been shown to adequately represent the performance of stratified thermal storages in cases when the charge and discharge flow rates into the storage are low and therefore mixing of the tank fluid is minimal [15].

The suitability of the simplified 1-D approach is based on the assumption that the temperature distribution through the thermal storage can be treated as 1-D, implying that temperature gradients exist in the vertical direction but are negligible in the horizontal direction [16]. If heat losses through the tank wall are high or if tank wall thermal conductivity is large then it would be expected that the associated heat transfer at the wall would result in nonuniform temperature distributions and lead to buoyancy-induced mixing of the storage tank. In addition, it has been suggested that actual heat loss rates from typical cylindrical thermal storages are higher than would be calculated by a simple 1-D approximate method used to estimate wall heat loss. The discrepancies would most likely be due to multidimensional effects that affect the heat loss rates from the storage and the diffusion of heat through the tank wall and the fluid. In stratified thermal storages, this may also lead to discrepancies in the tank temperature profile and lead to errors in heat loss prediction. The storage algorithm and the basic assumptions typically used in the computer modeling of solar storage heat losses (e.g., one-dimensional temperature profiles, minimal tank wall conduction, uniform wall heat loss) are described later, particularly in the context of a thermally stratified thermal storage.

3.1 Modeling Stratified Thermal Energy Storage

Previous research has shown that high degrees of stratification are possible in a correctly designed thermal storage system, [17]. Numerous models have been developed for liquid-based, sensible heat thermal storage [4]; however, most are simplified 1-D models. Some have been refined to account for the effects of mixing or the entrainment of fluid in the storage during charging or load draws. It is possible to develop a simple and efficient model of stratified thermal storage by dividing up the tank into N constant volume sections or "nodes," each assumed to be fully mixed and at a uniform temperature [13], as shown in Fig. 15.1. The choice of number of nodes N will determine the resolution to which the vertical temperature distribution can be modeled in the storage tank, that is, increasing N will allow for significant temperature gradients to be more accurately modeled [18]. For the special case of $N = 1$ the tank is modeled as a fully mixed tank and no stratification effects are possible [12].

The time–temperature history of a node can be predicted by performing an energy balance on each storage section, accounting for thermal losses to the surroundings and the influence of adjacent nodes (e.g., mass and energy flows, including conduction between the layers and vertical conduction through the tank walls (Fig. 15.2).

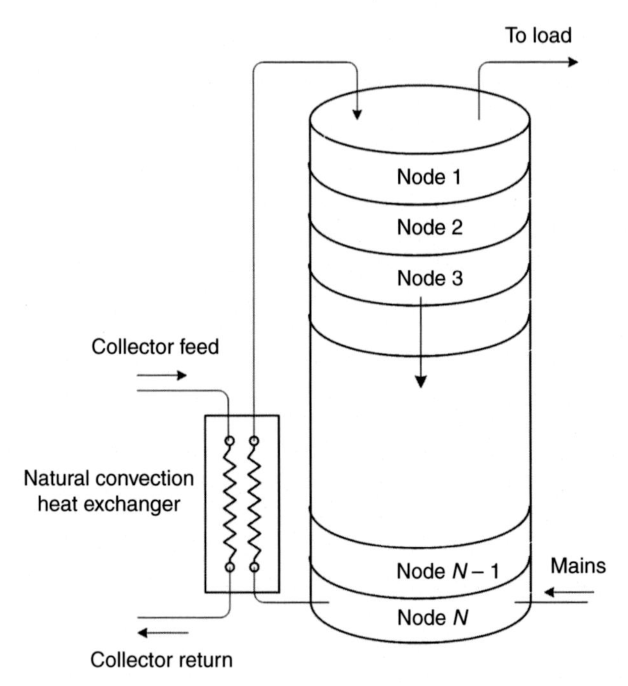

FIGURE 15.1 Storage tank divided into sections for the purpose of modeling thermal stratification [15].

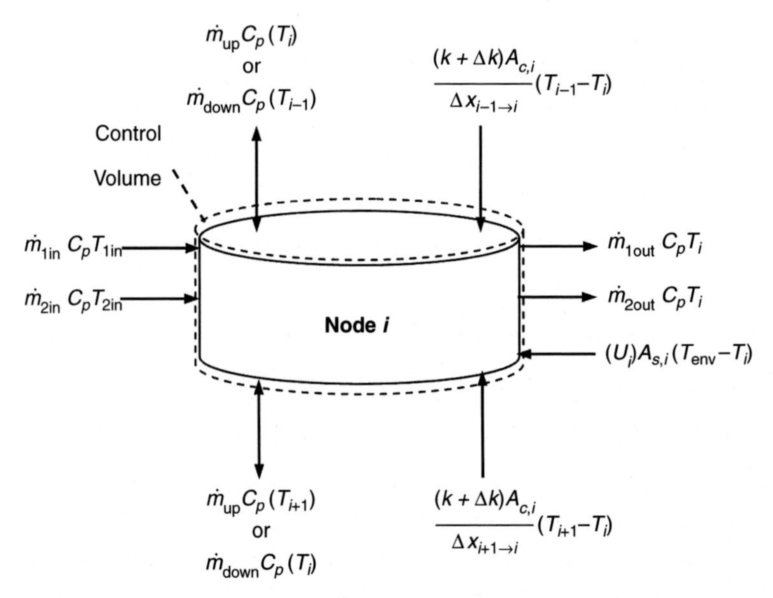

FIGURE 15.2 Control volume used to define the flow of mass into and out of node i where $1 < i < N$ [15].

In Fig. 15.2, $\dot{m}_{up}$ and $\dot{m}_{down}$ are the fluid flow rates up and down the tank, respectively; $A_{c,i}$ and $A_{s,i}$ are the cross-sectional and surface area of node i, respectively; k and Δk are tank fluid thermal conductivity and destratification conductivity (used to model destratification due to mixing at node interfaces and conduction along the tank wall) [6], respectively; C_p is the specific heat of the tank fluid; U_i is the node heat loss coefficient per unit area; $\dot{m}_{1in}$, $\dot{m}_{1out}$, $\dot{m}_{2in}$, and $\dot{m}_{2out}$ are the mass flow rates of entering and exiting fluids 1 and 2, respectively; T_{i+1}, T_i, T_{i-1}, T_{1in}, T_{2in}, and T_{env} are the temperatures located below, at, and above node i, the temperature of the entering fluid 1 and the entering fluid 2, and the temperature of the environment, respectively; and $\Delta x_{i+1\rightarrow i}$ and $\Delta x_{i-1\rightarrow i}$ are the center-to-center distance between node i and the node below and above it, respectively.

To estimate the temperature distribution and the heat loss characteristics of the vertical tank the energy and mass flows into and out of each storage node from adjacent nodes are estimated based on the node temperatures that existed at the beginning of the time step (Fig. 15.2). Any temperature inversions that result from these flows are eliminated by mixing appropriate nodes at the end of each time step [12]. To solve for the temperature distribution in the storage tank a set of N first-order, ordinary differential equations resulting from each node's energy balance [6] can be assembled, for example, an energy balance written about the ith tank node.

As the temperature of each node depends on the temperatures of the adjacent nodes and the temperature of the environment, it is necessary to simultaneously solve the system of equations. Computational efficiency and speed are therefore important, as an annual simulation may involve 200 000 time steps. In computer simulation programs (e.g., TRNSYS [13]), it is common practice to use standard numerical techniques to solve for the temperature distribution at each time step [6]. It has also been noted that simple storage models rapidly become more complicated as multiple inlet and outlet ports are added or auxiliary elements are placed in a storage tank [6,19]. As such, a number of variations have been introduced in current models, to better accommodate issues such as variable storage volumes, plume entrainment [5], and draw-off mixing [12]. Using a simple one-dimensional model as described earlier, it should be noted that the accuracy of the model relies on a number of assumptions being met. These are:

1. the flow of liquid within the tank is one-dimensional
2. the temperature and density of the fluid in each node is uniform and constant over the time step
3. the fluid streams from each node are considered fully mixed before they enter an adjacent node
4. the heat loss to the exterior of the tank and conduction in the tank walls are low enough that two- or three-dimensional temperature gradients do not form, promoting convection and de-stratification
5. the fluid velocities entering and exiting the storage tank are low enough that they do not promote extensive mixing within the storage tanks.

Not all of these assumptions are fully met in all real storages. For example, Shyu et al. [20] found that stratification decayed in a tank more rapidly than predicted for the theoretical rate when using the conductivity of water. The main reason for this is that the thermal conductivity of the wall material is typically higher than the conductivity of the water and this can promote convection motion along the wall and destratification within the tank. Given this, Shyu et al. [20] computed a table of "effective" conductivity values for various walls and insulation thicknesses to serve as a starting point for users.

4 SECOND LAW ANALYSIS OF THERMAL ENERGY STORAGE

While application of the First Law of Thermodynamics enables the determination of energy stored during a process (and the amount lost to the surroundings), the Second Law of Thermodynamics provides a mechanism for quantifying any degradation in the "usefulness" of the energy that occurs during the storing process [21]. To accomplish this, both exergy level and exergy efficiency have been widely used to evaluate the performance of TES systems [4]. Most recently, a study was conducted to examine the stored exergy of a stratified modular storage system when subjected to various charge and discharge strategies [15,22].

Traditionally, exergy is considered as a measure of the "quality" of energy or its potential to do work relative to a reference or dead state, usually representative of the surrounding conditions. Applying the First and Second Laws of Thermodynamics to a control volume with uniform properties the specific exergy of a substance, Ex, can be defined as:

$$Ex = (h - h_0) - T_0 \cdot (s - s_0)$$

where h and s are the specific enthalpies and entropies of the substance at its current temperature and pressure, and h_0 and s_0 are its enthalpies and entropies at a reference state; T_0 is the temperature of the reference state. It is highly desirable to develop TES systems that can store energy at its highest exergy level and to minimize the destruction of exergy associated with irreversible processes (i.e., entropy production). In a thermal storage, consisting of an effectively incompressible fluid (i.e., water), exergy destruction will primarily occur due to mixing and diffusion occurring during the charging, storage, and discharging processes. Exergy destruction during the storage of energy, over a period of time, occurs due to heat losses to the surroundings and the diffusion of heat through the fluid and the storage vessel. Exergy destruction also occurs during the charging and discharging of TES. Many indices have been proposed or are under development to quantify the Second Law performance of a TES system [23]; however, the performance of TES can be studied by observing the exergy level in the storage tank during the charging process. To avoid violating the First and Second Laws of Thermodynamics the maximum temperature in TES is determined by the maximum temperature occurring during the charge sequence. As such, high exergy will be achieved during charging if the bulk of the volume of the thermal storage can be brought as close as possible to the temperature of the charge fluid. In addition, higher degrees of temperature stratification in a storage should reduce exergy destruction associated with mixing and diffusion and are highly desirable in a storage. Therefore, to estimate the stored exergy values at any time t, (i.e., $Ex_{tank}(t)$), within TES, values of exergy in each of the nodes within the storage tank are summed:

$$Ex_{tank}(t) = \sum_{node=1}^{N} Ex_{node}(t)$$

5 SOLAR THERMAL ENERGY STORAGE SYSTEMS

Solar thermal energy systems have existed in a variety of forms for hundreds of years. The widespread use of solar energy to heat water for domestic consumption started in the early part of the 19th century; however, competition from low-cost fossil fuels reduced its popularity. It was not until the mid-1970s that it was reconsidered as an alternative to fossil- or nuclear-generated energy. During that period a number of configurations were developed in an effort to lower costs and improve performance. The developments that followed were often driven by local climatic, regulatory, and market conditions. As such, simple passive systems were developed for nonfreezing, hot climates and were widely used in the Mediterranean, Middle East, and Asia-Pacific regions. Different system approaches were used in Europe, Japan, North America, and Australia to produce systems that were freeze protected. However, virtually all systems include one or more solar collectors to capture and convert the Sun's energy into heat, a storage tank to store the available energy until it is required, and a circulation system to move a heat transfer fluid between the collectors and the storage tank. A simple conceptual diagram of a solar hot water–heating system is shown in Fig. 15.3.

TES is an integral component of a solar hot-water system that may significantly improve its efficiency and cost effectiveness by allowing better utilization of the solar hardware and matching of the solar resource to the load. Most

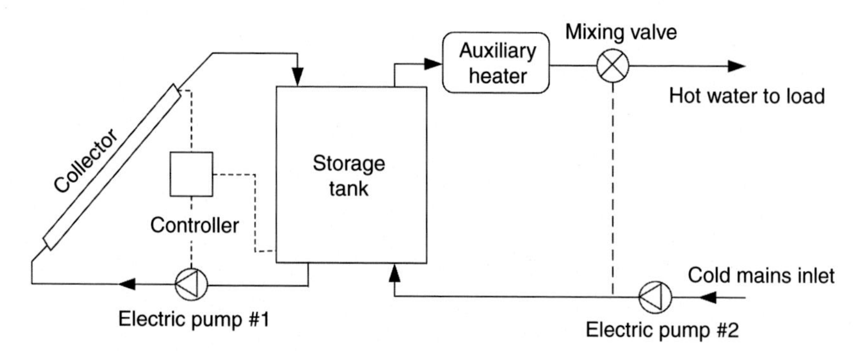

FIGURE 15.3 **Simple solar hot water–heating system [15].**

small- to medium-sized solar installations use diurnal storage, where energy is typically stored for 1 day or 2 days; however, weekly and seasonal storage is also used in certain applications. Primarily used to offset space heating loads, seasonal storage systems are designed to collect solar energy during the summer months and retain the heat in the storage for use during the winter months [24]. Characterized by their large capacity requirement (in the order of a hundred times the capacity of daily storage) [4], these systems typically run at a much higher cost and require a larger storage volume than short-term storages.

Much effort has been put into maximizing the performance of water storages and minimizing the cost of the storage vessels. Moreover, small hot-water heaters and storages (i.e., 180 and 270 L) have been produced in large quantity for the North American market and are readily available at low cost. Coupled to an external heat exchanger, they represent a cost-effective storage option for residential solar hot-water heaters. However, larger storage volumes in the range (500–1500) L are often required for multifamily residential units and small- to medium-sized commercial applications. Unfortunately, suitable storage vessels of this size are only produced in limited quantities, resulting in significantly higher costs per unit of storage volume [15]. In addition, these larger storage vessels are not well suited to retrofit situations where the storage vessel must be moved into a building space through existing door openings. Consequently, larger storages are often constructed onsite, and maintained at low pressure and vented to the atmosphere.

Finally, it has been shown that the thermal performance of water-based storage devices can be significantly improved by lowering the solar collector loop fluid flow rate, thus promoting an increase in thermal stratification in the storage tank(s) [1,10,25,26]. A study of "microflow" systems demonstrated that stratified storage delivered 37% more energy than a fully mixed storage of corresponding size [10]. This increase in efficiency can be attributed to two reasons: the low flow rate allows the hot water to enter and remain in a layer at the top of the tank thus allowing the energy in the tank to be available at a temperature that is closer to the desired load temperature; and, second, the resultant stratification allows cool fluid to circulate back to the solar collectors which increases the collector efficiency due to its lower inlet temperature [3]. In reality, the advantages of stratification will vary depending on the system configuration and distribution of the load throughout the day [27].

6 COLD THERMAL ENERGY STORAGE

Although sensible TES systems are predominantly used for space heating and domestic hot-water applications, cold thermal energy systems are becoming common. These systems are almost exclusively used in commercial and industrial applications as a method of shifting the energy consumption required for space cooling to the overnight period [28]. The ability to shift energy consumption to the overnight period can be beneficial for a number of reasons. The first and most significant is that most electrical jurisdictions in the world experience their peak electrical demand in the afternoon during the summer months. The ability to charge a cold TES system overnight and realize the cooling potential during the peak period helps reduce the peak load placed on the grid, and in turn could reduce the required electrical generating and transmission capacity required [29]. In addition to reducing the peak load the grid experiences, many power providers charge based on the demand, charging significantly higher electrical rates during peak periods in comparison with off-peak, overnight periods [29]. As such, a significant reduction in a building's utility costs can also be realized through implementation of a cold thermal storage system. In addition to utility cost reductions, utilization of chiller equipment at night can see an increase in performance and capacity, as exterior heat rejection temperatures are much lower than those seen during the day [28]. This combined with using the thermal storage capacity to meet peak cooling demand allows for smaller capacity chillers to be installed.

Cold thermal storage is obtained almost exclusively using water or water–glycol solutions as the storage medium, due to the ease of obtaining these fluids and their low costs [4]. Additionally, they integrate into and operate with almost all standard chillers, so custom units are not required. The principal decision required when designing and specifying cold storage units is whether to use only sensible thermal storage, or whether the latent potential of the materials will be utilized. In essence, will the thermal storage contain only the storage medium as a liquid, or will the formation of ice in some capacity be introduced to the thermal storage system. There are many ways that ice can be introduced into the storage system, whether it be as slurry, ice balls, builtup ice on coils, or encapsulated ice cubes [28].

There are a number of factors that influence the decision as to whether to use sensible storage or a more complex system that incorporates PCM. The largest factors are storage capacity and storage density (the amount of energy that can be stored in a unit of volume). Due to the enthalpy of fusion for water being approximately 350 kJ kg^{-1} a single kilogram of water delivering cooling at 15 °C is able to store a total of 412.7 kJ of cooling potential. The mass of water using only sensible storage kept at 3 °C (small buffer to ensure freezing does not occur) and cooling at 15 °C is only capable of storing 50.2 kJ of cooling potential. From this simple example, it can be seen that ice storage stores over 8 times the cooling potential compared to a water storage system, and depending on the storage temperature and the required temperature for cooling, this can approach 20 times the capacity [4,21]. A third option is to store the cooling potential using a glycol solution. Glycol solutions have the benefit of a much lower freezing point; however, they also have a slightly lower heat capacity compared with water. This allows lower temperatures to be obtained without the risk of freezing, but does not require the complex systems required for ice storage. In this case a single kilogram of 50/50 glycol–water solution by volume with a storage temperature of −10 °C is able to store 92 kJ of energy. A graphical comparison of the energy density for cold storage can be seen in Fig. 15.4.

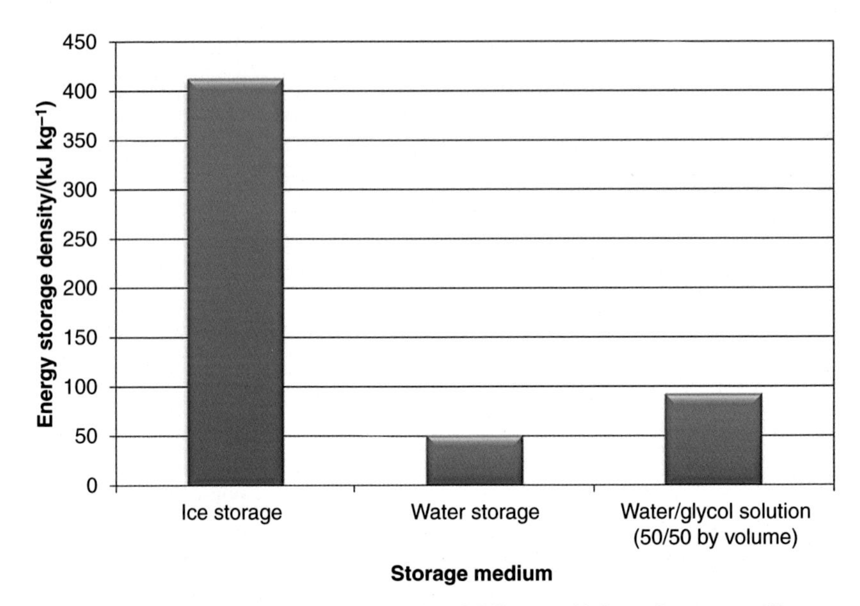

FIGURE 15.4 **Storage energy density capacity of different cold thermal storage media.**

Although energy storage density is much higher using ice storage the switch from sensible water storage to ice storage can have a negative effect on the performance of the refrigeration unit charging the thermal storage. This is caused by the fact that for water storage the evaporator temperature of the chiller is approximately 0 °C, while for an ice-based thermal storage system a colder evaporator temperature is required, typically –10 °C. At this reduced temperature, capacity drops to approximately 56% of that at 0 °C, while the coefficient of performance drops to 71% of that at 0 °C [4]. As a result, you typically use more energy and require a larger chiller to charge an ice storage system at the same rate as a water storage system making the capital purchase higher and increasing ongoing energy costs compared with the sensible storage system.

As a result of this degradation in performance at lower evaporator temperatures, ice storage should only typically be employed when the available space does not permit the use of sensible thermal storage due to lower storage density. In these cases a combination of sensible and latent storage can be employed, with the optimal combination found for building loads and usable space. Even with the degradation in performance, significant cost savings can be realized by building operators when incorporating cold thermal storage using either method.

7 SEASONAL STORAGE

While diurnal storage systems are designed to offset all or a portion of the daily heating and/or domestic hot-water demand, diurnal thermal storage has little to no effect on the seasonal performance of the heating system. Seasonal thermal storage systems meanwhile are used to meet the long-term, seasonal mismatch of available energy and energy demand. Seasonal TES is the storing of thermal energy, including heating or cooling potential, for the future long-term use of heating or cooling a building or for other extended periods of time [30]. When using ground source heat pump systems and solar thermal systems for space heating, often a thermal storage with an annual cycle time is required to maximize the energy efficiency and solar fraction of the system. The term solar fraction refers to the percentage of the overall load that is supplied by solar. The seasonal mismatch of energy availability and demand is the result of an overabundance of available heat, in the summer months, with little to no demand for use (with the exception of a small domestic hot-water requirement). Seasonal thermal storage systems allow for that excess heat to be collected when available in the summer months and stored for use over the winter period when heat is required [30]. Typically, this heat is collected from solar thermal collectors for storage or is the waste heat produced for air conditioning of buildings, as typically seen with ground source heat pump systems.

7.1 Applications

Solar thermal systems used for both space heating and domestic hot water are becoming more popular within the residential market; however, when using only diurnal storage these systems typically are unable to achieve solar fractions greater than 50% [31]. To increase the solar fraction of these systems a seasonal thermal storage system is required. When seasonal thermal storage systems are successfully implemented an annual solar fraction approaching 100% is obtainable [30]. To meet these high solar fractions, significant storage capacity is required, with seasonal thermal storage systems having a storage capacity typically between 100 and 1000 times greater per unit solar thermal collector than a diurnal thermal storage system [30]. Although storage capacities are significantly larger, solar thermal systems with seasonal storage systems typically have a capital cost double that of a similar system with only short-term storage [24].

Seasonal thermal storage is not only used with solar thermal heating systems, but is also commonly paired with heat pumps. Almost all liquid-to-liquid heat pump systems incorporate seasonal thermal storage, where source energy is extracted from the storage medium during the winter heating season and is converted to usable thermal energy by the heat pump. This energy can then be used either for space heating and/or domestic hot-water needs of a building. During the summer cooling period the waste heat from the air-conditioning process using the heat pump is deposited back into the seasonal storage medium, increasing its temperature and sensibly storing the heat for use in the reverse process during the winter [32].

Seasonal thermal storage systems are not only employed at the single dwelling or building level, but are now seeing widespread implementation at the community level. Particularly in Europe, however with some projects in North

America), solar district heating systems are becoming more prevalent and in some districts can be cost competitive with traditional heating sources including gas [33]. Solar district heating is the heating of a central thermal storage system, with the heat collected being distributed as needed to dwellings within the community. Solar collectors can be centrally located, or distributed throughout the community, but all supply heat to the central system. District heating systems use the largest seasonal thermal storage systems currently employed and can be in excess of 40 000 m^3 and contain in excess of 2000 m^2 of solar thermal collectors. Due to economies of scale and the central nature of these systems, total capital cost is typically only (20–30)% of that compared with putting individual solar thermal systems in each house fed by the district heating system [24].

7.2 Storage Methods

Seasonal thermal storage can be achieved by a number of methods and using many different media. As a result of the large storage capacity required to meet seasonal storage demand, and the relatively low cost of materials used for sensible storage, seasonal thermal storage systems predominantly use sensible thermal storage methods; however, some seasonal storage systems do still employ chemical or phase change systems. The following sections will outline different sensible storage methods used for seasonal thermal storage.

7.2.1 Large-Scale Tanks

Water tanks are one of the most favorable methods for seasonal thermal storage systems due to the numerous benefits of using water as the thermal storage medium. Water, compared with many other sensible thermal storage media, has much higher heat capacity. Additionally, water has the benefit of being easily pumped, allowing for the charge and discharge of the systems by pumping hot water directly into the tank, or out of the tank to the end heating use [4]. Alternatively, the water used for thermal storage can be pumped into heat exchangers, where heat can be transferred to other transfer media, most commonly glycol solutions, used in solar thermal and other heating systems. The use of tanks with defined and well-designed shapes and aspect ratios means thermal stratification can be more easily achieved and the benefits previously discussed with respect to diurnal storage tanks can be realized in larger, seasonal storage tanks.

Typically, large-scale water tanks are used as seasonal thermal storage for a single house, or smaller groups of houses, where the tanks are small enough to be built offsite and then installed in place. This allows easy integration during the construction process and connects directly to the solar thermal collector system and the space heating system in the house. While most tanks are built offsite and installed a number of projects in Europe have been constructed using extremely large, buried seasonal thermal storage tanks. These tanks are typically built onsite using poured concrete lined with either a stainless steel or plastic-based liner to prevent vapor diffusion through the tank walls. The outside of the tank is then insulated using waterproof material, typically consisting of glass-based or polyurethane-based products. These include a 4 500 m^3 tank in Hamburg and a 12 000 m^3 tank in Friedrichshafen (both in Germany) [34]. Large-scale thermal storage tanks are more often used in a large district heating system as intermediate storage between much larger thermal storage systems and the solar collector system. This method is used in the Drake Landing Solar Community in Okotoks, Alberta (Canada) where two 120 m^3 storage tanks are employed for short-term storage, between the borehole thermal energy system, the heating loads, and the solar thermal collector system installed throughout the community [35].

Seasonal thermal storage tanks are most commonly large, insulated cylindrical hot-water tanks with a vertical axis. The simplest and often the most inexpensive method for installing these tanks is to install them at ground level, with the tank projecting into the outdoor air. Although the simplest method a number of disadvantages exist. Most notably, with the tank exposed to ambient conditions, significant heat loss during the cold winter months is possible. To combat this heat loss, above-grade tanks must be well insulated, adding to the cost of installation. Additionally, above-grade tanks typically require larger pumps to overcome the hydraulic head present to pump the water into and out of the top of the tank, which can be significantly above the level of the equipment using the stored heat.

To combat many of the disadvantages associated with using an aboveground tank, one of the more popular methods for installing these tanks is to bury them. This method allows heat loss through the tank surface to the ambient air to be significantly reduced, as the ground temperature remains warmer through the winter months. This allows for a lower level of insulation requirement to achieve the same level of heat loss as the above-grade tanks. Additionally, this method provides the additional benefit allowing the space above the buried tank to be used (commonly a park or mechanical equipment is installed above the tank). When compared with an above-ground tank a buried tank is significantly more costly, with up to 30% of the total cost of the thermal storage system being the earth work required to excavate and bury the tank.

A third alternative, which is a hybrid of the two previous methods, involves the use of a bermed tank [36]. In this method the tank is partially buried, with the dirt removed to place the lower portion of the tank in the ground, then placed around the circumference of the tank and angled down toward the original grade line. This allows a larger portion of the tank to be either buried underground, or be soil backfilled, reducing heat loss from the tank and the insulation required to be installed. Additionally, it considerably lowers the height of the top of the tank in relation to the level grade and therefore decreases the pumping power required when compared with an aboveground tank (although still larger than required for a buried tank). The main benefit of using a bermed tank is that only minor excavation is required, and all the material remains onsite as the excavated material is used to create the berm around the tank. This significantly reduces the cost of the earth works to install the tank, and subsequently reduces the overall cost of the system.

7.2.2 Borehole Thermal Energy Storage

Borehole thermal energy storage (BTES) is one of the most common methods used for seasonal TES currently employed around the world. BTES involves using the ground as the storage medium, allowing heat to be added to the ground during the summer months, and extracted to meet the heating demands in the winter heating season. Boreholes are constructed by first drilling a hole to a depth ranging from (30 to 200) m in depth, depending on the ground conditions and composition [32]. Once the hole is drilled a U-tube heat exchanger is installed, which is a single pipe, typically made of polyethylene or high-density polyethylene, which is shaped into a U and inserted into the drilled hole. This allows the fluid to flow down to the bottom of the borehole and then return to the surface in a continuous loop. Once the piping is installed the remaining volume of the drilled borehole is filled with grout to provide structural support of the drilled hole and to increase thermal conductivity between heat transfer pipes and the ground. Although this is what is typically done in North America, water is commonly used as the filler material in European applications [37].

BTES can be implemented as a seasonal storage method for systems with a wide range of thermal capacities, from a single house right through to large-scale commercial buildings and district heating systems. The most common implementation of BTES systems is for the heating and cooling of individual residential houses, which is achieved by pairing BTES systems with a ground source heat pump. Typically, ground source heat pump BTES systems have a lower storage temperature and, as such, the heat from the ground cannot be used to directly heat the space or to meet the domestic hot-water demands of residents. As such, a heat pump is used to extract lower grade heat from the ground and transfer high-grade energy to the heat sink (e.g., house). During the summer cooling season, the heat pump is operated in reverse, extracting heat from the space and depositing it in the ground through the borehole. This process replenishes the heat in ground that has been extracted during the heating season and allows the heat to be stored for use in the next heating season. Ground source heat pumps are typically paired with one or two boreholes for residential use.

The use of seasonal BTES systems as part of large district heating systems and as a large-scale TES system for institutional and commercial applications are becoming increasingly popular. Unlike single residential systems that almost always require a heat pump to upgrade the energy stored in the ground, large-scale BTES can be used with a heat pump or as a standalone sensible thermal storage system, with the heat supplied by a large array of solar thermal collectors [32]. The size of the borehole thermal storage system can vary considerably based on the ground conditions, heating and domestic hot-water loads of the buildings or community, and the amount of energy required to be stored. Before these large-scale projects can be realized, ground testing to determine thermal conductivity and capacitance must be completed. The data obtained from these exploratory tests must then be implemented into detailed models of seasonal thermal storage systems, as well as the heat source (typically solar thermal collectors or industrial waste heat), heat pumps, and buildings to accurately obtain building loads. This completed model is then used to optimize the number and depth of the boreholes required to optimize the complete system.

Boreholes have been successfully used in many projects around the world in a number of different configurations. One of the most successful is the recently completed Drake Landing Solar Community previously mentioned [35]. This community, serviced by a district-heating system, contains 52 single detached homes, and achieves a heating solar fraction of in excess of 90% (meaning 90% of the heating demand is met using energy collected from the Sun). The community uses 800 solar thermal panels attached to the roofs of detached garages. These panels have a peak thermal production of 1.5 MW, which is fed into a central thermal storage system. The thermal storage system contains a short-term storage system consisting of a large water tank and a large BTES system used for long-term seasonal storage. The BTES contains 144 boreholes, each drilled to a depth of 37 m and spaced at 2.25 m on center. The system operates with 24 strings of 6 boreholes in series to maximize heat distribution within the system. The entire system has a total diameter of 35 m and the top is covered in insulation, a waterproof membrane, and then landscaped as a park to be used by members of the community. The BTES reaches temperatures in excess of 80 °C at the end of the summer months, and is then slowly released through the heating season through a district-heating network [38].

7.2.3 Aquifers

Aquifers are a method of storing thermal energy that uses naturally occurring groundwater as the storage medium. Aquifers are used predominately in Europe where their popularity has risen over the past few years. Using the Netherlands as an example, in 1995 only 29 aquifer thermal storage systems were operating, while in 2012 over 1800 systems were operating [39]. Aquifer storage systems come in many forms and configurations based on the desired system type and the ground properties of the proposed location. In general, aquifers can be either open or closed systems.

As an open system, groundwater, typically at temperatures which are significantly warmer than the outdoor ambient temperatures in the cold heating season, is used as a thermal energy source for heat pump systems. Groundwater is pumped out of the ground through a drilled well, passes through a heat exchanger with the heat pump, and then is dumped at the surface level, typically into a surface body of water [40]. Although this method uses groundwater as a thermal source, it is not a true thermal storage as it only removes energy from the ground, but does not actively replenish the energy removed, relying on the abundance of thermal energy present in the ground.

When using groundwater and aquifer thermal storage, a closed circuit is formed where water is pumped out of the ground, utilized in a heating or cooling process, and then returned to the ground. Heat is extracted during the heating season and then returned to the ground during the cooling season, replenishing the thermal energy available. Typical aquifer systems can be up to 200 m in depth and operate at relatively low temperatures, typical temperatures peaking at 20 °C and going as low as 5 °C when all energy is extracted [41]. As a result of the low operating temperatures, aquifer storage must be used in conjunction with a heat pump as the raw heat will not provide adequate heating.

Aquifer TES can be realized using two different configurations, with the appropriate method determined by the properties of the groundwater present at the site of installation. When groundwater flows a continuous regime is typically employed. Using this design, two wells are drilled, with one upstream of the second. Water is always extracted from the downstream well and returned upstream. This allows for a relatively constant temperature to be maintained throughout the year and has limited impact on groundwater properties. This method is much simpler in terms of control and piping installation; however, it has limited applications as the temperature remains relatively constant and there is almost no temperature difference between the two wells [40].

The more common arrangement is using a cyclic regime, where two wells are again used; however, they must be isolated from each other, meaning they are not fed by the same water mass. One well is used as hot storage and the second well maintained as cold thermal storage. During the heating season, heat is extracted from the hot well, used as the heat source for a heat pump, and the cold fluid deposited into the cold-side well. During the cooling season, cool water is used to extract heat using a heat pump and the hot fluid is deposited into the hot-side well. This works back and forth season to season, removing heat from one well and replenishing the thermal energy in the second well [40].

7.2.4 Rock-bed Thermal Energy Storage

Rock-bed thermal storage systems use rock as the storage medium, either as packed rock bed, small pebbles, or bricks. This storage method has a number of advantages when compared with other methods. These include the ability to operate at higher temperatures than other thermal storage methods, the systems require no drilling and can use air as the heat transfer fluid, reducing the costs associated with piping for fluids. Although some logistical benefits exist, packed rock-beds and pebble storage systems have a much lower energy density than water-based storage systems. As a result, to store the same amount of energy, a rock-based storage system must have a volume approximately three times greater than a water-based system, resulting in most seasonal rock-bed thermal storage systems having an immense volume (demonstration projects have been completed with rock-bed thermal storage systems in excess of 8000 m^3) [34].

One method being used to reduce the volume required is the hybrid gravel/water storage system. This is a compromise between the high initial capital costs of a large-scale water tank and the low cost and low energy storage capacity of rock-bed thermal storage systems [34]. When designing the system the required volume is determined through simulation based on the heating and cooling loads of the building, the heating method, and consequently the storage temperature (direct from the rock bed or incorporating a heat pump), and the desired portion of the load to be met by TES. In addition to determining the size of the storage the size of the individual rocks or pebbles is also significant. The smaller the rock size the more densely the rock bed is packed and the less free space present for the heat transfer fluid to pass through. The smaller the spaces between the rocks the greater the stratification that can be realized within the rock bed [42]. This allows for better system efficiencies as the top is much warmer to better meet the heating demand while the bottom remains cooler, allowing cooler return temperatures to the heating source.

8 CONCLUDING REMARKS

To make efficient use of a time-dependent source of energy, it is often necessary to store energy until it can be used to supply a particular load. For example, a storage system is particularly important for solar thermal systems as the availability of the solar resource varies over the day and season. To reduce cost and space requirements, thermal storages must have sufficient energy density, low energy losses, and efficient charge and discharge characteristics. To arrive at an efficient storage configuration for a particular application, it is necessary to conduct a detailed analysis of the thermodynamics, heat transfer, and fluid dynamics associated with that application.

Today, TES is considered an advanced energy technology [4]. Given the need to reduce greenhouse gas emissions and the increasing volatility of fossil fuels, the necessity to reduce energy consumption for heating and cooling becomes obvious. The use of thermal storage can alleviate the temporal energy mismatch between off-peak periods and building occupant demands (such as space conditioning and hot-water loads) by allowing energy to be stored, thus realizing significant energy savings and reducing the demand on fossil fuels during peak periods.

REFERENCES

[1] Lavan Z, Thompson J. Experimental study of thermally stratified hot water storage tanks. Sol Energy 1977;19:519–24.

[2] Hahne E, Chen Y. Numerical study of flow and heat transfer characteristics in hot water stores. Sol Energy 1998;64:9–18.

[3] Duffie JA, Beckman WA. Solar engineering of thermal processes. New York: John Wiley & Sons; 2013.

[4] Dincer I, Rosen M. Thermal energy storage; systems and applications, 2nd ed. Chichester, UK: John Wiley and Sons; 2011.

[5] Lightstone MF, Hollands KGT, Hassani AV. The effect of plume entrainment in the storage tank on calculated system performance. Proceedings of Solar Energy Society of Canada Inc., Ottawa, ON; 1988.

[6] Newton B. Modeling of solar storage sanks. M.Sc. thesis. Department of Mechanical Engineering, University of Wisconsin-Madison; 1995.

[7] Davidson JH, Adams DA. Fabric stratification manifolds for solar water heating. ASME J Sol Energy Eng 1994;116:130–6.

[8] Shah LJ, Andersen E, Furbo S. Theoretical and experimental investigation of inlet stratifiers for solar storage tanks. Appl Therm Eng 2005;25:2086–99.
[9] Andersen E, Furbo S, Fan J. Multilayer fabric stratification pipes for solar tanks. Sol Energy 2007;81:1219–26.
[10] Hollands KGT, Lightstone M. A review of low flow, stratified tank solar water heating systems. Sol Energy 1989;43:97–105.
[11] Hess CF, Miller CW. An experiment and numerical study on the effect of the wall in a thermocline type cylindrical enclosure. Sol Energy 1982;28:145–52.
[12] Kleinbach EM, Beckman WA, Klein SA. Performance study of one dimensional models for stratified thermal storage tanks. Sol Energy 1993;50:155–66.
[13] TRNSYS: a transient simulation program. Madison, WI: University of Wisconsin Solar Energy Laboratory; 2015.
[14] Nizami DJ, Lightstone MF, Harrison SJ, Cruickshank CA. Simulation of the interaction of a solar domestic hot water tank system with a compact heat exchanger. Proceedings of the Joint Conference of the Canadian Solar Building Research Network and the Solar Energy Society of Canada Inc., Fredericton, New Brunswick; 2008. p. 24–29.
[15] Cruickshank CA. Evaluation of a stratified multi-tank thermal storage for solar heating applications. Ph.D. thesis. Department of Mechanical and Materials Engineering, Queen's University, Kingston, ON, Canada; 2009.
[16] Cruickshank CA, Harrison SJ. Heat loss characteristics for a typical solar domestic hot water storage. Energy Buildings 2010;42:1703–10.
[17] Mather DW, Hollands KGT, Wright JL. Single- and multi-tank energy storage for solar heating systems: fundamentals. Sol Energy 2002;73:3–13.
[18] Cruickshank CA, Harrison SJ. Simulation and testing of stratified multi-tank, thermal storages for solar heating systems. Proceedings of the EuroSun 2006 Conference, Glasgow, Scotland, UK; 2006.
[19] Lightstone MF, Raithby GD, Hollands KGT. Numerical simulation of the charging of liquid storage tanks: comparison with experiment. ASME J Sol Energy Eng 1989;111:225–31.
[20] Shyu RJ, Lin JY, Fan LJ. Thermal analysis of stratified tanks. ASME J Sol Energy Eng 1989;111:55–61.
[21] Moran M, Shapiro H. Fundamentals of engineering thermodynamics. 6th ed. New York: John Wiley & Sons; 2008.
[22] Dickinson RM, Cruickshank CA. Exergy analysis of a multi-tank thermal storage for solar heating applications. Int J Exergy 2014;15:412–28.
[23] Haller MY, Cruickshank CA, Streicher W, Harrison SJ, Andersen E, Furbo S. Methods to determine stratification efficiency of thermal energy storage process – review and theoretical comparison. Sol Energy 2009;83:1847–60.
[24] Fisch MN, Guigas M, Dalenback JO. A review of large-scale solar heating systems in Europe. Sol Energy 1998;63:355–66.
[25] Phillips WF, Dave RN. Effects of stratification on the performance of liquid-based solar heating systems. Sol Energy 1982;29:111–20.
[26] Cristofari C, Notton G, Poggi P, Louche A. Influence of the flow rate and the tank stratification degree on the performance of a solar flat-plate collector. Int J Therm Sci 2003;42:455–69.
[27] Rosengarten G, Morrison G, Behnia M. A second law approach to characterizing thermally stratified hot water storage with application to solar water heaters. ASME J Sol Energy Eng 1999;121:194–200.
[28] Yau Y, Rismanchi B. A review on cool thermal storage technologies and operating strategies. Renew Sust Energy Rev 2012;16:787–97.

[29] Ontario Ministry of Energy. Achieving balance—Ontario's long term energy plan. Toronto, ON: Ontario Ministry of Energy; 2013. p. 1–13.

[30] Pinel P, Cruickshank CA, Beausoleil-Morrison I, Wills A. A review of available methods for seasonal storage of solar thermal energy in residential applications. Renew Sust Energy Rev 2011;15:3341–59.

[31] Edwards S. Sensitivity analysis of two solar combisystems using newly developed hot water draw profiles—M.A.Sc. thesis. Carleton University, Ottawa, ON; 2014.

[32] Gao L, Zhao J, Tang Z. A review on borehole seasonal solar thermal energy storage. Energy Procedia 2015;70:209–18.

[33] Flynn C, Siren K. Influence of location and design on the performance of a solar district heating system equipped with borehole seasonal storage. Renew Energy 2015;81:377–88.

[34] Xu J, Wang RZ, Li Y. A review of available technologies for seasonal thermal energy storage. Sol Energy 2014;103:610–38.

[35] Sibbitt B, McClenahan D, Djebbar R, Thornton J, Wong B, Carriere J, Kokko J. Measured and simulated performance of a high solar fraction district heating system with seasonal storage. Proceedings of the ISES Solar World Congress, Kassel, Germany; 2011.

[36] Rosen MA. The use of berms in thermal energy storage systems: energy-economic analysis. Sol Energy 1998;63:69–78.

[37] Kizilkan O, Dincer I. Borehole thermal energy storage system for heating applications: thermodynamic performance assessment. Energy Convers Manage 2015;90:53–61.

[38] Drake Landing Solar Community. http://www.dlsc.ca/

[39] Possemiers M, Huysmans M, Batelaan O. Influence of aquifer thermal energy storage on groundwater quality: a review illustrated by seven case studies in Belgium. J Hydrol: Regional Studies 2014;2:20–34.

[40] Lee KS. Underground thermal energy storage. Seoul, South Korea: Springer; 2013.

[41] Zeghici RM, Essink G, Hatog N. Integrated assessment of variable density–viscosity groundwater flow for a high temperature mono-well aquifer thermal energy storage (HT-ATES) system in a geothermal reservoir. Geothermics 2015;55:58–68.

[42] Singh H, Saini RP, Saini JS. Performance of packed bed solar energy storage system having large sized elements with low void fraction. Sol Energy 2013;87:22–34.

Part E

Chemical

Chapter 16

Hydrogen From Water Electrolysis

Greig Chisholm, Leroy Cronin
School of Chemistry, University of Glasgow, Glasgow, United Kingdom

1 INTRODUCTION

Renewable power cannot be stored yet the global growth of renewable energy sources and integration into the energy mix continues at a significant pace (Fig. 16.1) [1].

Despite great technological advancements, it is not possible to match the fairly predictable peaks and troughs of energy demand with the unpredictable peaks and troughs of energy supply from renewable sources. This issue necessitates the continued use of conventional and rapidly responding energy sources predominantly based on the combustion of hydrocarbon-based fuels to maintain energy supplies. As a result this approach can lead to scenarios wherein demand is low and is being met by conventional power sources coinciding with a peak in renewable energy production due to a particular weather pattern. While it could be instinctively assumed that the conventional power plant could simply be switched off, or at least turned down and energy demands met from renewable sources, in practice this is not possible as conventional fossil fuel power plants and nuclear plants cannot be quickly shut down and restarted, thus it is the renewable energy source that often must be curtailed such that all the energy they are capable of producing is not exploited. For example, in the United States up to 4% of wind power was curtailed in 2013 [2]. Integration of energy storage with renewable energy sources would enable all energy produced by a renewable energy source to be exploited, thus reducing the dependency on conventional, polluting sources of power. Indeed, it appears that flexible energy storage is the barrier to greater penetration and development of renewables.

Energy storage is already a fundamental concept in the global energy economy. Hydrocarbons are currently the most exploited source of energy, accounting for more than 85% of global energy production [3]. What we consider our most familiar fuels: gasoline, coal, and natural gas, to name but a few, are energy storage media themselves. Many millions of years ago, the energy of the

Storing Energy. http://dx.doi.org/10.1016/B978-0-12-803440-8.00016-6

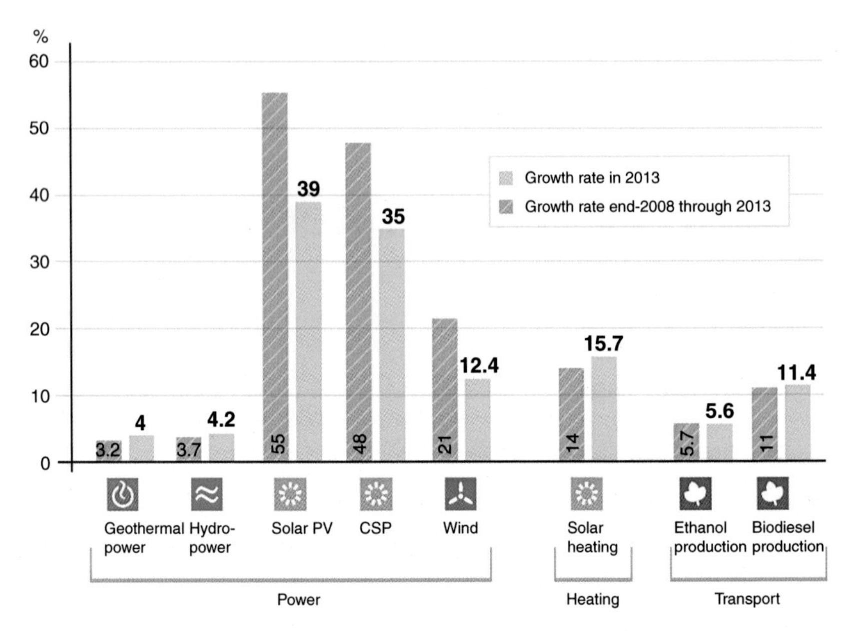

FIGURE 16.1 **Average annual growth rates of renewable energy capacity and biofuels production, end 2008–2013.**

Sun was consumed by plants enabling their growth and proliferation across the prehistoric Earth. These plants provided a rich and plentiful food source for our dinosaur ancestors and as these prehistoric plants and animals died and were covered by millennia of new growth, they were slowly transformed into the fossil fuels we exploit today. Thus energy storage is a concept that has existed for many millennia, but the fuels are literally fossils.

The challenges facing scientists and engineers developing new technologies for energy storage are twofold. First, one must consider the environmental impact of the energy storage medium. While hydrocarbons provide a rich source of energy the products of the combustion processes required to release the stored energy contribute to climate change, unless the carbon could be captured to produce a closed carbon cycle. Second, one must consider timescale. As stated earlier, the lifecycle of organic life and its subsequent decomposition into a usable fuel source is extremely slow, taking many millions of years. Thus when selecting a new energy storage medium the ideal candidate should be able to release stored energy without producing any pollutants or harmful side-products and should be able to store energy on an instantaneous timescale. Considering the fundamental requirements of energy storage, namely nonpolluting and quickly responding, hydrogen is a perfect medium for energy storage.

Hydrogen is not found in appreciable or exploitable concentrations freely on Earth and instead must be produced from other compounds. There are two principal routes to the production of hydrogen. Most commonly hydrogen is produced from natural gas via a process known as steam reforming. In addition to hydrogen, this process also produces carbon dioxide and is not a viable solution to the pollution-free production of hydrogen from excess renewable energy. Hydrogen may also be produced via electrolysis of water. In this process electricity (electro-) is used to break down (-lysis) water (H_2O) into its component parts of oxygen and hydrogen with no harmful or polluting side-products. If the electricity required to carry out this process is provided by a renewable source then a nonpolluting sustainable source of hydrogen is obtained. Release of the energy stored in hydrogen is an extremely clean process, producing only water as a byproduct and releasing large quantities of energy in doing so. Indeed on a weight-by-weight basis hydrogen produces almost four times more energy than the equivalent weight of gasoline [4].

2 HYDROGEN AS AN ENERGY VECTOR AND BASIC PRINCIPLES OF WATER ELECTROLYSIS

2.1 Hydrogen as an Energy Vector

Hydrogen is a gaseous element occurring as its diatomic gas H_2. For clarity when the chapter refers to hydrogen, unless otherwise noted, this is in reference to the diatomic molecule H_2. Since hydrogen does not naturally occur on Earth it must be formed by the decomposition of other molecules. Approximately 95% of the hydrogen produced around 50×10^6 t a^{-1} (around 50 million tonnes per annum) [5] is produced via steam reforming of natural gas and subsequent water–gas shift reaction (Scheme 16.1) with the remainder being formed via electrolysis. As can be seen from Scheme 16.1, the formation of hydrogen from natural gas results in the production of carbon dioxide. Given the finite supplies of natural gas and the greenhouse effect of carbon dioxide, production of hydrogen from this route does not address the needs of renewable energy storage. The production of hydrogen from water via electrolysis is a clean process, resulting in only oxygen being produced as a byproduct. If the electricity required to split the water into hydrogen and oxygen is supplied via a renewable energy source then the process is environmentally benign.

Hydrogen has two main applications as an energy vector. First, it may be combusted in a similar manner to natural gas. This is an extremely clean combustion process with water being the only product. A subset of this

Steam methane reforming

$$CH_4 + H_2O \longrightarrow CO + 3H_2$$

Water-gas shift reaction

$$CO + H_2O \longrightarrow CO_2 + H_2$$

SCHEME 16.1 Production of hydrogen from natural gas.

application is the storage of hydrogen in a conventional natural gas network, in a so-called power-to-gas process. This natural gas/hydrogen mixture is then combusted in an identical manner to unadulterated natural gas. It is worth noting that the coal gas that is used to supply gas to many homes and industries could contain up to 50% hydrogen. Natural gas supplies now largely supplant coal gas. In addition to combustion, hydrogen may also be exploited as an energy vector via catalytic recombination with oxygen. In contrast to the thermal energy produced in a combustion process, the catalytic reaction with oxygen produces electrical energy. This process is exploited by fuel cells to produce electrical energy.

2.2 History of Water Electrolysis

Water electrolysis was first demonstrated in 1789 by the Dutch merchants Jan Rudolph Deiman and Adriaan Paets van Troostwijk using an electrostatic generator to produce an electrostatic discharge between two gold electrodes immersed in water [6]. Later developments by Johann Wilhelm Ritter exploited Volta's battery technology and allowed separation of the product gases (Fig. 16.2) [7].

Almost a century later, in 1888 a method of industrial synthesis of hydrogen and oxygen via electrolysis was developed by the Russian engineer Dmitry Lachinov [8] and by 1902 more than 400 industrial water electrolyzers were in operation [9]. Early electrolyzers utilized aqueous alkaline solutions as their electrolytes, and this technology persists to this day. A more recent development in water electrolysis is the proton exchange membrane process, first described in the mid-1960s by General Electric as a method for producing electricity for the Gemini Space Program [10], and later adapted for electrolysis. Today a number of companies are active in the manufacture and development of electrolysis technologies with Proton, Hydrogenics, Giner, and ITM Power being leaders in the field.

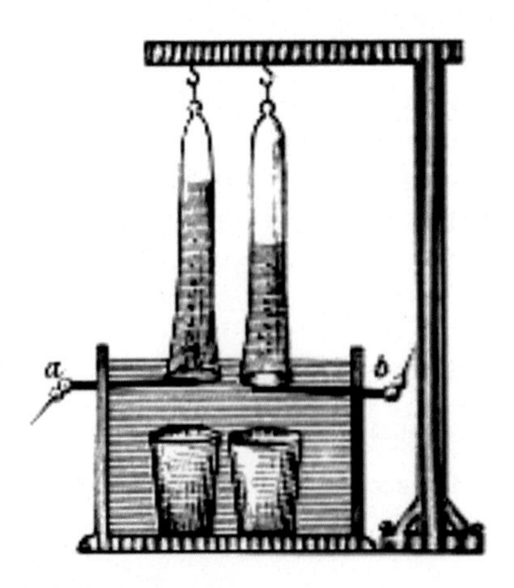

FIGURE 16.2 **Ritter's electrolysis apparatus.**

2.3 Electrochemistry and Thermodynamics

The electrolysis of water is thermodynamically disfavored and as such requires an input of energy to drive the process. In the case of the electrolytic splitting of water into hydrogen and oxygen, this energy input comes in the form of a potential difference between the anode and cathode of an electrochemical cell. By application of Gibbs energy equation (Eq. 16.1) and knowledge of the standard enthalpy of formation (286.03 kJ mol^{-1}) and ideal gas entropy (0.163 kJ mol^{-1} K^{-1}) of gaseous water we can calculate the Gibbs energy of water at 298 K (25 °C):

$$\begin{aligned} \Delta G^\circ &= \Delta H^\circ - T\Delta S^\circ \\ \Delta G^\circ/\left(\text{kJ mol}^{-1}\right) &= 286.03 - (298 \times 0.163) \\ \Delta G^\circ/\left(\text{kJ mol}^{-1}\right) &= 237.46 \end{aligned} \tag{16.1}$$

Gibbs energy

Using this Gibbs energy we can calculate the reversible voltage required to split water electrolytically (Eq. 16.2), where n is the number of electrons transferred and F is Faraday's constant. Two electrons are required to produce a single molecule of hydrogen, hence $n = 2$.

$$\begin{aligned} V_{\text{rev}} &= -\frac{\Delta G^\circ}{nF} \\ V_{\text{rev}}/V &= -\frac{237\ 460}{2 \times 96\ 485} \\ V_{\text{rev}} &= -1.23\ \text{V} \end{aligned} \tag{16.2}$$

Calculation of V_{rev}

This value of 1.23 V requires that all components be in the gaseous state, that is, that the water which is consumed in the reaction be vaporized. In a conventional electrolyzer where water is at atmospheric pressure and temperatures are typically below 80 °C, there is an additional energy input required as illustrated in Eq. 16.3. This gives rise to the thermoneutral potential of 1.48 V. The potential at which water splitting occurs can approach the reversible potential by carrying out the electrolysis at high temperature and pressure and approaches to this will be discussed in Section 3.

$$\begin{aligned} V_{\text{tn}} &= -\frac{\Delta H^\circ}{nF} \\ V_{\text{rev}}/V &= -\frac{286\ 030}{2 \times 96\ 485} \\ V_{\text{rev}} &= -1.48\ \text{V} \end{aligned} \tag{16.3}$$

Calculation of V_{tn}

Two further concepts are required to understand the electrochemistry of an electrolyzer: overpotential and efficiency. While the electrolysis of water is thermodynamically feasible at 1.48 V, in reality the reaction will occur prohibitively slowly and further potential must be applied to accelerate the reaction to a practical rate. This extra potential is known as overpotential and minimizing this additional potential is fundamental in producing efficient electrolytic cells. Overpotential is due to a combination of the very low conductivity of water and the high activation energy required to split water. The former may be addressed by incorporating salts, acids, or bases into the water to improve its conductivity. This is the case with an alkaline electrolyzer where the electrolyte is aqueous sodium hydroxide solution. The latter is addressed by the addition of suitable electrocatalysts, a concept considered in more detail in Section 3.

The efficiency of an electrolysis process can be considered in two ways. We can consider the Faradaic efficiency—the ratio of the number of moles of hydrogen produced versus the charge passed. If the process has a 100% Faradaic efficiency then every electron produced by the oxidation of water is transferred to a corresponding proton to produce hydrogen. Note that two electrons are required to produce a single mole of hydrogen (n = 2, Eq. 16.2). The actual Faradaic efficiency may approach 100%, but is always slightly lower due to parasitic electrochemical processes, for example, degradation of components of the electrochemical cell, and the crossover of hydrogen to the oxygen-producing side of the cell.

The energy efficiency of the cell or stack is measured by calculation of the energy available from the hydrogen produced by the cell, using the higher heating value of hydrogen (HHV) [11] and dividing this by the energy consumed by the cell as shown in Eq. 16.4, where I is the cell current, V is the voltage, and t is the time over which the hydrogen production was measured; η_{eff} is expressed as a percentage.

$$\eta_{\text{eff}} = \frac{\text{moles of hydrogen produced} \times HHV_{\text{H}_2}}{\text{I} \times V \times t} \times 100 \qquad (16.4)$$

Calculation of cell energy efficiency

Commercial electrolyzers often express their energy efficiency in terms of the whole system η_{sys}, including parasitic loads due to pumps, heaters, etc. In addition commercial systems are characterized by their specific energy consumption, that is, the energy required to produce 1 cubic meter of hydrogen under normal conditions. This is commonly expressed in kW h $(\text{Nm}^3)^{-1}$. Some typical values and the corresponding η_{sys} are given in Table 16.1.

The individual reactions taking place during electrolysis differ depending on the nature of the process taking place and will be discussed in full in Section 3.

TABLE.16.1 Efficiency Comparisons

Manufacturer	Product	System efficiency/ (kW h) (N m^3)$^{-1}$	η_{sys}/%	References
Hydrogenics	HyLYZER™	6.7	53%	[12]
Proton OnSite	Hogen® S-series	6.7	53%	[13]

3 HYDROGEN PRODUCTION VIA WATER ELECTROLYSIS

3.1 Water Electrolysis

The basic principles, chemistry, and thermodynamics of water electrolysis have been described in Section 2. Herein we discuss the device architectures, their advantages and disadvantages, and materials of construction. There are three main routes to the electrolysis of water: alkaline electrolysis, proton exchange membrane (PEM) electrolysis, and solid oxide electrolysis. Of these only alkaline and PEM electrolysis have been commercialized and while solid oxide electrolyzers show great technological promise, they are still the subject of considerable development [14]. The technological parameters defining the main types of commercial electrolyzers are shown in Table 16.2 [15,16]. Comparable information on solid oxide electrolyzers is not available due to the lack of commercial devices.

3.2 Alkaline Water Electrolysis

The first commercialized water electrolysis system was based on the principles of alkaline water electrolysis, and alkaline-based systems remain the most ubiquitously utilized water electrolysis systems [17]. A schematic of an alkaline water electrolyzer is given in Fig. 16.3. The cathode and anode are often manufactured from nickel [18] and the separator between the anodic and cathodic chambers is a polymer which is permeable to hydroxide ions and water molecules, for example, Zirfon Perl from Agfa. Alkaline electrolyzers have the advantage of technological maturity and relatively low cost compared with PEM electrolyzers.

The reactions taking place in an alkaline electrolyzer are given in Scheme 16.2.

There are a number of technological disadvantages associated with the alkaline electrolysis system: namely, low current density, limited ability to operate at low loads, and the inability to operate at high pressure. The latter two limitations are due to the crossover of gases possible through the separator. This will increase both with increasing pressure of hydrogen at the cathode and will also increase with reduced load where the oxygen production rate decreases and the hydrogen concentration in the oxygen stream can increase to dangerous levels (H_2 lower explosion limit >4%) [19].

TABLE 16.2 Comparison of Water Electrolysis Methods

	Alkaline electrolyzer	Proton exchange (PEM) electrolyzer
Cell temperature/(°C)	60–80	50–80
Cell pressure/(10^5 Pa) (bar)	<30	<30
Current density/(mA cm^{-2})	0.2–0.4	0.6–2.0
Cell voltage/V	1.8–2.4	1.8–2.2
Power density/(mW cm^{-2})	<1	<4.4
Efficiency (*HHV*)/(%)	62–82	67–82
Specific energy consumption stack/(kW h) $(Nm^3)^{-1}$ H_2	4.2–5.9	4.2–5.6
Specific energy consumption system/(kW h) $(Nm^3)^{-1}$ H_2	4.5–7.0	4.5–7.5
Partial load range/(%)	20–40	5–10
Cell area/(m^2)	>4	<0.03
H_2 production rate/(Nm^3 h^{-1})	<760	<10
Lifetime stack/h	<90 000	<20 000
Lifetime system/a (years)	20–30	10–20
Degradation rate/(μV h^{-1})	<3	<14

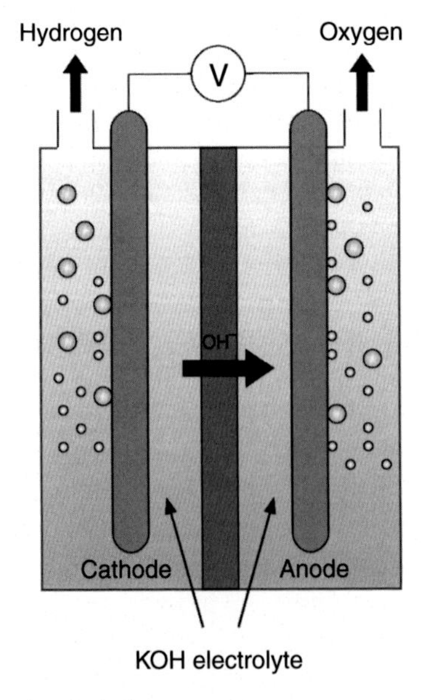

FIGURE 16.3 Alkaline water electrolyzer.

Cathode Reaction

$$H_2O_{(l)} + 2e^- \longrightarrow H_{2(g)} + 2OH^-_{(aq)}$$

Anode Reaction

$$2OH^-_{(aq)} \longrightarrow 2H_2O_{(l)} + {}^1/_2O_{2(g)} + 2e^-$$

SCHEME 16.2 Alkaline water electrolysis.

3.3 Proton Exchange Membrane Electrolysis

A schematic of a PEM electrolyzer is given in Fig. 16.4. Central to the operation of the PEM electrolyzer is the proton-conducting membrane. This membrane is made of a sulfonated fluorinated polymer. The most commonly used membrane is Nafion manufactured by DuPont. Various thicknesses of Nafion may be employed ranging from (25 to 250) μm. The exact thickness of membrane required for any specific application is determined by the conditions to be employed in the electrolyzer. Higher pressures and conditions likely to degrade the membrane at a greater rate, for example, low-load operation or frequent stop–starts tend to require a thicker membrane either to withstand the high differential pressure and minimize gas crossover or to provide a suitable margin for material loss due to degradation. The thinner the membrane the greater the efficiency of the electrolysis process. This is due to a reduction in the resistance of the membrane with decreasing thickness [20,21]. Thus a balancing act must be struck between the optimal electrochemical properties of the membrane and its suitability for a given set of process conditions. There is considerable ongoing research into the development of new membranes with improved properties including greater proton conductivity and improved mechanical strength [22,23].

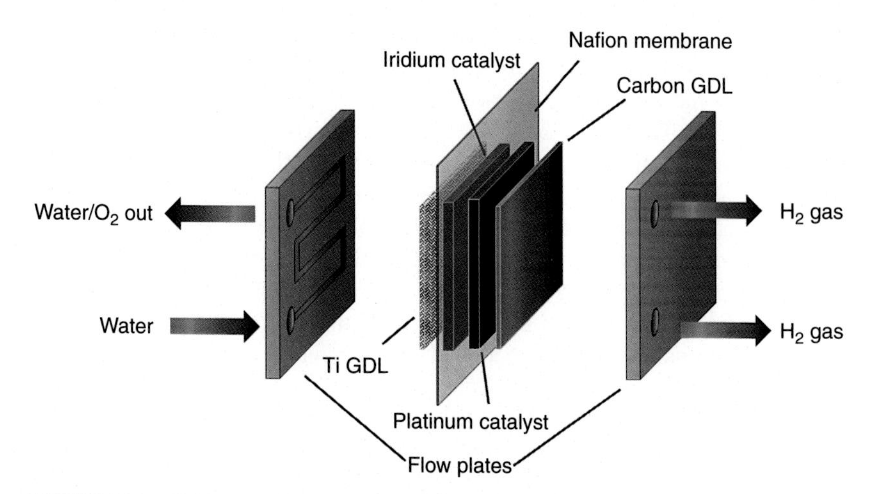

FIGURE 16.4 Cell layout diagram for a PEM Electrolyzer.

Cathode Reaction

$$2H^+_{(aq)} + 2e^- \longrightarrow H_{2(g)}$$

Anode Reaction

$$H_2O_{(l)} \longrightarrow {}^1/_2O_{2(g)} + 2H^+_{(aq)} + 2e^-$$

SCHEME 16.3 Proton exchange membrane water electrolysis.

As discussed in Section 2, the kinetics of water oxidation is extremely slow without the addition of suitable electrocatalysts. Due to the highly acidic nature of the PEM electrolysis process the choice of catalysts is limited to rare transition metals that are stable under acidic conditions, for example, rhodium, ruthenium, platinum, iridium, and their oxides [24]. The current state of the art for electrocatalysts in PEM water electrolysis is platinum at the cathode for proton reduction and iridium oxide at the anode for water oxidation. There has been a large amount of research undertaken to prepare new electrocatalysts for both water oxidation and proton reduction [25–28] to increase the efficiency of the process and to reduce cost. The combination of the membrane and the electrocatalysts is collectively known as membrane electrode assembly (MEA).

The water is fed into the anode of the electrolyzer and the product gases are conducted away from the reactive sites on the MEA via the flow channels of the bipolar flow plates. These flow plates are typically manufactured from titanium as this gives the mechanical strength and resistance to corrosion required. Stainless steel and graphite are also used in some systems though their application is limited by relatively poor corrosion properties and the poor mechanical strength of graphite compared with titanium. Despite titanium's near ubiquitous use, it is not the perfect material for bipolar plates. It is subject to oxidation and is very difficult to machine, adding to its cost. There is limited research into alternative materials. A recent publication described the use of three-dimensional printing to prepare flow plates though the difficulty of rendering these plates suitably conductive limits their current use to prototyping [29].

Between the flow fields of the bipolar plates and the MEA a conductive layer is added. This layer is known as the gas diffusion layer (GDL) and is added to improve the electrical connection between the bipolar plates and the MEA and also to insure the effective mass transport of both the reactant water and the product gases. The cathodic GDL is often made of carbon paper. This material is not suitable for use at the anode as the highly oxidative conditions at this electrode would quickly decompose the carbon paper. Instead, a titanium or similar inert metal mesh is inserted between the flow fields and the MEA.

The reactions taking place in a PEM electrolyzer are given in Scheme 16.3.

3.4 Solid Oxide Water Electrolysis

Solid oxide electrolysis differs from both alkaline and PEM systems in that the operating temperature is typically an order of magnitude greater in a solid

Cathode Reaction

$$H_2O_{(l)} + 2e^- \longrightarrow H_{2(g)} + O^{2-}_{(aq)}$$

Anode Reaction

$$O^{2-}_{(aq)} \longrightarrow {}^1/_2O_{2(g)} + 2e^-$$

SCHEME 16.4 Solid oxide electrolyzer water electrolysis.

oxide electrolyzer (SOE) spanning the range 800–1000 °C. At this temperature it should be apparent that the feed cannot be water and in this case the electrolyzer is fed with steam. The chemical reactions taking place in the SOE are given in Scheme 16.4.

Intuitively, the high operating temperatures of the SOE system would suggest that the efficiency of operation would be poor; however, this is not the case. The increase in thermal energy demand is compensated for by the decrease in the electrical energy demand and the overall energy demand of the system is largely insensitive to increasing the temperature. This is illustrated in Fig. 16.5 [30].

In the solid oxide electrolysis process, water in the form of high-pressure steam is reduced at the cathode to give hydrogen gas and oxygen anions. These anions migrate through the solid oxide electrolyte, where they are oxidized on the anode to produce oxygen gas. The electrons produced from the oxidation travel around the external circuit and supply the electrons for the water reduction. The product gases diffuse through the porous electrodes. A schematic of the SOE is given in Fig. 16.6 [31].

The SOE utilizes a solid electrolyte separating the anode and the cathode. In this regard the system has some similarity with the PEM system, where a solid

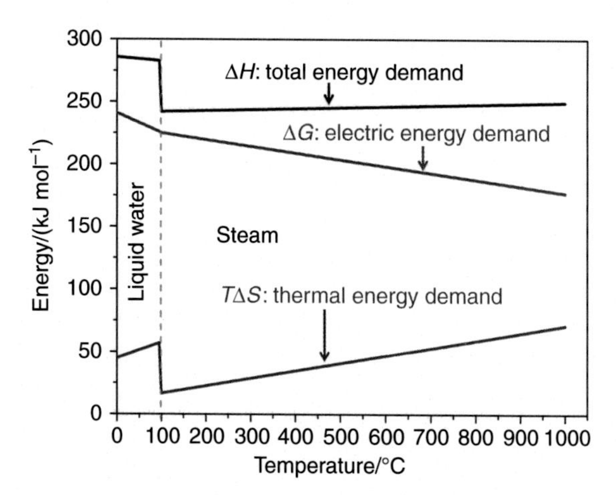

FIGURE 16.5 **Electrolysis energy demand.** *(Reproduced from Ref. [30] with permission from The Royal Society of Chemistry.)*

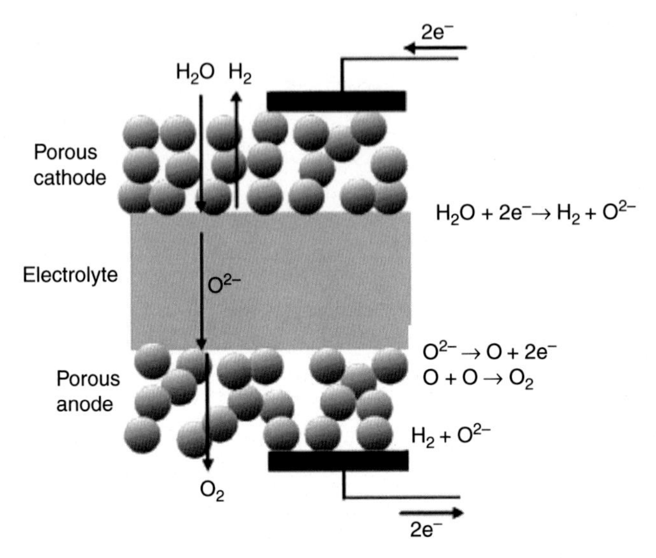

FIGURE 16.6 **SOE operation.**

electrolyte is also employed to provide ionic conductivity to the system. The current state-of-the-art electrolyte in a solid-oxide system is yttria-stabilized zirconia (YSZ) [30]. This material only has sufficient ionic conductivity at high temperatures, hence the requirement to operate at temperatures approaching 1000 °C. At this high temperature, there are numerous problems with cell integrity including poor long-term cell stability, interlayer diffusion, and fabrication and materials problems [32]. In addition to oxygen anion conductivity, it is also possible to operate an SOE on the principle of proton conductivity in a manner similar to a PEM electrolyzer [33]. In this system a different electrolyte must be adopted. Doped $BaCeO_3$ shows good conductivity in this system [34].

4 STRATEGIES FOR STORING ENERGY IN HYDROGEN

4.1 Properties of Hydrogen Related to Storage

Hydrogen was first identified as an element by Henry Cavendish in 1766 during a series of experiments wherein he added various metals to strong acids. He described the gas produced as inflammable air. The primitive apparatus used by Cavendish to first isolate and store hydrogen gas is shown in Fig. 16.7 [35]. Since this first isolation of hydrogen, considerable effort has gone into the development of suitable storage technologies for containing this gas. Storage by compression or via cryogenic liquefaction are the most developed technologies and form the basis of the technologies used in current automotive demonstration projects. A significant challenge associated with both pressurized and cryogenic storage is that the storage vessel itself contributes at least 90% to total system weight [36].

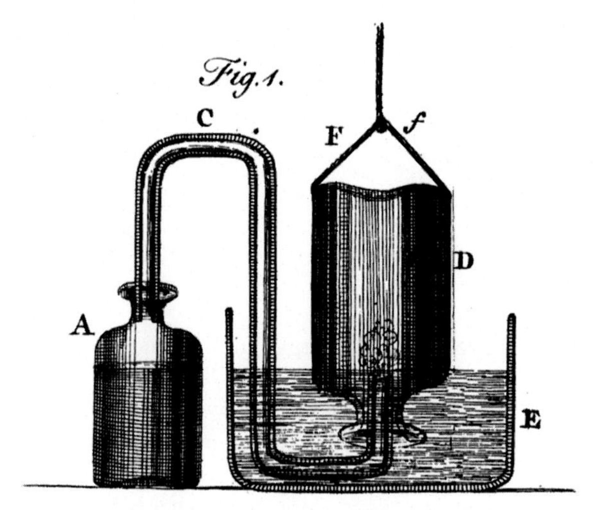

FIGURE 16.7 **Cavendish apparatus for isolation of hydrogen gas.**

Basic properties of hydrogen are given in Table 16.3.

Hydrogen has the highest energy density by weight of any fuel; however, due to its very low density, its volumetric energy density is very poor. Thus storage of large quantities of energy in the form of uncompressed hydrogen gas would require impractically large storage facilities. This requirement may be addressed by compression or liquefaction of the hydrogen, thus increasing its volumetric energy density.

4.2 Gaseous Hydrogen Storage

When considering a storage strategy for hydrogen it is important to consider the final application of the hydrogen and to match the pressure of the storage medium to that required by the application. The compression of hydrogen to a

TABLE 16.3 Hydrogen Properties

Property	Value
Melting point/K	13.99
Boiling point/K	20.27
Density/(g L^{-1})	0.0899
Higher heating value/(MJ kg^{-1})	141.80
Lower heating value/(MJ kg^{-1})	119.96
Gravimetric (specific) energy density/(MJ kg^{-1})	141.80
Volumetric energy density at STP/(MJ L^{-1})	0.0107

TABLE 16.4 DoE Hydrogen Storage Targets

Target	2009	2015	Ultimate
System gravimetric density/(mass%)	4.5	5.5	7.5
System volumetric density/(g L^{-1})	28	40	70
System fill time for 5 kg fill/(min)	4.2	3.3	2.5

suitable storage pressure consumes energy and consequently incurs an efficiency penalty. Compression of hydrogen to (350–700 $\times$ 10^5) Pa (350–700 bar) consumes energy equal to (5–20)% of the lower heating value of hydrogen. These pressure ranges are considered the optimum for transport applications using compressed hydrogen as they represent a balance between compression energy consumption, driving range of the vehicle being fueled, and the investment required in refueling stations and associated infrastructure [37]. Research and development into the storage of pressurized gaseous hydrogen is largely predicated on the emergence of a hydrogen vehicle and refueling sector. The US Department of Energy (DoE) has produced a series of targets for the storage of hydrogen (Table 16.4) [38].

The "ultimate" target represents the performance that is anticipated to be required for full penetration of hydrogen-powered vehicles into the light-duty market. In addition to the energy density requirements outlined earlier, there are also technical requirements related to the small size of the hydrogen molecule. This gives hydrogen a high diffusivity in many materials. In the case of storage of high-pressure hydrogen in steel containers, this diffusivity can result in embrittlement of the container leading to failure of the material [39–43]. The targets set by the DoE and the challenges of embrittlement have driven a great deal of research in the storage of hydrogen as a high-pressure gas.

In vehicular applications, two storage vessel technologies known as Type III and Type IV have grown to a dominant position in vehicular applications [44]. Type III storage vessels are composed of a pressure vessel made of a metallic liner fully wrapped with a fiber–resin composite [45,46]. Type IV storage vessels differ from type III vessels in that the pressure vessel is made of a polymeric liner as opposed to a metallic liner, and is again fully wrapped with a fiber–resin composite [45,46]. Type III storage vessels are considered to be the technically superior design due to the metallic liner. This liner gives the storage vessel a higher thermal conductivity which minimizes the temperature rise associated with the filling of the storage vessel with compressed hydrogen. High cost remains the most significant barrier to successful implementation of this technology, though public perception of the safety of compressed hydrogen storage may also need to be overcome [47,48].

4.3 Cryogenic Liquid Hydrogen Storage

Conversion of hydrogen to a cryogenic liquid requires a large input of energy to condense the gas. This energy input amounts to (30–40)% of the lower heating value of hydrogen and results in a corresponding loss of efficiency in any system relying on cryogenic hydrogen [49]. On this basis, it compares poorly with storage of hydrogen as a pressurized gas despite the increase in volumetric energy density to 8 MJ L^{-1}. In addition, storage containers for cryogenic hydrogen must be insulated and refrigerated to maintain the low temperature required and require frequent venting to allow evaporated hydrogen gas to escape.

4.4 Cryocompressed Hydrogen Storage

Cryocompression of hydrogen refers to several different storage methods that combine elements of compressed and cryogenic storage and includes liquid hydrogen, cold compressed hydrogen, or hydrogen stored as a mixture of saturated liquid and vapor. Common between methods is the storage of hydrogen in its cryo state and at pressures of $(250–350 \times 10^5)$ Pa (250–350 bar) [50]. Storage of hydrogen as a cryocompressed fluid serves to overcome the limitations of both compressed gas and cryogenic liquid storage allowing both an increase in volumetric energy density versus compressed gaseous hydrogen and also minimizing the evaporative losses associated with cryostorage at atmospheric pressure [51].

4.5 Hydrogen Storage by Physisorption

Nonmechanical means can be used to store hydrogen, for example, by physisorption onto porous supports. This process is reversible and the hydrogen may be adsorbed and desorbed over multiple cycles. The hydrogen is physically attached to the adsorbent via Van der Waals interactions and is limited to a monolayer of hydrogen. Sorbents with high surface areas therefore have the greatest capacity for hydrogen storage [52]. Cryogenic hydrogen absorption at temperatures below −195 °C can profoundly increase the level of physisorption of hydrogen [53,54].The absorbent properties of various materials are given in Table 16.5 [55].

4.6 Hydrogen Storage by Chemisorption

In contrast to hydrogen storage by physisorption, chemisorption of hydrogen requires that the hydrogen undergo a chemical transformation and become chemically bonded to another species e.g. in a hydride. The increased energy density in chemisorbed hydrogen is attributed to the shorter mean distance between hydrogen atoms in the hydride compared with the hydrogen molecules in a compressed or liquefied state (Fig. 16.8) [37]. The advantage of this approach is the very high energy density achievable compared with storage of hydrogen in compressed or liquefied form. In common with other hydrogen

TABLE 16.5 Properties of Selected Physisorption Media

Storage medium	Temperature/(°C)	Pressure/ (10^5 Pa) (bar)	Capacity/ (mass%)	Reference
Carbon nanotubes	27	1	0.2	[56]
	25	500	2.7	[57]
	−196	1	2.8	[58]
Graphene oxide	25	50	2.6	[59]
Polymers of intrinsic microporosity	−196	10	2.7	[60]
Hyper crosslinked polymers	−196	15	3.7	[61]
Covalent organic frameworks	−196	70	7.2	[62]
Zeolites	25	100	1.6	[63]
	−196	16	2.07	[64]
Metal organic frameworks	25	50	8	[65]
	−196	70	16.4	[66]
Clathrate hydrates	−3	120	4	[67]

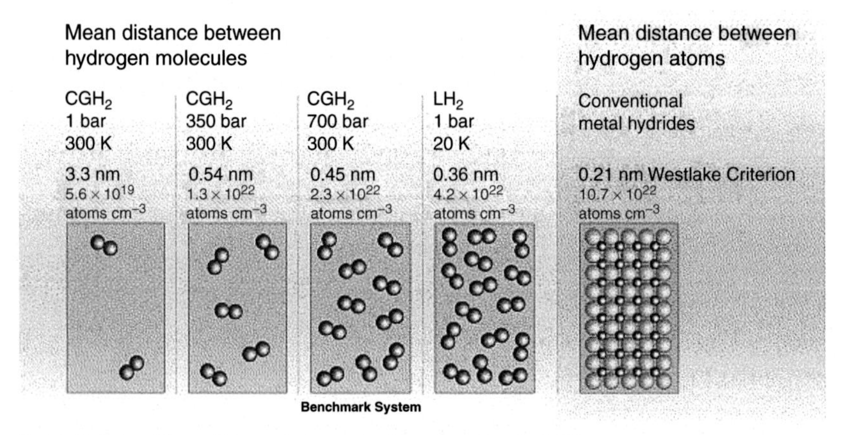

FIGURE 16.8 Mean distance between hydrogen molecules [37].

storage methods, storage via chemisorption has its share of technical challenges; namely, ease and cyclability of charging and discharging, storage capacity, and stability of the material after charging [55]. In particular, from a practical perspective if one was to consider a system wherein pressurized hydrogen was pumped at pressure onto a discharged hydrogen precursor, formation of the hydride would have a given heat of formation ΔH associated with it. If the tank is to be filled with hydrogen in the 150 s (2.5 min) filling time proposed by the DoE (Table 16.4) a great deal of heat would be required to dissipate in a very short time requiring a prohibitively large heat exchanger to be installed into the storage tank.

While there are clearly a number of technical challenges to be overcome in relation to the use of hydrides as a hydrogen storage medium, there are a number of candidates undergoing detailed development in an attempt to overcome these limitations. Current research is focused on metal hydrides based on alkali and alkaline earth elements such as MgH_2 [68] and complex hydrides such as $NaAlH_4$ [69], the borohydrides $LiBH_4$ and $NaBH_4$ [70], and $LiNH_2$ [71].

4.7 Power-to-Gas

Power-to-gas describes the injection of either hydrogen directly produced via electrolysis and suitably purified if required or methane produced via methanation of hydrogen into the natural gas network. This topic is described in more detail elsewhere in this volume (chapter 18) and is considered here only briefly for completeness. The adulteration of methane with hydrogen gas is well known historically. "Town gas" derived from coal comprising a mixture of hydrogen, carbon monoxide, and methane was supplied to industrial and domestic properties from the 1840 s and was only phased out in the United Kingdom with the introduction of natural gas sourced from the North Sea. The potential of the natural gas network, ubiquitous in all developed countries, has tremendous potential as a storage medium for energy. In the United Kingdom alone, more than 836 000 GW h of energy were consumed in the form of natural gas in 2011 [72]. Even small quantities of hydrogen gas stored in this network deliver a large amount of energy storage.

There are limitations to the storage of hydrogen in the natural gas network. Due to the potential for hydrogen embrittlement as discussed in relation to pressurized hydrogen storage mentioned previously, and the different burning characteristics of methane/hydrogen mixtures compared with pure methane [73], limitations are placed on the quantity of hydrogen in natural gas, for example, 0.1% by mole in the United Kingdom [74]. This limitation in the quantity of hydrogen that can be stored in the gas network can be overcome by conversion of the hydrogen to methane via reaction with carbon dioxide or carbon monoxide (Scheme 16.5) [75,76] in a process known as methanation.

$$CO + 3H_2 \longrightarrow CH_4 + H_2O$$

$$CO_2 + 4H_2 \longrightarrow CH_4 + 2H_2O$$

SCHEME 16.5 Methanation of hydrogen.

5 TECHNOLOGY DEMONSTRATIONS UTILIZING HYDROGEN AS AN ENERGY STORAGE MEDIUM

5.1 System Engineering

A block diagram of a generic energy storage system using hydrogen as the energy storage medium is given in Fig. 16.9.

The energy source for such an energy storage system is typically a renewable source such as wind or photovoltaics, though in principle any source of electrical energy may be used, even grid electricity if the economics are appropriate, for example, if very cheap grid electricity can be sourced during off-peak periods. The power-conditioning section of the system is of paramount importance and will determine whether electricity is taken from the energy source, the voltage of that electricity, and whether it should be sent directly to the electrolyzer or whether it should be stored in an intermediate energy store, for example, a battery. The use of a battery in the power-conditioning step can help to overcome limitations in the performance of an electrolyzer at low loads and to minimize start–stop cycling which can decrease the lifetime of the electrolyzer. Alkaline electrolyzers, in particular, have minimum loads typically around (20–40)% of full load [15]. PEM electrolyzers have much better low-load performance but at low loads gas crossover increases [77] as well as the degradation rate of the membrane [21]. The choice of electrolyzer type—alkaline, PEM, or SOE—will be based on the economic and technical aspects of the project. The characteristics of each type of electrolyzer are discussed in Section 3.

Hydrogen gas produced by an electrolyzer can be of varying purity depending on both the type of electrolyzer and the operating conditions employed. The two significant impurities found in the hydrogen gas stream from an electrolyzer are water and oxygen. Water is often removed by passage of the gas through a desiccant, for example, silica gel, or by pressure swing adsorption or a combination of both. Removal of oxygen is via a catalytic recombination process wherein the oxygen is recombined with hydrogen to produce water. This process is

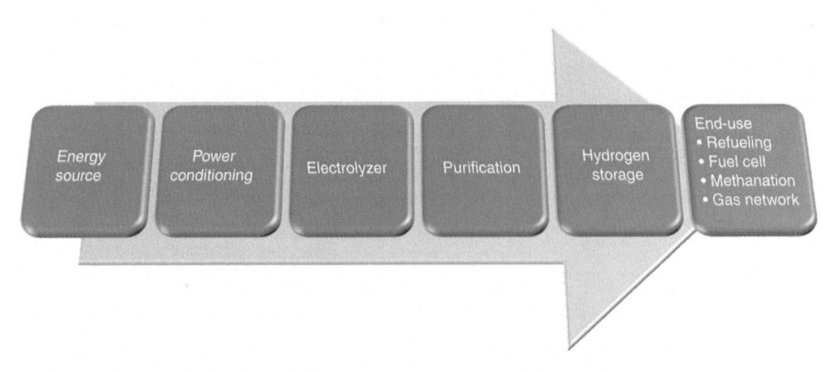

FIGURE 16.9 **Block diagram of hydrogen-based renewable energy storage.**

carried out by passage of the product gas through a palladium filter heated to (300–600) °C depending on the exact composition of the filter [78]. It goes without saying that this should be placed upstream of the final drying process to prevent reintroducing water to the product gas. The parasitic load from this balance of plant will reduce overall system efficiency. Contamination of the hydrogen with oxygen is typically at a lower level than the opposite case due to the greater size and lower mobility of the oxygen molecule within the membrane and the differential pressure between the hydrogen- and oxygen-producing sides of the membrane; however, despite being at a low level, oxygen contamination can reduce the purity of the product gas to a level where further purification is required, particularly for applications requiring high purity, for example, LED device manufacturing, silicon carbide epitaxy, and polysilicon manufacturing [79]. Hydrogen can be contaminated by up to 2.7% oxygen at high pressure [77]. While still outside the flammability limit for hydrogen/oxygen mixtures (3.9–95.8%), a safety margin must be considered and any reading greater than 2% is likely to lead to an alarm condition. In addition, this crossover of gases increases the likelihood of recombination of oxygen and hydrogen within the membrane of the electrolyzer which can lead to a reduction in the lifetime of the membrane and reduces the efficiency of the electrolyzer [80].

5.2 Renewable Energy Storage

The largest of the current renewable energy storage projects utilizing hydrogen that are currently in operation are facilities located in Germany at Grapzow [81] in the province of Mecklenburg-Vorpommern and in Falkenhagen [82] in Brandenburg. Both systems are coupled to a wind farm and utilize electrolyzers supplied by Hydrogenics and are in collaboration with the utility company E.ON. The 1 MW Grapzow electrolyzer is capable of producing 210 Nm^3 of hydrogen per hour. This hydrogen can be directed to either an internal combustion engine to generate electricity or injected into the local gas network. The Falkenhagen system consists of a 2 MW electrolyzer and can supply up to 360 Nm^3 of hydrogen for storage in the gas network. There are a number of comprehensive reviews of the coupling of electrolysis systems to renewable energy sources which give further details on a number of additional projects [83,84]. All projects at this stage can be considered as demonstration projects—showcasing technology and enabling further understanding of operational parameters in a live environment prior to scaleup to full commercial status.

The first hydrogen-based system for storing renewable electricity by means of electrolysis and subsequent hydrogen storage was realized in 1991 [85]. This project was based in Nuenberg vorm Wald in Germany and marked the beginning of that country's leadership in the adoption and demonstration of technologies essential for realization of hydrogen economy. This pilot plant operated from 1991 to 1999 and utilized electrical power generated from an array of photovoltaic panels with a capacity of 266 kWp (kilowatt-peak) (Fig. 16.10).

FIGURE 16.10 **Overhead view of the Nuenberg vorm Wald Plant [85].**

Over the lifetime of the project this demonstration plant utilized both alkaline (low and high pressure) and PEM electrolysis technologies (Fig. 16.11). This plant tested a number of subsystems connected to the electrolysis system including hydrogen and oxygen purification and storage systems, fuel cells, gas-fired boilers fueled with a hydrogen/methane mixture, and a liquid hydrogen–fueling

FIGURE 16.11 **Interior of the Neunberg Plant—DI Water tank and low pressure electrolysers (top right) and high pressure electrolyser (bottom right).**

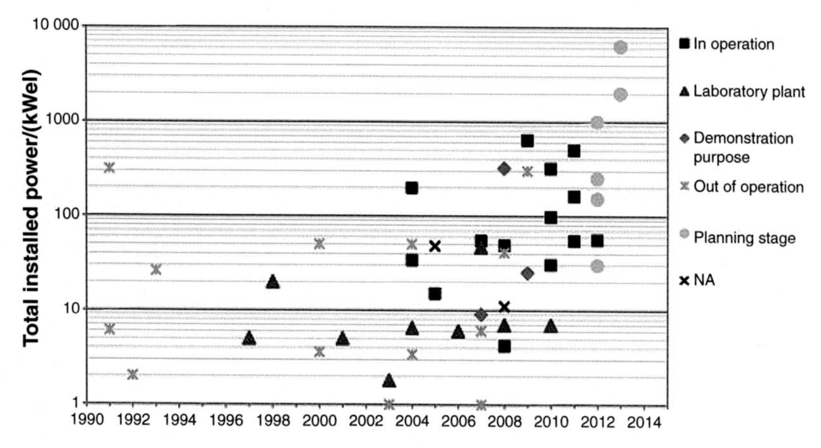

FIGURE 16.12 **Installed power of hydrogen storage demonstration plants.**

station used to fuel prototype liquid hydrogen–fueled cars. With regard to the latter, liquid hydrogen is no longer considered the technology of choice for exploiting hydrogen in transportation applications and has now been largely supplanted by pressurized hydrogen gas.

Since the pioneering work at Nuenberg (Fig. 16.10), many demonstration plants have followed. These were comprehensively reviewed by Gahleitner [84] in 2013. From this work it can be seen that that the number of demonstration projects both planned and in operation has not only increased in number but has increased in capacity (Fig. 16.12). Both are essential criteria for the eventual rollout of full-scale plants for the storage of energy as hydrogen gas.

6 EMERGING TECHNOLOGIES AND OUTLOOK

6.1 Electron-Coupled Proton Buffers and Decoupling of Hydrogen Gas Generation

An unavoidable consequence of electrolysis in a conventional electrolyzer is that the production of oxygen and hydrogen takes place in the same device separated only by a membrane. This in turn defines a number of limitations on the operation of a conventional system; namely, (1) extremely high levels of gas purity (>99.999%), specifically for electronic applications, cannot be directly obtained from the electrolyzer without requiring further purification [86]; (2) operation at low loads and at frequently changing loads reduces the lifetime of the membrane and further increases the potential for gas crossover [77]; and (3) production of hydrogen at high pressure requires high differential pressure across the membrane [87]. To maintain the integrity of the membrane a thicker membrane is used in these circumstances resulting in higher cell resistance and consequently lower efficiency. In this respect, it is interesting to see if water

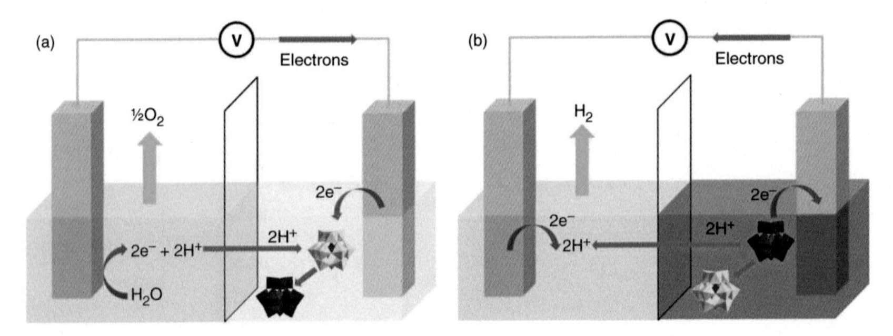

FIGURE 16.13 **ECPB mediated (a) oxygen and (b) hydrogen generation.** *(Reprinted with permission from McMillan Publishers Ltd: Symes MD, Cronin L. Nat Chem 2013;5:403–9, Copyright 2013.)*

electrolysis could be done in a fundamentally different way, perhaps mirroring the activation of water in photosynthesis which has two sets of coupled processes, that is, a light reaction which results in oxygen evolution and a dark reaction which harnesses the proton gradient and electrons for organic activation.

Recent work by Cronin et al. [88–90] has demonstrated that a new route to water splitting is possible and the limitations described earlier can be circumvented by decoupling the hydrogen and oxygen production via the introduction of a redox mediator, known as an electron-coupled proton buffer (ECPB) into the cathode of the electrolyzer cell. This mediator effectively intercepts the electrons formed by water oxidation, being itself reduced, and the consequent increase in negative charge is balanced by the protons formed (Fig. 16.13a) which are associated with the reduced mediator. The hydrogen evolution reaction can then be carried out in a separate device (Figure 16.13b) regenerating the oxidized ECPB ready for reuse.

The ability to act as an ECPB is determined by the position of the redox waves associated with the oxidation and reduction of the ECPB. To function as an ECPB the oxidations and reductions must occur at potentials between the oxygen- and hydrogen-evolving reactions. This allows the ECPB to "intercept" the electrons at a more positive potential than that required to combine with the protons to form hydrogen. This is illustrated in Fig. 16.14. The ECPB, in this case phosphomolybdic acid (solid line), demonstrates its electrochemical activity in the window between the oxidation of water and the reduction of protons (dashed line).

Cronin and coworkers have identified three different redox mediator types which show great promise in decoupling hydrogen and oxygen production. In the first example, the polyoxometalate, phosphomolybdic acid is used to decouple hydrogen and oxygen production resulting in two separate electrochemical steps [88]. In the second example the properties of the inorganic ECPB were extended to an organic system composed of potassium hydroquinone sulfonate [90]. The position of the redox waves of the ECPB defines whether or not the splitting of the voltages is largely symmetric, as with phosphomolybdic acid and potassium hydroquinone sulfonate, or asymmetric as in the final example

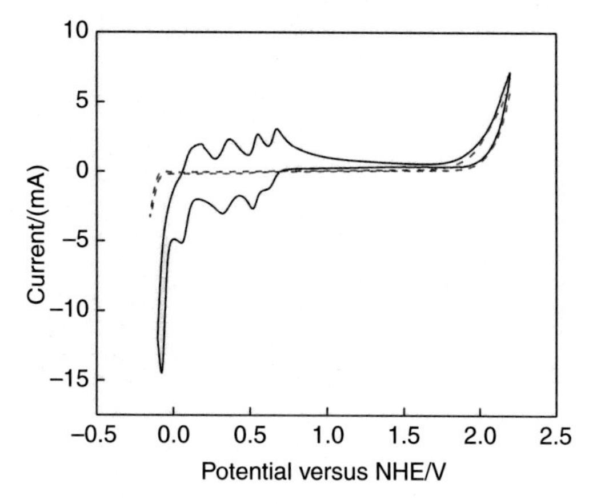

FIGURE 16.14 **Comparison of ECPB reduction potentials with water electrolysis potentials.**

of this technology. By choosing the polyoxometalate, silicotungstic acid, highly asymmetric behavior was discovered wherein only a single electrochemical input was required [89]. This electrochemical step produced the reduced form of the silicotungstic acid; however, it was found that, instead of a second electrochemical input to oxidize the acid and produce the hydrogen, this step could be successfully carried out by contact of the reduced ECPB with a suitable catalyst with no further electrical input (Fig. 16.15).

By separating the production of hydrogen from oxygen, devices based on ECPBs have the potential to overcome the low-load, high-pressure/efficiency, and purity limitations of conventional electrolyzers.

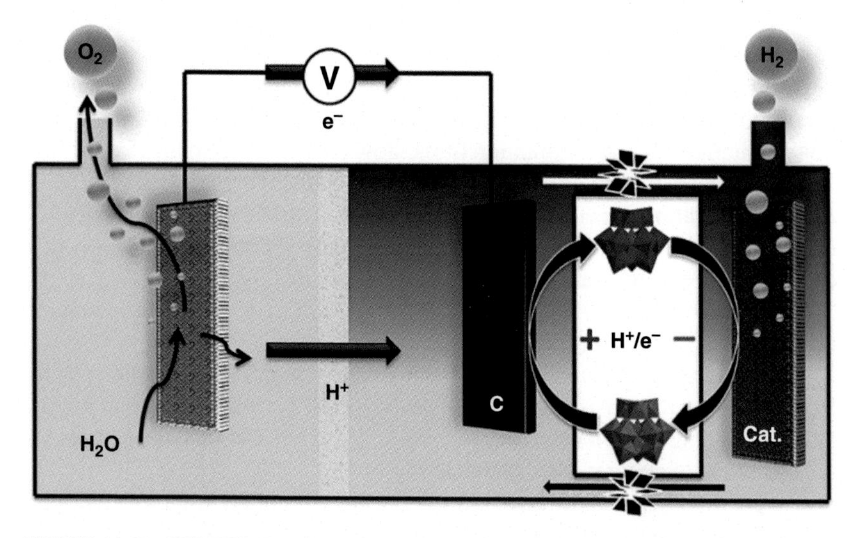

FIGURE 16.15 **ECPB Mediated electrolysis using a single electrochemical input.** *(Reprinted with permission from AAAS: Rausch B, Symes MD, Chisholm G, Cronin L. Science (80) 2014;345:1326–30.)*

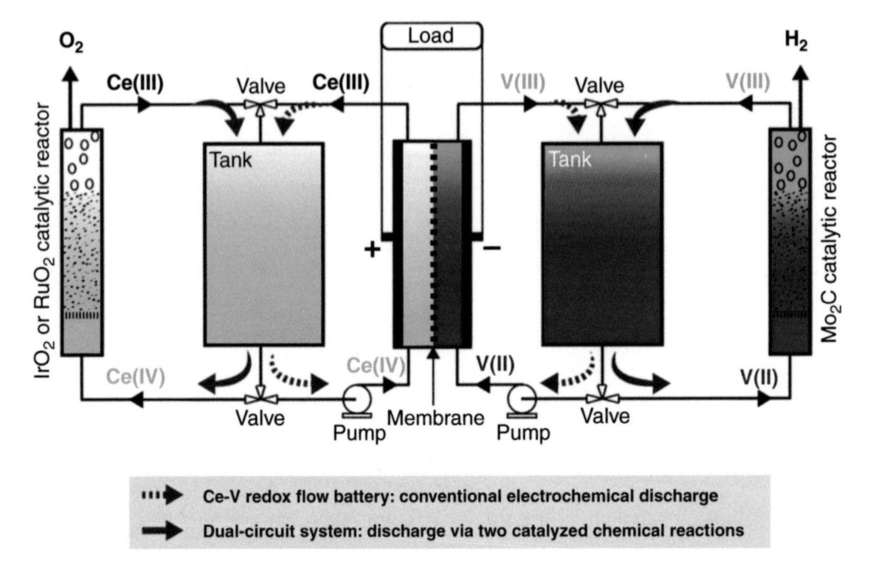

FIGURE 16.16 Ce–V Dual circuit flow battery, reproduced from reference [91] with permission from The Royal Society of Chemistry.

6.2 Flow Battery/Electrolyzer Hybrids

Redox flow batteries (see chapter 12) are a potential competitor technology to electrolysis as a route to economic and flexible energy storage and are discussed at length elsewhere in this volume. There is, however, a promising technology which serves to exploit elements of both flow batteries and electrolyzers to produce hydrogen from water. Amstutz et al. have recently developed a system based on indirect water electrolysis [91]. This system shares some commonalities with the ECPB concept described earlier in that the water-splitting reactions occur in two distinct liquid circuits of the device. This allows the water oxidation reaction to address a limitation of the flow battery wherein the flow battery has a fixed capacity and once this is reached no further energy can be stored. In having an additional water oxidation capability, hydrogen can be produced as an adjunct to the electrochemical energy storage of the flow battery. The system described is a cerium–vanadium (Ce–V) flow battery (Fig. 16.16).

Electrochemical charging of the flow battery proceeds via the following reactions during charging (Scheme 16.6).

Once charging is complete, or simultaneously with charging, the reduced V(II) species can be passed through a catalytic bed wherein it is oxidized to V(III) providing further V(II) for charging and also producing hydrogen gas via the following overall reaction (Scheme 16.7).

As previously noted the electrocatalyst most commonly used for hydrogen evolution in a conventional electrolyzer is platinum. The catalyst which is used in this system is molybdenum carbide (Mo_2C). It should also be noted that the

Cathode Reaction

$$V^{3+}_{(aq)} + e^- \longrightarrow V^{2+}_{(aq)}$$

Anode Reaction

$$Ce^{3+} \longrightarrow Ce^{4+} + e^-$$

SCHEME 16.6 Cerium–vanadium (Ce–V) flow battery reactions.

$$2H^+_{(aq)} + 2V^{2+}_{(aq)} \xrightarrow{Mo_2C} H_{2(g)} + 2V^{3+}_{(aq)}$$

SCHEME 16.7 Hydrogen production from Cerium–vanadium (Ce–V) flow battery.

production of hydrogen via the oxidation of V(II) does not require additional electrochemical input to produce the hydrogen gas.

The increase in renewable penetration in electricity generation and the increasing recognition that such penetration can only be managed via energy storage offer tremendous potential for the exploitation of established and nascent electrolysis processes in providing a solution to the energy storage dilemma. In addition, when one considers the slow but steady emergence of the hydrogen-fueled vehicle market the signs are extremely positive for a hydrogen economy and thus for the electrolysis industry required to support this.

REFERENCES

[1] REN21. Renewables 2014 global status report. Paris: REN21 Secretariat; 2014.

[2] Bird JCL, Wang X. Wind and solar energy curtailment: experience and practices in the United States. Golden, CO: National Renewable Energy Laboratory; 2014.

[3] BP statistical review of world energy, vol. 1. London: BP p.l.c.; 2014.

[4] Haynes WM, Lide DR. CRC handbook of chemistry and physics: a ready-reference book of chemical and physical data. Boca Raton, FL: CRC Press; 2011.

[5] HTAC. Report of the Hydrogen Production Expert Panel: a subcommittee of the Hydrogen & Fuel Cell Technical Advisory Committee. Washington, DC: Hydrogen & Fuel Cell Technical Advisory Committee; 2013.

[6] de Levie R. The electrolysis of water. J Electroanal Chem 1999;476:92–3.

[7] Berg H. Johann Wilhelm Ritter — the founder of scientific electrochemistry. Rev Polarogr 2008;54:99–103.

[8] Zhang J, Zhang L, Liu H, Sun A, Liu RS. Electrochemical technologies for energy storage and conversion. New York: Wiley; 2012.

[9] Kreuter W. Electrolysis: The important energy transformer in a world of sustainable energy. Int J Hydrogen Energ 1998;23:661–6.

[10] Pera MC, Hissel D, Gualous H, Turpin C. Electrochemical components. Chichester, UK: Wiley; 2013.

[11] Harrison KW, Remick R, Martin GD, Hoskin A. Hydrogen production: fundamentals and case study summaries. Oak Ridge, TN: Oak Ridge National Laboratory; 2010. 1.
[12] Hydrogen powered generator by electrolysis—HYLYZER 1 or 2. http://www.hydrogenics.com/hydrogen-products-solutions/industrial-hydrogen-generators-by-electrolysis/indoor-installation/hylyzer-1-or-2
[13] Hydrogen S series. http://protononsite.com/resources/brochures/hogen-s-series/
[14] Laguna-Bercero MA. Recent advances in high temperature electrolysis using solid oxide fuel cells: a review. J Power Sources 2012;203:4–16.
[15] Smolinka T, Günther M, Garche J. NOW-Studie Stand und Entwicklungspotenzial der Wasserelektrolyse zur Herstellung von Wasserstoff aus regenerativen Energien Kurzfassung des Abschlussberichts. Freiburg, Germany: Fraunhofer Institut für Solare Energiesysteme; 2011.
[16] Bertuccioli L, Chan A, Hart D, Lehner F, Madden B, Standen E. Fuel cells and hydrogen. Joint undertaking—development of water electrolysis in the European Union. Switzerland 2014.
[17] Ursua A, Gandia LM, Sanchis P. Hydrogen production from water electrolysis: current status and future trends. P IEEE 2012;100:410–26.
[18] Marini S, Salvi P, Nelli P, Pesenti R, Villa M, Berrettoni M, Berrettoni M, Zangari G, Kiros Y. Advanced alkaline water electrolysis. Electrochim Acta 2012;82:384–91.
[19] Schröder V, Emonts B, Janßen H, Schulze H-P. Explosion limits of hydrogen/oxygen mixtures at initial pressures up to 200 bar. Chem Eng Technol 2004;27:847–51.
[20] Ito H, Maeda T, Nakano A, Takenaka H. Properties of Nafion membranes under PEM water electrolysis conditions. Int J Hydrogen Energy 2011;36:10527–40.
[21] Chandesris M, Médeau V, Guillet N, Chelghoum S, Thoby D, Fouda-Onana F. Membrane degradation in PEM water electrolyzer: numerical modeling and experimental evidence of the influence of temperature and current density. Int J Hydrogen Energy 2015;40:1353–66.
[22] Goñi-Urtiaga A, Presvytes D, Scott K. Solid acids as electrolyte materials for proton exchange membrane (PEM) electrolysis: review. Int J Hydrogen Energy 2012;37:3358–72.
[23] Ayers KE, Capuano C, Anderson EB. Recent advances in cell cost and efficiency for PEM-based water electrolysis. ECS Trans 2012;41:15–22.
[24] Miles MH, Thomason MA. Periodic variations of overvoltages for water electrolysis in acid solutions from cyclic voltammetric studies. Electrochem Soc J 1976;123:1459–61.
[25] Gong M, Dai H. A mini review of NiFe-based materials as highly active oxygen evolution reaction electrocatalysts. Nano Res 2015;8:23–39.
[26] Siracusano S, Van Dijk N, Payne-Johnson E, Baglio V, Aricò AS. Nanosized IrO_x and $IrRuO_x$ electrocatalysts for the O_2 evolution reaction in PEM water electrolysers. Appl Catal B-Environ 2015;164:488–95.
[27] McKone JR, Marinescu SC, Brunschweig BS, Winkler JR, Gray HB. Earth-abundant hydrogen evolution electrocatalysts. Chem Sci 2014;5:865–78.
[28] Antolini E. Iridium as catalyst and cocatalyst for oxygen evolution/reduction in acidic polymer electrolyte membrane electrolyzers and fuel cells. ACS Catal 2014;4:1426–40.
[29] Chisholm G, Kitson PJ, Kirkaldy ND, Bloor LG, Cronin L. 3D printed flow plates for the electrolysis of water: an economic and adaptable approach to device manufacture. Energy Environ Sci 2014;7:3026–32.
[30] Bi L, Boulfrad S, Traversa E. Steam electrolysis by solid oxide electrolysis cells (SOECs) with proton-conducting oxides. Chem Soc Rev 2014;43:8255–70.
[31] Ni M, Leung M, Leung D. Technological development of hydrogen production by solid oxide electrolyzer cell (SOEC). Int J Hydrogen Energy 2008;33:2337–54.
[32] Minh NQ. Ceramic Fuel Cells. J Am Ceram Soc 1993;76:563–88.
[33] Kreuer KD. Proton Conducting Oxides. Annu Rev Mat Res 2003;33:333–59.
[34] Paria M. Electrical conduction in barium cerate doped with M_2O_3 (M = La, Nd, Ho) Solid State Ionics 1984;13:285–92.

[35] Cavendish H. Three papers, containing experiments on factitious air, by the Hon. Henry Cavendish, F. R. S. Philos Trans 1766;56:141–84.

[36] Rowsell JLC, Yaghi OM. Strategies for hydrogen storage in metal-organic frameworks. Angew Chem Int Edit 2005;44:4670–9.

[37] von Helmolt R, Eberle U. Fuel cell vehicles: Status 2007. J Power Sources 2007;165: 833–43.

[38] US Department of Energy, Targets for onboard hydrogen storage systems for light-duty vehicles. Washington, DC: US Department of Energy; 2009.

[39] Troiano AR. The role of hydrogen and other interstitials in the mechanical behavior of metals. Trans ASM 1960;52:54–80.

[40] Vennett RM, Ansell G. The effect of high-pressure hydrogen upon the tensile properties and fracture behavior of 304L stainless steel. Trans ASM 1967;60:242–51.

[41] Benson RB, Dann RK, Roberts LW. Hydrogen embrittlement of stainless steel. Trans AIME 1968;242:2199–205.

[42] Perng TP, Altstetter CJ. Comparison of hydrogen embrittlement of austenitic and ferritic stainless steels. Metall Trans A 1987;18:123–34.

[43] Lynch S. Hydrogen embrittlement phenomena and mechanisms. Corros Rev 2012;30: 105–23.

[44] Zheng J, Liu X, Xu P, Liu P, Zhao Y, Yang J. Development of high pressure gaseous hydrogen storage technologies. Int J Hydrogen Energy 2012;37:1048–57.

[45] Barthélémy H. Hydrogen storage — Industrial prospectives. Int J Hydrogen Energy 2012;37:17364–72.

[46] Hua TQ, Ahluwalia R, Peng JK, Kromer M, Lasher S, McKenney K, Law K, Sinha J. Technical assessment of compressed hydrogen storage tank systems for automotive applications. Int J Hydrogen Energy 2011;36:3037–49.

[47] Durbin DJ, Malardier-Jugroot C. Review of hydrogen storage techniques for on board vehicle applications. Int J Hydrogen Energy 2013;38:14595–617.

[48] Jorgensen SW. Hydrogen storage tanks for vehicles: recent progress and current status. Curr Opin Solid State Mat Sci 2011;15:39–43.

[49] Zheng J, Liu X, Xu P, Liu P, Zhao Y, Yang J. Development of high pressure gaseous hydrogen storage technologies. Int J Hydrogen Energy 2012;37:1048–57.

[50] Ahluwalia RK, Hua TQ, Peng JK, Lasher S, McKenney K, Sinha J, Gardiner M. Technical assessment of cryo-compressed hydrogen storage tank systems for automotive applications. Int J Hydrogen Energy 2010;35:4171–84.

[51] Aceves SM, Espinosa-Loza F, Ledesma-Orozco E, Ross TO, Weisberg AH, Brunner TC, Kircher O. High-density automotive hydrogen storage with cryogenic capable pressure vessels. Int J Hydrogen Energy 2010;35:1219–26.

[52] Armaroli N, Balzani V. The hydrogen issue. ChemSusChem 2011;4:21–36.

[53] Poirier E, Dailly A. Saturation properties of a supercritical gas sorbed in nanoporous materials. Phys Chem Chem Phys 2012;16544–51.

[54] Poirier E, Dailly A. On the nature of the adsorbed hydrogen phase in microporous metal–organic frameworks at supercritical temperatures. Langmuir 2009;25:12169–76.

[55] Dalebrook AF, Gan W, Grasemann M, Moret S, Laurenczy G. Hydrogen storage: beyond conventional methods. Chem Commun 2013;49:8735–51.

[56] Rzepka M, Lamp P, de la Casa-Lillo MA. Physisorption of hydrogen on microporous carbon and carbon nanotubes. J Phys Chem B 1998;102:10894–8.

[57] Jordá-Beneyto M, Suárez-García F, Lozano-Castelló D, Cazorla-Amorós D, Linares-Solano A. Hydrogen storage on chemically activated carbons and carbon nanomaterials at high pressures. Carbon N Y 2007;45:293–303.

[58] Jiang H-L, Liu B, Lan Y-Q, Kuratani K, Akita T, Shioyama H, Zong F, Xu Q. From metal–organic framework to nanoporous carbon: toward a very high surface area and hydrogen uptake. J Am Chem Soc 2011;133:11854–7.

[59] Aboutalebi SH, Aminorroaya-Yamini S, Nevirkovets I, Konstantinov K, Liu HK. Enhanced hydrogen storage in graphene oxide-mwcnts composite at room temperature. Adv Energy Mater 2012;2:1439–46.

[60] Budd PM, Butler A, Selbie J, Mahmood K, McKeown NB, Ghanem B, Msayib K, Book D, Walton A. The potential of organic polymer-based hydrogen storage materials. Phys Chem Chem Phys 2007;9:1802–8.

[61] Wood CD, Tan B, Trewin A, Niu H, Bradshaw D, Rosseinsky MJ, Khimyak YZ, Campbell NL, Kirk R, Stockel E, Cooper AI. Hydrogen storage in microporous hypercrosslinked organic polymer networks. Chem Mat 2007;19:2034–48.

[62] Furukawa H, Yaghi OM. Storage of Hydrogen, Methane, and Carbon Dioxide in Highly Porous Covalent Organic Frameworks for Clean Energy Applications. J Am Chem Soc 2009;131:8875–83.

[63] Li Y, Yang RT. Hydrogen storage in low silica type X zeolites. J Phys Chem B 2006;110: 17175–81.

[64] Dong J, Wang X, Xu H, Zhao Q, Li J. Hydrogen storage in several microporous zeolites. Int J Hydrogen Energy 2007;32:4998–5004.

[65] Stoeck U, Krause S, Bon V, Senkovska I, Kaskel S. A highly porous metal–organic framework, constructed from a cuboctahedral super-molecular building block, with exceptionally high methane uptake. Chem Commun 2012;48:10841.

[66] Farha OK, Yazaydına Ö, Eryazici I, Malliakas CD, Hauser BG, Kanatzidis MG, Nguyen ST, Snurr RQ, Hupp JT. De novo synthesis of a metal–organic framework material featuring ultrahigh surface area and gas storage capacities. Nat Chem 2010;2:944–8.

[67] Lee H, Lee J, Kim DY, Park J, Seo Y-T, Zeng H, Moudrakovski IL, Ratcliffe CI, Ripmeester JA. Tuning clathrate hydrates for hydrogen storage. Nature 2005;434:743–6.

[68] Zaluska A, Zaluski L, Ström-Olsen JO. Nanocrystalline magnesium for hydrogen storage. J Alloy Compd 1999;288:217–25.

[69] Bogdanović B, Schwickardi M. Ti-doped $NaAlH_4$ as a hydrogen-storage material – preparation by Ti-catalyzed hydrogenation of aluminum powder in conjunction with sodium hydride. Appl Phys A 2001;72:221–3.

[70] Fakiolu E. A review of hydrogen storage systems based on boron and its compounds. Int J Hydrogen Energy 2004;29:1371–6.

[71] Chen P, Xiong Z, Luo J, Lin J, Tan KL. Interaction of hydrogen with metal nitrides and imides. Nature 2002;420:302–4.

[72] MacLeay I, Harris K, Annut A. Digest of United Kingdom Energy Statistics (DUKES). London: Department of Energy & Climate Change; 2013.

[73] Ilbas MA, Crayford P, Yilmaz I, Bowen PJ, Syred N. Laminar-burning velocities of hydrogen-air and hydrogen-methane-air mixtures: an experimental study. Int J Hydrogen Energy 2006;31: 1768–79.

[74] Gas quality. http://www2.nationalgrid.com/uk/industry-information/gas-transmission-system-operations/gas-quality/

[75] Yaccato K, Carhart R, Hagemeyer A, Lesik A, Strasser P, Volpe AF, Turner H, Weinberg H, Grasselli RK, Brooks C. Competitive CO and CO_2 methanation over supported noble metal catalysts in high throughput scanning mass spectrometer. Appl Catal A-Gen 2005;296:30–48.

[76] Wang W, Gong J. Methanation of carbon dioxide: an overview. Front Chem Eng China 2011;5:2–10.

[77] Grigoriev SA, Porembskiy VI, Korobtsev SV, Fateev VN, Aupretre F, Millet P. High-pressure PEM water electrolysis and corresponding safety issues. Int J Hydrogen Energy 2011;36:2721–8.

[78] Burkhanov GS, Gorina NB, Kolchugina NB, Roshan NR, Slovetsky DI, Chistov EM. Palladium-based alloy membranes for separation of high purity hydrogen from hydrogen-containing gas mixtures. Platin Met Rev 2011;55:3–12.

[79] Succi M, Pirola S, Ruffenach S, Briot O. Managing gas purity in epitaxial growth. Cryst Res Technol 2011;46:809–12.

[80] Inaba M, Kinumoto T, Kiriake M, Umebayashi R, Tasaka A, Ogumi Z. Gas crossover and membrane degradation in polymer electrolyte fuel cell. Electrochim Acta 2006;51:5746–53.

[81] 140 MW wind park officially opens in Germany with energy storage facility using 1 MW power-to-gas system from Hydrogenics. http://www.hydrogenics.com/about-the-company/news-updates/2013/10/01/140-mw-wind-park-officially-opens-in-germany-with-energy-storage-facility-using-1-mw-power-to-gas-system-from-hydrogenics

[82] Largest power to gas facility in the world now operational with Hydrogenics technology. http://www.hydrogenics.com/about-the-company/news-updates/2013/06/14/largest-power-to-gas-facility-in-the-world-now-operational-with-hydrogenics-technology

[83] Yilanci A, Dincer I, Ozturk HK. A review on solar-hydrogen/fuel cell hybrid energy systems for stationary applications. Prog Energy Combust Sci 2009;35:231–44.

[84] Gahleitner G. Hydrogen from renewable electricity: an international review of power-to-gas pilot plants for stationary applications. Int J Hydrogen Energy 2013;38:2039–61.

[85] Szyszka A. Ten years of solar hydrogen demonstration project at Neunburg vorm Wald, Germany. Int J Hydrogen Energy 1998;23:849–60.

[86] Bright ideas—delivering smarter LED manufacturing through innovative gas technology. http://www.linde-gas.com/internet.global.lindegas.global/en/images/Linde_in_LED_brochure_WEB_1.117_39138.pdf

[87] Schalenbach M, Carmo M, Fritz DL, Mergel J, Stolten D. Pressurized PEM water electrolysis: efficiency and gas crossover. Int J Hydrogen Energy 2013;38:14921–33.

[88] Symes MD, Cronin L. Decoupling hydrogen and oxygen evolution during electrolytic water splitting using an electron-coupled-proton buffer. Nat Chem 2013;5:403–9.

[89] Rausch B, Symes MD, Chisholm G, Cronin L. Decoupled catalytic hydrogen evolution from a molecular metal oxide redox mediator in water splitting. Science 2014;345:1326–30.

[90] Rausch B, Symes MD, Cronin L. A bio-inspired, small molecule electron-coupled-proton buffer for decoupling the half-reactions of electrolytic water splitting. J Am Chem Soc 2013;135:13656–9.

[91] Amstutz V, Toghill KE, Powlesland F, Vrubel H, Comninellis C, Hu X, Girault HH. Renewable hydrogen generation from a dual-circuit redox flow battery. Energy Environ Sci 2014;7:2350–8.

Chapter 17

Thermochemical Energy Storage

Henner Kerskes
Research and Testing Centre for Solar Thermal Systems (TZS), Institute for Thermodynamics and Thermal Engineering (ITW), University of Stuttgart, Germany

1 INTRODUCTION

The development of our society toward a sustainable society with a modern energy system based on a high ratio of renewable energies together with high energy efficiencies brings thermal energy storage into a new focus. The integration of renewable energies, such as solar or wind power, with technologies which have high efficiencies, such as combined heat and power, require new storage technologies and present new challenges to those working in the field of heat storage. Compared with today's energy systems, compact and long-term storage processes will play an important role in future energy systems. High specific storage capacities and reduced heat losses are important technical aspects for future energy storage developments.

Numerous studies over the past few years have shown that thermochemical energy storage is a key technology to developing highly efficient short- and long-term thermal energy storage for various applications, such as solar thermal systems or cogeneration systems [1]. By storing energy in the form of chemical bonds of suitable materials, energy can be stored with almost no energy loss for long periods of time. At the same time, high energy storage density can be achieved. Both criteria are crucial for future energy storage applications.

Research activities in the field of low-temperature thermochemical energy storage (TCES) have developed strongly over the last few years—particularly in the field of material development and material optimization [2–5]. The main focus of this activity is on improving the chemical and thermal properties of materials such as increasing the energy storage density, enhancing the thermal conductivity, or improving cyclic stability. In addition, some attention has been paid to the design of the storage itself and its subcomponents such as the reactor. These topics are indispensable if the processes are to become commercially available.

Storing Energy. http://dx.doi.org/10.1016/B978-0-12-803440-8.00017-8

High-temperature heat storage systems are used to improve the energy efficiency of power plants and the recovery of process heat. They are also required for continuous power supply in solar thermal applications. Thermochemical reactions offer an option for high storage capacity even at high temperatures [6].

2 PHYSICAL FUNDAMENTALS OF THERMOCHEMICAL ENERGY STORAGE

More than 90% of all thermal energy storage processes used in a wide range of applications are sensible heat storage processes. For temperatures below 100 °C, water is mainly used as the storage material. For higher temperature applications, solid storage materials such as ceramics or liquids in the form of molten salts are available. The technology of sensible heat storage is well understood and much experience has been gained from the vast range of examples that are available. The situation is completely different for TCES systems and few proven examples exist in spite of its promising potential.

In TCES systems, reversible chemical or physical reactions are used for storing and releasing energy and as such are very different from storing sensible heat. Only in the case of gas-phase adsorption processes is there any similarity to the mechanism of latent heat or phase change energy storage. In Fig. 17.1 a comprehensive overview of the different thermal energy storage mechanisms is given. Thermochemical processes are divided into two main branches: sorption processes, which again can be divided into adsorption and absorption; and reversible chemical reactions, which in turn can be divided into solid–gas reactions and solid–liquid reactions.

2.1 Thermochemical Energy Storage (reaction)

Generally, thermochemical energy storage is based on the utilization of heat of reaction released by reversible chemical reactions. For example, a chemical compound of type A–B can be split reversibly into the components A and B by supplying heat.

In this process the supplied quantity of heat, denoted here by $\Delta_R H$, is used to break the A–B bond into independent species A and B. If the reverse reaction of

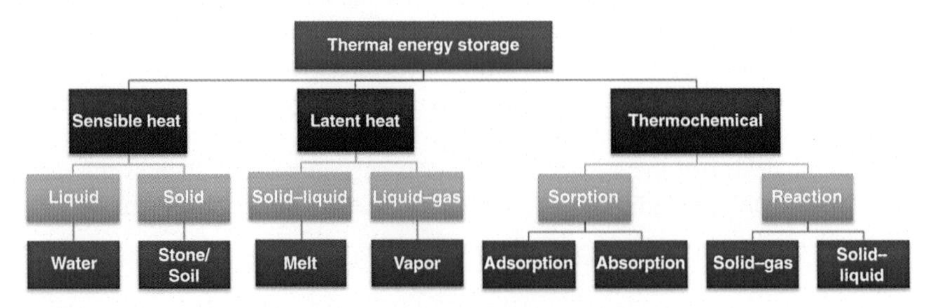

FIGURE 17.1 **Classification of thermal energy storage mechanism.**

TABLE 17.1 Examples of Reversible Solid/Gas Reaction for Thermal Energy Storage

Type of reaction	Reaction	Temperature range/°C
Dehydration of salt hydrates	$MgSO_4 \cdot 7H_2O \rightleftharpoons MgSO_4 \cdot H_2O + 6H_2O$	100–150
	$MgCl_2 \cdot 6H_2O \rightleftharpoons MgCl_2 \cdot H_2O + 5H_2O$	100–130
	$CaCl_2 \cdot 6H_2O \rightleftharpoons CaCl_2 \cdot H_2O + 5H_2O$	150–200
	$CuSO_4 \cdot 5H_2O \rightleftharpoons CuSO_4 \cdot H_2O + 4H_2O$	120–160
	$CuSO_4 \cdot H_2O \rightleftharpoons CuSO_4 + H_2O$	210–260
Deammoniation of ammonium chlorides	$CaCl_2 \cdot 8NH_3 \rightleftharpoons CaCl_2 \cdot 4NH_3 + 4NH_3$	25–100
	$CaCl_2 \cdot 4NH_3 \rightleftharpoons CaCl_2 \cdot 2NH_3 + 2NH_3$	40–120
	$MnCl_2 \cdot 6NH_3 \rightleftharpoons MnCl_2 \cdot 2NH_3 + 4NH_3$	40–160
Dehydration of metal hydrides	$MgH_2 \rightleftharpoons Mg + H_2$	200–400
	$Mg_2NiH_4 \rightleftharpoons Mg_2Ni + 2H_2$	150–300
Dehydration of metal hydroxides	$Mg(OH)_2 \rightleftharpoons MgO + H_2O$	250–350
	$Ca(OH)_2 \rightleftharpoons CaO + H_2O$	450–550
	$Ba(OH)_2 \rightleftharpoons BaO + H_2O$	700–800
Decarboxylation of metal carbonates	$ZnCO_3 \rightleftharpoons ZnO + CO_2$	100–150
	$MgCO_3 \rightleftharpoons MgO + CO_2$	350–450
	$CaCO_3 \rightleftharpoons CaO + CO_2$	850–950

turning products A and B into the compound A–B is avoided the energy stored in the chemical bonds can be stored without energy loss for any length of time.

An example of this type of reaction is the dehydration of salt hydrates, for example, magnesium hydroxide into magnesium oxide as shown in Eq. 17.1:

$$MgO + H_2O \leftrightarrow Mg(OH)_2 + \Delta_R H \qquad (17.1)$$

The energy required for the endothermic reaction can be provided by heat from any source. If the salt oxide is stored in a hermetically sealed space, then the energetic state can be kept without energy loss for an unlimited period of time. This means that heat losses only occur during charging and discharging. If at any later time the oxide is brought into contact with water or water vapor, hydration occurs and the quantity of heat supplied during dehydration is released.

Typical reversible solid/gas reactions with potential for thermal energy storage are listed in Table 17.1. These reaction types cover a temperature range from below 100 °C up to very high temperatures of over 800 °C. Besides the potential for high-energy storage density this is a reason this technology is attractive for a wide range of energy storage processes.

2.2 Prototype of the Combined Hot Water and Sorption Store

A related process, also important in the field of TCES, is the adsorption process. Eq. 17.2 describes the basic principle of the charging and discharging process:

$$A_{(s)} + xH_2O_{(g)} \leftrightarrow A \cdot xH_2O_{(s)} + \Delta_{ads}H \tag{17.2}$$

The basic reaction shown in Eq. 17.2 is very similar to the chemical reaction described in Eq. 17.1. During the energy-discharging process the solid reactant A is in contact with water vapor ($H_2O_{(g)}$). The solid adsorbs the water vapor to form the product $A \cdot x\ H_2O_{(s)}$ and the adsorption enthalpy ($\Delta_{ads}H$) is released in the form of heat. As this is a reversible adsorption reaction the same amount of heat has to be applied to decompose the product $A \cdot x\ H_2O_{(s)}$ into water vapor and the adsorbent A. It should be noted that the supply heat during desorption is typically at a higher temperature than the heat released during adsorption.

As in the case of chemical reactions there is no loss of energy during long-term storage, so long as the adsorbent is kept in a dry, water vapor–free environment.

The technology of adsorption is widely used in the chemical industry for a range of processes which include waste water treatment and gas purification. The process of adsorption is very well known and intensively investigated [7,8]. Therefore, only a brief introduction will be given which is necessary to understand the adsorption process for thermal energy storage and to understand the concepts discussed in Section 4.

The process of adsorption describes the attachment of molecules or ions from a gas phase or a liquid to a solid surface. Adsorption is a phase transfer process and can be explained as an enrichment of chemical species from a fluid phase onto the (inner) surface of a solid material. The solid surface provides energy-rich active sites which interact with the gas-phase species. By adsorbing these species the energy level of the surface is reduced to a thermodynamically more stable condition as energy is released. The solid is referred to as the adsorbent, the gas-phase molecule is called the adsorptive, and the adsorbed molecule is the adsorbate. Fig. 17.2 illustrates the adsorption process. As shown in Eq. 17.2, adsorption is an exothermal process and heating up the solid reverses the process, the binding forces are overcome, and the adsorbed species are released from the adsorbent back into the fluid. Thereby the adsorbent is raised to a higher energy level. This process is referred to as desorption.

The process of adsorption and desorption is an equilibrium controlled process. The amount of adsorbed mass per mass of adsorbent is referred to as loading X. Adsorption equilibrium depends on temperature and is defined by the partial pressure p^* of the adsorptive in the gas phase and the adsorbent loading

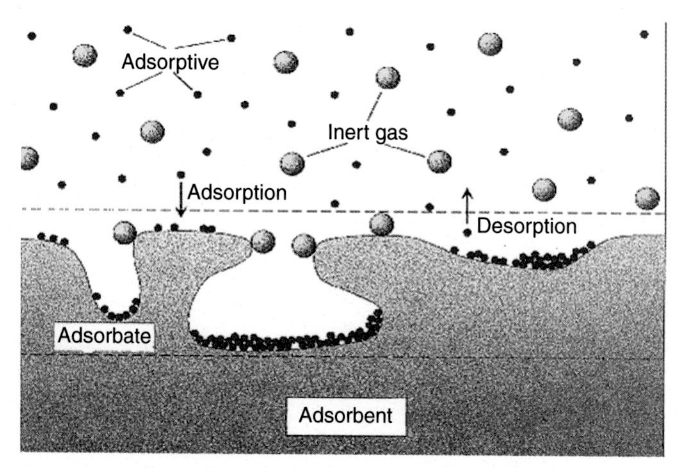

FIGURE 17.2 **Schematic representation of adsorption and desorption.**

X. The equilibrium can be expressed as a function $f(p^*, T, X)$. Typically, the solid loading *X* is given as a function of the partial pressure p^* at constant temperature *T*:

$$X = f(p^*)_T \tag{17.4}$$

This relation is called the adsorption isotherm at temperature *T*. In Fig. 17.3, adsorption isotherms of H_2O adsorption on zeolite 13X are depicted. Adsorption equilibrium is described by two dependencies:

- at constant temperature the adsorbed amount of H_2O increases with increasing water vapor pressure in the gas phase
- the higher the temperature the lower the equilibrium at constant pressure.

This behavior of energy storage defines the boundary condition for charging and discharging of thermochemical energy. For the discharging process it is beneficial to operate at low temperatures and high partial pressure to achieve a high amount of adsorbed mass. In a thermal process, desorption is achieved by increasing the temperature of the storage material. Furthermore, low partial pressure supports desorption.

In addition to adsorption capacity, the heat or enthalpy of adsorption is one of the most important parameters in heat storage. The enthalpy of adsorption is the difference in energy between the adsorption process and the desorption process.

The enthalpy of adsorption can be described as the sum of heat of condensation (for gas adsorption) and additional bonding forces which are weak Van der Waals forces or electrostatic forces. The value of these bonding forces is characteristic of the adsorbent material. For example activated carbon shows rather small bonding forces while zeolites have higher bonding forces. Due to the fact that the surface energy of adsorbent is inhomogeneous, preferential

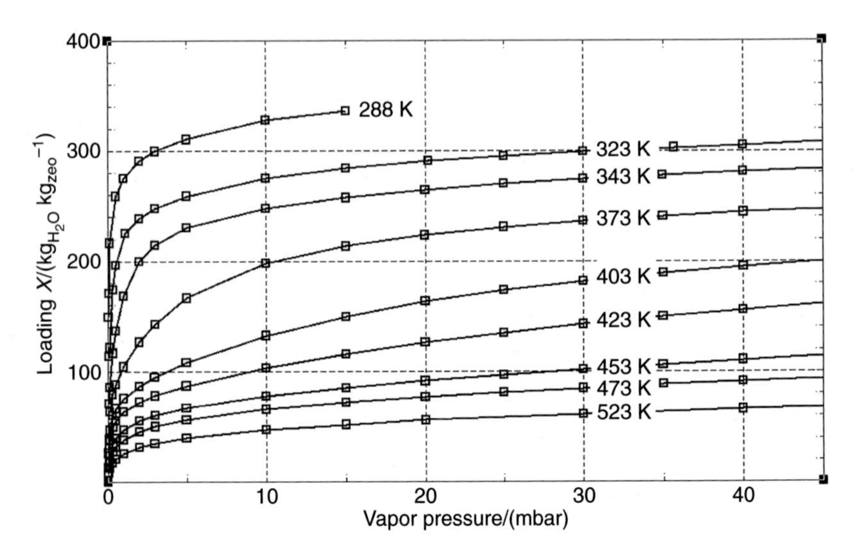

FIGURE 17.3 **Adsorption isotherms of water vapor on zeolite of type 13X [9].** Note: 1 mbar = 10^2 Pa.

sites of adsorption exist. The first molecules adsorb on the active sites with the highest bonding forces. Adsorption enthalpy is, in most cases, a function of the loading X. In Fig. 17.4 the heat of adsorption of water vapor on zeolite 13X is depicted as a function of loading X. Starting at a high energy state, enthalpy decreases continuously and ends up slightly above condensation enthalpy.

To take advantage of high energetic adsorption sites, almost complete desorption of the storage material is necessary. Compared with Fig. 17.3, very high temperatures or nearly zero water vapor pressure is mandatory to reach this energetic state. However, the adsorption processes of typical storage applications start at values of (0.05–0.1) g g^{-1} and, therefore, high energetic sites are not included in the process.

The thermal storage capacity provided by an adsorption process is the product of adsorption capacity and adsorption enthalpy. The specific mass amount of thermal energy stored can be expressed as:

$$q = \Delta X \cdot \Delta_{\text{ads}} h = (X_{\text{ads}} - X_{\text{ads}}) \cdot \int_{X_{\text{des}}}^{X_{\text{das}}} \Delta_{\text{ads}} h$$

and taking the mass of adsorbents into account the absolute value can be calculated as:

$$Q = m_{\text{ads}} \cdot \Delta X \cdot \Delta_{\text{ads}} h = m_{\text{ads}} \cdot (X_{\text{ads}} - X_{\text{des}}) \cdot \overline{\Delta_{\text{ads}} h}$$

where Q is stored energy; m_{ads} is the mass of adsorbents; X_{ads} is the amount of adsorbate after adsorption; X_{des} is the adsorbed amount after desorption; and $\Delta_{\text{ads}} h$ is mean adsorption enthalpy in the interval ($X_{\text{ads}} - X_{\text{des}}$).

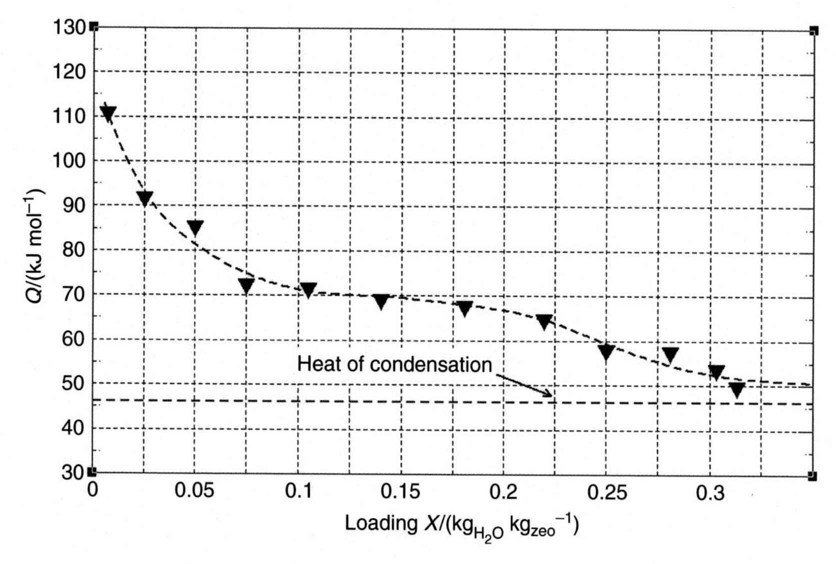

FIGURE 17.4 **Dependency of heat of adsorption on loading X for zeolite 13X [10].**

Storage density can be determined by dividing the value of Q by the volume of the storage material. Adsorption capacity and adsorption enthalpy are both dependent on the operating conditions during adsorption and desorption. Therefore, storage density must be specified in terms of the adsorption and desorption conditions.

3 STORAGE MATERIALS

Recently, there has been much interest shown in finding new materials for sorption or thermochemical storage for low- and high-temperature storage applications. This is largely due to the small thermal losses and the high storage density of thermochemical heat storage as well as the fact that it can be used for both low- and high-temperature applications. Even if the applications are very different there are general requirements a thermochemical storage material should fulfill. A high-performing storage material is characterized by:

- high adsorption capacity (water uptake)
- high heat of adsorption is required for high energy storage density
- fast reaction kinetic is desirable for high thermal power output during charging and discharging
- the desorption temperature should be at an appropriate level.

Compared with chemical reactions whose reaction kinetics strongly depend on the reaction temperature (Arrhenius law), some adsorption processes show high kinetics even at low temperatures. This makes the adsorption process attractive for low-temperature applications such as space heating and domestic

TABLE 17.2 Classification of Pore Size

Pore size/nm	Classification	Specific inner surface/($m^2\ g^{-1}$)
<2	Micropores	>400
$2 < d < 50$	Mesopores	10–400
>50	Macropores	0.5–2

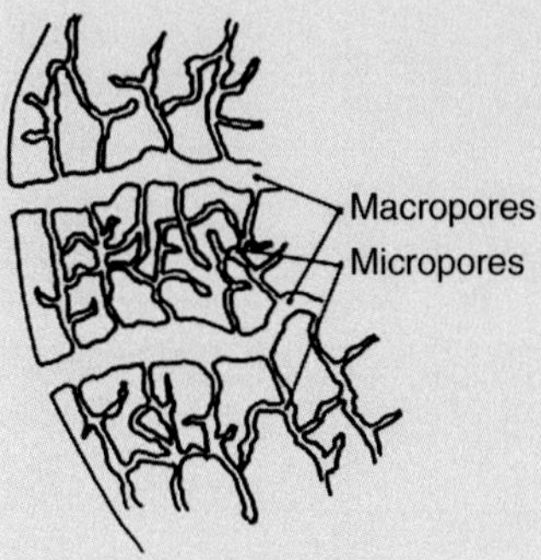

hot-water preparation where heat is required at temperatures below 100 °C. The processes of water vapor adsorption on zeolite and silica gel are the most studied processes for thermochemical energy storage [11–14].

3.1 Adsorption Materials

Adsorption materials are characterized by high porosity and a large inner surface. In technical applications the materials are used in different forms, for example, granules such as spheres or cylinders or similar forms with typical dimensions of (1–5) mm. Fig. 17.6 illustrates the pore structure inside a granule. The pores may be subdivided in macropores, mesopores, or micropores. A classification according to the pore diameter is given in Table 17.2. The diffusion of gas-phase molecules into the inner region of the particles occurs in the larger macropores and mesopores, while adsorption takes place in the micropores.

Classical adsorption materials are silica gel, zeolite, and active carbon. Their suitability depends upon the individual application. Active carbon is a cheap and robust adsorption material with rather small adsorption capacity and low adsorption enthalpy. It is often used for the adsorption process in purification, heat pump, or cooling applications. Silica gel and zeolite show better performance for energy storage because the product of adsorption capacity and adsorption enthalpy is much higher (compare Table 17.3). Most adsorption storage systems described in the literature use silica gel/water or zeolite/water as working pairs [15]. Compared with zeolite, silica gel shows a lower desorption temperature ($t < 120$ °C) which results from the lower adsorption enthalpy. From the technical point of view this may have advantages. On the other hand, the adsorption kinetics of silica gel are slower than for zeolites. Adsorption measurements carried out by Jähnig et al. [12] showed that the achievable temperature lift decreases with increasing loading of the silica gel.

TABLE 17.3 Characteristic Values of Thermochemical Storage Materials: Active Carbon, Silica Gel, and Zeolite

Characteristic values	Active carbon	Silica gel	Zeolite
Inner surface/(m^2 g^{-1})			650–750
Mean adsorption enthalpy/(kJ $kg^{-1}H_2O$)	~2400	~2600	~3500
Specific heat capacity c_p/(kJ K^{-1} kg^{-1})	0.709	0.9–1.0	0.8–0.9
Heat conductivity/(W m^{-1} K^{-1})	1.2–1.6	0.14–0.2	0.58

Temperature lift denotes the temperature increase occurring in the bed during adsorption. If temperature lift becomes too small for effective heat transfer the remaining adsorption capacity is lost.

An important advantage of zeolite is its high adsorption kinetics. The fast adsorption of gas-phase molecules in combination with high adsorption enthalpy yields high heat release. Steep adsorption fronts occur which is beneficial for process control. Even very low adsorptive concentrations can be completely adsorbed which makes zeolites perfect for working with moist air.

Of the wide range of different zeolite types, types A, NaX, and NaY have been tested successfully for energy storage. Type A zeolite (technically known as 4A and 5A) is a robust and inexpensive material with good hydrothermal stability. NaX (13X) and NaY-type zeolites show higher adsorption enthalpies than 4A and are very good candidates for thermal energy storage. Further improvements have been achieved by synthesizing novel binderless zeolites of type 4A and 13X [16]. This new class of zeolite is characterized by improved adsorption capacity, sufficient hydrothermal stability, and higher adsorption enthalpies.

In addition to classical adsorption materials, new materials are under development. The material class of aluminophosphate (AlPOs) and silicoaluminophosphate (SAPOs) represents a group of microporous materials with a zeolite-like crystalline structure. These materials are less hydrophilic than zeolite and may have the potential to fill the gap between zeolite and silica gel regarding optimization of adsorption strength and the ability to be desorbed at low temperatures. The structural types of ALPO-17 and ALPO-18 as well as SAPO-34 look very promising for heat storage with respect to the adsorption equilibrium. Experimental investigations [17] have shown that under moderate operating conditions the water uptake is much higher compared with zeolites. Even at desorption temperatures below 100 °C, significant adsorbed quantities of 25% in mass have been obtained. However, the application of AlPOs and SAPOs as heat storage material is at present not suitable due to cost reasons. The high production costs of the material are caused by the organic template which disappears during synthesis.

Very promising adsorption results has been obtained with another class of microporous materials called metal–organic framework (MOFs) materials. This material class is characterized by a huge internal surface and outstanding adsorption capacity. Water uptake values up to 1.4 g of water per gram of MOF have been measured [18]. MOFs exist in a wide range of different compositions. The large variety and adjustability of the pore structures enables tailored modification to improve the material with respect to heat storage. Due to these attractive properties the technology of MOFs is undergoing fast development. The component MIL-101 has been identified as material with excellent water adsorption properties. Important improvements have already been achieved concerning hydrothermal stability. Some MOFs are already commercially available but are still very expensive.

3.2 Salt Hydrates

Reversible solid–gas reactions offer the potential of even higher storage density than adsorption processes as a result of high enthalpies of reaction. Solid–gas reactions are easy to handle if the gas-phase reactant is present as a natural part of the environment, such as water vapor. In an open process neither condensing nor storing of the gas-phase reactant is necessary. This is the case for the reaction of salt hydration. Table 17.1 illustrates that the temperature range of reaction equilibrium is well suited for low-temperature applications. Furthermore, heat transformation is high and the theoretically achievable storage density is six to ten times higher than a hot-water store (temperature change $\Delta T = 50$ K). However, handling and practical use of these materials are complex. In recent research projects, different salt hydration systems have been investigated by several researchers. Favorably reviewed were sulfates such as magnesium sulfate ($MgSO_4$), chlorides such as calcium chloride ($CaCl_2$), or magnesium chloride and strontium bromide. [19]. Besides the theoretical potential some drawbacks do exist.

In the following, typical difficulties encountered during hydration and dehydration are discussed using the magnesium sulfate as an example. Magnesium sulfate monohydrate ($MgSO_4 \cdot H_2O$) has high potential as a chemical storage material. Taking into account the enthalpy of condensation of water vapor a theoretical storage density of 2.3 GJ $m^{-3}{}_{heptahydrate}$ (633 kW h $m^{-3}{}_{heptahydrate}$) can be obtained. This is about 11 times higher than the storage density of water storage with the same volume ($\Delta T = 50$ K).

Experimental investigations have been carried out by Bertsch et al. to analyze the reaction behavior of $MgSO_4$ in a fixed bed reactor [20]. The reactor had a length of 20 cm and a diameter of 3.5 cm and moist air was passed through it. The inlet conditions of a fixed bed reactor can be adjusted with regard to mass flow rate, temperature, and water vapor partial pressure. In experimental investigations it was found that hydration does not reach the heptahydrate state. Starting with magnesium monohydrate, water uptake was much smaller

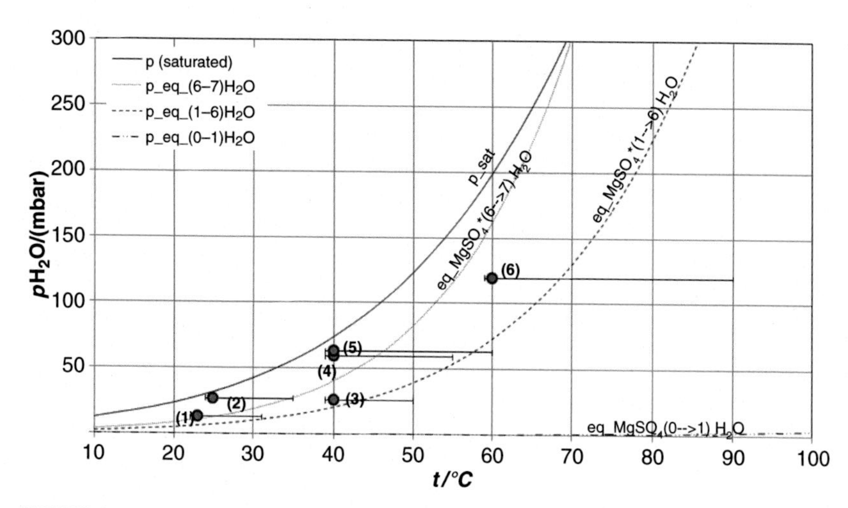

FIGURE 17.5 **Equilibrium curves of $MgSO_4$ and its hydrates and reactor inlet conditions of the air flow (point 1 to 6) for the different experiments.** Note: 1 mbar = 10^2 Pa.

than expected. The uptake of only three to four moles of water per mole of $MgSO_4 \cdot H_2O$ has been measured. There are different hydrates of magnesium sulfate, that is, mono-, bi-, tetra-, penta-, hexa-, and heptahydrate, even though not all of these hydrates are stable or crystalline. The equilibrium curves of mono-, hexa-, and heptahydrate are depicted in Fig. 17.5. To calculate equilibrium curves at atmospheric pressure, thermodynamic data have been taken from [21], [22], and [23]. Fig. 17.4 also depicts the reactor inlet conditions of the air flow (temperature, partial pressure of water vapor) of each experiment (black dots) and the maximum temperature lift achieved (bars to the right). In experiments 1, 2, and 3 the reactor inlet temperature was comparatively low. Furthermore, the inlet conditions were very close to the equilibrium of magnesium sulfate hexahydrate to heptahydrate. Both effects result in a slow reaction rate and a moderate temperature lift. In experiment 4 and 5 the higher inlet temperature favors hydration of the anhydrate and a higher reaction rate, a shorter hydration time, and a higher temperature lift was observed. The highest temperature lift of 30 K was obtained in experiment 6. However, a fully hydrated state of the magnesium sulfate cannot be achieved under these conditions as the reactor inlet condition is to the right of the equilibrium curve of hexahydrate to heptahydrate.

With increasing hydration of the salt a decreasing reactor temperature was observed in all experiments. This is due to the approach of reaction equilibrium as hydration of the salts proceeds. A decreasing reaction rate is the result and the lower enthalpy of reaction yields a small temperature lift. Furthermore, the maximum achievable reactor temperature is limited by the equilibrium of the intermediates. Magnesium sulfate monohydrate can absorb water at a higher temperature than a more hydrated salt such as magnesium sulfate hexahydrate. The storage densities achieved were below the theoretical value for all experiments.

Similar results have been reported by other researchers [24]. Their experiments show that the practical use of pure magnesium sulfate is difficult because of its low power density. They concluded that the application of magnesium sulfate as a thermochemical storage material is problematic. In addition, many authors report that $MgSO_4 \cdot 7H_2O$ melts or boils in its crystal water before degrading to lower hydrates [25,26]. This leads to crystal growth and bonding which hinders the hydration process by increasing diffusion resistance.

These findings are transferable to most other salt hydrates investigated for thermochemical energy storage. However, good experiences with pure strontium bromide as storage material were reported by Wyttenbach et al. [27].

3.3 Composite Materials

Much better performance has been achieved for highly dispersed salt on a carrier matrix [28,29]. The carrier matrix can be of passive or active type. By active, we mean that the supporting matrix takes part in the adsorption process.

In most cases zeolite is used as an active matrix for this purpose. A passive matrix can be any porous medium. Materials with a well-defined pore structure and pore size distribution have shown interesting results. Some of these storage materials already show good characteristics in terms of reaction kinetics, energy storage density, and mechanical stability. Composite salt and zeolite materials are prepared by impregnating commercially available zeolites (e.g., zeolite 13X or zeolite 4A particles) with a salt solution. Early experiments showed an increase in energy storage density compared with pure zeolite of approximately 20% [28,30–32]. The composite material showed similar behavior to that of pure zeolite—a high reaction rate associated with high temperature lift even at low water vapor pressures. To date these new composite materials have not been characterized in detail and the mechanism of interaction between salt and zeolite is not fully understood. Deeper insight into the reaction mechanism is necessary to develop composite materials which fulfill the expected targets which include tailored properties for energy storage.

Fig. 17.6 compares the experimentally achieved storage density of some thermal energy storage materials. Commercially available materials are silica gel, zeolite 4A, zeolite 13X, and a new type of binder-free zeolite (13XBF). In addition, the results of two noncommercially available composite materials are depicted. Composite A is made of a passive carrier matrix impregnated with $MgCl_2$ and reported by Zondag [33]. Composite B is zeolite 13X matrix impregnated with $MgCl_2$. The volume of the probes was 200 mL. All materials were desorbed at 140 °C in a dry air stream ($p^* < 10^2$ Pa or 1 mbar). Adsorptions were carried out at 30 °C and water vapor pressure of 20×10^2 Pa (20 mbar) For comparison reasons the storage density of water (sensible heat storage) for a change in temperature of $\Delta T = 50$ K is also depicted.

In recent years a wide range of thermochemical storage materials have been analyzed and tested and new materials are under development [34]. Good

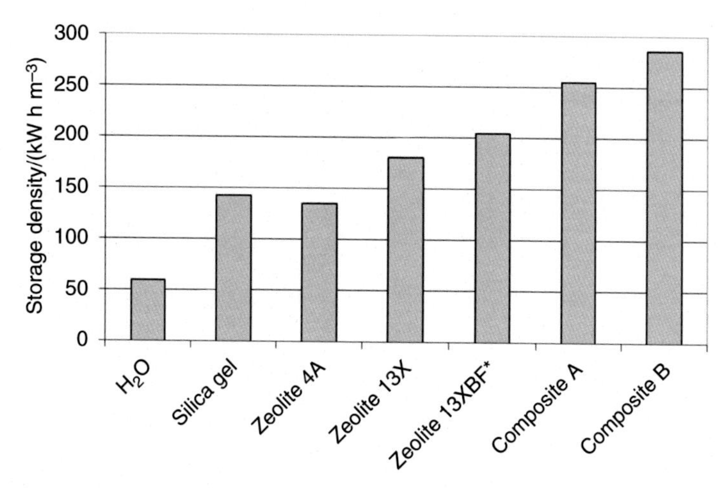

FIGURE 17.6 Comparison of storage materials. Operating conditions: desorption 140 °C in dry air; adsorption 30 °C and 2-kPa (20 mbar) water vapor pressure.

progress has been made regarding increasing performance, hydrothermal stability, and cost reduction.

4 THERMOCHEMICAL STORAGE CONCEPTS

The technology of TCES is expected to have high technical potential not only for compact short-term storage applications but also long-term and high-temperature storage applications. Besides the development of a wide range of new or improved storage materials, innovative storage concepts have been developed and tested under laboratory or pilot scale conditions [35,36]. At present no commercial thermochemical heat storage material is available on the market. Nevertheless, the positive outcomes of recent research activities point to expected market entry in the near future. In the section a discussion of the principal operation modes using an adsorption process for thermal energy storage is given.

The operation of thermochemical heat storage systems can be divided into open- and closed-system designs (Fig. 17.7). An open system works under atmospheric pressure and is in contact with the environment. This means the gas-phase reactant, in most cases water vapor, is extracted from the environment for the adsorption process and is released to the environment during desorption. By contrast, in a closed system the water vapor circulates in a hermetically closed loop typically under negative pressure. The evaporation and condensation of water has to be enforced by technical devices, and an additional water reservoir is necessary. From the design point of view the open system has the advantage of lower technical effort: no condenser, no evaporator, no water reservoir is necessary, it works at normal pressure and process control is not complex.

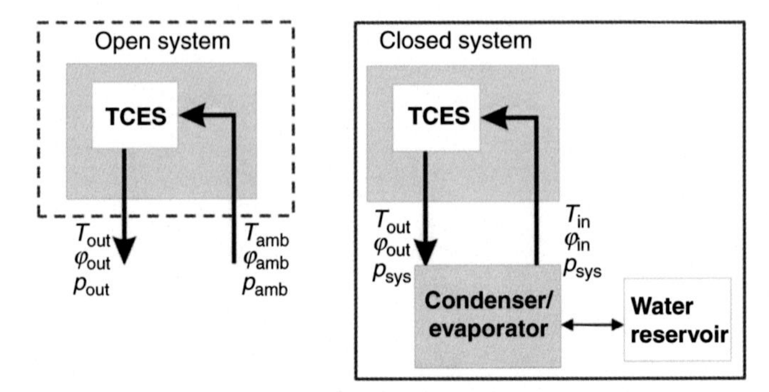

FIGURE 17.7 **Comparison of open- and closed-system concept.**

The open- and closed-system operation modes are presented in more detail to explain the difference and to discuss the advantages and disadvantages of both system designs.

4.1 Closed-System Operation Mode

The closed system consists of a vessel, containing the adsorbent, which is referred to as the adsorber. The adsorber is equipped with a heat exchanger for heat input during charging and to extract the heat of adsorption during discharging. Evaporation and condensation of the adsorptive can be performed in the same unit (heat exchanger) because the processes do not take place simultaneously. Furthermore, a water reservoir is needed to store the condensed water. The volume of the water reservoir should be dimensioned according to the maximum water uptake of the adsorbents (i.e., approximately 30% of the adsorber volume).

The process of TCES can be subdivided into three steps. These three steps are illustrated in Fig. 17.8 and are described as follows:

1. Desorption: charging the store
   During the charging process the adsorber is heated to a temperature high enough to almost dry the storage material. The heat Q_{des} has to be supplied to the storage material at a high temperature. Water vapor is released from the adsorbent and condensed in the condenser at a low temperature. The heat of condensation can either be used as a low-temperature heat source or has to be rejected to the environment. The liquid is stored in the reservoir.

   One of the most crucial points for attaining high storage density is achieving almost complete drying out of the adsorbent. The higher the temperature inside the adsorber and the lower the temperature of the condenser the better the desorption.

   To give an example let us assume zeolite 13X and water as a working pair. To attain a minimal water loading of 80 g kg^{-1} at a desorption temperature of 180 °C, the temperature of the condenser should be below 7 °C,

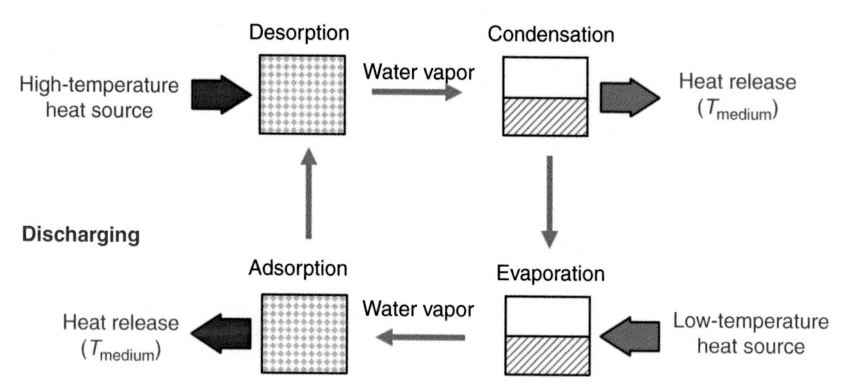

FIGURE 17.8 **Schematic sketch of closed adsorption process.**

which is the dew point temperature of water at 10^3 Pa (10 mbar) (compare with Fig. 17.3). In other words, good desorption requires either a very high desorption temperature or a very low condenser temperature. This depends on the application. After charging, the storage material and components will cool down to ambient temperature. If sensible energy cannot be used to meet the load it occurs as a heat loss.

2. Storage.
From the moment the system is at ambient temperature no further energy losses occur. As long as the storage material is hermetically sealed the energy of adsorption can be stored over an arbitrary time without any losses.
3. Adsorption: discharging the store.
Heat Q_{in} is supplied to the evaporator at a low temperature level to evaporate the liquid water. The resulting steam is adsorbed in the adsorber and the heat of adsorption is released. The adsorption material (adsorbent) is thus heated up. The released heat Q_{ads} can be used at a higher temperature level for heating purposes. This process can be driven as long as adsorption equilibrium is reached.

This means low-temperature energy for evaporation is supplied to gain useful heat at a higher temperature level during adsorption. In fact, from the thermodynamic point of view, adsorption energy storage is a heat pump cycle.

Due to the fact that adsorption is a completely reversible process the amount of heat supplied for desorption Q_{des} is exactly equal to the heat Q_{ads} gained back during adsorption. To fulfill the energy balance the heat of condensation must be equal to the energy of evaporation. In fact, the lower the adsorbent temperature and the higher the water vapor pressure in the evaporator the higher the maximum water loading of the adsorbent (compare Fig. 17.3).

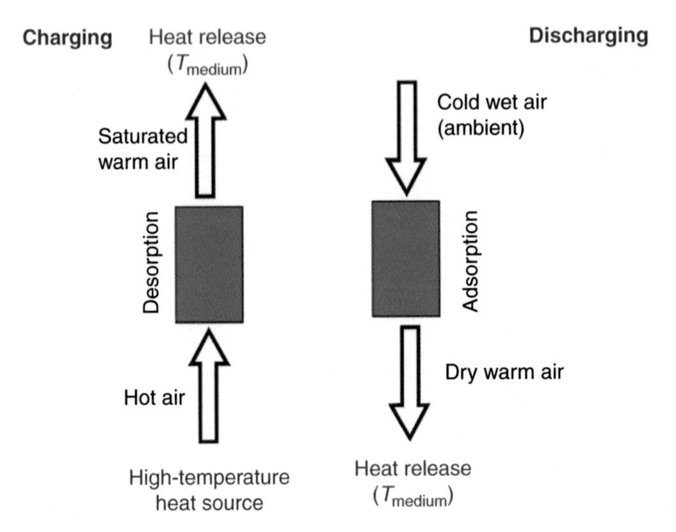

FIGURE 17.9 **Schematic sketch of open adsorption process.**

The challenge is to extract the heat from the adsorber. Furthermore, a low adsorbent temperature is a requirement to keep the adsorption process upright and to extract constant power from the store. The main heat transport mechanism is heat conduction. Unfortunately, the heat conductivity of adsorption materials like zeolite is very low (0.1–0.5 W m^{-1} K^{-1}). Therefore, the heat exchanger design plays an important role in storage performance.

The second power limitation aspect is the performance of the evaporator. The power extracted from the adsorption process equals the power of the evaporator multiplied by the ratio of adsorption enthalpy to evaporation enthalpy (for zeolite it is approximately a factor of 1.5). In Section 4.1 an example of a closed adsorption heat storage system is given.

4.2 Open-System Operation Mode

The open adsorption system is less complex in terms of apparatus und design. It consists mainly of an adsorber, a fan, and one or more heat exchangers. The open mode requires an adsorptive which is naturally present in the atmosphere, such as the water vapor within ambient air.

In Fig. 17.9 the operation of an open adsorption store is depicted schematically. Again the process can be divided into three steps:

1. Desorption: charging the store. For charging the store no heat exchanger inside the adsorber is required—just hot air blown through the store to heat up the storage material to the desired desorption temperature. Using zeolites, which are very efficient in combination with water vapor as the adsorptive, temperatures between (180 and 200) °C are favorable. The desorbed moisture

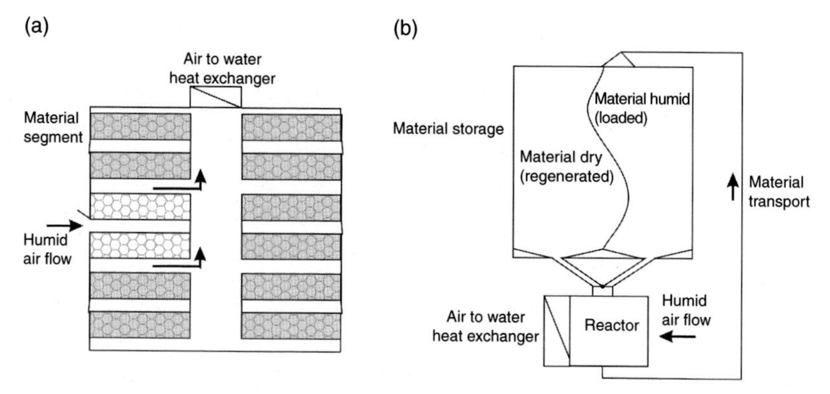

FIGURE 17.10 (a) Integrated and (b) external reactor concept for a thermochemical energy store.

is transported out of the adsorber by the air flow and released to ambient. Referring to Fig. 17.3, using ambient air heated to 200 °C (water vapor pressure 10×10^2 Pa or 10 mbar) will lead to a minimum water loading of 80 g kg^{-1}.

2. Storage. During the storing phase the storage material must be kept free of moisture.
3. Adsorption: discharging the store. Control of the discharging process is very simple. A fan is switched on to blow moist air through the storage system. Adsorption takes place, the solid is heated up by the release of adsorption enthalpy, and in turn the solid heats up the air flow, which now leaves the store very dry and at a higher temperature. This hot air flow can be used for heating purposes.

The open adsorption process is illustrated in Fig. 17.9. The adsorber, or more generally, the reactor is the key element of the system. The reactor is the apparatus where the chemical reaction/adsorption takes place. Two general approaches of reactor design can be found in the literature: the integrated reactor concept if the material reservoir is also the reactor (Fig. 17.10 left) and the external reactor concept if the material is stored separately from the reactor (Fig. 17.10 right).

4.2.1 *Integrated Reactor Concept*

In the integrated reactor concept the material is stationary inside the storage system. The main advantage is its simplicity. The type of reactor is very similar to a conventional fixed bed reactor. This is a known technology and many experiences of designing and operating of fixed bed reactors are available [37]. However, depending on the capacity of the store the size of the reactor may become very large. Reactors with large quantities of storage material ($m > 300$ kg) require subdivision of the volume for several reasons. An important aspect is the reduction in pressure drop when blowing moist air through the reactor. Low pressure drop is essential to limiting the electrical consumption of the fan/blower and is a necessity for high storage efficiency. Furthermore, dividing the storage mass into smaller parts is favorable and improves the heat and mass transfer inside the storage system. Reduction of the thermal mass heated up immediately reduces heat losses during desorption and adsorption. Additionally, this measure decreases the thermal response time of the system. This is an important improvement under transient operation conditions such as solar thermal applications. However, high temperatures ($t > 120$ °C) are needed for endothermic regeneration reaction/desorption. This implies the use of temperature-resistant materials throughout the entire TCES system.

In contrast to the external reactor design, material transport is not required. This implies less material stress as well as no technical or energetic effort for transporting the storage material.

4.2.2 External Reactor Concept

The external reactor concept separates the reactor from the storage material reservoir. Material transportation (e.g., vacuum conveying system) is required to move the material from the material reservoir to the reactor and vice versa. By separating the material reservoir and reactor the reaction is reduced to only a small part of the total storage material amount at any one time. Thermal heat capacities and heat losses, especially during the regeneration process, are reduced. Furthermore, the high temperature during material regeneration is restricted to the reactor only. As a result, the power of the storage system and storage capacity are decoupled. The design of the reactor defines the power of the system while storage capacity depends on the size of the storage vessel only. This insures a large degree of freedom in the ratio of power to capacity and enables good scalability. The storage vessel can be simple in design and no insulation is required.

Comparison of the integrated and external reactor concepts is given in [35].

5 SELECTED EXAMPLES

At present a number of concepts have been developed and some prototypes have been built, which shows dynamic development in this technology. This section describes selected examples of TCES concepts introduced in Section 4. This includes open and closed adsorption systems, systems using salt hydration as the storage mechanism, as well as high-temperature storage via dehydration of metal hydroxides. A detailed review of long-term sorption energy storage is given by N'Tsoukpoe [38] and Bales et al. [15].

5.1 Closed Adsorption Storage Systems

Thermochemical storage for solar space heating in a single-family house has been developed at the Institute for Sustainable Technologies, (Austria) (AEE

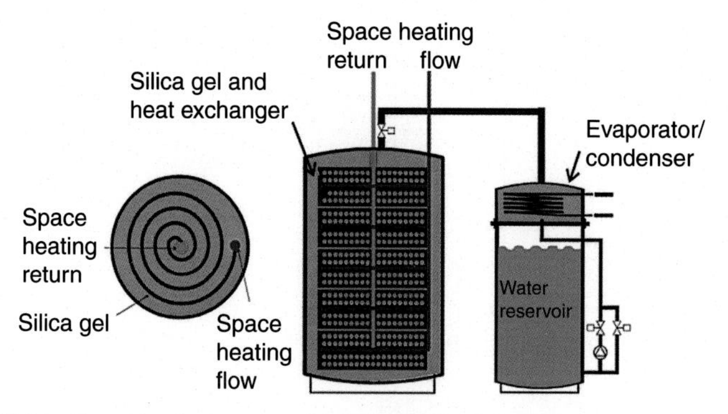

FIGURE 17.11 **Drawing of prototype of the MODESTORE project [12].**

Intec) [12]. In a project called MODESTORE a closed adsorption system has been developed. This system operates under vacuum conditions with silica gel and water as the working pair. It consists of a storage material vessel with an internal spiral heat exchanger, a water reservoir, and an evaporator/condenser unit. A schematic drawing of the sorption store is depicted in Fig. 17.11.

Charging the storage system (material desorption) is done by heating the storage material via the internal heat exchanger. Desorbed water vapor is condensed in the condenser and pumped into the water reservoir. During the discharging process (adsorption) water from the water reservoir is pumped into the evaporator to produce vapor. The vapor is adsorbed by the storage material and the heat of adsorption is released. The heat is transferred to the space-heating loop via the internal heat exchanger.

A pilot system of the sorption store was installed and monitored in a building located in Gleisdorf. The system consists of a 32 m^2 flat plate collector area, a 900 L water buffer store, two 500 kg sorption stores, a separate water store for hot-water preparation, and an auxiliary heater (wood pellets). Such a system has been successfully tested. However, experimental results remained below expected values for storage density. One main reason for the insufficient storage density originates from the working pair of silica gel and water vapor. The achieved temperature lift decreases with time during the adsorption process. A necessary temperature lift of at least 10 K was only reached during the first 35% of total storage capacity.

This research has been continued within the framework of the European project COMTES (Combined Development of Compact Thermal Energy Storage Technologies) [39]. The aim was to overcome these drawbacks using storage materials with higher performance and further improved system technology. A hydraulic scheme of the system is depicted in Fig. 17.12. A storage volume double the 1000 L adsorption store was realized in stainless steel vessels. Binderless 13X zeolite was used as the storage material which is characterized by high adsorption capacity and fast adsorption kinetics [9,40]. High-performance

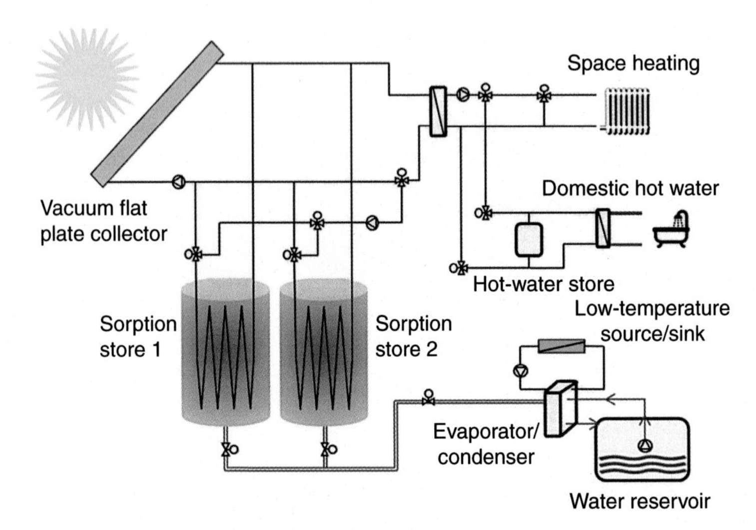

FIGURE 17.12 Scheme of the closed adsorption storage system developed within the project COMTES. *(Source AEE Intec.)*

vacuum flat plate solar thermal collectors serve as the heat source for desorbing the storage material during summer. The heat for evaporation is taken from the environment, using an air heat exchanger unit. In addition, the solar thermal collectors deliver low-temperature heat for evaporation in case of insufficient ambient conditions.

The focus has been heat and mass transfer inside the sorption unit. A heat exchanger design was developed which provides a high heat transfer rate. First experiments have shown that utilization of the storage material and the achieved temperature lift inside the store were successfully increased. The steam generated by the developed evaporator/condenser unit was continuously controlled and allowed constant power extraction of several kilowatts from the storage system. The demonstration plant was able to analyze thermal performance under realistic conditions. Fig. 17.13 shows the experimental setup and the solar thermal collectors installed at AEE Intec.

A similar closed adsorption storage system has been developed in another European project [41]. ZeoSys together with the Fraunhofer Institute for Interfacial Engineering and Biotechnology (IGB) has developed a compact adsorption storage system with the aim of storing industrial waste heat and heat from combined heat and power systems [41]. Basic investigations have been carried out in a small 1.5 L storage system where the performance of different sorption materials has been tested. Important process parameters, such as temperatures and pressure ranges, have been analyzed with respect to storage density and power output. Power is a crucial issue when using a solid storage medium with very low thermal conductivity such as zeolite ($\lambda \approx 0.1$ W m^{-1} K^{-1}). Several designs of heat exchangers have been tested in a medium-scale reactor of 15 L

FIGURE 17.13 **Experimental setup of 2 m^2 closed sorption storage at AEE-Intec (left) and vacuum flat plate collector for heat production (right).** *(Pictures AEE Intec.)*

volume. A good solution has been found with a heat exchanger configuration that showed a heat power rate more than 60% higher than measured with a parallel copper plate heat exchanger in the bulk. In the final step the improved system concept has been tested successfully in a fully functional storage system with 750 L of storage volume.

Both projects are promising examples for a technology that appears to have a number of advantages for heating applications as well as for industrial process heat applications.

5.2 Open Adsorption Storage Systems

Within the research project SolSpaces taking place at the University of Stuttgart (Germany), a new solar heating system has been developed with the aim of supplying the demand for space heating of energy-efficient buildings completely on a solar basis. The core element of the system is a sorption store, serving as seasonal energy store. In Fig. 17.14 the test building at the University of Stuttgart is shown. The building consists of a living space of 48 m^2 and has an annual heat demand of approximately 2,500 kW h.

The solar heating concept implemented is based on a solar thermal system in combination with a sorption heat store and is designed for very high solar fraction (up to 100%). Solar fraction is the amount of energy provided by the solar thermal system divided by the total energy required. Fig. 17.15 illustrates the operating mode of the solar heating system during summer and winter schematically. The sorption store is integrated between the controlled ventilation of the building and the indoor air leaving the space-heating zone. In this case the required moisture to achieve space heating by adsorption is supplied by indoor air. In winter, indoor air at room temperature (~20 °C) is allowed to pass through the sorption storage. The storage material adsorbs moisture from the air. The heat of adsorption is released and air flow heats up significantly above 20 °C. This warm air is then allowed to pass through the heat exchanger, where it heats up incoming fresh ambient air. The incoming air flow affects room heating. The power of the heating system is limited by the humidity of indoor air

FIGURE 17.14 **SolSpaces building at University of Stuttgart.**

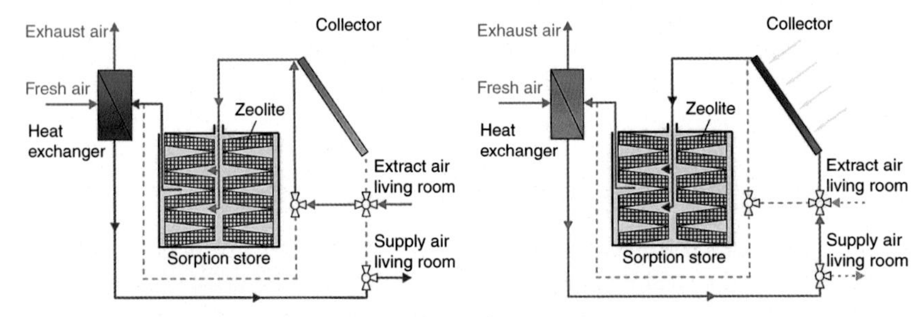

FIGURE 17.15 **Summer and winter operation mode of the solar thermal heating system with an open sorption store.**

and the air change ratio. Under typical operating conditions the system delivers power of (1–1.2) kW. This is sufficient approximately 85% of the time in the heating season.

In the summer months, when solar radiation is present in excess, the storage material is regenerated to charge the sorption store (desorption). Vacuum tube solar collectors are used to heat the ambient air. The hot air is then allowed to flow through the sorption store system to regenerate the storage material. The warm air leaving the sorption system is then passed through the heat exchanger to preheat incoming ambient air, before it enters the solar collectors.

A segmented storage system, subdivided into individual sorption units, has been developed. The sorption store has a cubic shape and a volume of 4.2 m^3. For efficient operation, it is necessary to divide the sorption store into individual segments. Due to segmentation a significant reduction in pressure drop is achieved. At the same time, the mass of the storage material involved in the adsorption/desorption processes is reduced. This leads to reduced heat losses

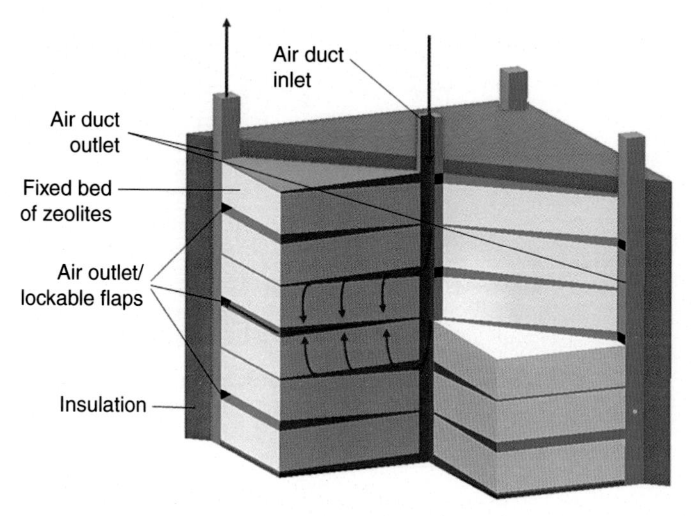

FIGURE 17.16 Concept of the SolSpaces segmented open sorption store.

due to smaller thermal capacity. Furthermore, regular and homogeneous flow distribution within each segment is an important requirement for the efficiency of the store. These considerations result in the storage concept, which is schematically depicted in Fig. 17.16. Four areas arise as a result of two vertical partitions on the diagonals, each of which is subdivided by horizontal planes.

At INSA de Lyon, the concept of an open adsorption system using exhaust air from the building as the water vapor source has been built. Hot air with a temperature between (80 and 150) °C will be provided in summer by air collectors for desorption. A new composite storage material, consisting of zeolite 13X as carrier matrix and 15% mass percent of magnesium sulfate, has been developed. Composite materials are characterized by simultaneous adsorption and hydration reaction. In a laboratory test reactor with 200 g of composite material a temperature lift of over 25 °C has been measured during the combined adsorption/hydration process (air flow rate of 8 L min^{-1} at 25 °C and relative humidity of 50%). A volumetric energy storage density of 166 kWh m^{-3} has been achieved. This is 65% of the theoretical storage density of the material (257 kW h m^{-3}) and a 27% increase compared with pure zeolite 13X (131 kW h m^{-3}). Microcalorimetry experiments revealed that energy density can be maintained over at least three charge and discharge cycles. Further research will focus on improving the carrier matrix and enhancing the energy storage density.

An open sorption system for long-term solar heat storage with magnesium chloride on a carrier matrix as storage material is under investigation at the Energy Research Centre of the Netherlands (ECN) [33,42]. A prototype reactor, containing 15 L of storage material has been successfully tested. The temperature for dehydrating the material was set to 130 °C. During hydration experiments a temperature lift of the airflow from (50 to 64) °C has been measured in

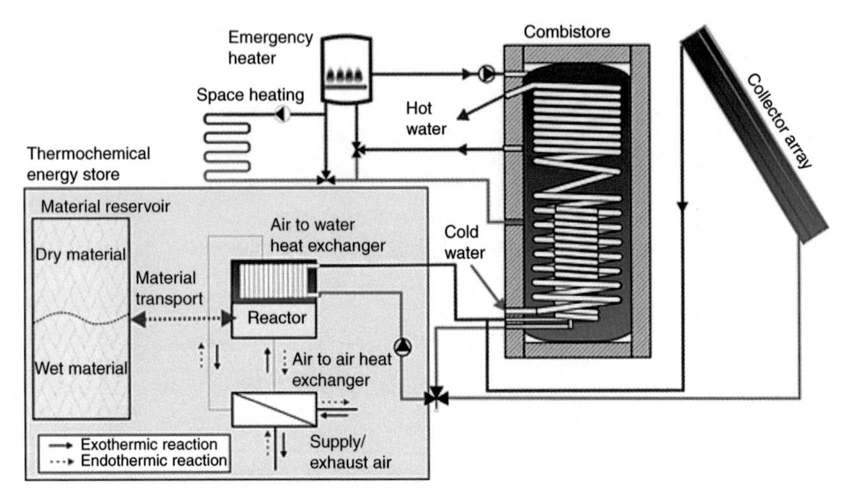

FIGURE 17.17 **Sketch of the external reactor design developed at ITW.**

the material bed [12 × 10^2 Pa (12 mbar) partial pressure of water vapour]. With further improvement of the reactor design and the heat exchangers it is expected that high energy density of approximately 280 kW h m^{-3} of the material can be technically reached.

The external reactor concept has been developed at the University of Stuttgart's Institute of Thermodynamics and Thermal Engineering (ITW). A new process design for solar thermal long-term heat storage has been developed [43] and a solar thermal combination system has been extended by incorporating a thermochemical energy store. In Fig. 17.17 a schematic drawing of the system concept is depicted. Similar to the concepts of the Solar Spaces project, it is an open adsorption/hydration system using ambient or exhaust air to provide humidity for the reaction.

The thermochemical energy store works as a low-power heating system and is connected to the combistore of the solar system via the collector loop heat exchanger. The thermochemical energy store consists of a material reservoir for the storage material and a reactor where heat and mass transfer take place during the reaction. The external reactor concept separates the storage material (a composite material of zeolite and salt) from the reactor. This has the advantage that the reaction is reduced to only a small part of the total storage material amount at a time. Thermal heat capacities and heat losses especially during the regeneration process are reduced. Furthermore, only the reactor has to withstand high temperatures whereas for the material reservoir low-cost materials can be used. Storage material transport between the material reservoir and reactor is done by a vacuum conveying system allowing very gentle material transport with low energetic expense.

The reactor is designed as a cross-flow reactor. The material enters the reactor from the top and runs gravity driven through the reactor. The air enters

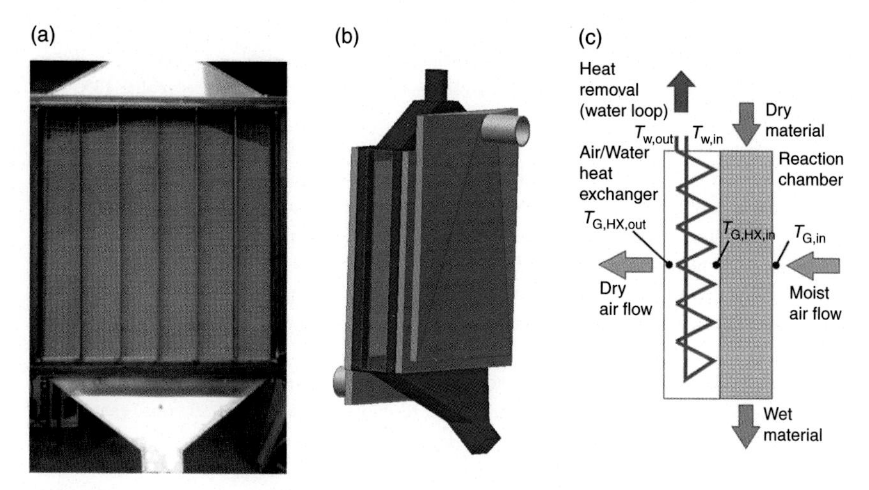

FIGURE 17.18 External cross-flow reactor of the CWS project: (a) pictures of reactor, (b) 3D illustration of the reactor, and (c) sketch of heat and mass fluxes of the reactor.

the reactor from the side and transports the humidity and heat into or out of the reactor. In heating mode, the heat released is transferred from the air flow to the water loop by an air-to-water heat exchanger. For material regeneration the air flow direction is reversed and the heat exchanger is used for transferring regeneration heat from the solar loop into the reactor via the air flow. A sketch of the reactor design and a picture of the laboratory prototype are depicted in Fig. 17.18. A detailed description of the reactor design and reactor operation mode is given in [44].

REFERENCES

[1] Aydin D, Casey SP, Riffat S. The latest advancements on thermochemical heat storage systems. Renew Sust Energ Rev, 41, 356–67.

[2] Posern K, Kaps C. Calorimetric studies of thermochemical heat storage materials based on mixtures of $MgSO_4$ and $MgCl_2$. Thermochim Acta 2010;502:73–6.

[3] Henninger S, Schmidt F, Henning H-M. Water adsorption characteristics of novel materials for heat transformation applications. Appl Therm Eng 2010;30:1692–702.

[4] Ristić A, Henninger S, Kaučič V. Two-component water sorbents for thermo-chemical energy storage—a role of the porous matrix. Proceedings of Innostock 2012, 12th International Conference on Energy Storage, Llleida, Spain; 2012.

[5] Aristov YI, Tokarev MM, Restuccia G, Cacciola G. Selective water sorbents for multiple applications, 2. $CaCl_2$ confined in micropores of silica gel: sorption properties. React Kinet Catal Lett 1996;59:335–42.

[6] Schmidt P, Bouché M, Linder M, Wörner A. Pilot plant development of high temperature thermochemical heat storage. Proceedings of Innostock 2012, 12th International Conference on Energy Storage, Llleida, Spain; 2012.

[7] Yang RT. Adsorbents: fundamentals and applications. Hoboken, NJ: Wiley-Interscience; 2003.

[8] Ruthven DM. Principles of adsorption and adsorption processes. New York: Wiley; 1984.
[9] Mette B, Kerskes H, Drück H, Müller-Steinhagen H. Experimental and numerical investigations on the water vapour adsorption isotherms and kinetics of binderless zeolite 13X. Int J Heat Mass Transf 2014;71:555–61.
[10] Jänchen J, Stach H. Shaping adsorption properties of nano-porous molecular sieves for solar thermal energy storage and heat pump applications. Sol Energy 2014;104:16–8.
[11] Jänchen J, Ackermann D, Weiler E, Hellwig U. Optimization of thermochemical storage by dealumination of zeolitic storage materials. 10th international conference on thermal energy storage. Pomona, NJ: EcoStock 2006; 2006.
[12] Jähnig D, Wagner W, Isaksson C. Thermo-chemical storage for solar space heating in a single-family house. 10th international conference on thermal energy storage. Pomona, NJ: EcoStock 2006; 2006.
[13] Kerskes H, Asenbeck S, Mette B, Bertsch F, Müller-Steinhagen H. Low temperature chemical heat storage – an investigation of hydration reactions. Effstock 2009, thermal energy storage for efficiency and sustainability.11th international conference on thermal energy storage. Stockholm, Sweden: Energi- och Miljötekniska Föreningen/EMTF Förlag; 2009.
[14] Cuypers R. MERITS: more effective use of renewables including compact seasonal thermal energy storage. Proceedings of Innostock 2012, 12th International Conference on Energy Storage, Llleida, Spain, 2012.
[15] Bales C. Laboratory tests of chemical reactions and prototype sorption storage units. Report of IEA solar heating and cooling programme—Task 32: advanced storage concepts for solar and low energy buildings. Paris, France: International Energy Agency; 2008.
[16] Jänchen J, Schumann K, Thrun E, Brandt A, Unger B, Hellwig U. Preparation, hydrothermal stability and thermal adsorption storage properties of binderless zeolite beads. Int J Low-Carbon Technol 2012;00:1–5.
[17] Schmidt FP. Optimizing Adsorbents for Heat Storage Applications. Dissertation, Freiburg im Breisgau, 2004.
[18] Henninger SK, Habib HA, Janiak C. MOFs as adsorbents for low temperature heating and cooling applications. J Am Chem Soc 2009;131:2776–7.
[19] Dawoud B, Aristov Y. Experimental study on the kinetics of water vapour sorption on selective water sorbents, silica gel and alumina under typical operating conditions of sorption heat pumps. Int J Heat Mass Transf 2003;46:273–81.
[20] Bertsch F, Mette B, Asenbeck S, Kerskes H, Müller-Steinhagen H. Low temperature chemical heat storage—an investigation of hydration reactions. Effstock 2009—11th international conference on thermal energy storage, 2009.
[21] Chou I-M, Seal II R. Determination of epsomite–hexahydrite equilibria by the humidity–buffer technique at 0.1 MPa with implications for phase equilibria in the system MgSO4–H_2O. Astrobiology. 2003; 3:3, 619–630.
[22] Knovel Database. Thermochemical properties of inorganic chemicals; 2008.
[23] Patnaik P. Handbook of inorganic chemicals. New York: McGraw-Hill Professional; 2002.
[24] Whiting G, Grondin D, Bennici S, Auroux A. Heats of water sorption studies on zeolite–$MgSO_4$ composites as potential thermochemical heat storage materials. Sol Energy Mat Sol Cells 2013;112:112–9.
[25] van Essen VM, Zondag H, Schuitema R, van Helden W, Rindt CCM. Materials for thermochemical storage: characterization of magnesium sulfate, First international congress on heating, cooling and buildings, Lisbon, Portugal; 2008.

[26] Emons H-H, Ziegenbalg G, Naumann R, Paulik F. Thermal decomposition of the magnesium sulfate hydrates under quasi-isothermal and quasi-isobaric conditions. J Therm Anal Calorim 1990;36:1265–79.

[27] Wyttenbach J, Tanguy G, Stephan L. Thermochemical seasonal storage demonstrator for a single-family house: design. Conference Proceedings Eurosun 2014, Aix-les-Bains, France; September 16–19, 2014.

[28] Hongois S, Kuznik F, Stevens P, Roux J-J. Development and characterisation of a new $MgSO_4$–zeolite composite for long-term thermal energy storage. Sol Energy Mat Sol Cells 2011;95:1831–7.

[29] Aristov YI, Gordeeva LG. Salt in a porous matrix adsorbent: design of the phase composition and sorption properties. Kinet Catal+ 2009;50:65–72.

[30] Chan KC, Chao CYH, Sze-To GN, Hui KS. Performance predictions for a new zeolite 13X/$CaCl_2$ composite adsorbent for adsorption cooling systems. Int J Heat Mass Transf 2012;55:3214–24.

[31] Aristov YI, Tokarev MM, Gordeeva LG, Snytnikov VN, Parmon VN. New composite sorbents for solar-driven technology of fresh water production from the atmosphere. Sol Energy 1999;66:165–8.

[32] von Beek T, Rindt C, Zondag H. Performance analysis of an atmospheric packed bed thermochemical heat storage system. Proceedings of Innostock 2012, 12th international conference on energy storage, Llleida, Spain; 2012.

[33] Zondag H, van Essen V, Bleijendaal LPJ, Kikkert B, Bakker M. Application of $MgCl_2$ H_2O for thermochemical seasonal solar heat storage. 5th international renewable energy storage conference (IRES 2010). SEMINARIS CampusHotel Berlin, Science & Conference Center, Berlin, Germany; 2010 November 22–24.

[34] Casey SP, Elvins J, Riffat S, Robinson A. Salt impregnated desiccant matrices for "open" thermochemical energy storage—selection, synthesis and characterisation of candidate materials. Energ Buildings 2014;84:412–25.

[35] Zondag HA, Schuitema R, Bleijendaal LPJ, Gores JC, van Essen M, van Helden W, Bakker M. R&D of thermochemical reactor concepts to enable heat storage of solar energy in residential houses. Paper presented at the 3rd International Conference on Energy Sustainability, July 19–23, 2009, San Francisco, CA. Proceedings of the ASME 3rd International Conference on Energy Sustainability. New York: American Society of Mechanical Engineers; 2009.

[36] Hauer A. Thermochemical energy storage for heating and cooling—first results of a demonstration project. Terrastock 2000, 8th International Conference on Thermal Energy Storage; 2000, p. 641–646.

[37] Eigenberger G, Ruppel W. Catalytic fixed-bed reactors. Ullmann's encyclopedia of industrial chemistry. Weinheim, Germany: Wiley–VCH Verlag; 2000.

[38] N'Tsoukpoe KE, Liu H, Le Pierrès N, Luo L. A review on long-term sorption solar energy storage. Renew Sust Energy Rev 2009;13:2385–96.

[39] van Helden W, Thür A, Weber R, Furbo S, Gantenbein P, Heinz A, Salg F, Kerskes H, Williamson T, Sörensen H, Isaksen K, Jänchen J. COMTES: parallel development of three compact systems for seasonal solar thermal storage: introduction. Proceedings of Innostock 2012, 12th international conference on energy storage, Llleida, Spain; 2012.

[40] Mehlhorn D, Valiullin R, Kärger J, Schumann K, Brandt A, Unger B. Transport enhancement in binderless zeolite X- and A-type molecular sieves revealed by PFG NMR diffusometry. Micropor Mesopor Mat 2014;188:126–32.

[41] Lass-Seyoum A, Blicker M, Borozdenko D, Friedrich T, Langhof T. Transfer of laboratory results on closed sorption thermo-chemical energy storage to a large-scale technical system. Energy Procedia 2012;30:310–20.

[42] Zondag H, Kikkert B, Smeding S, de Boer R, Bakker M. Prototype thermochemical heat storage with open reactor system. Proceedings of Innostock 2012, 12th international conference on energy storage, Llleida, Spain; 2012.

[43] Kerskes H, Mette B, Bertsch F, Asenbeck S, Drück H, Chemical energy storage using reversible solid/gas-reactions (CWS)—results of the research project, 1st International Conference on Solar Heating and Coolingfor Buildings and Industry (SHC 2012) 2012, 30, 294–304.

[44] Mette B, Kerskes H, Drück H, Müller-Steinhagen H. New highly efficient regeneration process for thermochemical energy storage. Appl Energy 2016;109:352–59.

Chapter 18

Power-to-Gas

Robert Tichler*, Stephan Bauer**
**Department of Energy Economics, Energy Institute, Johannes Kepler University Linz, Linz, Austria; **Innovation and Development, RAG Oil Exploration Company, Vienna, Austria*

1 INTRODUCTION

During the past (5–10) years a new energy storage system called power-to-gas has been developed, primarily in Europe and North America. Power-to-gas refers to the chemical storage of electrical energy with the energy stored in the form of gaseous substances such as methane or hydrogen. In the chapter the term "power-to-gas" is defined as the use of (excess) electrical energy from renewable energy sources to produce hydrogen in an electrolyzer and optionally also to use this hydrogen with carbon dioxide to synthesize methane or other hydrocarbons. Two general applications or process chains of power-to-gas systems may be differentiated:

- Using electric energy and carbon dioxide, synthetic methane can be produced. For this technology electrical energy is used from renewable energy sources. This energy is mainly, but not exclusively, excess energy that is produced from wind and photovoltaic sources and is stored in the form of methane. The conversion of hydrogen (H_2) and carbon dioxide (CO_2) into methane (CH_4) is carried out in specially designed facilities.
- Using electrical energy to produce hydrogen from electrical energy. Hydrogen can also be stored and used directly, especially in the transport sector. In addition, hydrogen can be added to natural gas and as such the hydrogen can be used in all energy segments (heat, electricity, transport) [1] (Fig. 18.1).

The initial intention for the implementation of power-to-gas plants is, on the one hand, to store intermittent electricity from renewable sources, thus avoiding the shutting down of generation plants in times of power surpluses and, on the other hand, to further integrate power generation in isolated regions. At the same time, systematic research and development of power-to-gas systems has uncovered and developed a wide range of systemic benefits, not only for power generation and transmission but also for energy systems, in general. The systemic approach to the power-to-gas form of energy storage system is the focus of the chapter. For details of electrolysis technologies see also chapter: Hydrogen from Water Electrolysis.

Storing Energy. http://dx.doi.org/10.1016/B978-0-12-803440-8.00018-X

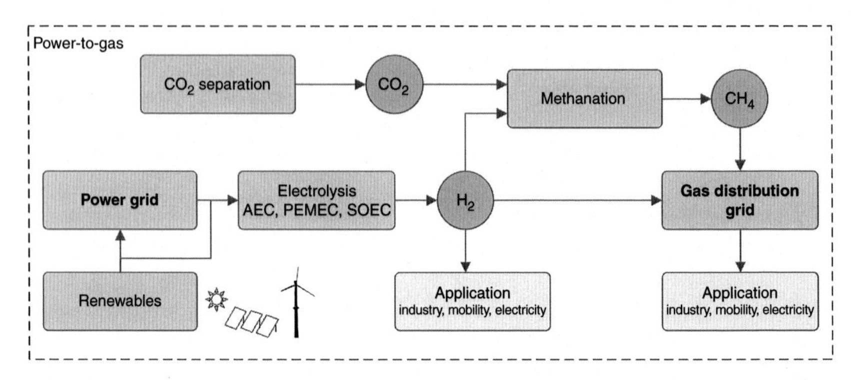

FIGURE 18.1 Process chain of power-to-gas. *AEC*, alkaline electrolyzer; *PEMEC*, proton exchange membrane electrolyzer; *SOEC*, solid oxide. *(Source: Reiter [2].)*

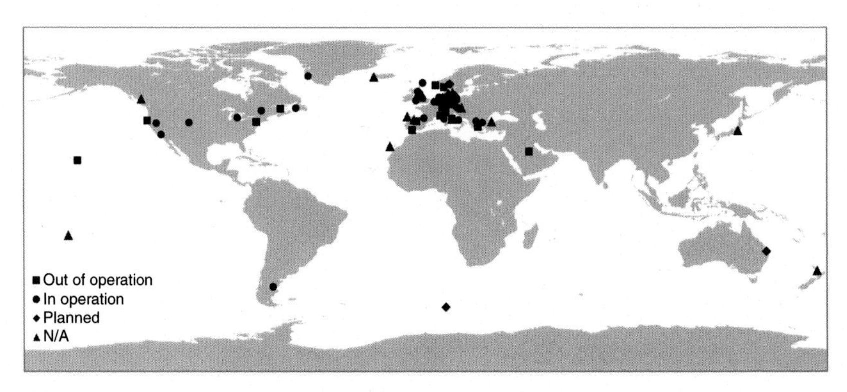

FIGURE 18.2 Overview of international power-to-gas pilot plants. *(Source: Reiter [2], based on information from Steinmüller et al. [5].)*

The increased integration of intermittent renewable energy sources such as wind and solar power is an increased challenge to the new flexible model of the energy system. At present, the main centers of research and development of power-to-gas systems are in European countries such as Germany, France, the United Kingdom, and Italy. Smaller investments are to be found in the Netherlands, Switzerland, and Austria and also in Canada and the United States (Fig. 18.2). In recent years, increased integration of renewable energies into the energy system, based on climate policy objectives and targets, has been adopted and implemented at national and multinational levels.

A sustainable and secure energy supply that is economically viable, environmentally sustainable, and socially responsible is a high priority in current energy politics. Due to the fact that the production of fossil fuels is shrinking and the necessity of preventing greenhouse gas emissions, necessary modifications to the energy supply mix have to be made and these changes must also take into account economic considerations.

This political framework based on economic challenges generates positive effects if optimally implemented, but could also cause new problems in the system. A new challenge in recent years is the rising proportion of intermittent electricity production, in particular through increasing production levels from wind and solar sources. Without solution strategies such an intermittent nature of renewable energy causes overcapacity and redundancy and thus increasing destabilization of power grids.

To handle inconsistent energy production from wind and solar sources in a resource-efficient way and thus address a significant challenge to the development of the energy system, the promotion of energy storage is necessary. Currently, there are limited numbers of implemented storage technologies for electrical energy. The only large-volume storage technology which is fully available at the moment (next to more flexible options such as load management) is pumped hydroelectric storage (see chapter: Pumped Hydroelectric Storage). In addition to all the advantages that this technology permits, pumped hydroelectric storage power plants are faced with restrictions such as dependence on topography. Therefore, it is necessary to develop alternative storage technologies and also to take advantage of specific technologies for different system tasks. One option is power-to-gas [3].

Energy storage systems such as power-to-gas will play a key role in integrating renewable energy sources with volatile production sources in addition to optimized power management. The availability of the correct amounts of energy at the proper time period presents a major challenge. Power-to-gas technology could be an important part of future storage portfolios because long-term storage as well as shifts in capacity between energy networks can be realized using power-to-gas, which allows for new possibilities in energy transmission [4].

Power-to-gas technologies and systems are currently (as of 2015) at a nascent stage in their development, with individual pilot and demonstration plants being implemented or designed in different sizes. Power-to-gas technology has specific advantages over other storage technologies. These include the possibility of long-term storage of energy, the ability to store enormous amounts of energy, the displacement of energy transport, and the binding of carbon dioxide. For a detailed explanation of the technological, energetic, economic, systemic, and ecological characteristics, reference is made to the literature, such as Steinmüller et al. [5], Sterner [6], Specht et al. [7], and Müller-Syring et al. [8].

In a nutshell, power-to-gas technology allows for long-term storage of electrical energy which does not already exist in another way within the current energy system, making the energy system more resource-efficient and flexible. This enables accelerated integration and continued realization of low- or zero-emission technologies such as wind power and photovoltaics, thereby promoting the achievement of both climate and energy policy objectives. Furthermore, power-to-gas allows new options in energy transport by load shifting from the power grid to the gas grid. This also reduces socioeconomic challenges such as social opposition to power grid expansions. In addition, power-to-gas technology has many potential strengths such as the incorporation of carbon dioxide in carbon capture and utilization, the provision of a new renewable energy source especially in the mobility market (no competition for acreage with food production), and the possibility of creating self-sufficient primary or backup energy supply systems (including fuel cells) for topographically remote regions or even buildings. The strengths of power-to-gas technology include, in particular, superordinate benefit for the entire energy system and for the economy as a whole [3].

The storage and conversion of energy are always accomplished with broad discussions and analysis about the technical efficiency factor. As expected the conversion of electricity into hydrogen with the option of synthesizing methane comes with reductions in the degrees of efficiency. However, power-to-gas technology is primarily implemented for storing electricity surpluses which cannot be integrated into the electricity system in other ways. Therefore, the introduction of power-to-gas technology into the general energy system can be regarded as an improvement.

2 DYNAMIC ELECTROLYZER AS A CORE PART OF POWER-TO-GAS PLANTS

As discussed earlier, power-to-gas technology can be implemented using (excess) electrical energy from renewable energy sources to produce hydrogen in an electrolyzer and optionally synthesize the hydrogen produced with carbon dioxide into methane or other hydrocarbons. The main focus is therefore on electrolytic supply of hydrogen without additional primary energy sources used in biomass or fossil raw materials outside electricity production. The electrolyzer thus represents the main technological component in a power-to-gas system. It uses electric power for the cleavage of water into hydrogen and oxygen according to the following reaction equation:

$$2\,H_2O \rightarrow 2\,H_2 + O_2 \tag{18.1}$$

Depending on the electrolyte a distinction between alkaline electrolysis cells (AEC), proton exchange membrane electrolysis cells (PEMEC), and solid oxide electrolysis cells (SOEC) has to be made. A more detailed description of the state of the art, latest developments, and characteristics of the different electrolytic types can be found, for example, in Ursua et al. [9] or Smolinka et al. [10] or in chapter: Hydrogen from Water Electrolysis.

Alkaline electrolyzers use an aqueous alkaline electrolyte and provide the most widely used technology with lowest investment costs [11]. AEC systems are considered robust and are already available in high-performance units. Challenges remain, especially in dynamic modes, as efficiency and hydrogen quality are severely affected in partial load operation [10]. Alkaline electrolyzers also require a large amount of space compared with PEM electrolyzers [12]. Further developments of conventional alkaline electrolyzers include improvements related to increased gas pressure. Problems arise, however, when there are leaks in the system as this can cause a discharge of corrosive electrolytes [13].

PEM electrolyzers use a polymer electrolyte membrane and are significantly more compact than alkaline electrolyzers. With better startup performance, faster response to load changes, and higher quality hydrogen (fewer impurities), PEM-EC are better suited to dynamic operations. According to Smolinka et al. [10] there are still challenges in terms of the lifespan of the membrane and high-investment costs due to the use of noble metal catalysts such as platinum. In addition, the available power sizes of PEM electrolyzers are still significantly lower than those in alkaline electrolyzers [14].

SOEC electrolysis technology has the greatest need for development. Operation of this electrolyzer is carried out with the use of high thermal energy from an external heat source. The high temperatures result in accelerated reaction kinetics, and hence high-priced noble metal catalysts can be avoided [9]. By using heat the required power is reduced and electrical efficiency is increased. However, challenges exist with regard to material stress due to the high temperatures and, moreover, the SOEC require a large amount of space due to the design of their complex system [12]. The main technical parameters of alkaline and PEM electrolyzers are shown in Table 18.1.

Both alkaline and PEM electrolyzers achieve a system efficiency between (60–70)%. Key challenges for both electrolysis technologies are mainly dealing with fluctuating power input (from renewable energy sources), the present associated lower efficiency, and the hydrogen quality in partial load operations. Highly dynamic operation also has negative effects on the stability of the system. To make power-to-gas technology economically attractive the currently high investment costs for electrolyzers need to be reduced significantly. While PEM electrolyzers respond better to rapid load changes and are well suited to

TABLE 18.1 Typical Characteristics of Alkaline and PEM Electrolyzers

	AEC	PEMEC
Available nominal power/MWel	Several	Up to 1
Performance range/%	20–100 of the available nominal power	0–100 of the available nominal power
Operating pressure/10^5 Pa (bar)	1–30	Up to 100
Operating temperature/°C	60–90	~80
Duration/a (years)	10–20	6–15
Space requirement	PEMEC are by a factor of 5–10 smaller than AEC	

Source: Tichler et al. [3], based on information from Ursua et al. [9], Smolinka et al. [10], and Maclay et al. [15].

dynamic operations, efficiency and hydrogen quality are severely compromised for alkaline electrolyzers operating at partial load. Although alkaline pressure electrolyzers have improved dynamic behavior, problems arise from leaks in the system and an associated outlet for corrosive electrolytes. Key challenges for both electrolysis technologies lie in dealing with fluctuating power input (from renewable energy sources) and the resulting lower efficiency and life expectancy of electrolyzers (connected to degradation). To make power-to-gas technology economically attractive the currently high investment costs for electrolyzers have to be significantly reduced [3].

The development of new technologies always entails a significant learning curve, which usually leads to a reduction of investment costs with increasing installed capacity or quantity produced. The learning rate is to be determined separately for each technology; a learning rate of 20% has been found to be typical for many components. A learning rate of 20% means a reduction in the specific investment costs of 20% at double the cumulative installed power. For photovoltaic systems, for example, the learning rate is approximately 20%, and for wind turbines the learning rate is estimated to be between (10–15)%. Also learning rates between (15–25)% are estimated for fuel cells [16]. It has been predicted that the future cost development of electrolysis technologies [17] will have a learning rate of 18%. By comparison, the specified learning rate for the production of hydrogen by steam reforming of natural gas (steam methane reforming) is around 11%. When applied to the cost of development of water electrolysis a learning rate of 18% implies that if an increase of 50 times the cumulative installed capacity of PEM and alkaline electrolyzers (compared with current installed capacity) can be achieved by 2025, investment costs of electrolyzers will be reduced by 67%.

3 METHANATION PROCESSES WITHIN POWER-TO-GAS

As already mentioned, power-to-gas systems allow for the option of producing methane by reducing carbon dioxide using the hydrogen generated. Thus, the term "power-to-gas" always implies a hydrogen path and the option of an additional methane path.

In general, there are numerous sources and separation technologies available for the provision of CO_2 for this methanation process, some of which are described in IPCC [18] or Li et al. [19]. A large amount of CO_2 can be obtained from the combustion of fossil fuels or renewable resources in power plants. Postcombustion, precombustion, oxyfuel, or chemical-looping technologies are used for the separation of the CO_2. Depending on the separation technology a certain amount of heat from the power plant is needed, thus increasing the consumption of primary energy. This leads to an overall decrease in efficiency of the power plant by (7–10)%. Furthermore, between (20–40)% more primary energy per generated kilowatt-hour is needed [20,21].

Carbon dioxide can also be obtained from industrial processes such as cement or lime production and from various fermentation processes with purity

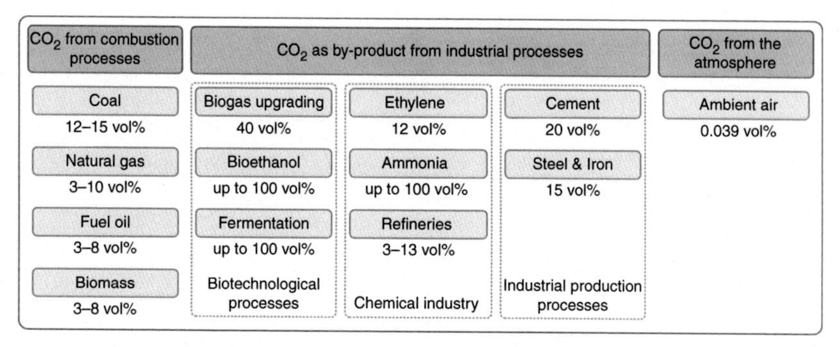

FIGURE 18.3 **CO_2 sources for utilization in power-to-gas.** *(Source: Reiter [2], based on information from Metz et al. [26].)*

levels depending on the type of process involved [22]. For the separation of carbon dioxide, various physical and chemical adsorption– and membrane–separation processes can be used (see Ryckebosch et al. [23] for a detailed description). Biogenic sources of carbon dioxide from anaerobic digestion (biogas) or from biomass gasification are possibly the most suitable sources for use in power-to-gas plants [21] (Fig. 18.3).

In theory, CO_2 can also be separated from ambient air but due to the low concentrations in the atmosphere (c. 400 ppm) the energy expenditure necessary for this process is high which leads to high costs. Depending on the separation technology, between (320–440) kJ mol^{-1} of CO_2 are needed [24,25].

Once the hydrogen production chain is in full swing and there is a surplus of hydrogen, carbon dioxide need not only be used for methane synthesis, it can also be used for the production of synthetic gas and subsequently liquid hydrocarbons (a process called power-to-liquid). For the conversion of synthetic gas into liquid hydrocarbons there are various methods, ranging from methanol synthesis, Fischer–Tropsch synthesis, oxosynthesis, and fermentation. In addition to the production of synthesis gas as a feedstock for liquid hydrocarbons (power-to-fuel or power-to-liquid), liquid hydrocarbons can also be synthesized directly from CO_2 and hydrogen (or water); for instance, the production of methanol from CO_2 and H_2 as described by Olah [27]. However, these processes and products are not associated with our original definition of power-to-gas and will not be further discussed.

Reverting back to the basic methanation process of the power-to-gas system in which methane (CH_4) is produced from hydrogen and carbon dioxide in a catalytic process called the Sabatier process; see Eq. (18.2). While CO methanation is an already proven technology in coal gasification, CO_2 methanation is still in development. The achievable efficiencies are around 80% (and fast approaching the efficiency of CO methanation), but there are still challenges in terms of thermal management and long-term stability of the catalyst [3].

$$CO_2 + 4H_2 \rightarrow CH_4 + 2H_2O \tag{18.2}$$

TABLE 18.2 Parameters of CO_2 Methanation

	CO_2 Methanation
Performance range/%	80–110 of the available nominal power
Operating pressure/10^5 Pa (bar)	6–8
Operating temperature/°C	180–350
Efficiency/%	70–85
Space requirement	Depending on the plant size, a doubling of capacity is not accompanied by a double space requirement
Development status	Demonstration stage

Source: Tichler et al. [3], based on information from Sterner [6], Breyer et al. [22] and Cover et al. [28].

Typical parameters of CO_2 methanation are shown in Table 18.2. For the methane synthesis different reactor systems are available, which can be divided into two-phase and three-phase systems. In two-phase systems in which fixed, coated honeycomb, and fluidized bed reactors are used, the gaseous reactants and the catalyst are fixed in position. In three-phase systems (e.g., bubble column) a liquid heat transfer medium is used [3].

4 MULTIFUNCTIONAL APPLICATIONS OF THE POWER-TO-GAS SYSTEM

This section is primarily based on Tichler and Steinmüller [4]. The long-term storage of electricity is a unique feature of power-to-gas among electricity storage options. Beyond this, power-to-gas systems offer far more than only the storage of electricity. As a consequence, power-to-gas should not be considered exclusively as a storage option; it can be useful for other systemic functions, such as an alternative option for the transportation of energy. The variety of technical options within power-to-gas system show an extremely wide range of specific uses and technological characteristics.

Generally, five identifiable benefits for the energy system can be stated under our definition of power-to-gas and in turn different solution strategies for different applications can be realized [29]:

- to provide long-term electrical energy storage and the associated improvement in management of highly intermittent and inconsistent electricity production
- the shift of energy transport from the power system to the gas system and the associated lower intensity needed in expansion of the power infrastructure
- the ability to raise the share of renewable energy in the transport sector through the use of synthetic methane (and of hydrogen) from renewable sources [30]

- the creation of self-sufficient energy solutions in topographically difficult and remote regions for all relevant energy segments: electricity, heat, and transport
- the use of carbon dioxide as a raw material (and the resulting reduction in greenhouse gas emissions).

The list of basic capabilities implies various forms of process chains, different business models with different technologies, and different benchmarks in the energy system. This makes for compact analysis of current and expected business forms in terms of the compatibility of the system or of competitive technologies. As a consequence, economic evaluation of specific applications of the power-to-gas system and associated competing systems or alternative solutions is necessary.

In the following section a multitude of possible applications of power-to-gas plants are listed. These applications do not include ratings for economic viability, legal implementation, or even a discussion or analysis of optimal operation or the technology involved. The applications are written in such a way that each point has a concrete benefit and specific intention for a market participant for the construction and operation of a power-to-gas plant.

Various applications of a power-to-gas plant for the implementation of a particular benefit to market participants of a specific energy market include the following [4]:

- A power grid operator implements a power-to-gas plant to substitute for investments in electricity grid extension for transmission networks, which would otherwise need increasing energy transport volumes between supply-and-demand centers. Thus, the transport of energy can be transferred to the gas system. Thus, the main intention of establishing a power-to-gas plant is to reduce infrastructure investment costs in the electricity network and reduce social frictions which result from local opposition to new transmission lines.
- A power grid operator implements a power-to-gas plant in combination with technology for reconversion—such as a fuel cell—for private households, enterprises, or technical systems in topographically remote regions to substitute for investment in an expensive power grid connection and to guarantee year-round supply. The main purpose of establishing a power-to-gas plant is also to reduce infrastructure costs in the electricity network.
- A power grid operator implements a power-to-gas plant to solve the load management problem of the power system (especially at the distribution system level) in times of high production of electrical energy from intermittent, regional renewable energy sources through the storage of electrical energy and to optimize system costs to reduce the current balance.
- A gas network operator implements a power-to-gas plant to achieve higher utilization of gas networks by shifting the transport of energy from the electricity to the gas network. The primary purpose of establishing a power-to-gas plant here is expansion of capacity in network operation.

- A potential hydrogen service provider implements a power-to-gas plant to achieve higher utilization of the network by shifting the transport of energy from electricity to hydrogen power—equivalent to the previous case of the gas network operator.
- A wind and/or photovoltaic system operator implements a power-to-gas plant to avoid cessation of renewable energy production so that production continues to operate in the wind turbine or photovoltaic system at any time regardless of energy demand and transportation capacity Thus, the overall efficiency of the system and annual full load hours can be increased.
- A wind and/or photovoltaic system operator implements a power-to-gas plant to take advantage of available energy production for intermediate storage to price-to-sell at optimum times in the electricity market. This can be done to optimize the current sale.
- A biogas plant operator implements a power-to-gas plant to increase the use and retention of carbon dioxide by producing synthetic methane and to increase overall efficiency.
- A gas storage operator implements a power-to-gas plant to achieve higher utilization of gas storage at specific times through additional natural gas production.
- A gas trader implements a power-to-gas plant to bring a new additional renewable gas product to market that can be sold.
- An electricity producer/trader implements a power-to-gas plant to bring a new additional renewable product to market that can be sold.
- A fuel producer/trader implements a power-to-gas plant to bring a new additional renewable product to market that can be sold.
- An industrial plant (chemical industry) implements a power-to-gas plant to offer a new renewable chemical/material product.
- An electricity producer/trader implements a power-to-gas plant to use renewable electricity generated in a topographically remote area with high potential for renewable energy generation (e.g., Sahara, Patagonia) and to transport this energy to demand centers (e.g., through gas pipelines).
- A service station operator implements a power-to-gas plant to offer a new renewable hydrogen product and to make the supply of hydrogen independent of grid electricity.
- An industrial plant with a commitment to CO_2 allowances implements a power-to-gas plant to bind the otherwise emitted carbon dioxide to synthetic methane, thereby increasing the efficiency and production capacity of existing resources.
- An industrial plant with a commitment to CO_2 allowances implements a power-to-gas plant to replace fossil fuels with renewable sources, thereby reducing the cost of CO_2 allowances.
- A power producer implements a power-to-gas plant to provide additional negative balancing energy and be able to generate revenue in the balancing market.

- A power producer implements a power-to-gas plant to provide a positive balance of energy and thus avoid replacement investments in alternative underworked plants.
- The automotive industry implements a power-to-gas plant to reduce CO_2-equivalent emissions of the fleet and thus comply with legal requirements and to offer new products.
- The operator of a public transportation fleet (e.g., bus, tram, train) implements a power-to-gas plant to reduce CO_2-equivalent emissions of the fleet and to insure mobility with renewable energy sources.
- A private household or a business implements a power-to-gas plant (in combination with a fuel cell) to produce energy completely for their own needs and therefore to become a self-contained system to insure self-sufficiency and act as a status symbol.
- A private household or a business implements a power-to-gas plant for intermediate storage of power which allows optimization of costs in the case of flexible rates.
- A producer of an alternative gaseous energy carrier (biogas, coal gas) operates a power-to-gas plant to modify existing gas quality and thus allow for feeding into the natural gas grid.
- Regions prone to strong potential topographic impact of electricity grid expansion (or topographical interventions by large conventional energy storage, such as pumped storage power plants) implement power-to-gas plants to meet transportation or storage needs by shifting the burden from the electricity to the gas network. This can reduce social frictions from developments which deface the landscape and/or settlement areas.
- The "public sector" operates power-to-gas plants to increase the share of renewable energy sources, as well as the overall efficiency of the energy system (by reducing the number of shutdowns of generating plants) [4].

It can be seen then that the power-to-gas process allows for a variety of applications. Of course, the characteristics of a business compatible with the respective benchmarks are also constituted very differently.

As a consequence, power-to-gas can generally be described as a very flexible system in terms of its variety of applications and different forms in national and international energy systems. The different business models that can be developed based on the specific capabilities of the system also imply specific and varied cases for different market players. This expression of a multifunctional use of power-to-gas in the future energy system also has as a consequence a wide economic impact on the technologies involved. The real purpose of the development of the power-to-gas system, however, stems from the challenge of a rising share of intermittent and volatile generation sources and the necessary option for additional energy storage that allows for long-term storage in times of excess production [4].

In addition, it is worth considering the respective advantages and disadvantages of energy carriers, hydrogen, and synthetic methane from power-to-gas plants [4,31]. The advantages of hydrogen over synthetic methane from power-to-gas plants are:

- hydrogen has lower production costs than synthetic methane
- without additional caching modules the production of hydrogen allows for an overall more dynamic mode of operation
- the production of hydrogen involves lower conversion losses, therefore the efficiency is higher than that for the production of synthetic methane
- standalone systems for energy storage can be implemented more easily with hydrogen than with synthetic methane
- no carbon dioxide source is required for hydrogen production, therefore production is location independent
- the combustion of hydrogen produces virtually no direct emissions, whereas synthetic methane does [4,31].

However, there exist several advantages to synthetic methane over hydrogen from power-to-gas plants such as:

- storage of synthetic methane is less elaborated than the direct storage of hydrogen—direct storage of hydrogen is expensive and technologically much more sophisticated
- the use and transport of synthetic methane may have recourse to an existing infrastructure, while pure hydrogen networks exist only in a few regions
- the production and use of synthetic methane has fewer restrictions for end-users in terms of technology compatibility and the granting of guarantees
- Using synthetic methane is less complex due to it having a similar calorific value to conventional natural gas—there are minor variations in accounting systems in relation to higher hydrogen shares in natural gas
- the place of supply in the natural gas grid—for storage and transportation of energy—is more problematic for hydrogen feed-in, as one has to be mindful of the exact mixing—synthetic methane on the other side can be fed without any problems, as long as the standards are complied with
- the production of synthetic methane solves the problem of potential dependence on other market participants, who have already added the maximum amount of hydrogen to the gas network [4,31].

Overall assessment in determining unique advantage can only be done on an individual basis case by case. A flexible power-to-gas system allows for the use of both energy sources [31].

The broad range of applications and above all the necessary application of long-term storage of electricity by converting electricity to hydrogen and (with carbon dioxide) to methane via power-to-gas technologies correspond to general forecasts of future long-term substitution within primary energy carriers. Energy-carrying gas and its different characteristics is represented in many forecasts for future power supply as the essential transitional technology or immanent energy carrier on the road to a "hydrogen economy" [4]. Furthermore, the loss of importance of liquid energy sources such as oil and solid fuels like wood, coal, and uranium is predicted by Hefner [32] in the global context. As a consequence the "age of energy gases" is expected to start in this decade as shown by Hefner with respect to future development of the global composition of energy suppliers [32].

5 UNDERGROUND GAS STORAGE IN THE CONTEXT OF POWER-TO-GAS

A necessary part of power-to-gas technology is to develop a storage capacity for the gas produced from fluctuating renewable energy in the form of underground gas storage facilities. The well-established and, in many parts of the world, comprehensive existing gas infrastructure can be used for such energy storage. Existing gas pipeline networks can in fact already offer a certain degree of flexibility due to pressure elasticity, a buffer that electricity grids cannot provide. The actual energy reservoirs in the system, however, are underground gas storage facilities. According to the International Gas Union (IGU) report [33], the total working gas volume in the world adds up to more than 390×10^9 m^3 (390 billion cubic meters) of natural gas (Fig. 18.4).

Today, underground gas storage facilities are technologically mature and form the backbone of secure energy supplies. Underground gas storage facilities were initially developed to balance out seasonal fluctuations between transport capacities and consumption. Over the past few years, these energy reservoirs have also been increasingly utilized as trading instruments. Chapter: Traditional Bulk Energy Storage—Coal and Underground Natural Gas and Oil Storage provides an overview of underground gas storage technology. According to today's

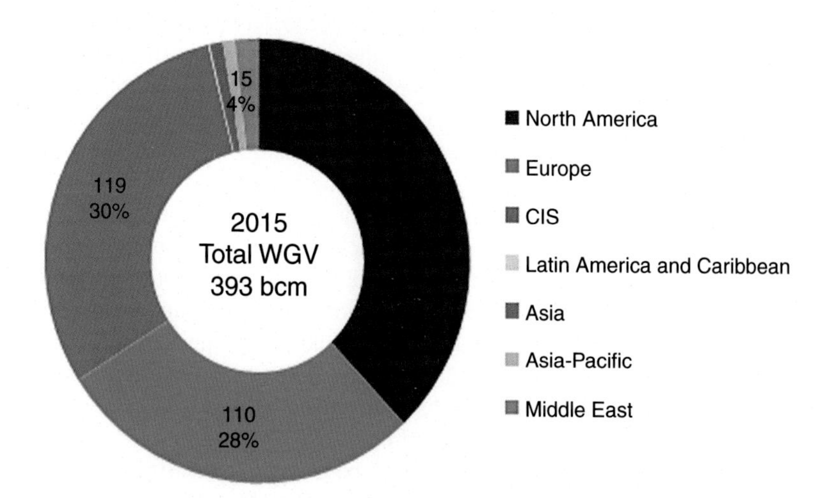

FIGURE 18.4 **Underground Gas Storage: total working gas volume.** The figures relate to 10^9 m^3 (billion cubic meters). *(Source: own figure based on information in Ref. [33].)*

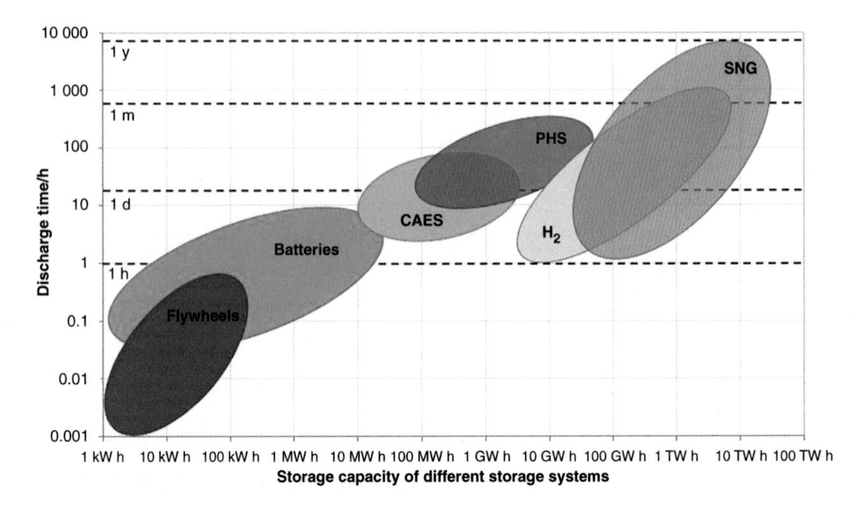

FIGURE 18.5 **Withdrawal time and storage capacities of different energy storage systems.** *CAES*, compressed air energy storage; *PHS*, pumped hydroelectric storage, H_2; *SNG*, underground gas storage of hydrogen or synthetic natural gas. *(Source: own figure based on information in Ref. [34].)*

considerations, underground storage of renewable energy is so attractive because it is the only technology that can be used to store vast quantities of energy over long periods of time and, furthermore, has the unique advantage of being able to balance seasonal fluctuations in energy generation. Fig. 18.5 provides a comparative logarithmic size overview of the storage capacities of various energy storage technologies.

When storing pure methane (synthetic natural gas, SNG), there are no technical problems anticipated in underground gas storage, as methane is the predominant element in natural gas. However, it is necessary to examine particular prevailing national regulatory and legal framework conditions for gas storage, which in many cases have up to now only been defined for natural hydrocarbons. Efforts should be made to amend or supplement these regulations to also include synthetic hydrocarbons. A different story altogether is the storage of pure hydrogen (see chapter: Larger Scale Hydrogen Storage).

Another possibility to be considered is the case where hydrogen is added to natural gas or, due to an incomplete reaction in the methanation process, hydrogen is mixed in with the synthetic methane. The addition of hydrogen into the natural gas infrastructure in controlled amounts is being pursued to achieve greater efficiency in the power-to-gas process and to save investment and running costs involved in the methanation process. Existing infrastructure should continue to be used and, moreover, this path should be pursued wherever there is no existing infrastructure for pure hydrogen.

There have been ongoing efforts in Europe for quite some time to examine the existing natural gas infrastructure in terms of its hydrogen compatibility and to define permissible limit values. On the one hand, gases containing large amounts of hydrogen were in circulation during the city gas era; on the other hand, customer devices available today are unable to handle a certain level of hydrogen content in natural gas. In a first initiative, spearheaded by the NATURALHY project, the pipeline network was examined [35]. In particular, in Germany there were additional studies geared toward examining the whole spectrum of hydrogen compatibility. In these studies it was emphasized that underground gas storage facilities should be examined more carefully for possible reactions with hydrogen [36]. The German Society for Petroleum and Coal Science Technology (DGMK) subsequently commissioned a detailed literature research which resulted in two studies: "Influence of hydrogen on underground gas storage" and "Influence of bio-methane and hydrogen on the microbiology of underground gas storage." Both studies indicate numerous elements of risk and potential for damage, all of which, however, were derived from related inquiries involving CO_2 research or were based on pure hydrogen applications. Real-life experiments or operating experience based on hydrogen–natural gas mixtures do not exist. It has been pointed out that underground gas storage facilities, due to their varying geology, geochemistry, and operating conditions, should each be viewed individually [37,38] These findings, among many others, were summarized in a European Gas Research Group (GERG) study [39]. An actual project involving an in situ field experiment for underground storage of renewable energy in the form of hydrogen–methane mixtures is currently being carried out in Austria by a consortium led by RAG Rohöl-Aufsuchungs AG under the title "Underground sun storage" [40,41]. A project of a similar nature is being completed in Argentina by HYCHICO. Final results of these projects are expected at the end of 2016.

ACKNOWLEDGMENT

The authors would like to thank Gerda Reiter for her valuable and constructive input during the preparation of the chapter.

REFERENCES

[1] Tichler R. Volkswirtschaftliche Relevanz von Power to Gas für das zukünftige Energiesystem. IEWT 2013, Erneuerbare Energien: überforderte Energiemärkte? Vienna; 2013.

[2] Reiter G. Power-to-gas. In: Garland N, Samsun RC, Stolten D, editors. Data, facts and figures on fuel cells. Wiley-Verlag; 2015.

[3] Tichler R, Lindorfer J, Friedl C, Reiter G, Steinmüller H. FTI-Roadmap Power-to-Gas für Österreich, nachhaltig wirtschaften. Vienna, Austria: Bundesministerium für Verkehr, Innovation und Technologie; 2014. p. 50

[4] Lehner M, Tichler R, Steinmüller H, Koppe M. Power-to-gas: technology and business models. Heidelberg, Germany: Springer Verlag; 2014.

[5] Steinmüller H, Tichler R, Reiter G et al. Power-to-Gas—eine Systemanalyse. Markt- und Technologiescouting und –analyse. Linz, Austria: Energieinstitut an der Johannes Kepler Universität Linz; 2014.

[6] Sterner M. Bioenergy and renewable power methane in integrated 100% renewable energy systems: limiting global warming by transforming energy systems. Dissertation, Kassel University, Germany; 2009.
[7] Specht M, Zuberbühler U, Baumgart F, Feigl B, Frick V, Stürmer B, Sterner M, Waldstein G. Storing renewable rnergy in the natural gas grid. Methane via power-to-gas (P2G): a renewable fuel for mobility. Proceedings of the 6th conference on gas powered vehicles—the real and economical CO_2 alternative, Stuttgart, Germany, 2011; October 26–27.
[8] Müller-Syring G, Henel M, Krause H, Rasmusson H, Mlaker H, Köppel W. et al. Power-to-gas: Entwicklung von Anlagenkonzepten im Rahmen der DVGW-Innovationsoffensive. Gas/Erdgas 2011; November. 770–777.
[9] Ursua A, Gandia LM, Sanchis P. P IEEE 2012; 100:410–426.
[10] Smolinka T, Günther M, Garche J. NOW-Studie "Stand und Entwicklungspotenzial der Wasserelektrolyse zur Herstellung von Wasserstoff aus regenerativen Energien" Kurzfassung des Abschlussberichts. Freiburg, Germany: Fraunhofer Institut für Solare Energiesysteme; 2011.
[11] Holladay JD, Hu J, King DL, Wang Y. Catal Today 2009;139:244–60.
[12] Carmo M, Fritz D, Mergel J, Stolten D. Int J Hydrogen Energy 2013;38:4901–34.
[13] Zeng K, Zhang D. Prog Energy Combust Sci 2010;36:307–26.
[14] Graves C, Ebbesen SD, Mogensen M, Lackner KS. Renew Sust Energy Rev 2011;15:1–23.
[15] Maclay JD, Brouwer J, Samuelsen GS. Int J Hydrogen Energy 2011;36:12130–40.
[16] Neij L. Energy Policy 2008;36:2200–11.
[17] Schoots K, Ferioli F, Kramer GJ, van der Zwaana BCC. Int J Hydrogen Energy 2008;33: 2630–45.
[18] IPPC. Carbon dioxide capture and storage. Cambridge, UK: Cambridge University Press; 2005.
[19] Li B, Duan Y, Luebke D, Morreale B. Advances in CO_2 capture technology: a patent review. Appl Energy 2013;102:1439–47.
[20] Rubin ES, Mantripragada H, Marks A, Versteeg P, Kitchin J. Prog Energy Combust Sci 2012;38:630–71.
[21] Trost T, Horn S, Jentsch M, Sterner M. Z Energiewirtsch 2012;36:173–90.
[22] Breyer C, Rieke S, Sterner M, Schmid J. Hybrid PV–wind–renewable methane power plants. EU PVSEC Proceedings European Photovoltaic Solar Energy Conference, Hamburg, Germany; 2011.
[23] Ryckebosch E, Drouillon M, Vervaeren H. Techniques for transformation of biogas to biomethane. Biomass Bioenerg 2011;35:1633–45.
[24] Gattrell M, Gupta N, Co A. Energy Convers Manage 2007;48:1255–65.
[25] Pearson RJ, Eisaman MD, Turner JWG, Edwards PP, Jiang Z, Kuznetsov VL, Littau KA, Di Marco L, Tylor SRG. P IEEE 2012; 100(2):440–460.
[26] Metz B, Davidson O, de Coninck H, Loos M, Meyer L. Carbon dioxide capture and storage, IPCC. Cambridge: Cambridge University Press; 2005.
[27] Olah G. J Org Chem 2009;74:487–98.
[28] Cover AE, Hubbard DA, Jain SK, Shah KV, Koneru PB, Wong EW. Review of selected shift and methanation processes for SNG production. Houston, TX: Kellogg Rust Synfuels Inc; 1985.
[29] Tichler R, Gahleitner G. Power to Gas—Speichertechnologie für das Energiesystem der Zukunft. Linz, Austria: Energieinstitut an der Johannes Kepler Universität Linz; 2012.
[30] Gahleitner G, Lindorfer J. Alternative fuels for mobility and transport: harnessing excess electricity from renewable power sources with power to gas. ECEEE summer study 2013. Toulon-Hyères, France: European Council for an Energy Efficient Economy; 2013.

[31] Reiter G, Tichler R, Steinmüller H et al.Wirtschaftlichkeit und Systemanalyse von Power-to-Gas-Konzepten. DVGW Technoökonomische Studie von Power to gas Konzepten. Bonn, Germany: Deutscher Verein des Gas; 2014.

[32] Hefner R III. The age of energy gases. China's opportunity for global energy leadership. Oklahoma City: The GHK Company; 2007.

[33] Goryl L, Lorenc V. Report Study Group 2.1: UGS database. IGU world gas conference, Paris, France; 2015.

[34] Specht M, Brellochs J, Frick V, Stürmer B, Zuberbühler U, Sterner M, Waldstein G. Erdöl Erdgas Kohle 2010;126:342–6.

[35] Altfeld K, Pinchbeck D. Admissible hydrogen concentrations in natural gas systems. Gas for energy 3/2013. Munich, Germany: DIV Deutscher Industrieverlag; 2013.

[36] Müller-Syring G, Henel M, Köppel W, Mlaker H, Sterner M, Höcher T. Abschlussbericht: Entwicklung von modularen Konzepten zur Erzeugung, Speicherung und Einspeisung von Wasserstoff und Methan ins Erdgasnetz. DVGW-Projekt G1-07-10. Bonn, Germany: Deutscher Verein des Gas; 2013.

[37] Reitenbach V, Albrecht D, Ganzer L. Influence of hydrogen on underground gas storage: literature study. DGMK-Project 752. Hamburg, Germany: Deutsche Wissenschaftliche Gesellschaft für Erdöl, Erdgas und Kohle; 2014.

[38] Wagner M, Ballerstedt H. Influence of bio-methane and hydrogen on the microbiology of underground gas storage: literature study. DGMK-Project 756. Hamburg, Germany: DGMK Deutsche Wissenschaftliche Gesellschaft für Erdöl, Erdgas und Kohle; 2013.

[39] Nadau L. Underground storage: literature survey of hydrogen and natural gas mixture behavior. GERG admissible hydrogen concentrations in natural gas. Belgium: GERG; 2013.

[40] Bauer S, Gubik A, Pichler M, Loibner A, Scherr K, Schritter J, Mori G, Vidic K, Tichler R. DVGW Energie Wasser-Praxis 2014;65:50–4.

[41] Bauer S, Pichler M. Underground sun storage: a study on properties of hydrogen admixture in porous underground-gas-storage facilities by means of an in-situ experiment. World Gas Conference, Paris, France; 2015.

Chapter 19

Traditional Bulk Energy Storage—Coal and Underground Natural Gas and Oil Storage

Fritz Crotogino
R&D Department, KBB Underground Technologies GmbH, Hannover, Germany

1 INTRODUCTION

The purpose of the chapter is to dispel the widely held belief that the need to store large volumes of energy to balance out production and demand is a special aspect which primarily only affects an energy system based on fluctuating wind and solar power. This opinion has been frequently used as an argument against the establishment of renewables as part of energy transition. According to the same lobbyists, there are hardly any suitable storage technologies available to hold large volumes of electrical energy.

This argument overlooks the fact that as the demand for energy rose exponentially as a result of industrialization in Europe and the United States, in particular, instead of being extracted in the vicinity of areas with the highest demand, fossil fuels in the form of coal, oil, and natural gas increasingly had to be transported long distances by rail, ship, or pipeline—leading to the knock-on effect of having to construct storage facilities close to the main centers of demand. The specific reasons for this included:

- The fact that coal, oil, and natural gas tend to be produced at a more or less steady rate, which is frequently out of step with varying demand at a range of timescales (such as seasonally dependent gas consumption for heating purposes).
- The transport chain is designed for continuous transport rates with as little fluctuation as possible to keep the enormous investment costs as low as possible, and so as not to have to design them to cope with demand peaks (e.g., oil pipelines several thousands of kilometres long from Siberia to Central Europe).
- The discontinuous demand from, for example, a power plant, which is dependent on time of day, week, day, and season.

Storing Energy. http://dx.doi.org/10.1016/B978-0-12-803440-8.00019-1

- Reserves to compensate for stoppages along the transport chain.
- Reserves for periods of unusually high demand, for example, during longer intense periods of frost.

If one then casts an eye on a future energy system relying on renewables, one sees many parallels with respect to the demand for storage space, including aspects such as the transport of wind power from coastal areas to inland regions, the general balancing out of production and demand, as well as the need for reserves to compensate for unforeseen technical or political events.

Another, and actually more important, purpose for the chapter is the fact that the crucial technologies to be used in future for grid-scale storage of electrical energy from renewables are largely based on technologies which have been tried and tested for several decades for the storage of liquid and gaseous fossil fuels in underground geological formations.

The timing of the production of primary energy sources and the actual demand for electricity rarely coincide perfectly for energy systems based on either conventional fossil fuels or tomorrow's renewable energy sources. Both require large storage capacities to balance out production and demand at the grid scale.

The major advantage of supplying power in an energy system based on fossil fuels is, however, that the primary energy sources are *chemically* bound up in coal, oil, and gas *before* being converted into electricity in a power plant. First, this means very high volumetric energy storage densities—a medium-sized car can drive up to 1000 km on one tank of petrol. Second, fossil fuels can easily be stored on the surface or below ground. This makes it possible to generate electrical energy in tune with demand, which reduces the need for power storage capacities. Actual storage is primarily provided here by the primary fossil fuels themselves. This is highlighted by the fact that Germany, for example, has stockpiles of natural gas which could satisfy demand for approximately 4 weeks, but only has storage capacities for less than one hour of electrical energy: a ratio of around 5000/1!

Therefore, while the previous energy system primarily depended on primary fossil fuels or energy sources, a future energy system largely based on wind and solar power[1] produces electrical power as the primary energy source. Outsourcing the storage function to fossil fuels is therefore ruled out in this case. This means that power first has to be converted into a storable medium to be able to store the large volumes of energy necessary (in the gigawatt hour to terawatt hour range)—if one excludes batteries which are primarily suitable for smaller capacities.

The following sections look at the surface storage of large volumes of coal, the storage of oil underground in artificially constructed salt caverns, and the storage of natural gas in various geological formations.

---

1. This naturally does not apply to solar- thermal power plants, which, however, only play a very minor role in the overall energy mix.

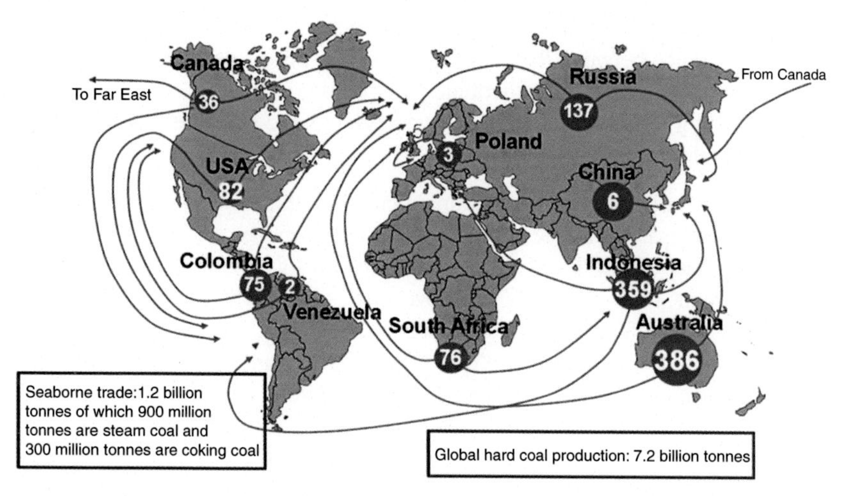

FIGURE 19.1 **Main trade flows in seaborne hard coal trade in 2014.** *(Source: VDKi. © Verein der Kohlenimporteure eV. [1])*

2 COAL

Today, coal forms the basis for one-quarter of global energy consumption and forms over one-third of power production. The most important coal-exporting countries are Indonesia, Australia, Russia, the United States, Colombia, South Africa, and Canada, while the most important importing countries are China, Japan, India, South Korea, Taiwan, Germany, and the United Kingdom (Fig. 19.1) [1]. This makes it clear that considerable effort is usually required to transport coal to far-flung centers of consumption, such as Europe, and that this must be maintained at the most continuous level possible to optimize the investment costs for rail and ship transport. Various storage facilities therefore have to be integrated along the transport chain from the mine producing the coal to the consumer—primarily power plants and the steel industry. The tasks typically undertaken by storage facilities include:

- Balancing out continuous coal production and discontinuous rail and long-distance ship transport to import terminals in distant sales markets.
- Balancing out discontinuous rail and inland ship transport from import terminals to power plants and other major consumers.
- Stockpiling at the sites of major consumers to bridge temporary stoppages along the transport chain to end–consumers, for example, complete stoppage of inland waterway transportation in extreme winters.
- Stockpiling at power plants to satisfy long-term above-average demand for coal-generated power during long periods of calm in countries with a high proportion of wind or solar power.

- Stockpiling at power plants, in general, to generate so-called residual load—the difference between the capacity of priority power generated by fluctuating wind turbines and photovoltaic plants and grid load.
- Stockpiling different qualities of coal held by large consumers to create optimal coal mixtures.

Large stockpiles of coal are complemented by the enormous total capacity of transport ships themselves. European import seaports, for example, Rotterdam, have coal stockpiles with a capacity of up to several million tonnes—energy content[2] of c. 8000 GW h for every 10^6 t)—which are needed for rapid handling and therefore are not used as reserve storages.

The capacity of stockpiles held at power plants typically range from a few tens of thousands of tonnes to up to a few hundreds of thousands of tonnes corresponding to (80–800) GW h. One of the largest power plants in Germany has coal stockpiles with a theoretical capacity of over 350 000 t (2 800 GW h): this amount would be adequate for a period of approximately 4 weeks during an emergency.

However, these figures are not very suitable for estimating the capacity of reserve stockpiles because differentiation is required between active and passive stockpiles. Active stockpiles are used for coal deliveries and supplying the power plant. Passive stockpiles alone can be used as long-term storage. Most locations do not have any passive stockpiles, and many types of coal are unsuitable for long-term storage because they have a tendency to self-ignite. Therefore, they represent a high level of risk to the environment, operations, and invested capital!

Different types of coal storage are used in practice:

- A heap stockpile is an open stockpile of solid fuel which is therefore exposed to the weather. This type of storage is very simple and easy to implement for large volumes
- Bunker and shed stockpile

The reason for using closed storage areas is usually to avoid polluting the environment with emissions of dust, and to prevent water contamination. Fire protection expenditure is much higher than in the case of open heaps. Potential capacities are of up to 10 000 t—in other words, much lower than open storage areas.

3 OIL

Oil is still the paramount energy source worldwide, ahead of coal and natural gas. In a similar way to the situation affecting coal, oil must in most cases be transported long distances, in this case by pipeline and/or ship from major oil fields to the main centers of consumption.

---

2. Conversion based on a caloric value of 8 kW h kg^{-1}.

FIGURE 19.2 **Oil tanker unloading terminal and tank farm of NWO in Wilhelmshaven (Germany).** *(© Nord-West Oelleitung GmbH (photo design Klaus Schreiber) [2].)*

Storage is therefore a very important element in the oil transport chain. Large storage systems are required for maritime transport at export terminals, import terminals (in particular), and where international pipelines interface with transport and distribution networks. However, the largest demand for storage is the holding of strategic reserves to guarantee supplies of oil during periods of crisis.

The most well-known features of oil storage facilities are the surface oil tanks shown in Fig. 19.2 (an aerial photograph of a tanker unloading) together with the terminal and tank farm at NWO, Wilhelmshaven (Germany) [2], which forms the interface between incoming tanker loads and long-distance pipelines. Twenty six tanks are available for interim storage, each holding around 30 000 m^3 (324 GW h[3]), as well as nine tanks with capacities of around 100 000 m^3 (1 080 GW h). The total capacity of the tank farm is 1.6×10^6 m^3 (17 300 GW h). The main purpose of this tank farm is to act as a buffer so that oil tankers can be unloaded within a few hours, before the oil is transported onward in long-distance pipelines or held for long-term storage in caverns, see later discussion.

Today's standard tank farms place high priority on protecting the environment and, in particular, the hydrosphere, for example, a containment wall around each tank which can hold back the oil in the unlikely event of a leak.

3. Conversions are based on a density of 900 kg m^{-3} and a caloric value of 12 kW h kg^{-1}.

However, the tanks can only be protected to a certain extent against external influences.

An alternative to surface tank farms are man-made cavities constructed in underground geological formations below the groundwater horizon. The surface facilities of these underground storage caverns consist largely of a pump station and the connecting pipelines. As a result of the storage being deep underground, there is far less risk to soil, drinking water, or groundwater contamination. Furthermore, underground storage has the advantage that in the event of technical malfunctions or terrorist attacks there is no surface storage plant or structure that can be damaged, destroyed, or result in dangerous leaks. Other advantages include much lower costs and a much smaller footprint.

The main underground storage options are artificially constructed salt caverns, followed to a lesser extent by rock caverns constructed by conventional mining, and in a very few individual cases old abandoned mines.

3.1 Salt Caverns

A special characteristic of salt caverns is the possibility of constructing the caverns from the surface by drilling a well down into the salt formation. Water is pumped down this well after installing cemented casing strings and temporary pipes. The water dissolves the rock salt to create the cavity with the required shape. The brine produced as a result of dissolution of the salt is displaced to the surface where it is disposed of in an environmentally friendly way, or used as a raw material. Fig. 19.3 is a schematic diagram of a completed oil cavern in operation. For a more detailed description of salt storage caverns see Section 4 on natural gas and salt caverns.

3.1.1 Operation of Oil Storage Salt Caverns

Two methods are available for the operation of caverns storing liquids:

- Using pumps in a similar way to a tank: a submersible pump is installed in a withdrawal pipe (string) above the cavern sump to remove the storage product. The cavity above the top of the product is filled with air. This method is rarely used in the case of salt caverns because having atmospheric pressure in an air-filled space is only possible at low depths because of the need to maintain the stability of the caverns.
- Using brine displacement (Fig. 19.3): the oil is stored in equilibrium with the brine from the former solution-mining process. Because brine is heavier than crude oil the oil sits above the brine and no mixing occurs. When brine is pumped into the cavern via the brine string, it displaces the crude oil through the annulus to the surface and then into the delivery pipelines. During filling the cavern oil is pumped through the annulus into the cavern, where it displaces brine via the brine string, which is sent to a surface brine storage reservoir and eventually reused in the storage and distribution process.

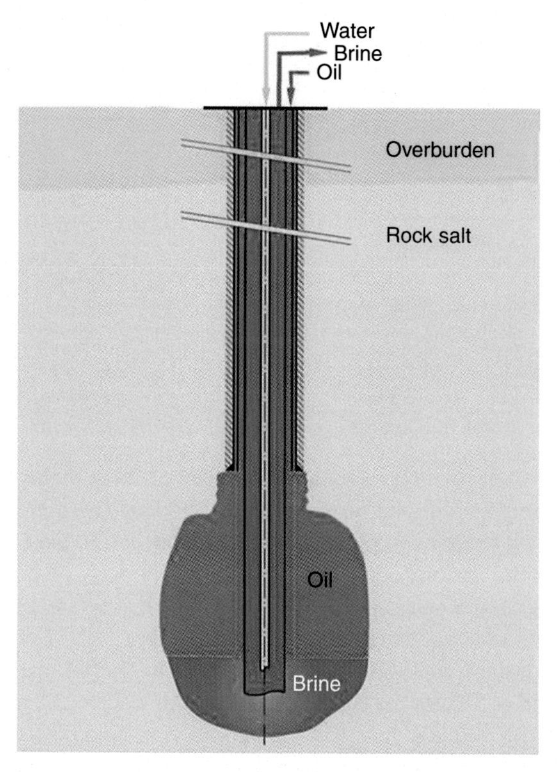

FIGURE 19.3 **Crude oil cavern with brine displacement.** *(© KBB Underground Technologies GmbH.)*

The disadvantage of this method is that if it is to be used regularly, a brine pond is required on the surface with the same volume as the cavern. In practice, brine buffer ponds are largely only used for liquid gas storage caverns because of the more frequent cycling of the operation.

In the case of strategic crude oil storage for emergency use, it is possible to use water instead of brine to displace the oil. However, the use of water increases the volume of a cavern by c. 15% every time the oil is removed. This is due to the solubility of the salt and is the reason the caverns in the German strategic reserve are only designed for a maximum of five cycles (Table 19.1).

3.1.2 Important Oil Storage Facilities Realized in Salt Caverns

In practice, commercial operating storage facilities for crude oil are generally built on the surface primarily because the construction of tank farms is not dependent on the availability of suitable salt formations and because the expense involved in planning, authorizing, and operating surface brine ponds is fairly considerable. One of the exceptions is the Louisiana Offshore Oil Project

TABLE 19.1 Typical Data for an Oil Storage Cavern in a North German Salt Dome

Average depth/m	1 400
Volume/m^3	400 000
Max. diameter/m	50
Height/m	600
Max. rate (in/out)/(m^3 h^{-1})	400
Energy content, calorific value/GW h	4 320
Max. capacity (in/out)/GW	4.3

(LOOP) in the vicinity of New Orleans, Louisiana (United States), which serves as the largest import terminal on the Gulf of Mexico. There are also two very large strategic oil reserves worldwide—in Germany and in the United States—using salt caverns and scheduled for use during periods of crisis. The main reasons for using the salt cavern option in these cases are the much lower investment costs and much higher security and safety.

In Germany the Crude Oil Storage Association (EBV) operates 4 coastal sites with a total of 58 caverns and a total volume of 12.5×10^6 m^3, corresponding to around 135 000 GW h or 135 TW h. This is around 60% of the reserves. The oil stored in these reserves is exclusively intended for maintaining supplies during a crisis and can therefore not be used for government price regulation purposes, for instance. The Netherlands and Portugal lease storage capacity for strategic reserves from other cavern operators [3,4].

The Strategic Petroleum Reserve (SPR) of the US government comprises a complex of four sites with deep underground storage caverns created in salt domes along the Texas and Louisiana Gulf coasts. The caverns have a capacity of 116×10^6 m^3 corresponding to around 1.3×10^6 GW h or 1,300 TW h and store emergency supplies of crude oil [5]. Unlike the conditions existing in Germany the US government can also use storage facilities for price regulation.

3.2 Rock Caverns

Conventionally mined rock caverns are underground cavities constructed by using conventional mining techniques (shaft sinking, excavation of cavities by blasting or cutting). Mined rock caverns can be constructed in a certain range of geological formations which need to allow for the construction and operation of large, long-term stable caverns. These formations were initially considered intrinsically impervious to water and hydrocarbons. Later designs in fractured rock used the head of water as a water curtain (hydrodynamic containment) to contain the hydrocarbons (see later).

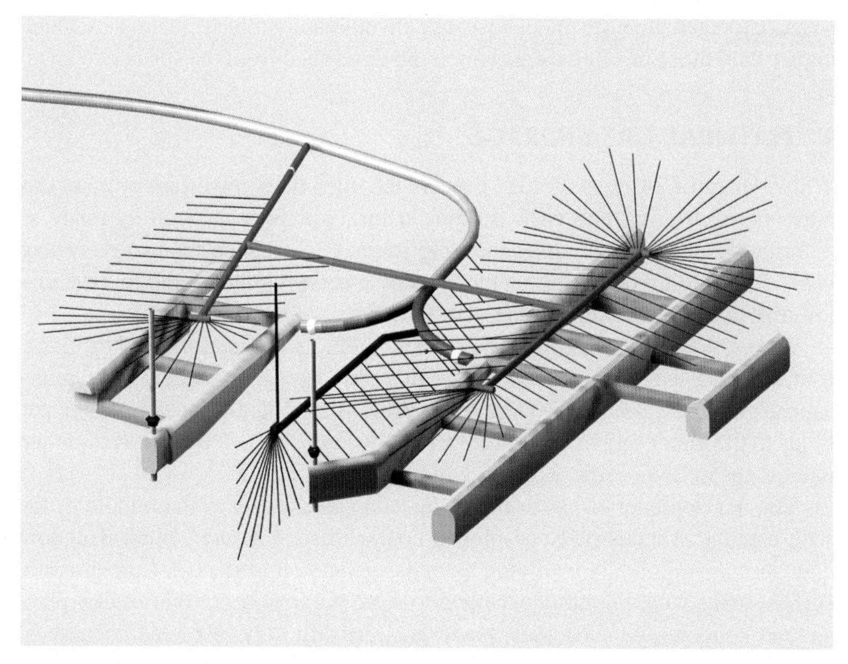

FIGURE 19.4 Conventionally mined hard rock cavern with a water curtain. *(© Geostock Entrepose.)*

Rock caverns have been developed primarily for the storage of liquid hydrocarbons like oil, gasoline, and liquid petroleum gases (Fig. 19.4). Most of these developments were started in the United States after World War II. Other countries such as the Scandinavian countries, France, and South Korea followed in the time between 1960 and the 1980s, and in the 1990s East Asian countries such as China, Japan, and India followed suit. All these countries have abundant suitable sound rock available. A recent example is Singapore's Jurong rock cavern project launched in 2014. The caverns are located at a depth of 130 m and are designed for liquid hydrocarbons such as crude oil, condensate, naphtha, and gas oil. The first phase comprised five 340 m long caverns with nine storage galleries providing 1.470×10^6 m^3 of storage, and 8 km of tunnels costing some US$$1.3 \times 10^9$ (US$1.3 billion). The second phase is planned to double this capacity [6].

Caverns using hydrodynamic containment are based on the principle that the requisite strong and competent rocks typically have fractures and fissures and are therefore not impervious to liquids (and gases). The fractures and fissures, on the other hand, lead to an inflow of water if groundwater is present, or if water is artificially provided and if the storage pressure is kept below the water column pressure. The pressurized water flow then prevents the escape of the storage product to the hydrosphere. The small amount of inflowing water is collected in the cavern sump, pumped to the surface, and disposed of after eliminating traces of the hydrocarbon product.

As expected the investment costs of conventionally mined caverns are much higher than those of salt caverns which are developed from the surface.

4 NATURAL GAS STORAGE

With a share of 24% (as of 2013), gas is the third most important primary energy source after oil and coal. Burning natural gas is considered by many as a "bridging technology" along the long journey to creating an energy system based largely on renewables—which is why gas is still considered to have major growth potential in the foreseeable future.

The world's largest natural gas producer by far is Russia, followed by Qatar, Iran, China, Saudi Arabia, and Algeria. In a similar way to coal and oil in many aspects, natural gas also has to be transported very long distances from the gas fields to the major centers of consumption—via pipelines on land or as liquefied natural gas in LNG carriers crossing the major oceans.

The development of gas infrastructures began as early as the middle of the 19th century in industrially developed countries such as the United Kingdom and Germany—initially involving town gas generated from coal. Low-pressure surface tanks were initially developed to store the town gas, followed by pressurized tanks whose capacities, however, were still very restricted because of the low operating pressures and their relatively small volumes.

Town gas was replaced by natural gas in the middle of the 20th century after the discovery of major fields in the Netherlands, Russia, and in the North Sea—particularly in the British and Norwegian sections. This transition changed the previously locally structured gas industry into a regional, national, and increasingly international industry based on an international network of interconnected pipelines with major integrated storage facilities.

The first experience with storing gas underground in depleted oil fields and, particularly, depleted gas fields as well as in water-bearing reservoirs (so-called aquifer formations) and in artificially constructed salt caverns had already been acquired during the town gas era. Deep underground geological formations are particularly attractive for the storage of gas because it can be stored at high pressures, unlike the large tanks on the surface which are limited to relatively low pressures. Another advantage of underground gas storage is that very large geometrical storage volumes can be created. The combination of high pressures and large volumes are the factors which enable gas storage facilities to achieve extremely high energy storage capacities, even though the density of the gases, even after compression, is relatively low compared with liquid or even solid fuels.

The operation of underground gas storage facilities usually involves injecting the gas into the storage cavern under pressure, and withdrawing it at a later date by releasing the pressure. This means that underground storage is not only possible in open cavities—such as artificially constructed caverns (see sec. 3.1)—but also in natural reservoirs consisting of a matrix of countless interconnected microscopic pores. A special feature of storage operations involving compression and pressure release is that operational injection and production pressures must stay within a range between a minimum and a maximum operating pressure. The storage capacity achievable within this range is called the working gas. Residual gas which remains in the storage below the minimum pressure is called the cushion gas. Depending on the storage technology the ratio of working gas to cushion gas ranges approximately from 2:1 in gas caverns to around 1:1 in pore storage sites. Expenditure for the cushion gas needs to be considered as part of the investment costs.

The following describes the most important storage options for natural gas in detail. Large-scale storage discussed in chapter: Larger Scale Hydrogen Storage describes the storage options available for hydrogen largely based on options developed over many decades for the storage of natural gas.

4.1 Depleted Oil and Gas Fields

Oil and gas fields are the product of long-dead residues of plants and small animals locked into rocks many millions of years ago and buried for long periods of time at great depths under high pressures and high temperatures. Because of the low permeability of earlier laid-down rocks compared with the overlying more permeable rocks, the hydrocarbon fluids generated later migrated upward. In some cases the hydrocarbons were prevented from rising up further by the presence of overlying impermeable rock layers (cap rock) which then contained the hydrocarbons in so-called traps. This gave rise to the existence of today's oil and gas fields (reservoirs) in the porous horizons beneath cap rocks (Fig. 19.5).

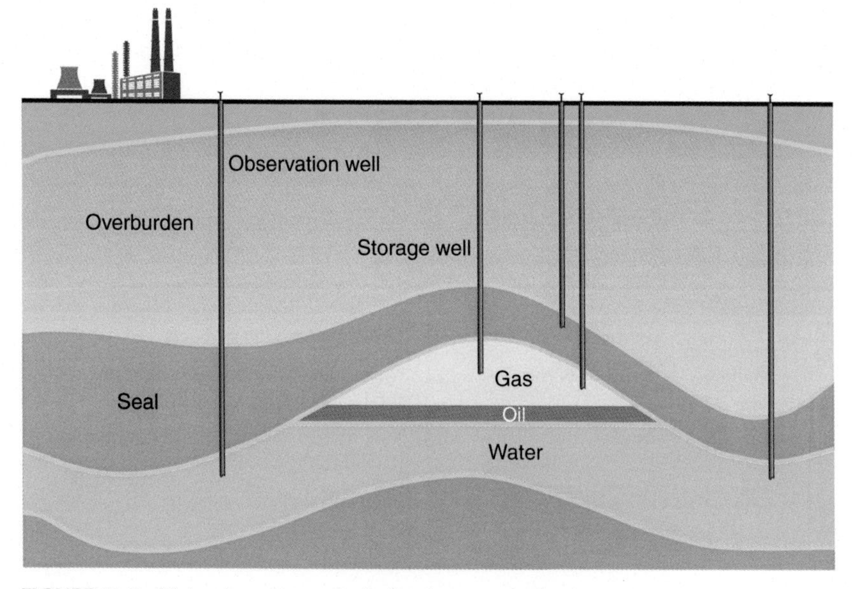

FIGURE 19.5 **Natural gas storage in depleted gas reservoir.**

Oil and gas does not naturally collect in large open cavities, but in a matrix of countless tiny interconnected pores within a "reservoir". Numerous wells normally have to be drilled to release the oil and gas in the reservoir discovered by exploration because flow resistance in the matrix of pores over excessively long distances in the field would overly restrict the production rates that could otherwise be achieved.

Oil and gas fields, which have been largely depleted, can be converted into natural gas storage spaces if certain conditions required for subsequent storage operations are satisfied. In some cases, existing wells can be utilized further for the injection of gas. The major benefit of depleted oil and gas fields is that the tightness of the reservoir has already been proven by its existence over geological time periods. Because of flow resistance in the pore matrix associated with the system, depleted oil and gas fields are primarily suitable for seasonal storage operations, that is, injection in the summer half of the year and withdrawal in the winter half of the year—and are therefore less suitable for flexible operations with frequent cycles and high production and injection rates. In some cases these disadvantages can be compensated for to a certain degree by using increasingly utilized horizontal drilling technology.

Although oil fields are in principle just as suitable for conversion to gas storages as a depleted gas field, gas fields tend to be used much more frequently in practice because the higher hydrocarbon fractions from the oil in a depleted oil field have negative effects on natural gas storage operations. Moreover, not every depleted oil and gas field can be converted into a gas storage facility. Crucial prerequisites are adequate permeability and porosity of the reservoir rock to facilitate the production and injection rates necessary for storage operations and suitable depths because excessive depths are associated with uneconomically high storage pressures (Table 19.2).

TABLE 19.2 Advantages and Disadvantages of Gas Storage in Depleted Oil and Gas Fields

Advantages	Disadvantages
Use of existing reservoir	Low deliverability per well
Excellent knowledge of reservoir (geology, performance)	Maximum (0.5–1)% daily withdrawal of total capacity
Low investment costs for conversion	Maximum 1–2 turnovers per annum low deliverability and injectivity
Possibility of reusing existing wells	High percentage of cushion gas

4.2 Aquifer Storage

The geological sequence of rocks in the Earth's crust consists of impermeable as well as porous, permeable formations. The porous, permeable formations are known as reservoirs, or when they are filled with groundwater (which is generally the case), as aquifer formations or simply aquifers.

In certain cases, such mostly saline groundwater-filled nondrinking water aquifers lying at great depths can also be used for the construction of gas storage facilities, by injecting gas into the formation via wells to displace the water laterally and downward. The starting position is more challenging than in the case of an oil or gas field because the tightness of the top of the reservoir has not been confirmed over long periods of time by the trapping of light hydrocarbons. The first condition that has to be met to verify the suitability of an aquifer for the construction of a gas storage facility is therefore confirmation that the overlying rock layer is impermeable to gas. Another requirement is the presence of a dome-shaped structure below the cap rock to enable subsequently injected gas to remain trapped within a large but controllable volume—in other words, the gas does not simply flow away through the sides in an uncontrolled manner. If these conditions are satisfied the reservoir still has to have adequate porosity and permeability to enable subsequent storage operations to be realized at the necessary injection and production rates.

Unlike the situation encountered in a depleted oil and gas field, which has already benefited from extensive exploration in the runup to the start of production operations, the geology of an aquifer, its structure and lithology, and its reservoir-engineering properties, are generally largely unknown. This means that a very extensive step-by-step exploration campaign is necessary to verify the suitability of the aquifer for the installation of a gas storage facility.

Exploration involves seismic methods to explore the structure of formations; several exploration wells to extract rock samples to determine the material parameters and to carry out pump tests to evaluate intrinsic reservoir properties; and then simulation of the storage operations based on these data. The exploration phase can take several years and is associated with the risk of ultimately not being able to confirm suitability.

If the exploration phase is completed successfully the next step is the drilling and completion of numerous wells because flow resistance in the pore matrix has to be overcome in just the same way as in an oil and gas field. If the distances between individual wells in the reservoir are too great, this would severely restrict achievable production and injection rates (Table 19.3).

We conclude this section by pointing out that a large number of aquifer storage facilities have in the past, been set up worldwide despite the aforementioned disadvantages. However, in certain areas, such as in central France, no suitable depleted reservoirs or salt formations are available. Finally, because of the costly exploration phase, as well as technical problems, only a few additional aquifer storage facilities have been set up in Europe over the past few years.

TABLE 19.3 Advantages and Disadvantages of Gas Storage in Aquifer Formations

Advantages	Disadvantages
Use of existing reservoirs	Need for extensive, costly, and time-consuming exploration phase
No hydrocarbon residues in reservoir, which might contaminate storage product	Low deliverability per well
	Maximum (0.5–1)% daily withdrawal of total capacity
	Maximum 1–2 turnovers per annum
	Low deliverability and injectivity
	High percentage of cushion gas

We conclude this section by pointing out that a large number of aquifer storage facilities have been set up worldwide despite the aforementioned disadvantages. This is due to the depleted oil and gas fields that are available and the fact that there are large regions with storage needs, such as in France, where there are no suitable salt formations, oil fields, or gas fields in most industrial regions. Because of the costly exploration phase, as well as technical problems, only a few additional aquifer storage facilities have been set up in Europe over the past few years.

4.3 Salt Caverns

Salt caverns are artificially constructed cavities in salt formations. The special properties of rock salt allow:

- cost-efficient construction of cavities with very large volumes by solution mining through a well drilled from the surface
- stable operations over long periods of time without any internal structural installations
- safe containment of liquids and gases under high pressure
- storage of very large volumes of liquids and gases at high pressures and, therefore, large amounts of energy.

Construction involves drilling a well deep into the rock salt with a diameter of $d < 1$ m (Fig. 19.6). Several pipes—so-called casings—are installed in this well in a telescope-like arrangement and bonded with cement to the surrounding rock to make the casings gas tight. Two additional casings are suspended in the well to construct the cavern during the solution-mining process. Water is injected into the inner pipe string to dissolve the salt. The brine which this produces is displaced to the surface through the inner annulus or reverse. A protective fluid (blanket) is injected into the outer annulus to prevent uncontrolled upward

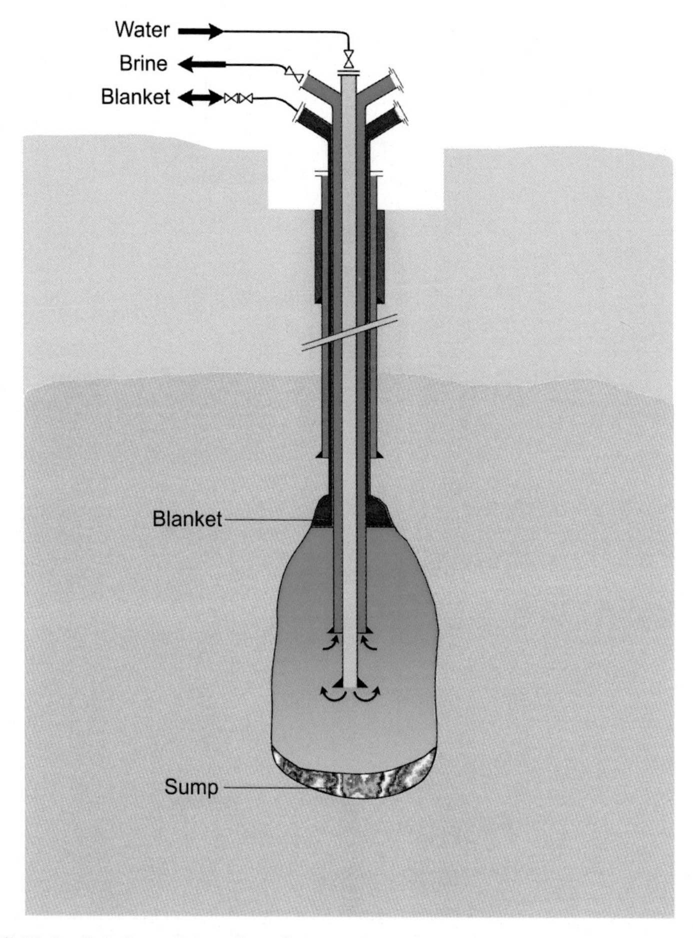

FIGURE 19.6 **Solution mining of a salt cavern.** *(© KBB Underground Technologieos GmbH.)*

solution of the salt formation. The brine produced by solution mining must be disposed of in an environmentally compatible way, for example, by pumping it into the sea. Alternatively, it can be used as a raw material for salt production or by the chemical industry.

Once the cavern has reached its final volume a test is carried out to confirm the tightness of the cemented casings. After a successful test the gas production string is then installed by sealing off the annulus from the inner cemented casing and filling it with a protective liquid (Fig. 19.7). In the unlikely event of a leak into this casing, this leaking gas would be immediately detected because this would cause the gas in the fluid to immediately rise upward. This crucial safety feature is supplemented by a subsurface safety valve which closes automatically if the cavern head at the surface becomes damaged. However, before this

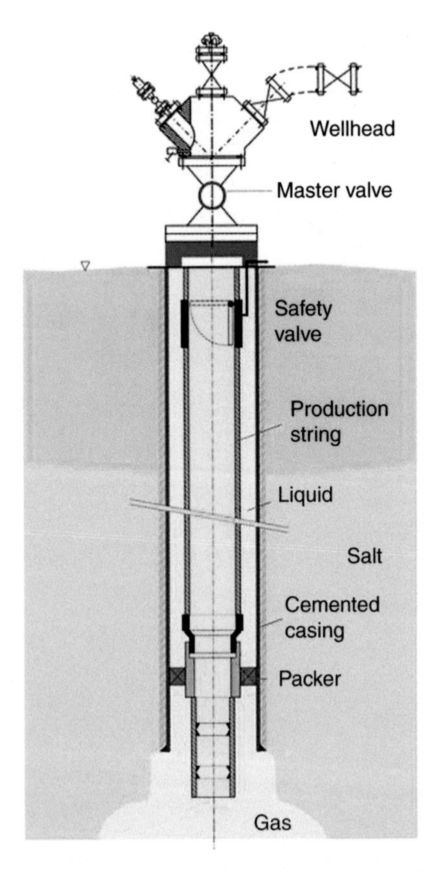

FIGURE 19.7 **Completion of a gas cavern.** *(© KBB Underground Technologies GmbH.)*

valve can be installed, the remaining brine in the cavern has to be displaced to the surface via a "brine displacement string" by the storage gas injected.

Salt caverns for storage have typical geometrical volumes from several 10^5 m^3 (100 000 m^3) to maximum 10^6 m^3, and maximum pressures of 200×10^5 Pa (200 bar). The minimum pressure is around one-third of this, which leads to a favorable working-gas-to-cushion-gas ratio of 2/1.

Salt caverns are particularly suitable for flexible operations with frequent cycles and high injection and production rates because no pressure losses occur within the essentially open storage volume, as would be the case in a rock matrix with pore storage. Rock salt is also impermeable to and does not react with conventional gases—however, a certain amount of saturation with water vapor must be expected from the residual brine remaining in the sump of the cavern.

These properties are particularly important for the storage of hydrogen (see chapter 20: Larger Scale Hydrogen Storage) which is very reactive, and which has to remain extremely pure, particularly for its future use as a fuel in hydrogen fuel cells.

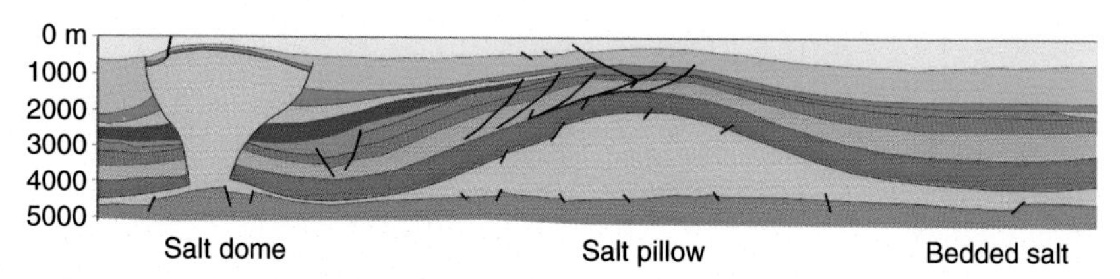

FIGURE 19.8 **Classification of salt formations.** *(© KBB Underground Technologies GmbH.)*

A suitable salt formation must be available to construct and operate storage caverns with economically viable volumes. Roughly speaking the following main types occur in nature: salt dome, salt pillow, and bedded salt (Fig. 19.8).

The first requirement for a suitable formation is adequate vertical thickness and lateral extent. Adequately thick salt horizons are also required above, below, and adjacent to each cavern to guarantee the stability and tightness of the rock salt surrounding the caverns. With respect to subsequent storage pressures the formation must also be adequately deep but not located at excessive depths. In addition, the salt formation must not contain too large a proportion of insoluble constituents because this could jeopardize the creation of cavities with adequate net volumes. And, finally, the suitability of a location largely also depends on the ability to utilize or to dispose of the brine generated by the solution-mining process: The creation of 1 m^3 of cavity generates c. 8 m^3 of brine; this means that around 4×10^6 m^3 of brine are generated when constructing a cavern with a typical volume of 5×10^5 m^3 (500 000 m^3)—this brine can be used by industry or disposed of in an environmentally compatible way, for example, into the sea or into deep saline aquifers (Table 19.4).

In Table 19.4, m^3(std) refers to standard cubic meter, which is defined here as the gas mass within a volume of 1 cubic metre under a pressure of

TABLE 19.4 Typical Gas Cavern Dimensions

V_{geom}/m^3	5×10^5	Geometrical volume
d_{roof}/m	1000	Roof depth
d_{sump}/m	1350	Cavern sump depth
p_{min}/MPa	6.0	Min. operating pressure
p_{max}/MPa	18.0	Max. operating pressure
m_{work}/kg	4.8×10^7	Working gas (mass)
m_{cush}/kg	2.5×10^7	Cushion gas (mass)
V_{work}/m^3 (std)	6.1×10^7	Working gas (standard volume)
V_{cush}/m^3 (std)	3.2×10^7	Cushion gas (standard volume)

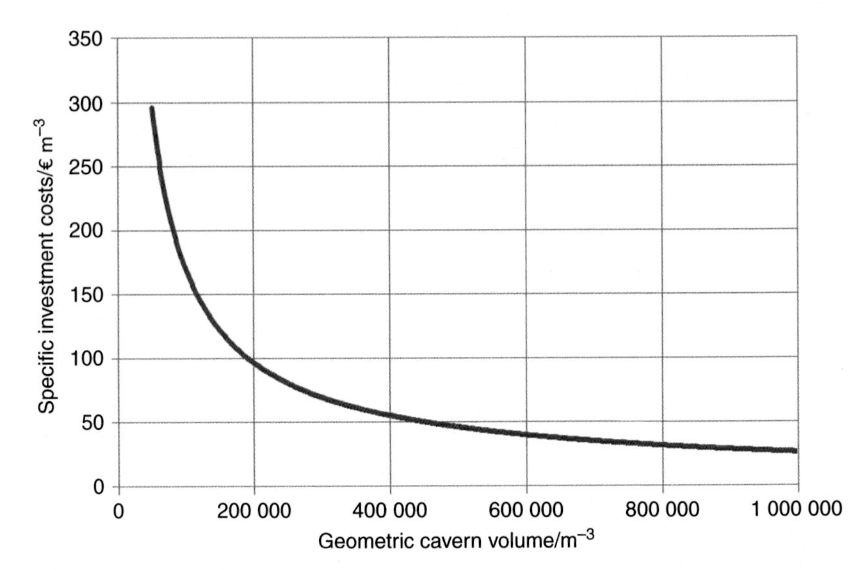

FIGURE 19.9 Specific gas cavern investment costs. *(© KBB UT.)*

1.013 × 10^5 Pa (1.013 bar) and at a temperature of 273.15 K. Standard conditions are subject to minor national differences; standard cubic meters are commonly used in the oil and gas industry.

The investment costs for storage caverns are typically associated with large starting costs for one-off investments in the necessary geological exploration and infrastructure of the well, and then relatively minor costs for the creation of the actual cavity. Fig. 19.9 gives an indication of the specific costs depending on the volume. The usual volumes of several 100 000 m^3 are associated with costs of around (30–50) € m^{-3} (geometrical); these costs also depend on whether the cavern is constructed on a greenfield site or a brownfield site, where caverns already exist. These costs do not include filling the caverns with the cushion gas.

The time required for constructing storage caverns can be divided up into the planning phase, the main approval phase, and the construction phase. The time required to construct the cavern itself largely depends on the solution-mining process, which in turn depends on the maximum possible water injection rate (Table 19.5).

When planning underground energy storage facilities in future, it is important to remember that the time required for realization can easily involve 10 years, even under favorable conditions. The key tasks of natural gas storage in salt caverns are:

- seasonal flexibility
- peak shaving
- strategic stock

TABLE 19.5 Advantages and Disadvantages of Gas Storage in Salt Caverns

Advantages	Disadvantages
High safety due to only one well per storage cavern	Need for exploration phase
Low geological risk	Several years construction time
High flexibility, maximum 10–12 turnovers per annum	Need to dispose large quantities of salt brine
High deliverability and injectivity/high rates	
Low percentage of cushion gas	
No reactions between storage gas and rock salt	

- balancing
- shift of gas flows
- asset-backed trading.

5 CONCLUSIONS

In the traditional world of fossil fuels, bulk storage in underground geological formations has played and still plays a vital role for many decades to minimize transport costs and balance out supply and demand. Future bulk storage for fluctuating wind and solar energy in the form of compressed air, hydrogen, or green methane storage will be largely based on technologies successfully developed for the storage of natural gas. This also applies to the various purposes for which storage facilities are built.

REFERENCES

[1] German Coal Importer Association (VDKi). http://english.kohlenimporteure.de/press-details/press-release-no-1-2015.html
[2] NWO. https://www.nwowhv.de/c/index.php/en/facilities/tank-farm
[3] LOOP. https://www.loopllc.com/home
[4] EBV. http://www.ebv-oil.org/cms/pdf/EBV_Informationsbroschuere_dba.pdf
[5] US Office of Fossil Energy. www.energy.gov
[6] Jurong International. http://www.jurong.com/index.php?option=com_portfolio&task=detail&type=17&Itemid=65

Chapter 20

Larger Scale Hydrogen Storage

Fritz Crotogino
R&D Department, KBB Underground Technologies GmbH, Hannover, Germany

1 HYDROGEN ECONOMY—FROM THE ORIGINAL IDEA TO TODAY'S CONCEPT

Hydrogen has long attracted a great deal of interest because of its outstanding properties, such as clean, CO_2-free combustion, and the ease with which it can be stored compared with electrical energy. It was already known in the 19th century that fossil fuels would eventually become exhausted, and that a potential solution lay in the use of the inexhaustible energy of the Sun, water and wind. In 1874, for instance, Jules Verne in *The Mysterious Island* went into raptures: "I believe that water will one day be employed as a fuel, that hydrogen and oxygen which constitute it, used singly or together, will furnish an inexhaustible source of heat and light, of an intensity of which coal is not capable. I believe, then, that when the deposits of coal are exhausted, we shall heat and warm ourselves with water. Water will be the coal of the future." The author does need to be corrected slightly, however, because although hydrogen is an excellent source of energy it is completely absent in nature as an energy source because of its high level of reactivity.

The term hydrogen economy was coined in the United States back in the 1970s to describe a system in which the well-known problems associated with the use of fossil fuels could be solved by using hydrogen as a secondary energy source, an energy carrier, and as a storage medium.

Today, in the age of the *Energiewende*,[1] the term hydrogen economy has dropped out of the headlines. We now tend to talk about an energy economy based in the future on renewables. However, hydrogen plays a very similar and much more specific role in this system as the ideal medium to integrate fluctuating wind and solar power, which is generated independent of demand, and particularly in terms of the transport of large capacities, and storage of large volumes of energy at the grid scale.

The initial discussion mainly considered the deployment of hydrogen for the purposes of storing excess electricity and turning it back into electricity when required, which was initially criticized because of the associated conversion

1. German expression for the transition from fossil and nuclear towards renewable energies.

Storing Energy. http://dx.doi.org/10.1016/B978-0-12-803440-8.00020-8

losses. The breakthrough for green hydrogen only arrived when the following aspects were recognized:

- The use of green hydrogen as a fuel in the mobility sector, as well as a raw material for the chemical industry, was an energetically much more efficient alternative to reconverting hydrogen into electricity.
- Storage capacities up to the terawatt-hour range required in future could not be realized using mechanical storage such as pumped hydro or compressed air energy storage (CAES) because of low volumetric energy storage densities, but could only be achieved using chemical storage based on hydrogen or green methane.

Hydrogen deployment gained another major boost against the background of energy transition and the power-to-gas concept initially advocated by the German gas industry: this involves the natural gas industry making its gas grid available to add green excess electricity in the form of hydrogen to natural gas, which ultimately means transporting green energy via existing pipelines and storing it in existing huge underground storage facilities. However, because the amount of hydrogen which can be mixed with natural gas is now known to have certain limits, scientists developed a supplementary alternative by synthesizing green hydrogen with carbon dioxide to form green methane. This gas can be mixed in any ratio with natural gas or even replace it in the long term.

In conclusion, power-to-gas can be considered a major constituent of the future energy system based on renewables. However, it is unclear today, what the future holds for the three green hydrogen options: (1) distribution and deployment of pure hydrogen in a separate hydrogen grid, including storage; (2) mixing hydrogen with natural gas; or (3) methanization followed by mixing with or even completely replacing natural gas. Fig. 20.1 [1] is a schematic

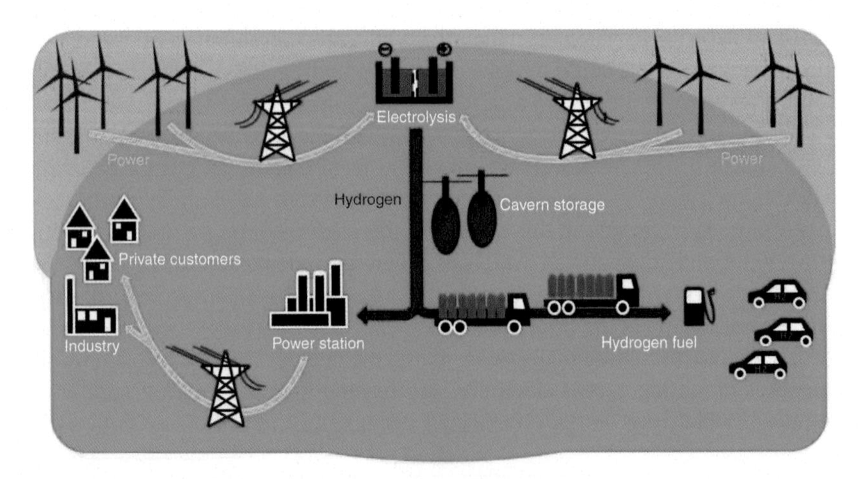

FIGURE 20.1 **Wind/solar–hydrogen system.** *(© NOW [1])*

diagram of current concepts for the future wind/solar–hydrogen system including storage.

2 WHY USE HYDROGEN STORAGE TO COMPENSATE FOR FLUCTUATING RENEWABLES?

2.1 Storage Demand at Various Timescales

There is a fundamental difference between the electricity grid and the gas grid: production and consumption in the electricity grid have to match at all times because the electricity grid itself has basically no storage capacity at all. The fact that today's electricity supply system—primarily based on fossil fuels—operates flexibly and reliably despite the varying demand at a range of timescales is primarily because of the enormous capacities of storage facilities for the primary fuels—coal, oil, and natural gas—which power plants can make use of at any time in accordance with demand levels.

In contrast, tomorrow's energy system will be mainly based on primary *renewable* energy sources and largely on fluctuating wind and solar power which is available in the form of electrical energy after being converted by wind turbines or solar panels. The resulting power production is primarily dependent on the weather, time of day, and time of year, and does not correlate at all with power consumption, which itself fluctuates according to daily, weekly, and seasonal timescales. With the exception of backup power plants using fossil fuels, balancing out the production of and the demand for green power in the existing electricity grid can only currently fall back on relatively low storage capacities, mainly in the form of pumped hydro power plants and batteries in the future. By way of comparison, Germany has storage capacities for natural gas corresponding to around 200 TW h, but only 0.04 TW h of pumped hydro capacity—a ratio of 5000/1! In other words, reserves of natural gas capable of satisfying demand for several weeks compared with a storage capacity for electricity which would be used up in less than an hour!

The differences between the amount of power available from renewables and the demand for electricity can be classified as follows in a strongly simplified way as far as power storage is concerned:

- Short-term deviations in the minutes to hour range, for example, deviations in actual power generation from the day-ahead forecast or day–night deviations.
- Medium-term deviations such as atypical weather conditions, for example, a period of calm lasting several days.
- Seasonal deviations resulting from annual differences in power generation from wind and solar power plants. Although wind and solar power theoretically compensate for each other when averaged out over the year the main period of generation is in the winter half.

2.2 Estimate of Future Storage Demand

Experts have long agreed that the transition to renewables will lead to considerable demand for storage. The real question though is how much storage is actually required. The question is not easy to answer because various options are available to balance out production and demand:

- Increasing the installed generation capacity of wind and solar power generators: to produce adequate electricity, even during periods when output from wind farms and solar energy plants is low, installations must be erected with a capacity which far exceeds demand during a full load period. The higher the generation capacity installed, the smaller the need for storage facilities.
- Upgrading the storage capacity and the input and output capacities of storage facilities: the higher the storage capacity, the lower the necessary installed generation capacity of wind and solar power modules.
- Availability of backup power plants burning fossil fuels: the lower the amount of reserve generation capacity, the higher the demand for storage capacity and performance.
- Grid upgrade: dependent on the actual amount of grid upgrading and the associated connection to remote regions, local production differences reflecting local weather can be balanced out, which results in lower demand for storage capacity.

Optimizing this complex system is not only an economic, but also a political issue. This particularly concerns the willingness to tolerate the use of backup fossil fuel power plants over a longer period of time. It is therefore not surprising that there is a wide difference between forecasts for the future storage demand for renewables (Fig. 20.2); the data presented are based on results published [2] and show the forecasts of various German institutes for Germany. The findings cover a huge range from (4–40) TW h. One reason for forecast variations is the different time horizons of the various studies (percentages refer to the proportion of renewables in the power generation mix).

One fact remains undisputed, however: storage demand forecast for an industrial country such as Germany with around 80 million inhabitants will be in the long term in the low terawatt-hour range independent of the specific study.

2.3 Which Storage Technologies Support Capacity in the High Gigawatt-Hour Range?

The following options are available in the foreseeable future for the storage of electrical energy at the grid scale:

- Pumped hydro power plants in which excess power is used to pump water from a lower to a higher pond, and vice versa.

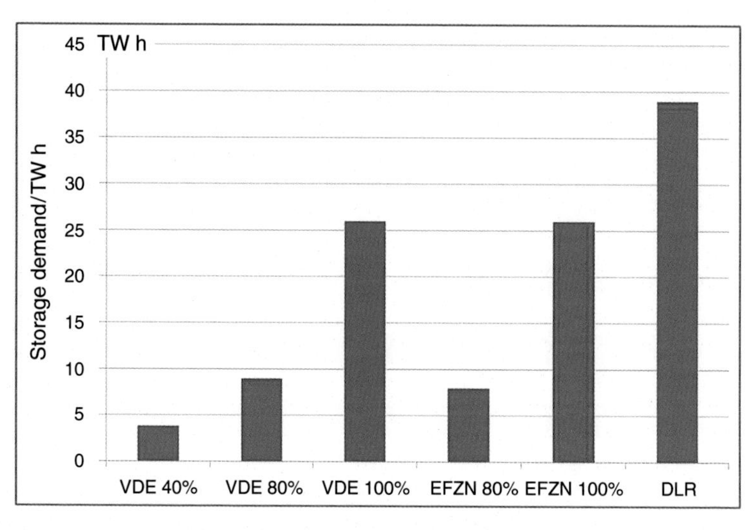

FIGURE 20.2 **Forecasts for future storage demand in Germany.**

- CAES power plants, in which excess electricity is used to compress air in underground storages. The compressed air is then used to drive a turbine connected to a generator to produce electricity on demand.
- Hydrogen storage: electricity is used here to electrolyze water and split it into hydrogen and oxygen. After compression the hydrogen can then be stored in a cavern, for instance. The hydrogen can then be used when needed to generate electricity in a gas power plant or directly used for fuel cell vehicles or chemical feedstock.
- Methane storage for methane synthesized by methanation from hydrogen and carbon dioxide—both from renewable sources. These storage facilities are basically identical to conventional natural gas storage facilities. A major reason for such storage is the understanding that final clarification is still required to determine whether pore storage is suitable for storing hydrogen underground—doubts currently exist because of the possibility of in situ reactions taking place. The solution might be the methanation of hydrogen.

Pumped hydro storage facilities and CAES power plants are based on the principle of *mechanical* storage, which is characterized by low volumetric storage capacities. Hydrogen and methane systems, on the other hand, are based on *chemical* storage. The volumetric energy storage capacity of chemical storages is around two orders of magnitude higher than that of mechanical storages, as shown in Fig. 20.3. Two values are given for methane and hydrogen: the *hv* value refers to the calorific value, and the *cc* value refers to the amount of energy produced after using the gas to create electricity in a combined cycle gas power plant at an assumed efficiency (η) of 60%.

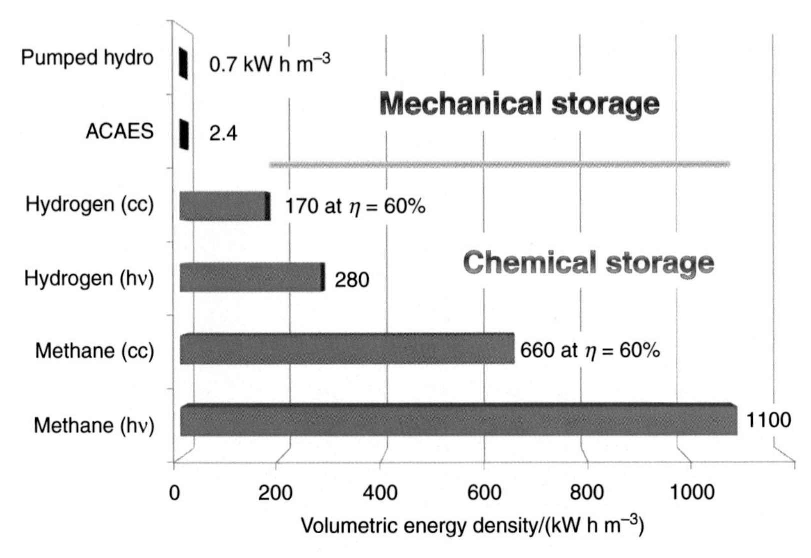

FIGURE 20.3 **Volumetric energy storage capacity.**

As just pointed out, in Fig. 20.3 *hv* refers to the calorific value, and *cc* to the amount of energy produced after burning the gas in a combined cycle gas power plant at an assumed efficiency (η) of 60% to create electricity. Estimated energy densities are based on gas storage pressures between (60–180) $\times 10^5$ Pa (60–180 bar) and the difference in levels for pumped hydro ponds of 300 m.

It is obvious that storage demand in the terawatt-hour range in real life can only be realized by the chemical storage of hydrogen or green methane. Fig. 20.4 is intended to provide an approximate overview of the capacity and discharge time of the various large-scale storage options. Discharge

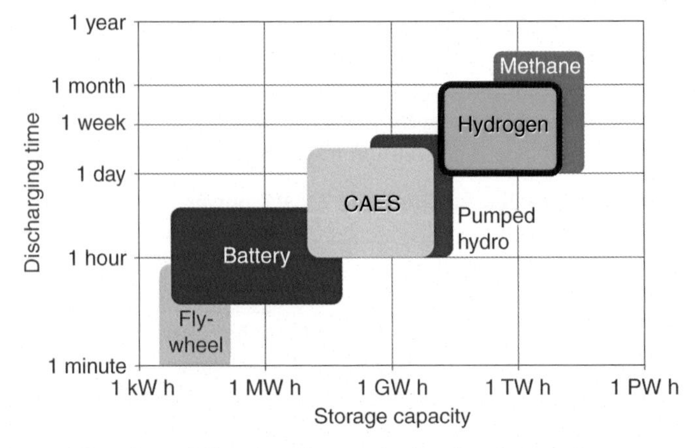

FIGURE 20.4 **Capacity and discharge time ranges of various large-scale storage options.**

time means the maximum duration of power production at maximum power output.

3 HYDROGEN IN THE CHEMICAL INDUSTRY

Although the renewable energy community has a very positive opinion of the future use of hydrogen, one should not overlook public fears with respect to the future deployment of hydrogen. It is therefore important in a chapter on the planned bulk storage of green hydrogen within the context of a wind/solar–hydrogen system to point out that the industry and competent authorities worldwide already have many decades of experience in the safe handling of large volumes of hydrogen.

Around 50×10^6 t a^{-1} of hydrogen is currently produced around the world, of which 2×10^6 t a^{-1} is produced in Germany [3].

The traditional hydrogen market comprises three main players:

- Merchant companies which trade hydrogen.
- Captive producers which produce hydrogen for their direct customers or their own use. Major uses are for oil refining, where hydrogen is used to hydrotreat crude oil as part of the refining process, producing ammonia for fertilizer, food production (e.g., hydrogenation), treating metals, etc.
- Chemical plants producing hydrogen as a byproduct resulting from processes like chlorine-alkali electrolysis and ethylene production.

In Europe (as of 2004), 80% of the total hydrogen was consumed by two main industrial sectors: refineries (50%) and the ammonia industry (32%), which are both captive users [4]. The lion's share of the hydrogen produced is processed at the production site. The remaining share is largely transported by pipeline. This mainly involves high-pressure pipelines, of which there are c. 2000 km in the United States (as at 2006) and over 300 km in two pipeline grids in Germany.

As for industrially used hydrogen, this is primarily extracted from natural gas by reforming. Fig. 20.5 shows a steam reformer plant operated by Linde AG.

Large chemical engineering facilities with sophisticated safety systems have been developed and installed to produce the aforementioned volumes of hydrogen. Large storage capacities have been successfully realized in underground salt caverns for many years in the United Kingdom and the United States. These do not differ in any significant way from storage facilities for high-pressure natural gas (see Section 5.3). One great advantage of using hydrogen for storing energy is that the safe and economical handling of large volumes of hydrogen is standard engineering practice worldwide. So, when hydrogen is used in future as a secondary energy carrier in an energy system largely based on renewables, use can be made of this comprehensive experience built up over many decades—with a well-developed set of codes and standards governing production, storage, and deployment.

FIGURE 20.5 **Steam reformer plant.** *(© Linde AG.)*

4 OPTIONS FOR LARGE-SCALE UNDERGROUND GAS STORAGE

4.1 Overview

The storage of hydrogen in geological formations does not differ significantly from the well-established, widespread technologies implemented for the storage of natural gas (see chapter: Traditional Bulk Energy Storage—Coal and Underground Natural Gas and Oil Storage) and is therefore largely based on the many years of experience acquired from these operations. The following sections therefore only recapitulate the conventional technologies used for the storage of natural gas with a focus on large-scale storage in geological formations. A detailed discussion of the specific technical and geotechnical issues related to hydrogen storage follows in Section 5, including a description of concrete experience acquired to date.

For the time being, natural gas will remain one of the main sources of energy. Annual consumption of around 2.6×10^{12} kg or 3.3×10^{12} m^3(std) in 2012 [5] corresponds to an average value of 3800 GW.[2,3] The main centers of

---

2. m^3(std) refers to standard cubic meter, which is definded as the gas content of 1 m^3 at standard pressure (1.00 atm, 1.013 bar) and standard temperature (273.15 K); the exact numbers for standard conditions are subject to minor national differences; standard cubic meters are common in the oil and gas industry.

3. Conversion based on caloric value of 10 kW h m^{-3}.

gas production are concentrated in only a few regions around the world and are often located long distances from the centers of consumption and often situated in politically unstable countries. The international gas industry therefore had to tackle the following aspects right from the beginning:

- balancing out primarily continuous gas production and seasonally fluctuating demand
- creating reserves to bridge supply shutdowns for technical or political reasons
- establishing reserves for extremely cold periods.

The new aspect of gas trading was added after deregulation of the energy markets. The gas industry has invested large amounts of money in this aspect in recent decades for the construction of a suitable gas grid, including large-scale storage facilities. In Europe alone, there are approximately 2.2×10^6 km of natural gas pipelines, of which c. 2×10^5 km are high-pressure pipelines. There are also 150 underground storage facilities with a working gas capacity of approximately 97×10^9 m^3(std) or 970 TW h [6].

It became clear early on for many reasons that very large quantities of gas would be better stored under the ground in geological formations rather than on the surface in pressurized tanks. Only storage facilities located deep underground, and covered by several hundred meters of rock, allow the use of high pressures up to 2×10^7 Pa (200 bar) and above as well as allowing large storage volumes and therefore enormous storage capacities, accompanied by relatively low specific investment costs. Fig. 20.6 visualizes a range of storage options in underground geological formations.

Early on, use was largely made of depleted hydrocarbon reservoirs, gas, and sometimes also oil fields, which were converted into gas storage facilities by injecting gas into the reservoirs when required and withdrawing it again later when needed. The basic suitability of such storage facilities had already been verified by the existence of oil and/or gas fields under high pressure over geological time periods before production began.

Sometime later, use was also made of natural water-bearing horizons (aquifers) for the storage of natural gas, which made operators independent of locations only associated with hydrocarbon deposits. The gas pumped into the aquifers displaces the water and the gas can be withdrawn again whenever required.

The third major large-scale storage option, which was mainly developed in the last 40 years in Europe and North America, is man-made salt caverns. These caverns are constructed from the surface by injecting water into the rock salt via a borehole to leach out a cavity by solution mining.

These three dominant concepts are supplemented by the subsequent use of abandoned mines and rock caverns excavated using conventional mining techniques. Some abandoned mines have very large open underground workings. The crucial prerequisite for their subsequent use as storage for compressed gas is of course the tightness of the surrounding rock: in practice, only very few rock

FIGURE 20.6 **Large-scale storage options for natural gas.** *(© KBB UT.)*

types are geologically suitable for gases, and these are frequently interbedded in underground workings with other less impervious rock types. Moreover, mines are generally not designed from the start to cope with the stresses involved in the cyclic pressure changes associated with the operation of large-scale gas storage. The author is only aware of one project presently in operation in which a very small former salt mine was converted into a natural gas storage; this is the Burggraf-Bernsdorf storage facility in Germany.

An interesting concept successfully tested in a pilot project in Sweden was the construction of a rock cavern in stable hard rock, in this case granite, which was lined with a gas-tight stainless steel skin [7]. In other words, the surrounding rock mass is used to create a stable cavity which can withstand the forces generated by the excavation. The actual barrier which keeps the gas in place, however, is the steel lining. This concept could possibly be further developed for the storage of hydrogen, and would then make possible gas storage in regions where geological conditions do not allow the construction of storage facilities in depleted oil or gas fields, aquifers, or rock salt formations.

In all storage options the storage process almost exclusively involves the compression of gas up to a maximum pressure. Withdrawal is then allowed down to a minimum permissible pressure. The capacity available in storage operations between these maximum and minimum pressures is called the working gas, while the volume which cannot be extracted below the minimum pressure is called the cushion gas. One of the quality criteria for underground storage is therefore the working-gas-to-cushion-gas capacity ratio, which generally varies between 2/1 for caverns and 1/1 for reservoirs. This is particularly true for hydrogen, since cushion gas is a major item of investment costs.

The key advantages of underground compared to aboveground gas storage are:

- very high intrinsic safety and security against external influences because of the deep location below the surface
- enormous storage capacities thanks to the combination of large geometrical volumes and high pressures
- small surface footprints
- minor effects on the environment
- low specific investment costs compared with surface gas storage.

4.2 Depleted Oil and Gas Fields

Most of the gas storage facilities around the world have been constructed in depleted gas fields (Fig. 20.7), and to a minor extent also in depleted oil fields. The main advantage of this option is that natural storage space already exists, the imperviousness of the storage formation has been confirmed by the trapping of oil and gas over geological timescales due to a sealing formation, and a great

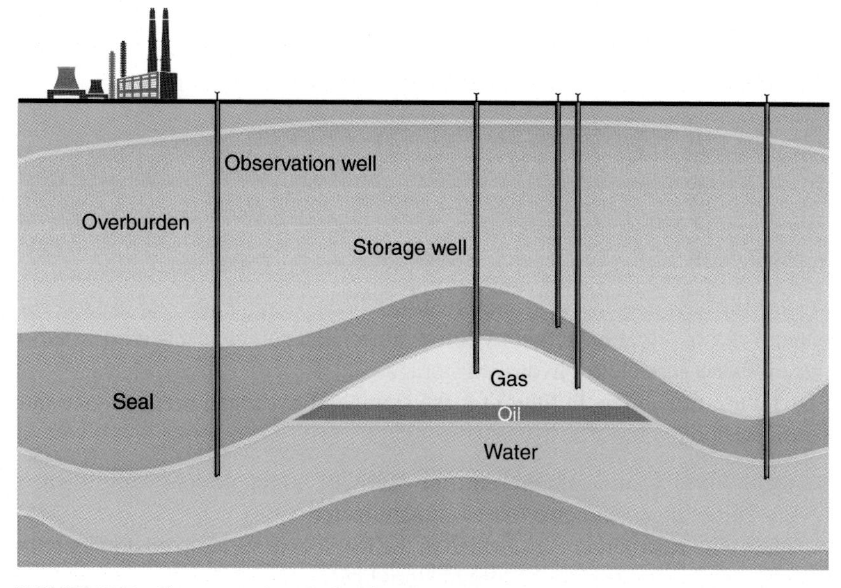

FIGURE 20.7 **Gas storage in a depleted gas reservoir.**

deal of information on the underground reservoir properties has been gathered during exploration and production phases.

Because oil and gas are located in a fine matrix of pores—unlike one large open cavity in a salt cavern—the fluid moving through the storage site has to overcome large flow resistances. Pore storage is therefore most suitable for seasonal applications, with only a few cycles at relatively low production and injection rates. Another characteristic is the relatively high proportion of cushion gas.

The use of depleted oil and gas fields in future for the storage of hydrogen needs to take into consideration the fact that hydrogen could:

- become contaminated with hydrocarbon residues
- react with the minerals present, for example, sulfur, which would deplete stored hydrogen and give rise to contamination
- react with microorganisms, which in the worst case scenario, could block the pore spaces or lead to the loss of stored gas.

Section 5 contains a detailed description of previous experience with the storage of hydrogen as well as a current discussion concerning the future use of depleted oil and gas fields for hydrogen storage.

4.3 Aquifer Storage

Aquifer storage is similar in many ways to a depleted hydrocarbon reservoir, in that the hydrocarbon reservoir shown in Fig. 20.7 was in most cases originally a water-filled reservoir within which hydrocarbons accumulated at a later date after migrating upward from lower horizons. The main difference in using an aquifer formation for storage is that the basic suitability of each individual aquifer still has to be confirmed. Requisite conditions include the following:

- that the top of the aquifer is sealed by an impermeable layer of rock
- the presence of a dome-shaped structure to hold the injected gas in a defined space
- high permeability of the aquifer with adequate pore space.

Establishing these facts usually requires a considerable amount of time and expense for exploration activities before a decision can be reached on whether the aquifer is suitable for hydrogen storage.

The use of aquifers in future for the storage of hydrogen needs to take into consideration the fact that hydrogen could:

- react with the minerals present, for example, sulfur, which would deplete stored hydrogen and give rise to contamination
- react with microorganisms, which in the worst case scenario could block the pore spaces and lead to the loss of stored gas.

An advantage of aquifers over depleted hydrocarbon fields is the fact that there is no risk of contaminating the hydrogen with hydrocarbon residues. The reason numerous aquifer gas storage facilities have been developed worldwide, despite the many disadvantages compared with depleted oil and gas fields, is that suitable oil and gas fields are not present in many regions.

As already pointed out, Section 5 contains a detailed description of previous experience with the storage of hydrogen as well as a current discussion concerning the future use of aquifer formations for hydrogen storage.

4.4 Salt Caverns

Unlike depleted hydrocarbon reservoirs and aquifer formations, salt caverns are artificially constructed cavities in deep-lying salt formations (Fig. 20.6). They are constructed by drilling a well into the salt and then pumping in water to dissolve the salt and create a cavity. Rock salt has a number of very favorable properties which make it suitable for constructing gas caverns:

- The cavity—the cavern—can be constructed from the surface. This means avoiding the high costs of sinking an access mine shaft and for underground tunneling work, as well as the associated costs for equipment and labor.
- Rock salt is impervious to gases and liquids as long as the internal pressure is lower than the formation pressure. Gas caverns are therefore operated at a maximum operating pressure of around 80% of the initial formation pressure. Unlike a surface pressure tank a hydrogen cavern could therefore never explode!
- Rock salt enables using very large, unclad cavities which will be stable for very long periods of time. In other words, unlike the conditions existing in other rock types, no engineering measures are required to stabilize the cavity.
- Rock salt is inert to storage media—a crucial aspect for the storage of hydrogen, in particular.

The main disadvantage, however, is the need to dispose of large volumes of saturated brine in an environmentally friendly way.

Before constructing a salt cavern the salt formation first needs to be investigated to assess its suitability. In general, the amount of additional exploration activity required is fairly minor because many salt structures have already been known for a long time from past exploration measures for salt extraction and, particularly, as a result of oil and gas exploration.

4.5 Comparison of Storage Options

Table 20.1 lists the key properties of the main underground gas storage options.

TABLE 20.1 Compilation of Key Properties of Main Underground Gas Storage Options

	Depleted gas reservoir	Aquifer formation	Salt cavern
Use of existing underground space	Yes	No	In case of former brine production caverns
Exploration effort	Low	High	Medium
Annual turnover frequency	Low (1–2)	Low (1–2)	High (10–12)
Deliverability/Injectivity	Low	Low	High
Ratio working/cushion gas capacity	1/1	1/1	2/1
Positive experience with hydrogen storage	No	No	Yes

5 UNDERGROUND HYDROGEN STORAGE IN DETAIL

5.1 Special Criteria for Storage Operations with a Focus on Fluctuating Injection or Withdrawal

Although the strategy in future is for most of the power generated by wind turbines and solar panels to be used directly—to keep the losses associated with conversion as low as possible —some of the power which cannot be accommodated by the grids at a particular time, or for which only very low prices can be achieved, should be stored temporarily. For a wind/solar–hydrogen system this unfortunately means that it is the irregular portion of the power production which has to be fed into the electrolyzer to split water into oxygen and hydrogen. A knock-on effect is that strongly fluctuating power input into the electrolyzer gives rise to similarly fluctuating hydrogen output and thus input to the storage.

The mode of operation which is prone to frequent and strong changes in flow rates is a serious challenge, not only for electrolyzers, but also:

- for compressors (taking into account safety measures) which inject hydrogen into the storage
- for the short-term and long-term stability of the storage host rock because the fluctuation in flow rates causes high pressurization and depressurization rates, as well as pressure fluctuation stresses.

It is forecast, however, that at some point in the future the demand for green hydrogen, for example, as a fuel for mobility or raw material for industry, will no longer be covered by excess wind or solar energy alone. This will then give rise to more uniform input for electrolyzers and caverns.

5.2 Standard Engineering Practice

5.2.1 Experience from the Underground Storage of Town Gas

Town gas, the predecessor of natural gas, was used as a fuel in many applications up until the 1970s. This gas, which is produced from coal, contains methane, nitrogen, and carbon monoxide, but also up to 50% hydrogen [8,9].

Engineers in the town gas era also had to cope with the problem of balancing out fluctuations between production and demand. When the capacities of gas tanks installed on the surface were no longer adequate, town gas with a high proportion of hydrogen was then stored in depleted gas fields and oil fields, as well as in aquifer formations. This took place in Germany in three depleted oil or gas fields and eight aquifers. Town gas was also stored in aquifers in other countries like France. In addition, town gas was stored in the first gas cavern ever constructed in Germany as well as in an abandoned salt mine.

The many years of experience with the underground storage of town gas with a high proportion of hydrogen provide an important source today for practical experience, which is of great significance for the future deployment of the various storage options available for pure hydrogen. Care must be taken here to take into consideration the specific properties of all of the components of town gas, as well as their interactions, when analyzing the cause of any problems which may have arisen at the time.

Hydrogen sulfide (H_2S) accumulated in stored gas as a result of the hydrolysis of carbonyl sulfide from town gas, the reduction of sulfate by bacteria, or released directly from the storage rock. This caused problems in meters installed on the surface. In the case of hydrocarbon gas fields the injected gas mixed with hydrocarbon residues in the depleted fields and was not particularly problematic because town gas already contained a significant amount of methane.

Unlike depleted oil and gas fields the imperviousness of the cover rock to gas has to be verified at aquifer storage sites before commissioning a gas storage facility. In one case, however, impervious problems resulted in subsequent shutdown despite positive testing prior to commissioning. Gas volume losses also occurred in aquifer storage, in particular, as a result of gas solution, phase dispersion, and biochemical alteration [8].

5.2.2 Standard Engineering Practice for Hydrogen Caverns

Practical experience in the construction and operation of hydrogen caverns has been gathered over a period of c. 40 years in the (petro) chemical industry which makes use of these caverns. These include:

- three small, shallow caverns in Teesside (United Kingdom)
- three large, deep caverns at various locations in the Houston, Texas, area in the United States.

Published data on these caverns is listed in Table 20.2. The three small caverns on Teesside (United Kingdom) act as a buffer in the shared network of several chemical and petrochemical producers and consumers. The caverns have

TABLE 20.2 Metrics of Hydrogen Caverns in the United States and the United Kingdom [10]

	Teesside (UK)	Clemens Dome, Texas (USA)	Moss Bluff, Texas (USA)	Spindletop, Texas (USA)
Salt formation	Bedded salt	Salt dome	Salt dome	Salt dome
Operator	Sabic Petrochem.	Chevron Phillips Chemical Comp.	Praxair	Air Liquide
Commissioned	1972	1986	2007	information not available
Geometrical volume/m^3	210 000	580 000	566 000	906 000
Mean cavern depth/m	365	1 000	1 200	1 340
Pressure range/10^5 Pa (bar)	45	70–137	55–152	68–202
Net energy stored/GW h	27	81	123	274
Amount of H_2/t	810	2 400	3 690	8 230
Net volume/m^3 (std)	9.12×10^6	27.3×10^6	41.5×10^6	92.6×10^6

fairly small dimensions and lie in just a 50 m thick salt layer. Unlike the usual practice of operating gas caverns by compressing or decompressing the gas, these three caverns are operated at constant pressure by displacing the gas with brine held in surface ponds. The hydrogen stored in the caverns is generated by reforming natural gas.

The caverns in the United States were also constructed for the petrochemical industry. The caverns are used to cover operational shutdowns in generating or consuming installations and, therefore, insure smooth production operations. The caverns are connected to a hydrogen pipeline grid with a total length of several hundred kilometers.

The dimensioning and shape of the US hydrogen caverns basically corresponds to that of modern natural gas caverns. However, because certain safety components are not used, such as an additional production tubing string or a subsurface safety valve, they would not be authorized if built in this way in Europe. To be constructed in Europe, they would have to be adapted to the usual specifications stipulated for the storage of natural gas.

Published data gained in performing integrity tests and impervious (tightness) tests in one of the US hydrogen caverns showed values comparable with those of state-of-the-art natural gas caverns.

5.3 Dimensioning and Operational Metrics of Future Hydrogen Storage Caverns

Table 20.3 summarizes the data for a future hydrogen cavern matching today's standard engineering practice in the natural gas cavern industry. Figures for storage capacity and rates are still approximate because the actual figures will depend on a range of parameters, especially the mode of operation and, thus, gas temperatures.

5.4 Outlook

The fact that hydrogen caverns have been successfully operated for many years in the United Kingdom and the United States does not necessarily mean that the technologies used for the planning, authorization, and operation of future green hydrogen caverns can be taken over without any significant additional research and development. This is because of the much higher engineering specifications currently in place for the construction of new gas

TABLE 20.3 Data for Model Hydrogen Cavern

Geometrical volume /m^3	500 000
Average depth/m	1 200
Production string diameter/m	0.20
Minimum operating pressure/10^5 Pa (bar)	60
Maimum operating pressure/10^5 Pa (bar)	180
Cushion gas amount/m^3 (std)	29×10^6
Cushion gas mass/kg	2.6×10^6
Cushion gas energy/GW h	87
Working gas amount/m^3 (std)	46×10^6
Working gas mass/kg	4.1×10^6
Working gas energy/GW h	141
Max. injection + withdrawal volume rate/m^3(std) h^{-1}	160 000
Max. injection + withdrawal mass rate/kg h^{-1}	14 135
Max. injection + withdrawal energy rate/MW	480

caverns in Europe as well of increasing opposition by the public to the introduction of new technologies when used in underground engineering projects, cf. carbon capture and storage and fracking.

National and European institutions have launched major subsidy programs to kick-start the technology because of the future importance of bulk hydrogen storage in salt caverns for successful implementation of energy transition. Insuring public acceptance also plays an important part in this strategy.

After the successful conclusion of several desktop studies in recent years on the geological, technical, and economic aspects of future wind–hydrogen systems involving underground storage, new research and development projects are focused on testing the technology in practice in a pilot plant. The aim is to plan, construct, and operate an installation in the megawatt range, which is supplied with wind power and generates hydrogen with the help of an electrolyzer. This hydrogen will then be compressed and injected into a small cavern with a volume of, say, 10^4 m^3. To insure that the simulation is as realistic as possible the provision of green power will be linked to prices in the energy market and will also take into consideration seasonal differences. The operation of such a pilot plant will provide information on the following aspects, in particular:

- the choice of materials and their successful use in practice
- the extremely fluctuating operation of plant components including the cavern itself
- dimensioning the plant with respect to extremely high safety levels
- verifying safety during operation
- investment and operation costs as well as hydrogen generation costs.

REFERENCES

[1] NOW (German National Organization of Hydrogen and Fuel-cell Technology). Integration von Wind-Wasserstoff-Systemen in das Energiesystem (Integration of wind-hydrogen systems in the energy system). Berlin, Germany: NOW GmbH; 2014.

[2] Fichtner Group. Erstellung eines Entwicklungskonzeptes Energiespeicher Niedersachsen (Preparation of a development concept for energy storages for the state of Lower Saxony). Hannover, Germany: Innovationszentrum Niedersachsen; 2014. http://www.energiespeicher-nds.de/fileadmin/Studien/Fichtner_Studie.PDF

[3] H2ydro guide, the hydrogen-guide. http://www.hydrogeit.de/wasserstoff.htm

[4] Roads2HyCom. European Hydrogen Infrastructure Atlas.

[5] BP statistical review of world energy, Jun 2014. bp.com/statisticalreview

[6] Eurogas statistical report 2005 quoted after Gert Müller-Syring: *Wasserstoff im Erdgasnetz/Migrationspfade für Biowasserstoff* (Hydrogen as part of the natural gas grid/migration paths for bio hydrogen), lecture, Leipzig, 2010

[7] Tengborg P. Storage of highly compressed gases in underground lined rock caverns—more than 10 years of experiences. Proceedings of the world tunnel congress, Foz do Iguaçu, Brazil; 2014.

[8] Schmitz S. Einfluss von Wasserstoff als Gasbegleitstoff auf Untergrundspeicher (Influence of hydrogen included in town gas on underground storages). DBI-Fachforum Energiespeicherkonzepte und Wasserstoff, Berlin; 2011.

[9] Pichler M. Assessment of hydrogen–rock interactions during geological storage of CH_4–H_2 mixtures. Master thesis. Montanuniversitaet Leoben, Austria; 2013.

[10] DLR. Plan-DelyKaD: Studie über die Planung einer Demonstrationsanlage zur Wasserstoff-Kraftstoffgewinnung durch Elektrolyse mit Zwischenspeicherung in Salzkavernen unter Druck (Study about planning a demonstration plant for hydrogen fuel production via electroysis combined wth intermediate storage in salt caverns under pressure). Stuttgart, Germany: BMU (Federal Ministry for Economic Affairs and Energy); 2015.

Part F

Integration

Chapter 21

Energy Storage Integration

Philip Taylor*, Charalampos Patsios, Stalin Munoz Vaca**, David Greenwood**, Neal Wade****
**Institute for Sustainability, Newcastle University, Newcastle upon Tyne, United Kingdom; **School of Electrical and Electronic Engineering, Newcastle University, Newcastle upon Tyne, United Kingdom*

1 INTRODUCTION

The chapter seeks to cover the essential aspects of the network integration of electrical energy storage (EES) systems. The chapter covers energy storage policy and markets, energy storage planning and operation, demonstration projects involving network integration of energy storage and energy storage modeling. The chapter finishes by drawing conclusions about the current state of energy storage deployment and future requirements for research, development, and deployment. Interest in electrical ESS is increasing as the opportunities for their application become more compelling in an electricity industry with a backdrop of aging assets, increasing distributed generation, and a desire to transform networks into smart grids.

Energy storage is not in itself new; electrical ESS have been in use since at least 1870 when Victorian industrialist, Lord Armstrong, built one of the world's first hydroelectric power stations at Cragside in Northumberland (United Kingdom). In hydroelectric schemes, the penstock valve regulates the conversion of potential energy held by water in an upper reservoir into electrical energy by a turbine-generator set. The storage capacity of a scheme is determined by the volume of water available in the reservoir and the power output by the rating of the generator.

It was identified as early as 1959 that to make best use of renewable energy resources with a meteorologically dependent output a storage element to the overall system would increase the energy yield. As well as increasing yield the ability to add dependability to renewable resources has been widely investigated. Despite this long track record a number of research challenges remain to be overcome if energy storage is to become ubiquitous and live up to its potential.

The current global implementation of energy storage in power systems is relatively small but continuously growing with approximately 665 deployed projects recorded as of 2012 [1]. Worldwide grid energy storage capacity was

Storing Energy. http://dx.doi.org/10.1016/B978-0-12-803440-8.00021-X

estimated at 152 GW (including projects announced, funded, under construction, and deployed), of which 99% are attributed to pumped hydro schemes and the remaining installations are new nontraditional storage systems (such as batteries and flywheels). ESS are designed to meet performance criteria spanning timescales from milliseconds to days and even seasons and from watt-hours (W h) to gigawatt-hours (GW h).

A number of commercial and regulatory barriers exist to the large-scale takeup of electrical ESS in power systems, including [2]:

- the high implementation costs for energy storage in many cases make it uneconomical compared with conventional solutions;
- the unbundled electricity system (prevalent in many countries) which lacks transparency means stakeholders cannot determine the full value of energy storage;
- the undetermined asset class for energy storage, which functions as both generation and demand, means potential investors cannot always realize the benefits of storage; and
- low electricity market liquidity, changing market conditions, and a lack of common standards and procedures.

Electricity distribution networks have entered a period of considerable change, driven by several interconnected factors: aging network assets, installation of distributed generators, carbon reduction targets, regulatory incentives, and the availability of new technologies [3,4]. In this climate the use of distributed storage has reemerged as an area of considerable interest. The end of this period of transition will be signaled by the successful establishment of the technology and practices that must go together to create what is termed the Smart Grid. The UK Electricity Networks Strategy Group (ENSG) provide a useful definition of the term [5]: a smart grid as part of an electricity power system can intelligently integrate the actions of all users connected to it—generators, consumers, and those that do both—in order to deliver efficiently sustainable, economic, and secure electricity supplies.

It is clear that the integration of electrical ESS into electrical networks is a key enabler for smart grids and decarbonization of the electricity industry. The chapter describes the key issues which must be considered and addressed when attempting to integrate energy storage into electrical networks.

2 ENERGY POLICY AND MARKETS

Governments, utilities, regulators, and other electricity stakeholders are all interested in the role of ESS in providing solutions in evolving and future power systems due to their versatility in providing power and energy capacity. As policies, electricity markets, and regulatory frameworks are constantly evolving, so is ESS, which although in its infancy will mature in the years ahead. It is estimated that the global demand for ESS will be £72 billion by 2017 [1], and in the United Kingdom, for example, bulk ESS has been projected to provide annual benefits of £120 million by 2020, £2 billion by 2030, and over £10 billion by 2050 to integrate low-carbon technologies (LCTs) to the grid (with similar achievable benefits for distributed ESS) [6]. The investment potential in the United Kingdom can be applicable to power sectors in numerous countries facing similar issues. Nevertheless, the unconventional operation and different functions of ESS complicates their operation under the current regulatory and market structures.

2.1 Background

2.1.1 Regulation

To increase competitiveness, provide higher quality services to consumers, and drive down costs in the power sector, deregulation was introduced for generation and supply functions [7]. In a restructured and deregulated electricity system, generation and supply functions are generally classed as competitive while the transmission and distribution (T&D) networks are regarded as natural monopolies and are regulated [8]. Regulation is used as a tool to drive down the cost of electricity and ensure a low electricity tariff for customers, provide a return on investment for electricity network stakeholders involved with T&D, and provide incentives to T&D companies to improve both network and operating efficiencies to the benefit of customers.

2.1.2 Electricity Market

Wholesale electricity markets usually operate as a centralized market (power pool) or decentralized market (bilateral contracts) [9]. The markets in a liberalized electricity system are futures, spot (day-ahead and intraday), balancing, ancillary services, and retail. In the wholesale forward market, short-term contracts are carried out in the spot market (day-ahead and intraday markets) and long-term contracts are made in the futures market, which covers trades for a week up to a year. To maintain grid frequency and system stability, supply and demand has to be balanced constantly in real time due to the lack of storage capacity in power systems. System balancing is carried out via the balancing and ancillary services market to account for shortfalls in the spot market.

2.2 Business Models for Using Energy Storage

A major issue affecting the wider implementation of ESS is the higher costs they add to the already expensive T&D networks or renewable energy systems (RES) deployments, which often renders them uneconomical if used for a single application when compared against alternative conventional solutions. Thus, developing a viable business model for the provision of multiple functions is important for the success of ESS.

In an unbundled power sector, ESS could be used (if regulation permits) for competitive (deregulated) services in the wholesale energy market (day-ahead and intraday), balancing and ancillary services markets, and capacity markets to maximize value across the electricity value chain. According to Pomper [10], ESS owners and providers can be categorized into six types: merchant providers, transmission system operators, distribution system operators, customer group, or contract storage providers. The ownership types, regulatory frameworks, and location of the ESS all influence the business model, which can either be regulated and/or competitive.

ESS used under the regulated business model would provide a guaranteed revenue source as this would be fixed based on contractual terms for services provided or if owned by a regulated network operator would lead to guaranteed cost recovery. Conversely, the deregulated or competitive business model can be used to participate competitively in the electricity markets and can additionally be used to provide regulated services without major interference [10,11].

2.3 Review of National Policies, Regulation, and Electricity Market Arrangements Supporting Storage

Europe. ESS investment in Europe covers over 20% of the ESS market worldwide [1,12,13]. The European Commission (EC) developed the Strategy Energy Technology Plan (SET-Plan) for developing and implementing an EU energy technology policy for the transition to a low-carbon economy [14]. The aim of the SET-Plan is to change the EC's approach toward investing in research, development, and demonstration (RD&D) activities for a low-carbon economy and it includes materials for ESS pathways for energy storage in the United Kingdom [15].

United Kingdom. The UK government has a 15% renewables target by 2020 and plans for the increased electrification of transportation and heating by 2030. Thus, the government has identified ESS, interconnection, and demand response as crucial in enabling the United Kingdom to reach its targets for transforming the electricity system by the year 2050 [16]. However, there is no clarity on the future role of ESS in the United Kingdom and consequently no specific regulation for ESS. There are no specific license conditions for ownership and operation of ESS, which functions as a load or generator. At present, ESS is considered as a generator under license conditions [17].

Scandinavia. ESS has been identified as a pivotal element in the Danish 2050 energy vision [18]. The present regulation in Denmark treats ESS as load, hence ESS is liable to grid charges for load. In Norway, there are grid charges for pumped hydro storage as load or generator with an additional charge for energy consumption during peak periods [17]. At present, there is no specific regulation or electricity market change for ESS in Norway and Denmark but there are future plans for using storage.

Germany. Germany's electricity market is Europe's largest [19]. ESS is considered to be a key component in the country's move toward a reliable, economically stable, and efficient power system. A report on European regulatory aspects for electricity storage [20] concluded that the lack of regulations, opportunities, and mechanisms to support the competitive use of ESS is affecting the uptake of ESS in Germany.

Spain and Italy. There is no specific regulation for ESS in Spain and legislative initiatives have been restricted to the Canary Islands where compensation is realized through regulated capacity and energy payments. In Italy the rapid increase in RES led to new legislative initiatives and proposals to be passed. The Legislative Decree 28/11 implementing directive 2009/28/EC calls on Terna, the Italian transmission system operator (TSO), to identify network reinforcements, including ESS, to enable energy from RES to be fully dispatched [21].

Australia. Australia has high carbon emission reduction targets as the country has the highest per capita GHG emissions in the Organization for Economic Co-operation and Development (OECD) and one of the highest globally [22]. There is currently a target of 20% electricity production from RES by 2020 (as illustrated in Fig. 21.1), which is expected to help reduce GHG emissions by 5% [23–26]. Nonetheless, the increase in RES has not yet brought about an interest in ESS, which currently do not participate in the energy market. Due to the lack of support, experience, and uncertainty of its future role, utilities do not include ESS in network plans and are unsure on how to recover costs for investing in ESS.

China. The Chinese electricity sector is vertically integrated with state-owned monopolies but there are currently plans to reform the power sector by unbundling transmission and distribution and deregulating the electricity market [27].

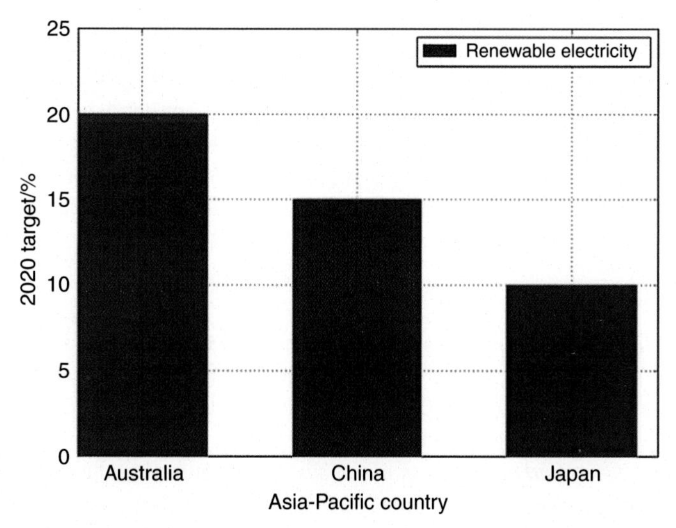

FIGURE 21.1 Renewable energy sector (RES-E) 2020 percentage targets for energy consumption in Australia, China, and Japan. *(Source: [21].)*

The market for ESS use is motivated by the need to increase the efficiency of the grid by the integration of RES. As the Chinese electricity market is not competitive, policies would be the main drivers for developing ESS. However, the current lack of national policies supporting ESS is a major barrier to national uptake of the technology [28,29].

Japan. The power generation and retail sectors of the power industry are liberalized and controlled by 10 vertically integrated power companies [30]. After the Fukushima nuclear disaster the government's interest in the use of ESS in providing security of supply has increased. There is a short- to medium-term target for 15% ESS capacity to be deployed on the grid [31].

United States. The investment in ESS is growing and was encouraged by the Energy Independence and Security Act of 2007. This identified the use of advanced electricity storage and peak-shaving technologies as a means of modernizing the grid to maintain reliable and secure electricity infrastructure and to meet growth in demand [13,32]. Since ESS do not fall under the conventional functions of generation, transmission, or distribution, the Federal Energy Regulatory Commission (FERC) individually addresses issues with the classification of ESS for use on the grid [30]. A major challenge for FERC is developing and adapting markets in deregulated states and creating proper evaluation frameworks in regulated states to allow ESS technologies to have economic value for the range of benefits that they can provide [33,34].

2.4 Regulation, Electricity Markets, and Their Impact on Storage Implementation

The regulatory barriers restricting the uptake of ESS are to a large extent dependent on the extent of unbundling practised. In countries where unbundling has not been fully realized and there is vertical integration, it is easier for utilities to deploy ESS across the power system to support the grid and to also use ESS commercially in the electricity market. On the other hand, in an unbundled system, the benefits derived from implementing ESS are more challenging to determine and accomplish because there are multiple actors involved from generation to consumers with different goals, practices, and regulation systems in place.

2.4.1 Storage Regulatory Barriers

Renewable integration policies. ESS could be used to store excess energy from RES and substitute for grid expansion. However, there is little incentive for investment in ESS because of the high priority and financial compensation provided to renewable generators to curtail excess energy.

Transmission and distribution use of system charges. Regulation determines whether T&D use of system charges should apply to ESS when used to provide services on the grid. Presently, ESS used on the grid is subject to T&D charges as a generator, consumer, or both, depending on the country.

Undetermined asset classification. ESS are multifunctional and can serve as a generator, transmission, or distribution asset, or as an end-user, depending on the required end-goal. Consequently, ESS asset classification is undetermined under present regulatory conditions and this directly affects eligibility for ESS asset ownership, grid tariffs, and cost recovery for regulated assets.

Lack of framework and incentives for storage service provision to network operators. There are no incentives or rewards in place for improved power quality, and power quality benefits are difficult to quantify [35]. The benefits ESS provide by improving capacity utilization of the grid and increasing the efficiency of centralized generators are also difficult to quantify.

Unwillingness to take risks or innovate. Although some ESS technologies are established, most technologies (other than Pumped Hydro Storage (PHS)) are still developing for use on the grid. Hence the lack of experience and high investment costs make it a risky venture.

Lack of standards and practices. Most ESS with the exception of PHS are relatively new and developing technologies (e.g., compressed air energy storage, hydrogen storage) with minimal deployments. This has resulted in a lack of necessary standards and practices to carry out thorough economic assessment, system design, and deployment.

Policies for other competing technologies or solutions. Current policies favor established technologies (e.g., interconnections, gas peaking power plants) over the use of ESS which have limited operational experience holding back the growth of ESS implementations.

Investment dilemma. The difficulty in determining the wide range of benefits across the grid makes it difficult to quantify the overall value of an ESS investment. This affects the profitability of investing in ESS and is especially the case for independent ESS owners.

Energy storage not being considered as part of RES schemes. The production of electricity from ESS connected to the grid may or may not be from RES. This creates a difficulty in trying to include the benefit of ESS under RES subsidies.

No benefit for controlled and dispatchable RES. Generally, generation-based support mechanisms (market premiums or feed-in tariffs) and priority dispatch are part of regulatory frameworks and policies to increase the uptake of RES. However, these mechanisms do not include and compensate for the controlled dispatch of renewable energy to meet demand and supply variations on the grid [20].

2.4.2 Storage Market Design Barriers

Limitations on service participation. ESS owners may be unclear of the future state of charge (SOC) of ESS when they are participating in the balancing or reserves market. This may prevent ESS owners from being involved in the spot market and in providing grid support services because it would be difficult to guarantee use in the balancing or reserves market if ESS are used for several services.

Lack of market liquidity. A liberalized electricity market that promotes competition favors a liquid wholesale market. However, bigger generators engage in bilateral contracts to mitigate issues that arise as a result of volatile prices in the spot market, which is currently affecting Distributed Generation (DG) operators in the EU [36]. This leads to low electricity market liquidity which is an entry barrier for ESS owners because it limits access and results in an unreliable market [37].

Market operation requirements and market fees. Satisfying the requirements of the spot market could be difficult if ESS are being used in multiple energy markets. Confirmation is required in close to real time for the provision of balancing and other ancillary services, which conflicts with wholesale market requirements, in which participants confirm their position ahead of real time in futures, day-ahead, or hour-ahead markets.

Decline in spread of peak and off-peak energy prices. The spread of energy prices during peak and off-peak periods provides an avenue for ESS owners to gain revenues from energy arbitrage. The price spread is affected by factors including the demand and generation mix and unpredictable fuel and carbon dioxide prices [38,39].

Unfair advantage provided to regulated utilities. The use of ESS by natural monopolies could complicate electricity market operation as it can provide regulated network operators with a way to influence the electricity market price and provide a biased advantage, which goes against the principles of unbundling.

Market price control mechanisms. Price control mechanisms enacted in different countries may affect the revenues ESS can make from arbitrage. As ESS may often operate for shorter periods during the year, compared with conventional generation, the opportunity to recover investment costs during periods of volatility in the markets is important. A price cap will create uncertainties that will significantly affect the business case for investing in ESS.

Wholesale and retail price market distortion. T&D operators have the potential to distort the electricity market by participating in the wholesale and retail markets while also obtaining regulated revenue on ESS, as a network asset, placing them at an advantage against other ESS or generation owners.

Undifferentiated remuneration for reserve and other ancillary services. ESS are compensated in the same way as traditional regulation service providers despite the additional benefits that their accuracy, high responsiveness, and rapid ramp rate can provide.

Penalties for not meeting scheduled energy dispatches. Using storage under a business model where it provides regulated and competitive services would be difficult to control. This is crucial as market rules place financial penalties on operators if an ESS is contracted to provide reserve services or electricity in the wholesale market but does not have enough available energy due to it being used for other services.

Value assessment from market operations. The method of assessing the potential revenues from ESS providing services in different electricity markets is complex because of the associated risks and uncertainties of changing market conditions.

Size requirements for ancillary services markets. There are limitations placed on the minimum duration and size of generation that can participate in providing regulation services. There are also limitations on the energy capacity in reserve markets. This limits participation from operators with smaller sized ESS.

3 ENERGY STORAGE PLANNING

Planning the use of energy storage in electrical networks is an important task which involves offline analysis to determine the optimal rating, capacity, location, voltage level, and service provision for ESS. Network operators are interested in the costs and benefits of different technologies to manage their assets.

Reverse power flow and voltage rise are related events which are most extreme when generation is highest and demand is lowest [40]. These both become more problematic as the penetration of rooftop PV increases and can limit the amount of distributed generation that should be installed in a network area [41]. Consumers connected to the LV network may have no direct problems as a consequence of reverse power flow, but this can affect voltage regulation on the medium voltage network including additional cycling of tap-changing transformers [42]. Reverse power flow is currently evident in the UK distribution network, as metered data from an LV transformer shows (Fig. 21.1). Voltage must not exceed 10% above or 6% below 230 V under UK regulation [43].

It is estimated £32 billion of investment is needed to mitigate the effects of distributed generation in the UK electrical network. To manage the network without directly interfering with generation or customer demand, network operators can either reduce network impedance (reconductor), add discretionary loads, demand-side management, or energy storage [42]. Indeed, under the new price control scheme (RIIO: revenue, incentives, innovation, and outputs) there is a financial incentive for distribution network operators (DNOs) to invest in new technologies and techniques such as energy storage [44]. In a competitive industry, there is a need to assess the cost implications of these innovative technologies relative to traditional mitigation methods. Energy storage is widely considered to be a technically viable solution to the problems expected in the distribution network, eg, in [45]; however there are few industrial distributed storage projects, costs are high, and DNOs do not necessarily have the experience to plan for new technologies.

EES technologies can generally be split into two broad categories. Utility or bulk-scale energy storage, such as pumped hydro and compressed air, are capable of delivering several megawatts of power over (1–8) h and due to cost and geological restrictions are suitable for transmission applications [46]. Distributed storage systems typically deliver smaller amounts of power for a similar period to utility storage but can be scaled in terms of rating, location, and capacity [47].

As summarized in [45], storage can offer a number of benefits to DNOs including voltage support, power flow management, restoration, network management, and compliance with regulatory requirements. Reverse power flow and voltage rise can be managed using storage to absorb power from the generators. Peak shaving (PS) requires discharge of stored energy into the network to reduce the loading on transformers and cables. To reduce reverse power flow or to peak-shave requires a specific amount of energy to be supplied or absorbed. Assuming thermal limits are not exceeded, power/energy specifications of the storage are unaffected by its location in the network.

However, the power required to manage a voltage problem is directly affected by the location of the storage unit [48,49]. As such, to reduce the capacity and power required to solve a voltage problem, storage should be distributed at many nodes within the network. It is often suggested that locating the storage within the network (in properties or on the street) will allow the greatest ability to reduce the power needed to solve the voltage problem. In real networks, however, there may be hundreds of nodes (busbars, customer connection points) where an energy storage unit could be connected. If multiple storage units are proposed the number of feasible combinations of energy storage increases rapidly; for example, there are 2.25×10^{16} ways of locating 10 storage units among 200 feasible locations.

3.1 Heuristic Techniques

Due to the complexity of this problem heuristic approaches are often considered for locating distributed storage. A number of papers consider heuristic approaches in the location, sizing, or operation of energy storage. In [50] a Tabu search approach is used for sizing energy storage by considering unit commitment. In [51] the authors use a genetic algorithm to locate superconducting magnetic energy storage to maximize the voltage stability index. In [19] three cost-based heuristics are shown for managing voltage rise in LV networks and the authors find that deterministic approaches are not as good as stochastic methods because they are unable to search the entire problem space. In [52] a genetic algorithm is used to locate and size a single energy storage unit to achieve benefits in reducing loss, voltage deviation, and costs. In [53] a genetic algorithm is combined with a sequential quadratic programming approach to locate capacitors and energy storage in a medium-voltage (MV) smart grid. In [47] a multiobjective algorithm is used to locate and size storage units in a 34 bus, 24 kV network. Objectives include reduction of storage power and capacity, minimizing the probability of voltage deviations, maximization of arbitrage revenue, and minimization of lost ancillary service opportunities. The heuristic used in [47] builds on work in [54] where wind, PV, and CHP units are located in a distribution network using a genetic algorithm.

Although global and local search methods have been applied to distribution networks in the literature, further consideration is needed into how their application can provide relevant results to DNOs in relation to distributed energy storage. This particularly applies in the area of planning given uncertainty and a lack of control of the location of distributed generation.

3.2 Probabilistic Techniques

When a primary substation reaches its capacity limit the standard solution is to reinforce the network with additional circuit capacity. Under the right conditions the required additional peak capacity can be provided from EES [55], real-time thermal ratings (RTTR) [56], or a combination of both. Probabilistic methods for calculating the size of an EES system for a demand peak-shaving application are described in the following text. The impact of both power and energy capacity are considered, along with the reliability of the energy storage.

3.2.1 Peak Shaving

Peak shaving (PS) refers to the reduction of electricity demand at times of peak consumption. Electricity demand varies throughout the day; in the United Kingdom this peak typically occurs in the early evening. In the majority of cases, peak demand only occurs for a small fraction of the time [57], but the generation, transmission, and distribution systems are constrained by it. A PS scheme attempts to reduce the demand peak either literally, through demand-side response [58] or offsetting the demand by supplying power locally through distributed generation or energy storage [59]. It is necessary to determine how large an EES system, in terms of both power rating and energy capacity, should be used to offset peak demand. EES has several potential advantages over conventional reinforcement; it can help solve other problems, for example, overvoltage on the local network, it does not require the same long planning consultation, nor does it result in the same long-term lock-in; it can participate in balancing, reserve, and frequency service markets.

Fig. 21.2 shows the demand over 24 h at a primary substation (33 kV/11 kV) in the United Kingdom. Between 1700 and 2000 the demand exceeds the overhead line limit, so the EES would need to provide PS for those 3 h. The peak exceeds the demand by around 4 MV A, so the EES would need a converter rated at 4 MV A to successfully offset the peak. The total area between the demand curve and the line rating is around 8 MW h; this is the energy capacity that would be required for the EES to meet the PS demand.

Electricity demand is highly variable, depending on time of day, day of week, month and season, as well as physical variables such as temperature and whether it is light or dark. Consequently, PS time, power, and energy will all vary from day to day. The ability to predict this demand is a crucial aspect of using EES for PS because sufficient energy will need to be stored in the EES prior to the PS event.

For this example system, increasing the power rating of the EES only provided benefits up to around 4 MW, while energy capacity continued to provide

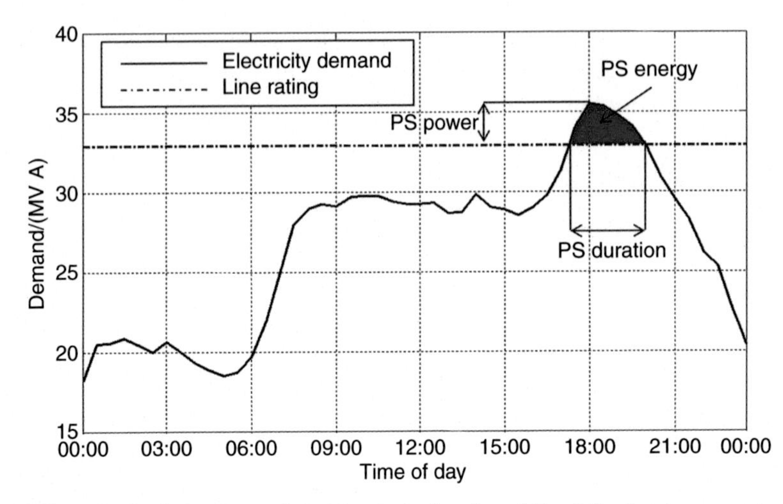

FIGURE 21.2 **Peak shaving needs to take place when the existing infrastructure cannot support the demand peak.** A storage system would need to meet both power and energy requirements to successfully offset the peak.

improvements up to 20 MW h. The addition of RTTR proved more beneficial in a load growth scenario because demand is likely to be above the static rating for a greater proportion of the time. RTTR also reduces utilization of the EES for PS, either freeing it to perform more commercial services or extending its operational lifetime. This extension is particularly useful given that the combination of RTTR and EES is likely to solve a network problem for considerably longer than either would in isolation.

3.3 Planning Storage for Security of Supply

EES can bring benefits on occasions when the network is disrupted. Under these circumstances the number of customers who can be resupplied at any time, and especially at times of peak load, is likely to be limited by the thermal ratings and consequent maximum capacity of certain critical sections of the network. Increasing this number would constitute an improvement in quality of service and may be required by the national industry regulator. In some circumstances, this requirement could involve costly capital expenditure, for example, to reconductor an overhead line or underground cable to a higher static thermal rating.

However, if supplementary energy supplies are made available from EES, then the number of customers supplied during such an event can also be increased. This can be achieved by smart deployment of a battery storage system.

This methodology is illustrated by a case study, described in the following section, which also calculates the supply shortfall that would be expected under faulted conditions. The potential impact of EES is then outlined, and the potential of different-size EES systems to reduce the risk of customer disconnection is calculated. The final section draws conclusions from the case study, in particular as regards the potentially wider use of EES technology and the associated methodology.

3.3.1 Case Study

Primary substations "A" and "B " together serve over 16 000 customers in the north of England, with a present peak demand of over 34 MW. They are supplied by two independent teed 33 kV circuits, as shown in Fig. 21.3. These supply circuits each consist of underground cable for the first 2.9 km, followed by 1.6 km of overhead line to the tee. These sections of overhead line are the most critical as regards static ratings.

In the event of a fault on one of the two circuits the remaining circuit would be required to carry the full load to both primaries. The critical section of both circuits is 175 mm^2 aluminum conductor steel reinforced (ACSR) overhead line, with static ratings of 30.8 MV A (winter), 28.6 MV A (spring/autumn), and 24.7 MV A (summer). Analysis of actual half-hourly load data

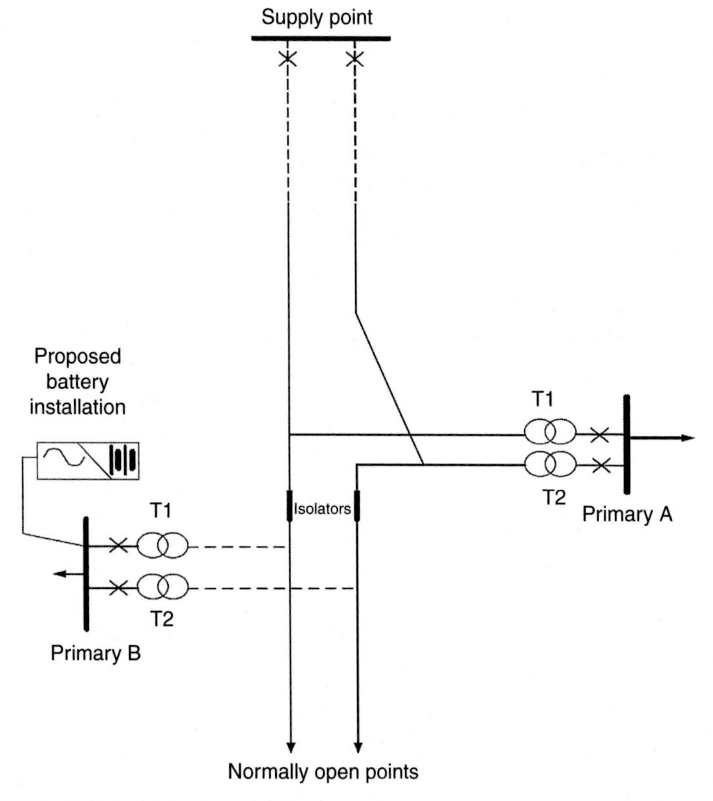

FIGURE 21.3 Schematic of supply circuits.

for the 12 months from Aug. 2011 to Jul. 2012 indicates that the summation of load at both primaries reached peak values of 33.28 MV A in winter (Jan. 18), 29.29 MV A in spring (Apr. 25), and 27.86 MV A in summer (May 1), all of which are in excess of the single-circuit static rating.

3.3.2 Potential Impact of Electrical Energy Storage

The following analysis assumes the connection of a 5.0 MW h, 2.5 MV A EES system to the 11 kV busbars at Primary "B." The ability of the EES system alone to secure the shortfall is a function of both the power rating of the converters and the energy storage capacity of the unit. For example, on Jan. 18, 2019, the expected worst day of winter, the shortfall would last from 1700 until 2030 and would be in excess of 2.5 MV A from (1730–1900) inclusive, a total of 2 h, as shown in Fig. 21.4. With peak converter power of 2.5 MV A the shortfall could be fully secured for 0.5 h, then partly secured for 2.0 h, then fully secured for a further 1.5 h.

The total energy shortfall on Jan. 18 is 8.75 MW h, which is almost double the assumed battery capacity of 5.0 MW h. As regards energy, only the first 1.8 h of shortfall could be secured.

For smaller power shortfalls the converter can be operated at below capacity and thus support an energy shortfall of longer duration.

3.3.3 Effect of Electrical Energy Storage System Size

Analysis of power and energy shortfalls on the 146 days per year on which an energy shortfall could be expected for part of the day in the event of $(n-1)$ loss of a single circuit has been carried out, and the results are shown in Table 21.1. The duration of the shortfall period, which is not shown in Table 21.1, ranges

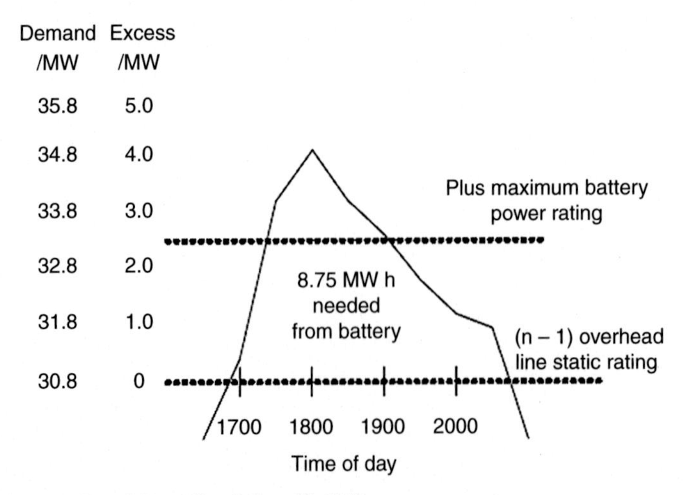

FIGURE 21.4 **Shortfall predicted, Jan. 18, 2019.**

TABLE 21.1 Number of Days with Specified Energy and Peak Power Shortfalls

	Peak power shortfall/MV A						
Energy shortfall/MW h	0.0–0.5	0.5–1.0	1.0–1.5	1.5–2.0	2.0–2.5	2.5–3.0	Over 3.0
0.0–1.0	22	18	3				
1.0–2.0	1	5	21				
2.0–3.0			3	5	2		
3.0–4.0			2	4	7		
4.0–5.0			2	1	7	1	
5.0–6.0			2	1	5	1	
6.0–7.0			1	3	2	3	1
7.0–8.0				1		2	3
8.0–9.0				2		1	2
9.0–10.0				1		1	
Over 10.0				1		6	3

over those 146 days from a minimum of 0.5 h up to a maximum of 13 h. It is assumed that the time between shortfalls (at least 11 h with load levels well below peak) allows the EES to be fully recharged for any combination of converter size and storage capacity.

The numbers in the body of Table 21.1 indicate the number of days on which energy and peak power shortfalls fall within a given range. So, for example, there were 21 days within the projected year 2018–19 when, in the event of an $(n - 1)$ fault, the energy shortfall would be between (1.0–2.0) MW h and the peak power shortfall would be between (1.0–1.5) MV A.

Table 21.1 can be used to evaluate the effectiveness of a converter and battery of the size installed at the trial site (2.5 MV A and 5.0 MW h). By totaling cells, it can be seen that on 103 of the 146 days this EES system would be able to meet the whole shortfall. On 19 of the remaining days its converter power rating would be sufficient, but not its energy storage capacity. On one of the remaining days the energy storage capacity would be sufficient, but not the converter power rating. On the remaining 23 days, neither would be sufficient. On this basis Table 21.1 can be used to reach decisions about EES system size.

4 ENERGY STORAGE OPERATION

Strategies are needed to operate energy storage in a live network situation to ensure the specified control objectives are met. The required complexity can range from a predetermined schedule that remains unchanged under all network, market, and forecast conditions to a fully automated system that reads all of these and adapts under a sophisticated decision-making process. A storage system might act in isolation, in coordination with other storage systems, or in combination with other interventions, such as demand-side response, real-time thermal ratings, or onload tap changers. The fundamental energy limit of any storage media leads to the conclusion that forecasting is a key feature of any storage installation. This may be implicit in the sizing calculation that is made when specifying the system at the design stage—the operating strategy then assumes sufficient energy will be available based upon the design. Alternatively, a forecasting function may be explicitly incorporated in the operation strategy which regulates the duty to be performed by the storage system throughout its lifetime.

This section explores a number of energy storage operation strategies that work in isolation and in concert with other devices to provide a range of network control services.

4.1 Balanced and Unbalanced Power Exchange Strategies

Integrating renewable energy into LV networks brings a number of challenges to existing distribution networks, particularly steady-state voltage rise and in some cases voltage unbalance. EES systems can play an essential role in facilitating renewable energy integration by mitigating voltage rise and unbalance problems. Three-phase voltage measurements from a low-voltage network in the customer-led network revolution (CLNR) project are shown in Fig. 21.5. It can be seen that the three-phase voltages are always higher than rated voltage (1 pu, i.e., one per unit) and the maximum voltage reaches 1.075 pu, which is close to the 1.1 pu

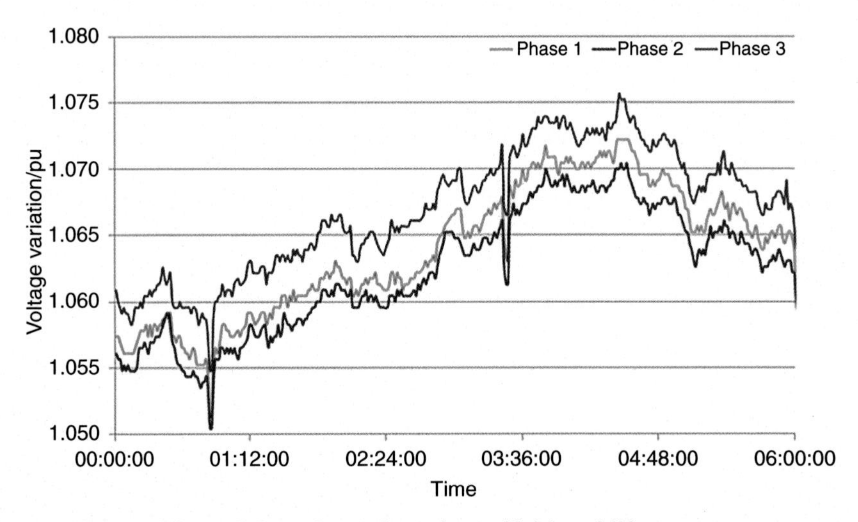

FIGURE 21.5 Measured three-phase voltages from midnight to 0600.

statutory voltage limit in the United Kingdom and Europe. It can also be seen that the three-phase voltages are not balanced; the maximum unbalance factor was found to be 0.45% against the limit of 1.3% in the United Kingdom.

Two EES voltage control strategies have been evaluated: balanced power exchange control strategy (BPECS) and unbalanced power exchange control strategy (UPECS). In BPECS the power flows on each of the phases of the EES are the same. In UPECS the power flow on each phase of the EES can be controlled independently. The test network for evaluating the controllers is shown in Fig. 21.6 with the domestic loads on the first branch of Feeder 1, to which a total of 15 households are connected, modeled in detail, while the loads on the remaining branches of this feeder and other feeders are modeled as aggregated load.

Simulation results demonstrating the operation of the EES with UPECS are shown in Fig. 21.7. The three-phase voltages at N2, which is the remote end of the feeder, are illustrated in Fig. 21.7a. The power import/export of the PV generation system and EES system are illustrated in Fig. 21.7b.

It can be seen that the three-phase voltages increase after PV generation starts injecting its maximum output power into the network at time $t = 5$ s and the voltage of Phase C rises above the statutory voltage limit. At time $t = 10$ s the EES starts importing power from the network which decreases the voltage below the limit. This indicates that the EES can mitigate violation of the voltage limit effectively.

Table 21.2 shows the required EES converter power rating using each of the voltage control strategies. It can be seen that the EES converter with UPECS requires a lower three-phase power rating than that with BPECS. This result demonstrates that the ability to control power exchange independently on each of the EES phases reduces the required energy storage capacity and converter power rating. Further information on these strategies can be found in [60].

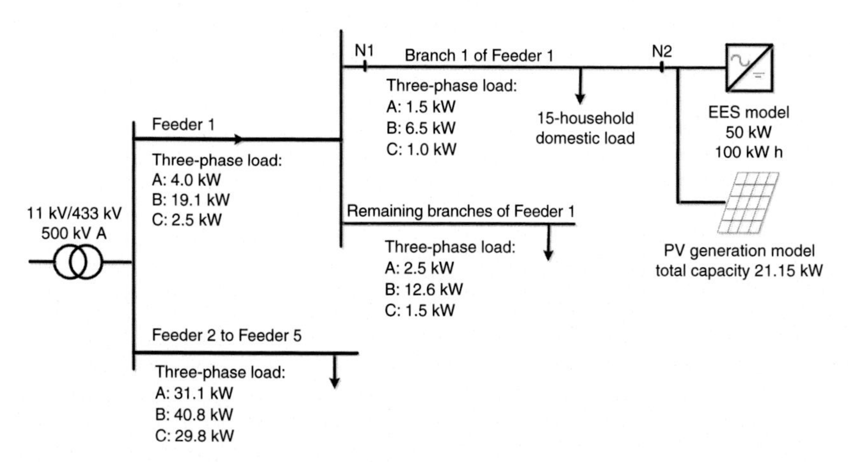

FIGURE 21.6 Low-voltage section of simulation network.

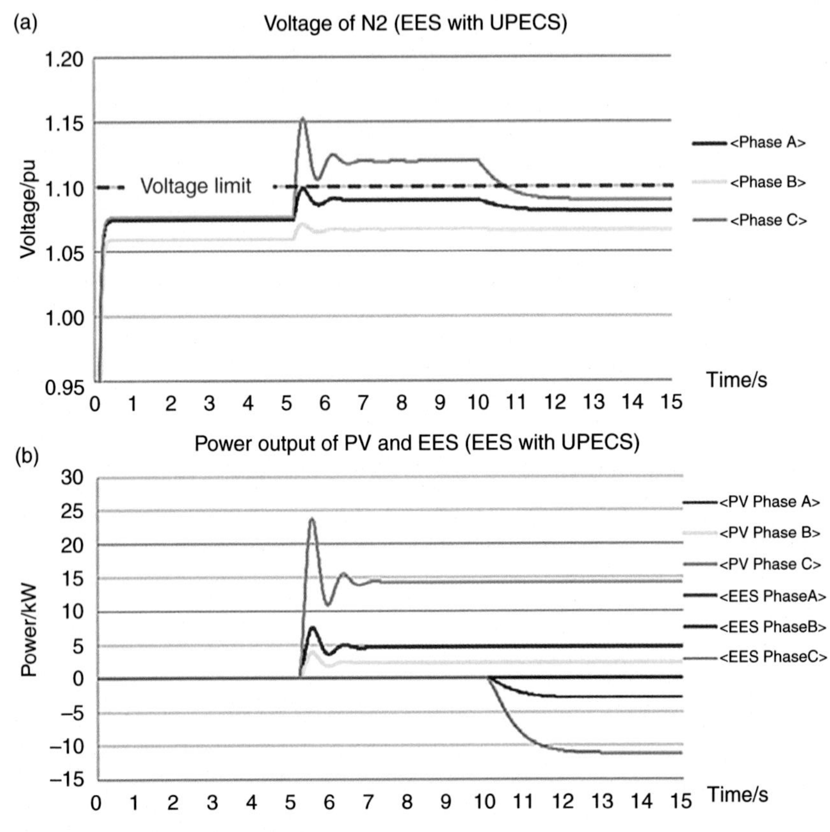

FIGURE 21.7 **Simulation results.**

TABLE 21.2 Required EES Converter Power Rating

	Phase A	Phase B	Phase C
EES with BPECS/kW	11	11	11
EES with UPECS/kW	3	0	11

4.2 Combining Energy Storage and Demand Response

The application of demand-side response (DSR) can yield numerous network benefits, such as reduction of the generation margin and improvements to the investment and operational efficiencies of both transmission and distribution systems [58]. In [61] it was demonstrated how DSR can also be used to solve distribution network voltage problems. Trials from the CLNR project showed that DSR was not always immediately available. In one case, on receiving a call from the DNO, the customer started a diesel generator to supply power to meet their demand. The customer load was thus reduced by over 800 kW for 4 h. It is important to note that there was a delay of approximately 20 min before customer consumption was actually reduced.

Energy storage under direct control of the network operator can be called into operation very rapidly. There are examples of energy storage and DSR being used together for optimizing network capacity [62] and cost reductions [63].

Under a scenario with a high penetration of electric vehicles (EVs) and heat pumps (HPs) within a localized cluster of 230 domestic customers, the steady-state voltage limit is violated. A control scheme using EES in conjunction with DSR has been devised to solve the problem.

The demand curve with a high EV and HP scenario and the resulting voltage profile are shown in Fig. 21.8. The voltage at the remote end of the longest LV feeder drops below the statutory limit (0.94 pu in the United Kingdom) during the nighttime peak period, early morning, and afternoon peak time.

To mitigate violation of steady-state voltage limits the collaborative control system will instruct the EES to operate first and export real power into the LV network to increase the voltage at the remote end. The collaborative voltage control scheme will simultaneously call for DSR. When operation of the DSR is confirmed and the steady-state voltage is within limits the collaborative voltage control system will instruct the EES to reduce real power export and thus conserve the limited storage capacity.

In Fig. 21.9 the voltage profile between 2330 and 0400 is seen to drop below the 0.94 pu limit, at approximately 0015 in the morning. The EES then injected 10 kW of real power into the grid to bring the voltage back above the limit and, at the same time, the DSR command was issued. After 20 min the consumption of the DSR customer started to reduce but did not reach a stable level until 0100. At this time the voltage of the network was close to the statutory limit, therefore the collaborative voltage control scheme decided to maintain the output of the

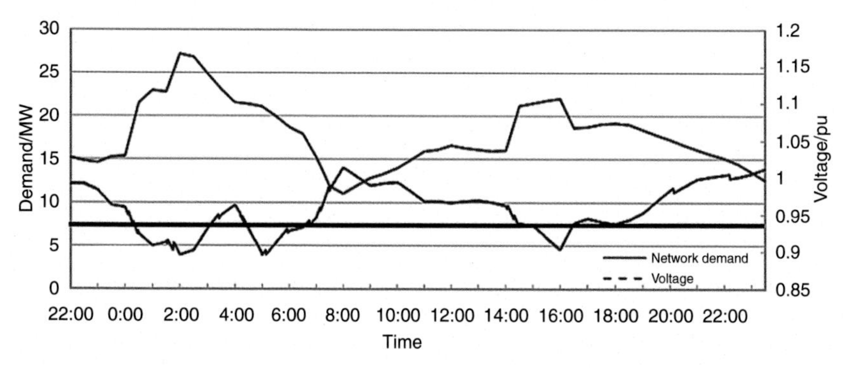

FIGURE 21.8 **Demand and voltage profiles during 1 day.** Thick black line indicates minimum allowable voltage.

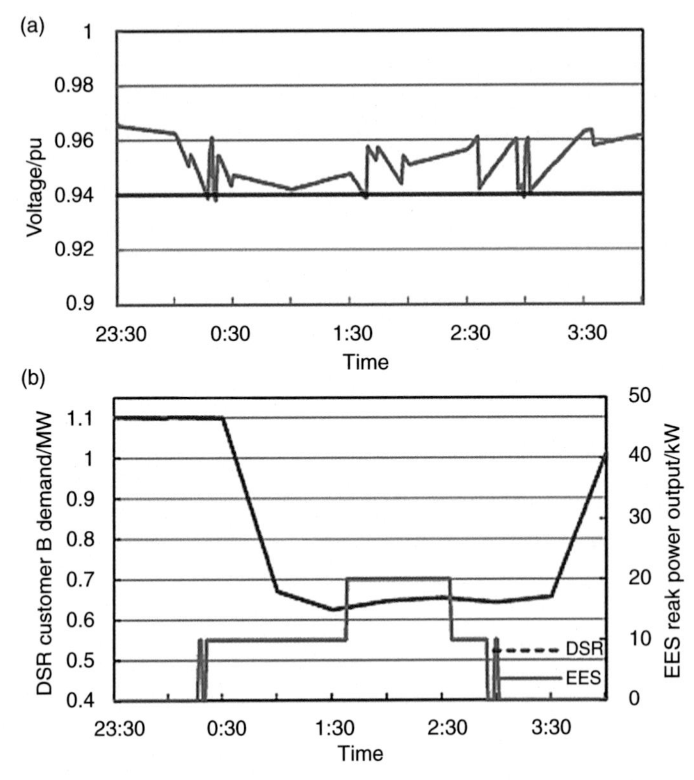

FIGURE 21.9 Voltage control during early morning period. (a) Voltage profile and (b) DSR and EES responses.

EES to prevent a further voltage problem. However, around 45 min later, when the voltage again went below the limit, the EES started to inject more power into the network to keep the voltage above the limit.

It can be seen that use of the two techniques in collaboration offers benefits beyond the use of a single technique. The use of DSR on its own would result in a voltage problem sustained for approximately 20 min; this is avoided due to the fast response of the EES. The capacity of the EES required is reduced because the DSR system can remove or reduce the need for storage intervention after 20 min. Given that EES technology is currently expensive and the cost of DSR is lower than the cost of EES, this is a valuable contribution. Further information on this study can be found in [64].

4.3 Coordination of Multiple Energy Storage Units

Using more than one ESS in a coordinated control system can offer a number of potential benefits, both in the provision of the service and the operational flexibility of the storage units. To examine the relative merits of coordinating storage systems the ideas of decentralized, centralized, and coordinated control need to be examined.

In decentralized control each unit uses local measurements to control its charging/discharging function. This type of control strategy does not require a wider communication scheme and so is in some respects robust, reliable, and cost effective compared with centralized control [65]. However, due to no communication between units, support cannot be received from other units if either an extreme SOC or a power limit is reached or in the event of complete unit failure.

In centralized control the charging and discharging control actions for each unit are determined in a central controller. This approach requires online information of the network state and high computation speed [66]. A significant drawback of this control approach is cost, since it requires a fast, high-reliability communications network. In the event of communication failure each unit would not be able to respond to a voltage excursion.

In coordinated control the control strategy combines the positive features of both centralized and decentralized control [67,68]. The distinctive features of this control are robustness with respect to intermittency and latency of feedback and tolerance to connection and disconnection of network components.

A study has compared the relative strengths of these approaches to control [69]. In Fig. 21.10 battery energy storage system (BESS) decentralized controllers

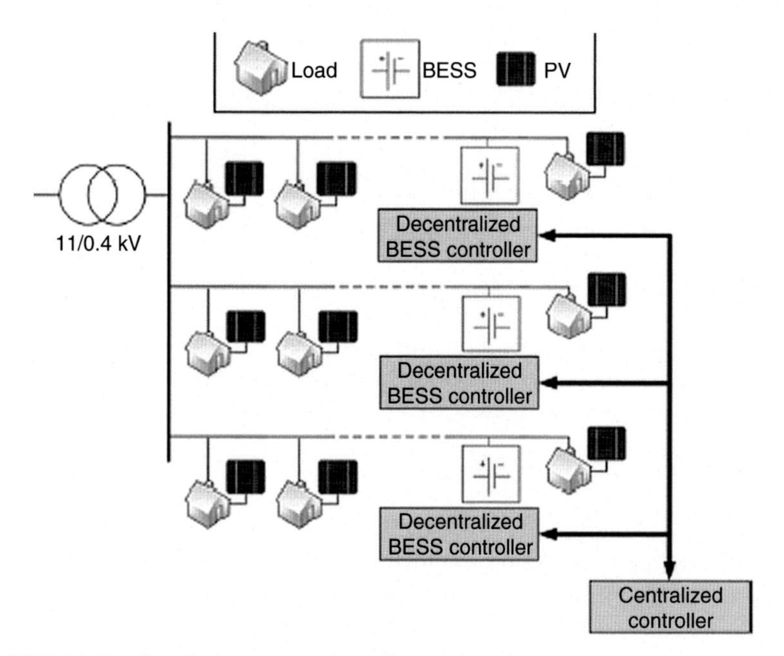

FIGURE 21.10 Coordinated control of multiple energy storage units.

act on local measurements of voltage, combined with a voltage sensitivity factor, to determine active and reactive power set points. The voltage sensitivity factor (VSF) is a measure of how large an impact that variations in power will have on voltage at a given node. The centralized controller determines which BESSs should be used to solve voltage problems by considering the remaining battery cycle life, energy storage availability, and their VSF. This analysis concluded that:

- By sharing power and energy between the BESSs the scheme is able to solve real-time voltage problems that cannot be solved with independently controlled BESSs with the same power and energy capacity.
- The rated power and energy of BESS units at locations with the most severe requirements are reduced, hence the largest unit is smaller when compared with a unit in the same location with noncoordinated control.
- The even sizing of BESS units offers advantages in maintenance and economies of scale in manufacturing.
- There is greater potential for this proposed method to adapt to changes in location of extra PV generation, albeit to the limit where extra capacity would then be required. This is not the case with noncoordinated control.
- The addition of an aging model more evenly utilizes the BESSs and consequently reduces the cost of battery replacements for the storage operator, both in terms of battery replacement and maintenance requirements.

4.4 Summary

This brief summary of storage operation studies shows how there are many considerations to be made in the provision of storage control. A single storage unit will have to provide one or more services in the areas of network control or market operations. To do so, reliance on direct or indirect forecasting of the duty to be provided is required. There can be benefits in sharing the duty required to provide a service between multiple storage units or alternative technologies such as DSR.

5 DEMONSTRATION PROJECTS

This section describes three demonstration projects on UK distribution networks. In each case, we present the ESS installed and explain what it was intended to accomplish, we present trial results and summarize the outcomes of the project.

5.1 Hemsby Energy Storage

This was the first installation of large-scale energy storage on a distribution network in Great Britain [70–72]. A picture of the storage site is shown in Fig. 21.11, and the local network, close to Great Yarmouth in the United Kingdom, is shown in Fig. 21.12. The ESS is located at a normally open point at

FIGURE 21.11 **Dynamic energy storage [73].**

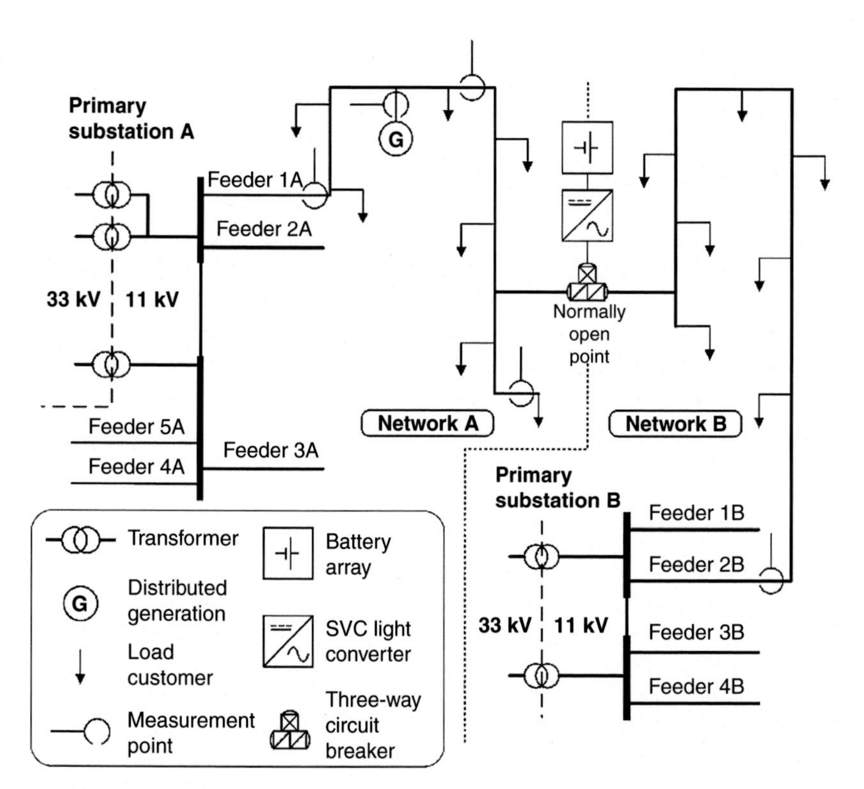

FIGURE 21.12 **Primary and secondary instrumentation points in demonstration network 11 kV feeder.**

the remote ends of Feeder 1A and Feeder 2B. A 2.25 MW windfarm is located midway along Feeder 1A. The area is not on the gas grid, so there is a significant electric heating load. Feeder 1A has peak demand (2.4 MW) occurring between 0100 and 0200, and Feeder 2B has a contrasting daytime peak (4.6 MW) when holiday parks have high occupancy.

5.1.1 Energy Storage System

The storage system was nominally rated as a 200 kW h/200 kW network, and the storage medium selected was lithium-ion batteries. The ESS could operate in four quadrants, simultaneously exchanging real and reactive power with the network in either forward or reverse direction. The converter was rated at 850 kV A which permitted 600 kW and 600 kV Ar to be transferred simultaneously. The battery modules were able to deliver 600 kW of power, but were generally limited to 200 kW (1 C).

5.1.2 Automatic Control

To carry out a substantial series of trials, it was necessary to operate the ESS under automatic control. The control algorithm developed could read required variables, process them in a flexible manner, and issue set points for implementation on the ESS. The core control algorithm could be configured to respond to any of the values from instrumentation points on the network; this was called an "event," with associated "location," "threshold" (of the measurement), and "action" (power set points).

As an extension to the basic event, location, threshold, action process described earlier an algorithm was developed to track the time and magnitude of peak power flow occurring on the feeder. This function was required for this demonstration project because daily and seasonal variability in feeder power flow lead to either limited use or frequent saturation of the ESS capability.

Further to the event-driven algorithms described previously a scheduling system was developed to provide a facility to run repeated set point combinations on the ESS. This provided the capability to carry out sequences of tests on the ESS, designed to verify the operating parameters of the system.

5.1.3 Trial Results and Validation

The effect of the peak-shaving algorithm, described earlier, in conjunction with EES in reducing power flow on the 11 kV feeders is illustrated in Fig. 21.13. It can be seen that, due to the scale of the EES relative to the feeder demand, the effect of the EES system on the load profile is relatively small.

Operation of the EES and the impact that this operation has on estimated SOC of the battery is illustrated in Fig. 21.14.

SOC negative overshoot is due to the SOC battery terminal voltage estimation technique adopted by the system [74]. When estimating SOC under dynamic conditions there is a significant voltage drop due to internal impedance, whereas when export power is reduced to zero, there is no current flow

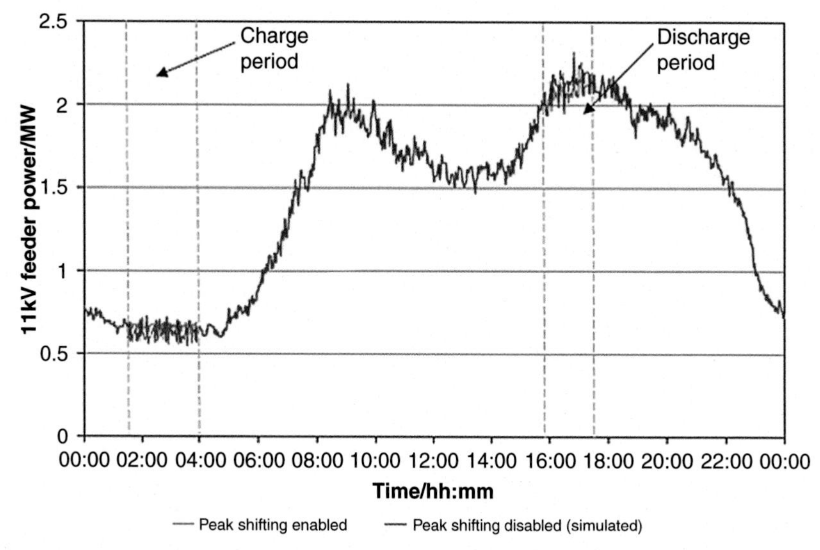

FIGURE 21.13 **11 kV feeder power flows with and without peak shaving enabled (Aug. 7, 2013).**

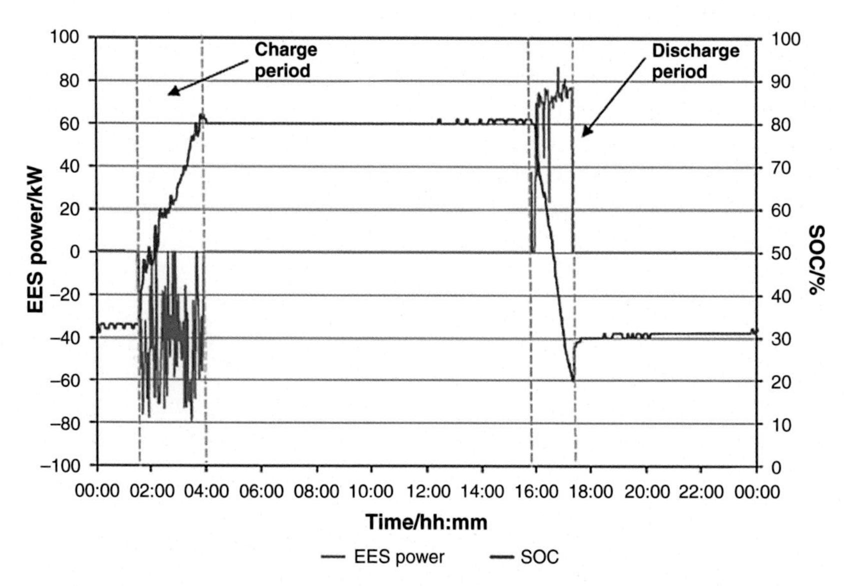

FIGURE 21.14 **Real power import/export and SOC of EES system (Aug. 7, 2013).**

in the battery circuit and no voltage drop. Charging/discharging rates, battery age, state of health (SOH) [75], and environmental conditions (e.g., ambient temperature) have been shown to have significant impact on this internal impedance value [76]. Consequently, the instantaneous online SOC estimator is significantly lower at the end of discharging than the no-load measurement (as shown in Fig. 21.14).

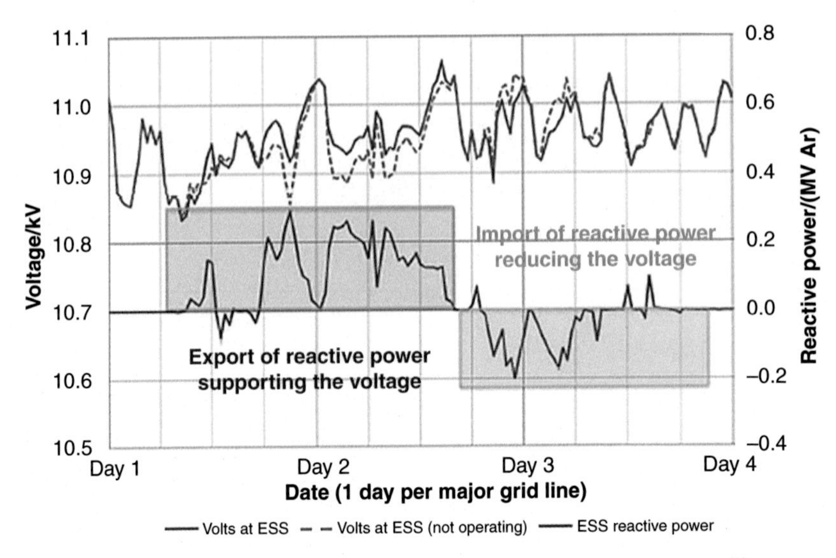

FIGURE 21.15 **Reactive power exchanges with network to support voltage stability.**

Results measured from the network in Fig. 21.15 show the action of automatic voltage control. To make a comparison with the effect that would have been seen without the ESS in operation a load flow was carried out to simulate voltage at the point of common coupling (PCC), both with and without the ESS. For this early test, limited data acquisition was available on the network, so voltage at the PCC was simulated from measurements at the primary substation. This situation limits the clarity of the effect of automatic control, but the raising of low voltages and trimming of high voltages can be seen.

5.1.4 Outcomes

This project demonstrated automatic, algorithmic control of an ESS connected to the remote end of an 11 kV distribution feeder on a UK distribution network. The primary purpose of this project was to explore practical and technical features of planning, deploying and operating energy storage on an 11 kV distribution network.

5.2 Energy Storage in the Customer-Led Network Revolution

ESS were installed at multiple voltage levels as part of the CLNR project, which was the largest smart grid demonstration project ever conducted in the United Kingdom. Large-scale storage, comprising 5 MW h/2.5 MV A of lithium-ion cells, was located on the 6 kV Northern Power Grid distribution network at Rise Carr near Darlington in the north of England. The storage facility and a simplified circuit diagram of the local network are shown in Fig. 21.16 and Fig. 21.17. In addition to this a number of smaller ESS (100–200) kW h were installed on

FIGURE 21.16 **A picture of the storage at Rise Carr.**

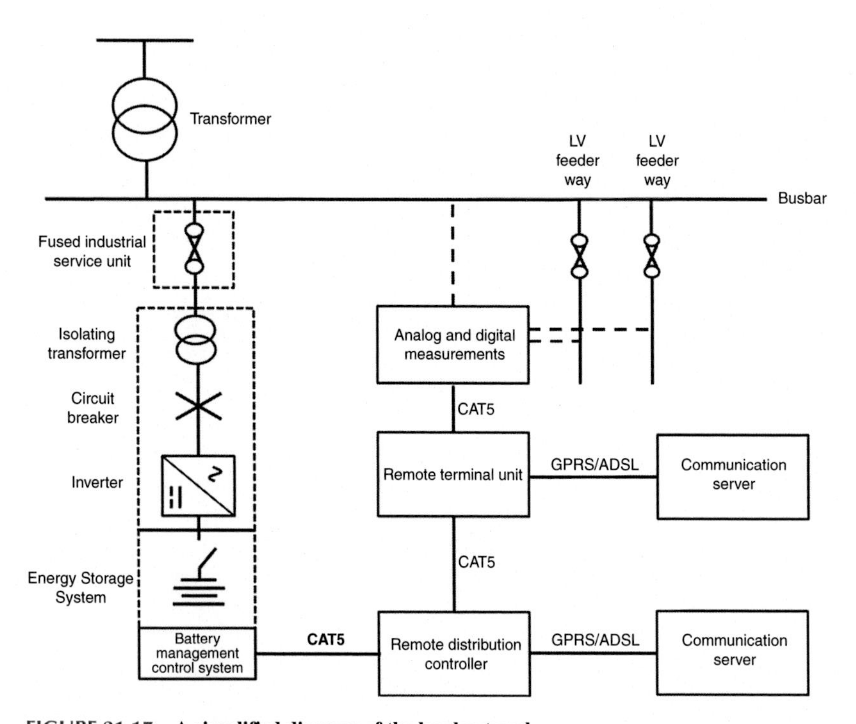

FIGURE 21.17 **A simplified diagram of the local network.**

rural distribution networks. The aim of these installations was to demonstrate the benefits of individual ESS in coordinated energy storage responses, to test network interventions with multiple functions, and to interact with other network interventions such as demand response and real-time thermal ratings.

5.2.1 Trial Results

The ESS at Rise Carr is primarily required to reduce the loading on the 11 kV/6 kV transformer. This transformer operates using a variable ampacity based on the local weather. Fig. 21.18 and Fig. 21.19 show power being supplied from the ESS to reduce demand to below the transformer thermal limit. Fig. 21.18 shows the current in the transformer with and without the action of the ESS, the transformer current, and the current supplied by the ESS. Fig. 21.19 shows the impact of delivering this service on the SOC of EES.

Fig. 21.20 shows real power voltage control of a LV feeder-located energy storage unit at Mortimer Road (EES3, 50 kW/100 kW h). This is autonomous voltage control with upper and lower limits of 0.420 and 0.408 kV, respectively. Positive power implies real power export from the EES unit to the grid.

5.2.2 Outcomes

The CLNR EES demonstrations showed that EES can work at a variety of voltage levels and provide a mixture of network services in conjunction with other network interventions. The ESS was deployed on a variety of networks (representative of around 80% of the UK distribution network) and was of a variety of scales.

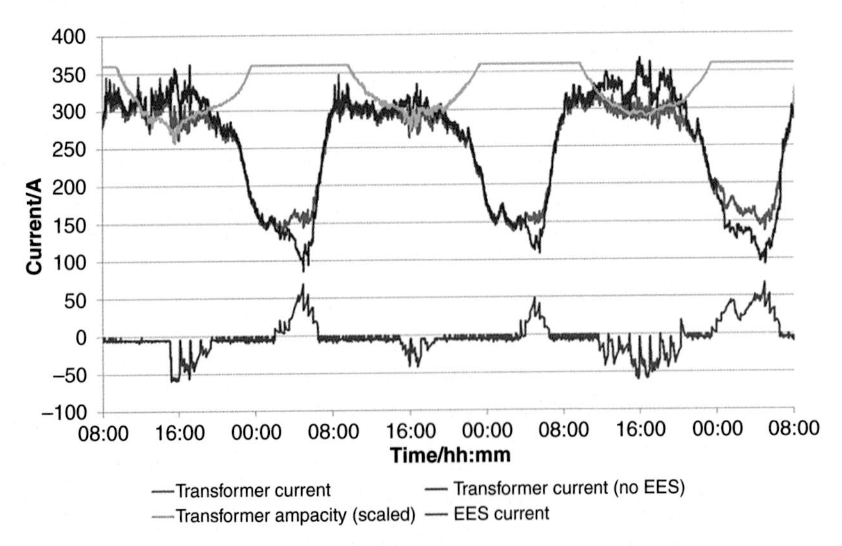

FIGURE 21.18 **Real-power response for power flow management at Rise Carr.**

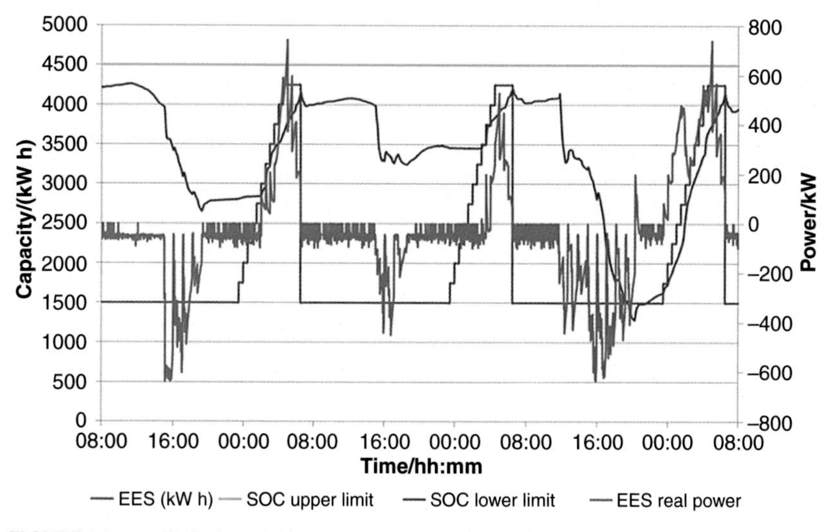

FIGURE 21.19 **EES state of charge during active power response.**

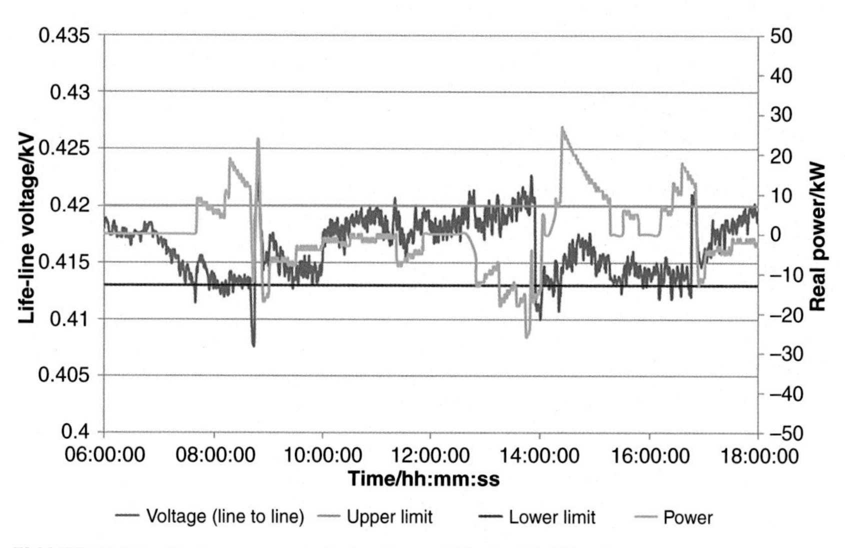

FIGURE 21.20 **Real power control of voltage at Wooler St. Mary's.**

5.3 Smarter Network Storage

Installing EES for just one application is typically insufficient to make the capital investment worthwhile. The principal goal of smarter network storage (SNS) was to install large-scale EES for a variety of system benefits to maximize its value. The project installed what, at the time, was the largest battery in Europe—10 MW h/6 MW/7.5 MV A of lithium-ion storage, pictured in Fig. 21.21—at a primary substation (33/11 kV) in Bedfordshire (England). The primary purpose of the EES was to defer the need to invest in new network infrastructure by reducing peak demand on the network, shown in Fig. 21.22. To offset investment costs, revenue was gained from contracts to provide frequency response and short-term operating reserve, as well as being offered in a tolling contract to energy traders [55].

The aim of the project was to demonstrate that EES on this scale can serve as a network asset, while providing a return on investment. A forecasting and

FIGURE 21.21 The site of the storage (left) and the battery racks (right).

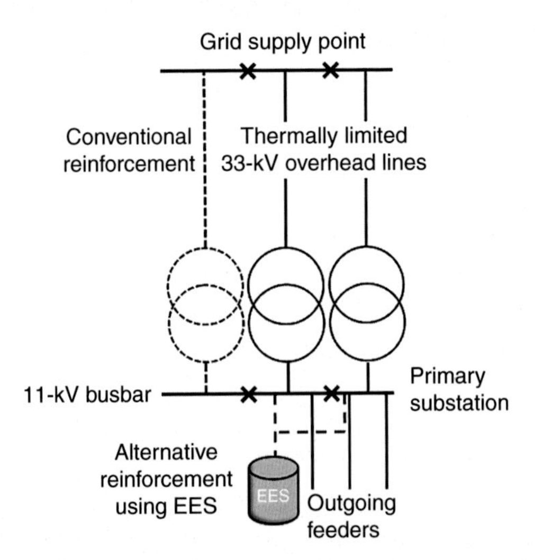

FIGURE 21.22 Comparison between the conventional reinforcement option and the alternative using EES.

service scheduling system was developed to allow the ESS to participate in commercial services while maintaining available power and energy for the primary demand peak-shaving service.

5.3.1 *Trial Results*

At the time of writing, the SNS project is still ongoing, but only single-service trials have been carried out. The results of some of these trials are presented here, but ultimately the intention is to demonstrate a portfolio of services that are automatically scheduled and delivered and maximize the lifetime value of the ESS.

Fig. 21.23 shows the ESS carrying out its primary peak-shaving function. Peak demand is brought below the circuit thermal limit between 1815 and 1915, meaning that in the event of single-circuit outage the power supply could continue without any additional intervention. This is significant in that the ESS is providing a necessary reinforcement service, indicating that the industry has moved beyond demonstration projects and into ESS being used to solve real network problems.

Fig. 21.24 shows the ESS providing a dynamic frequency response service. When system frequency deviates from the nominal value by more than 0.05 Hz the ESS automatically provides a power response proportionate to the deviation. This, combined with similar responses from other ancillary service providers, brings the frequency back within its limits. In the example shown the ESS is providing a bidirectional response, allowing it to act on both high- and low-frequency events.

5.3.2 *Outcomes*

Many of the outcomes of SNS are yet to be realized because the project is, at the time of writing, still in early stage trials. However, the project is already

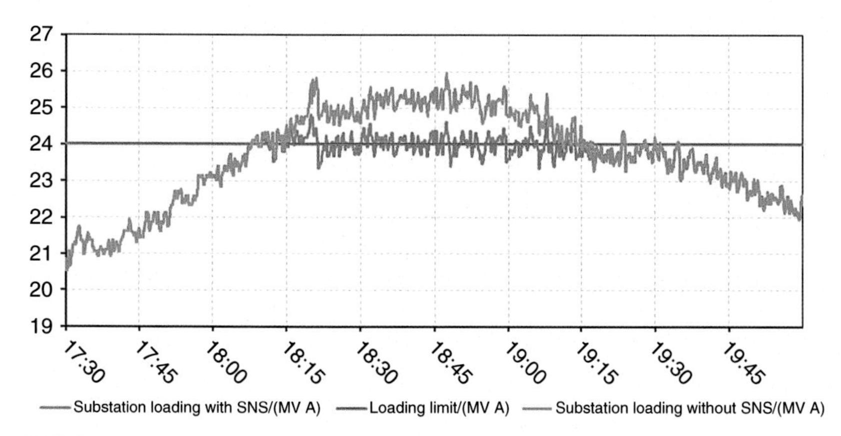

FIGURE 21.23 **Peak demand reduction using the ESS.**

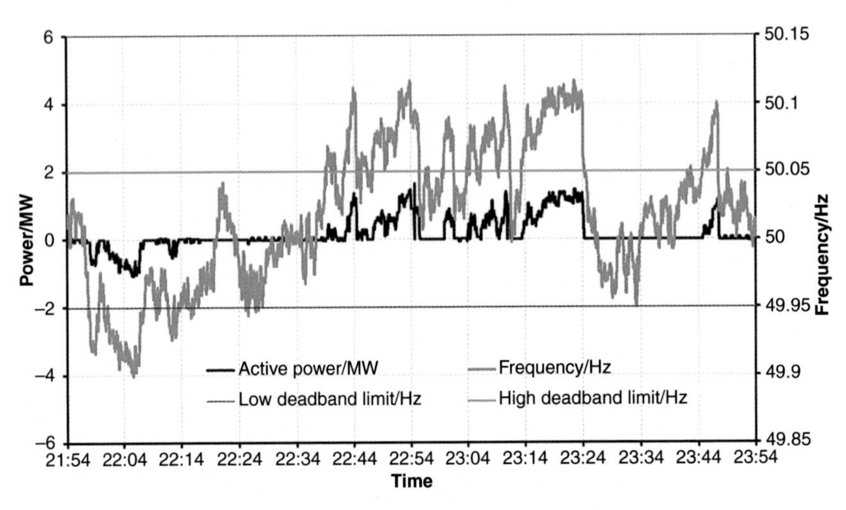

FIGURE 21.24 **Dynamic firm frequency response provided by the EES.**

demonstrating that it is possible to use ESS to solve real network problems, and to participate in ancillary service markets while doing so. Over the lifetime of the project, evidence will be created on the cost-effectiveness of this solution, as well as the suitability of existing markets and the regulatory framework to support this kind of approach.

6 INTEGRATED MODELING APPROACH

A grid-scale ESS comprises three main components: the energy storage medium, a power electronic interface, and a high-level control algorithm which chooses how to operate the system based on internal and external measurements. The literature contains many examples of isolated modeling of individual energy storage mediums, power electronic interfaces, and control algorithms for energy storage. However, when assessing the performance of a complete ESS, the interaction between components gives rise to a range of phenomena that are difficult to quantify if studied in isolation. An integrated electrothermochemical modeling methodology seeks to address this problem directly by integrating reduced-order models of battery cell chemistry, power electronic circuits, and grid operation into a computationally efficient framework [77]. Physics-based battery and power converter models are implemented with sufficient detail to account for energy losses and track battery degradation through solid–electrolyte interphase (SEI) layer growth. This BESS model is coupled with a grid model. The same grid control objectives can be met through different battery operating set points, resulting in considerable differences in battery degradation and roundtrip efficiency.

The proposed benefits stemming from this approach are:

- to enable online mapping of asset characteristics and operating costs
- to assist in cost-effective sizing of grid-connected BESS
- to lead to more efficient and cost-effective operation of BESS
- to allow development and testing of novel control architectures in a realistic and detailed environment
- to facilitate incorporation of the model in real-time simulation platforms.

6.1 Methodology

Fig. 21.25 is a block diagram of the integrated model showing details of individual blocks as well as flows of power and information. Battery roundtrip energy efficiency is calculated over a cycle that begins and ends at the same SOC.

6.1.1 Lithium-Ion Battery Model

Lithium-ion batteries can be modeled at varying degrees of complexity; 3D, 2D, 1D, and 0D. The 0D model, known as the single-particle model (SPM), has been adopted in this work. This has advantages in terms of low computational cost, but is typically only valid up to approximately 2C (the current required to empty the cell in half an hour) operation; network services are typically below this rate, so the model's use is justified. The battery chemistry modeled is based on lithium–polymer cells with an operating voltage window of (4.2–2.7) V.

Battery voltage varies significantly with SOC, as shown in Fig. 21.26. Thus, selecting a suitable SOC range and designing the electronics around that voltage range is important when designing BESS.

The SPM accounts for nonlinear electrochemical performance by treating each electrode as equivalent to a single sphere [78]. The modeling approach for the SPM is based on work by Chaturvedi et al. [79] and Ramadass et al. [80] and has been adopted in this work unless otherwise stated.

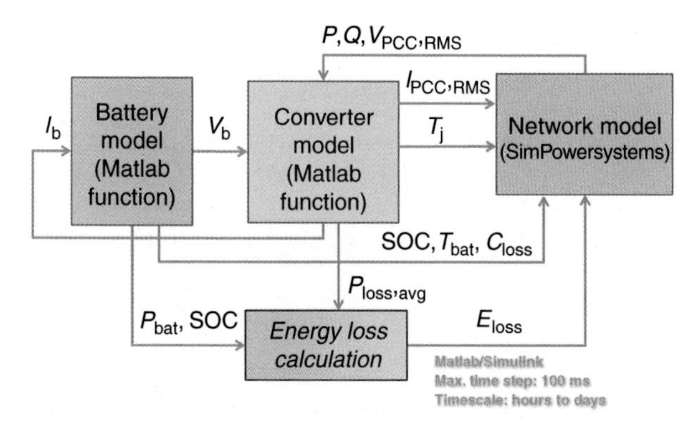

FIGURE 21.25 **Block diagram of the proposed integrated model and details of individual blocks of the whole system.**

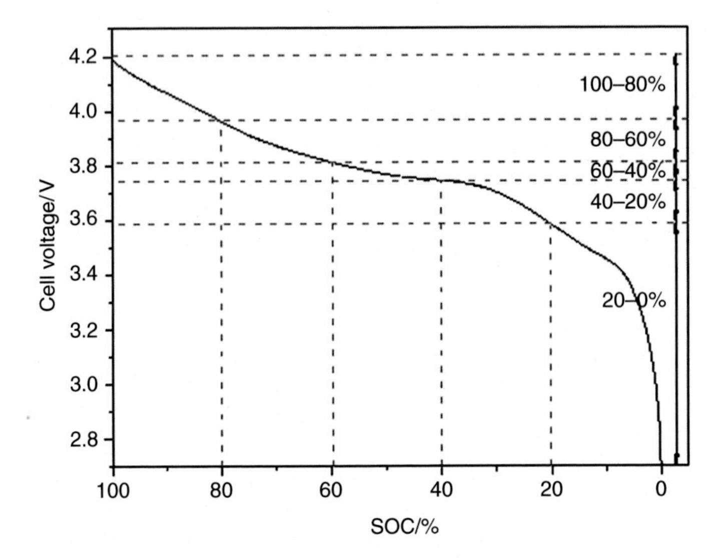

FIGURE 21.26 **Measured cell voltage as a function of SOC for a 4.8 A h Kokam lithium-ion battery.**

6.1.2 Converter Model

In this study the two-stage topology presented in Fig. 21.27 is used; it allows significant variation in battery voltage, and hence more complete charge extraction, without compromising system efficiency.

The converter is represented by an "average mode," based on semiconductor datasheet information, coupled to a piecewise linear approximation of a single 50 Hz fundamental period of the voltages and currents occurring in one leg of a three-phase bridge. Balanced three-phase waveforms are assumed during simulation, so power losses only need to be calculated for a pair of semiconductors. The model accounts for conduction and switching losses and losses in the filters of both the DC/DC and DC/AC converters.

6.1.3 Network Model and Control Algorithm

The section of network used to test the integrated model is based on a real LV network in the United Kingdom and is shown in Fig. 21.28. A scenario is examined

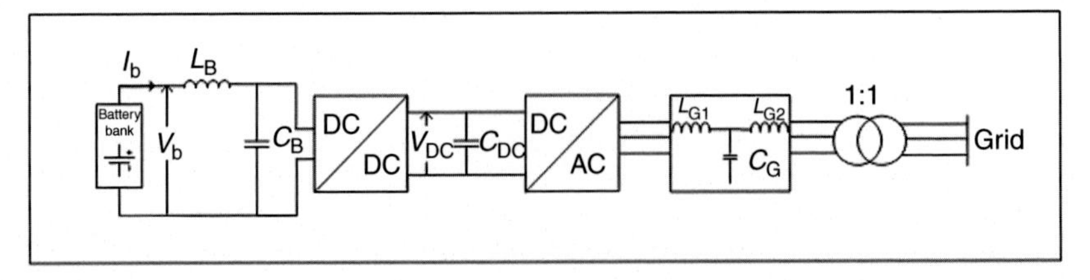

FIGURE 21.27 **Two-stage topology comprised of battery bank, DC to DC and DC to AC conversion stages.**

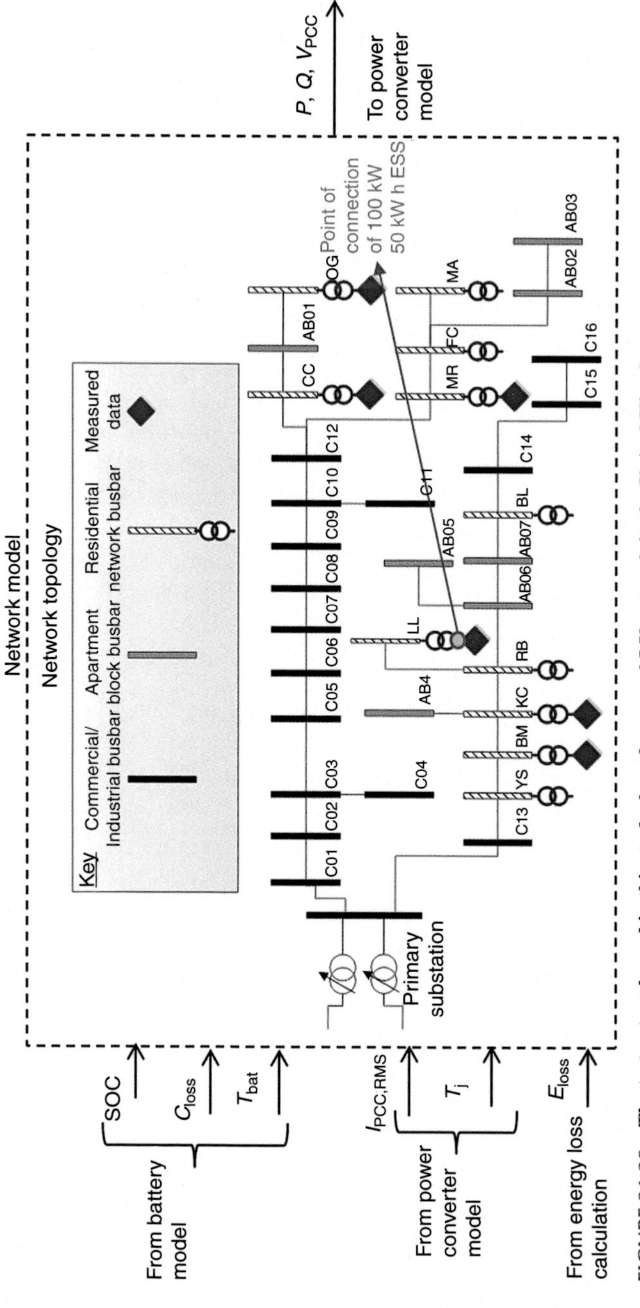

FIGURE 21.28 **The test network used in this study, based on a real LV network in the United Kingdom.**

with a 100 kW 50 kW h BESS connected directly to the secondary substation, as shown in Fig. 21.28, rated at 315 kV A. At present the loading limit of the transformer is not exceeded, but loading may increase such that network reinforcement is required. Under this scenario the BESS flattens the high evening load peaks and reduces transformer loading below the thermal thresholds while also providing ancillary services.

The BESS connection to the grid uses current sources responding to active and reactive power set points. Power set points along with the voltage at the PCC are passed to the converter model, which calculates battery current and outputs this to the battery model, which calculates battery parameters.

6.1.3.1 BESS control algorithm

The control algorithm's primary function is to follow the transformer load and respond by flattening it by discharging the battery when demand exceeds 300 kW. The algorithm issues power commands that equal the difference between the load and the 300 kW threshhold. The algorithm also accounts for SOC and current limitations; these are calculated online in the battery model and aim to maintain individual cell voltage between (2.7–4.2) V limits. The BESS will normally float at a preselected SOC; it will be allowed to charge, either prior to the evening peak or during the evening.

Fig. 21.29 displays the typical operation of the BESS, set to float at 30% SOC.

6.2 Results and Discussions

The system was simulated, using the integrated modeling approach, for an equivalent time window of 24 h. An autumn load profile, scaled such that the evening peak exceeds the transformer loading limit, was used in this study.

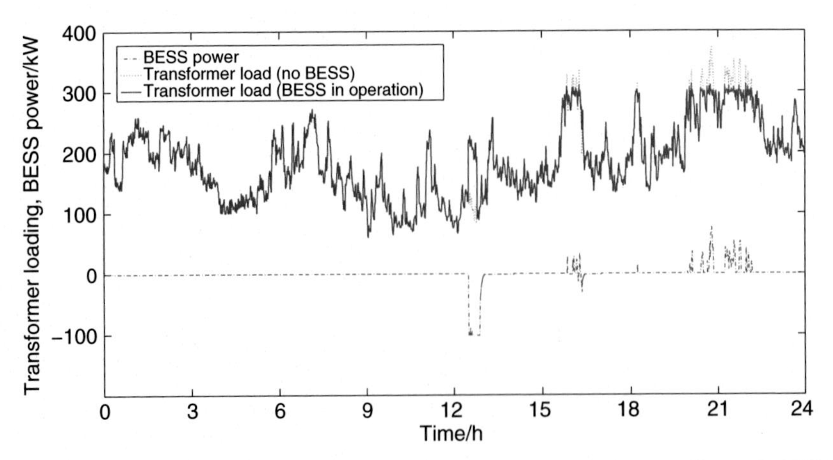

FIGURE 21.29 Typical operation of the BESS, showing transformer loading and BESS power.

Fig. 21.30 shows BESS operation for two different combinations of floating SOC as well as SOC limits:

1. Floating SOC 100%, maximum SOC 100%, minimum SOC 20%.
2. Floating SOC 30%, maximum SOC 80%, minimum SOC 0%.

Fig. 21.30 shows the power flowing through the transformer, the BESS power, SOC, battery capacity loss, and energy absorbed and released during operation of the BESS. As can be seen in Fig. 21.30d, capacity loss varies significantly depending on the floating SOC, with lower SOCs resulting in lower degradation.

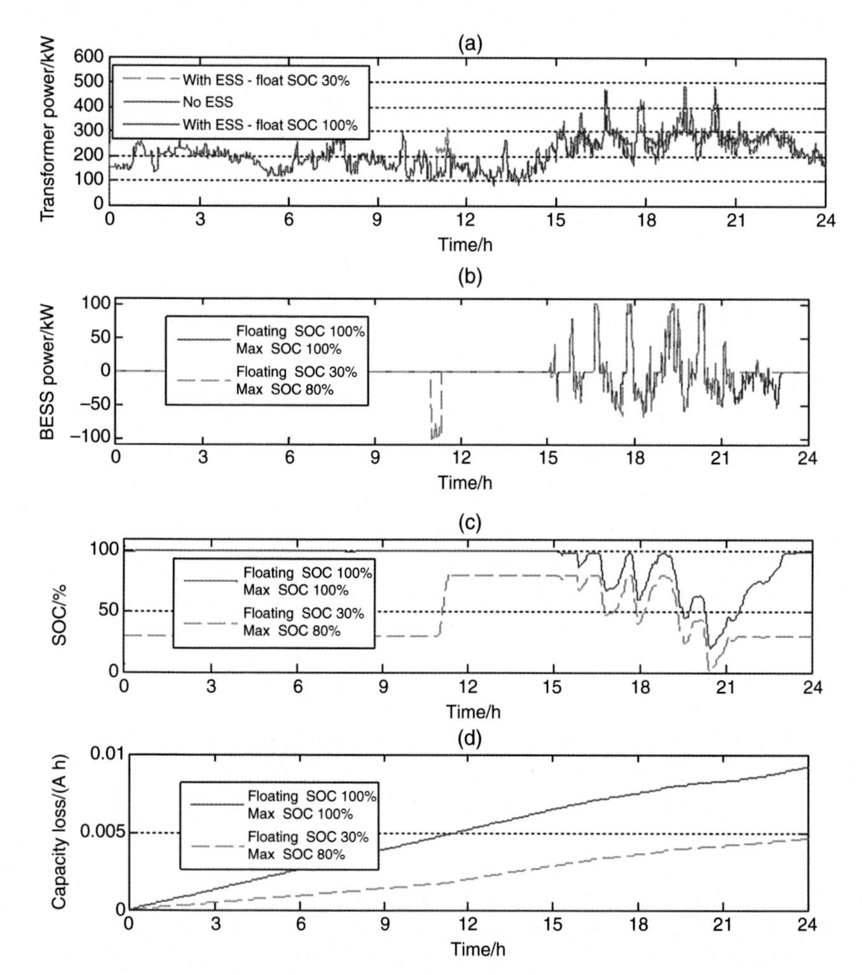

FIGURE 21.30 BESS operation for two combinations of floating SOC and SOC limits, showing (a) power flowing through the transformer; (b) BESS power; (c) battery SOC; and (d) battery capacity loss.

The energy absorbed and released during operation of the BESS is depicted in Fig. 21.31a for combination 1. The energy dissipated in the batteries through internal irreversible losses is 33.4% of total losses. Fig. 21.31b shows the dependency of battery efficiency on the depth of discharge. Battery efficiency decreases as the depth of discharge is increased

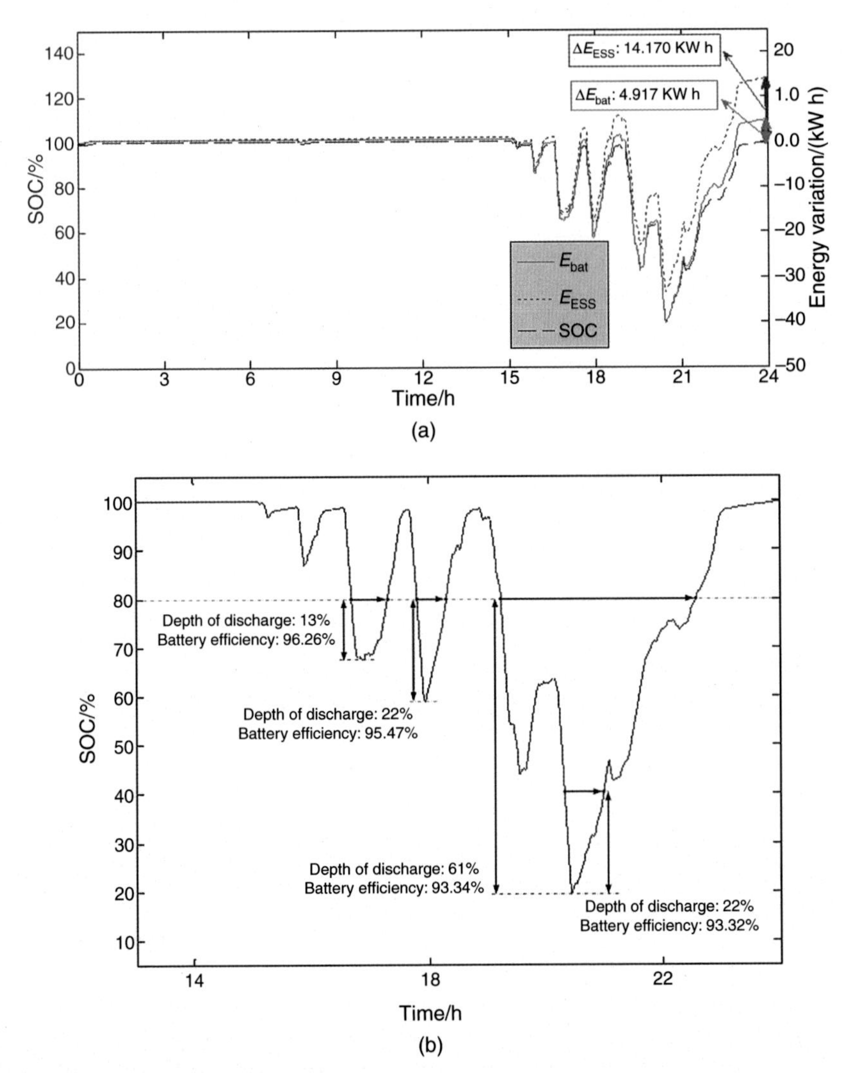

FIGURE 21.31 Roundtrip energy losses and battery efficiency: (a) energy absorbed and released during operation of the BESS, and calculated total roundtrip energy losses ΔE_{ESS} and battery roundtrip losses ΔE_{bat}, respectively, over time for a 24 h cycle; and (b) SOC over time for a 24 h cycle marking battery roundtrip efficiency calculated for different depths of discharge and SOC values.

TABLE 21.3 Combinations of Different Floating SOC and SOC Limits

		Max.–Min. SOC/%		
Float SOC/%		Max. 100–Min. 20	Max. 90–Min. 10	Max. 80–Min. 0
	100	V	X	X
	90	V	V	X
	60	V	V	V
	30	V	V	V

V, valid combination; X, nonvalid combination.

for cycles with equal SOC set points. In the case of cycles with similar depths of discharge, efficiency is higher in the cycles with the highest SOC set points.

Simulations were then undertaken to study the effects of floating SOC and SOC limits on degradation and efficiency. The scenarios considered are shown in Table 21.3.

Results from these simulations are shown in Fig. 21.32. Floating at a higher SOC results in a higher rate of capacity loss. A 100% increase in capacity loss can be noted for the two extreme cases, that is, floating at 100% and 30% SOC.

Fig. 21.32 shows energy losses for the same combinations. Converter losses are not greatly affected by the floating SOC, depending instead on system utilization. Fig. 21.33 reaffirms that the main contributor to losses in the BESS is the battery interface and that higher depths of discharge will result in reduced battery efficiency.

These results imply that the choice of both the floating SOC and the SOC limits are critical in terms of efficiency and degradation. Even though the same energy was delivered to the grid under the same power by several combinations, capacity loss can be significantly different depending on the load profile and SOC restrictions.

The previous conclusions suggest that accurate load prediction could lead to optimized BESS operation, where load demand can be met and capacity loss minimized. The results also suggest that the targets of increasing roundtrip efficiency and minimizing battery degradation are contradictory in terms of establishing SOC set points. However, these vary nonlinearly and thus require models of sufficient fidelity to capture this. It would not have been possible to establish these conclusions without adopting such an integrated modeling approach.

The model incorporates accurate calculations of the key battery and power converter properties that impose important constraints on the services provided to the grid. The latter can be taken into consideration in control algorithms

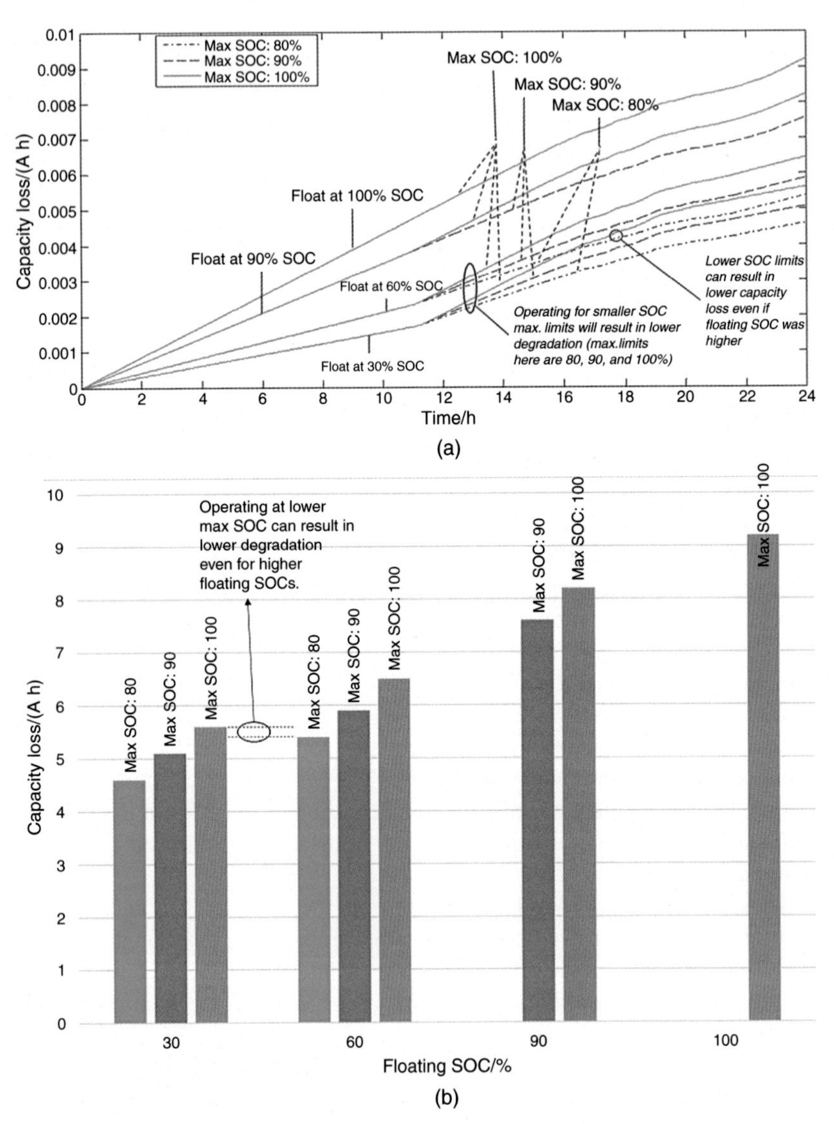

FIGURE 21.32 Battery degradation for 24 h of operation for different combinations of floating SOC and maximum SOC limits: (a) capacity loss over time and (b) total capacity loss at the end of the 24 h operation for a 4.8 A h cell.

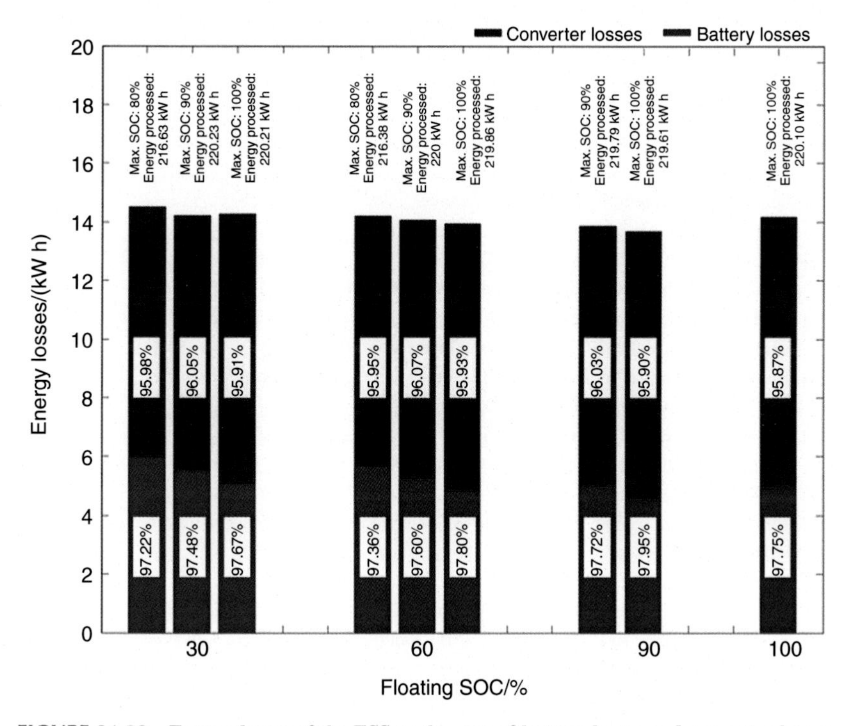

FIGURE 21.33 **Energy losses of the ESS as the sum of battery losses and converter losses.** The efficiency of the converter and the battery is calculated based on total energy processed during the day. Energy losses are presented for different combinations of floating SOC and maximum SOC limits.

targeting objectives that can range from voltage control to arbitrage and renewable energy time shift. This methodology could be applied to any other energy storage technology and thus acts as a platform for future work in this area.

REFERENCES

[1] The Electricity Advisory Committee, 2012. http://energy.gov/sites/prod/files/EAC%20Paper%20-%202012%20Storage%20Report%20-%2015%20Nov%202012.pdf

[2] Anuta OH, Taylor P, Jones D, McEntee T, Wade N. Renew Sust Energy Rev 2014;38: 489–508.

[3] Bouffard F, Kirschen DS. Energy Policy 2008;36:4504–8.

[4] Scott J. Power Eng 2004;18:12–3.

[5] Electricity Networks Strategy Group, 2009. A smart grid vision. http://webarchive.nationalarchives.gov.uk/20100919181607/http:/www.ensg.gov.uk/assets/ensg_smart_grid_wg_smart_grid_vision_final_issue_1.pdf

[6] Strbac G, Aunedi M, Pudjianto D, Djapic P, Teng F, Sturt A, Jackravut D, Sansom R, Yufit V, Brandon N. Report for Carbon Trust, 2012. http://www.carbontrust.com/media/129310/energy-storage-systems-role-value-strategic-assessment.pdf

[7] Ritschel A, Smestad GP. Energy Policy 2003;31:1379–91.

[8] Jamasb T, Pollitt M. Energy Policy 2007;35:6163–87.

[9] Barroso LA, Cavalcanti TH, Giesbertz P, Purchala K. CIGRE/IEEE PES International Symposium, 2005, New Orleans, LA. p. 9–16. DOI: 10.1109/CIGRE.2005.1532720.

[10] Pomper DE.. Silver Spring, MD: National Regulatory Research Institute; 2011. http://nrri.org/

[11] Ruester S, He X, Vasconcelos L, Glachant J-M. Think final report. Florence, Italy: European University Institute; 2012. http://think.eui.eu

[12] Pike Research, 2012. http://www.navigantresearch.com/

[13] Ponsot-Jacquin C, Bertrand J-F. 2012. http://www.ifpenergiesnouvelles.com/Publications/Available-studies/Panorama-technical-reports/Panorama-2013

[14] Taylor P, Bolton R, Stone D, Zhang X-P, Martin C, Upham P. Report for the centre for low carbon futures, York, 2012. http://www.lowcarbonfutures.org/

[15] European Commission, Brussels, 2011. https://ec.europa.eu/research/industrial_technologies/pdf/materials-roadmap-elcet-13122011_en.pdf

[16] UK Government, Department of Energy and Climate Change, 2011. https://www.gov.uk/government/publications/planning-our-electric-future-a-white-paper-for-secure-affordable-and-low-carbon-energy

[17] European Commission, Brussels, 2013. https://ec.europa.eu/energy/sites/ener/files/energy_storage.pdf

[18] Danish Government, 2011. http://www.kebmin.dk/sites/kebmin.dk/files/news/from-coal-oil-and-gas-to-green-energy/Energy%20Strategy%202050%20web.pdf

[19] Germany Trade and Investment, 2013. http://www.gtai.de/GTAI/Navigation/EN/Invest/Industries/Smarter-business/Smart-energy/germanys-energy-concept,t = the-german-electricity-market--a-brief-overview,did = 622868.html

[20] GROW-DERS Project. http://www.growders.eu/papers/CICED%202010%20paper.pdf

[21] Rangoni B. Utilities Policy 2012;23:31–9.

[22] Australian Government, Department of Industry, Innovation, Climate Change, Science, Research and Tertiary Education. Australia's sixth national communication on climate change, 2013. http://unfccc.int/resource/docs/natc/aus_nc6.pdf

[23] Skyllas-Kazacos M. ECS Trans 2013;50:1–5.

[24] REN21, 2012. http://www.ren21.net/Portals/0/documents/Resources/GSR2012_low%20res_FINAL.pdf

[25] Australian Government, Climate Change Authority, 2012. http://www.climatechangeauthority.gov.au/reviews/2012-renewable-energy-target-review

[26] Arif MT, Oo AMT, Ali ABM. J Renew Energ 2013;2013. article ID785636, 15 pages.

[27] Regulatory Assistance Project, 2008. http://www.raponline.org/featured-work/power-sector-regulatory-market-reform-china

[28] Abele A, Elkind E, Intrator J, Washom B. University of California, Berkeley School of Law; 2011. http://www.energy.ca.gov/2011publications/CEC-500-2011-047/CEC-500-2011-047.pdf

[29] Royal Academy of Engineering, 2012. http://www.raeng.org.uk/publications/reports/future-of-energy-storage-technologies-and-storage

[30] Schwartz D. London: Law Business Research Ltd; 2012. http://thelawreviews.co.uk/titles/1204/energy-regulation-markets-review/

[31] Roberts BP, Sandberg C. P IEEE 2011;99:1139–44.

[32] US Government. US Code. Washington, DC: US Government Printing Office; 2010. https://www.law.cornell.edu/uscode/text/42

[33] Bhatnagar D, Loose V. A perspective for state electric utility regulators. SAND2012-9422. Albuquerque, NM: Sandia National Laboratories; 2012. http://www.sandia.gov/ess/publications/SAND2012-9422.pdf

[34] California Public Utility Commission, 2010. http://www.clean-coalition.org/site/wp-content/uploads/2012/11/R1012007-Clean-Coalition-Comments-on-Proposed-Decision.pdf

[35] Driesen J, Green T, van Craenenbroeck T, Belmans R. Power Engineering Society Winter Meeting, 2002, New York. Piscataway, NJ: IEEE; 2002. vol. 1, p. 262–7.

[36] International Energy Agency, 2002. http://library.umac.mo/ebooks/b13623175.pdf, https://www.iea.org/

[37] UK Government, Department of Energy and Climate Change, 2012. https://www.gov.uk/government/uploads/system/uploads/attachment_data/file/197633/liquidity_measures_ia.pdf

[38] Miller N, Manz D, Roedel J, Marken P, Kronbeck E. Power and Energy Society general meeting, 2010, Minneapolis, MN. Piscataway, NJ: IEEE; 2010. p. 1–7.

[39] Genoese M, Genoese F, Most D, Fichtner W. Seventh international conference on the European energy market, 2010, Madrid, Spain. p. 1–6.

[40] Zahedi A. Renew Sust Energy Rev 2011;15:866–70.

[41] Barton JP, Infield DG. IEEE Trans Energy Convers 2004;19:441–8.

[42] Passey R, Spooner T, MacGill I, Watt M, Syngellakis K. Energy Policy 2011;39:6280–90.

[43] HMSO. UK, 2002. p. 13–14. http://www.legislation.gov.uk/uksi/2002/2665/contents/made

[44] OFGEMLondon, UK, 2010. https://www.ofgem.gov.uk/ofgem-publications/64031/re-wiringbritainfs.pdf

[45] Wade N, Taylor P, Lang P, Jones P. Energy Policy 2010;38:7180–8.

[46] Denholm P, Margolis RM. Energy Policy 2007;35:2852–61.

[47] Geth F, Tant J, Haesen E, Driesen J, Belmans R. IEEE Power and Energy Society general meeting, 2010, Minneapolis, MN. Piscataway, NJ: IEEE.

[48] Kashem MA, Ledwich G. Electr Power Syst Res 2007;77:10–23.

[49] Crossland A, Jones D, Wade N.22nd international conference and exhibition on electricity distribution, Stockholm, 2013.

[50] Chakraborty S, Senjyu T, Toyama H, Saber AY, Funabashi T. IET generation. Transm Distrib 2009;3:987–99.

[51] Xiaohua H, Guomin Z, Liye X. IEEE Trans Appl Superconduct 2010;20:1316–9.

[52] Chang Y, Mao X, Zhao Y, Feng S, Chen H, Finlow D. J Power Sources 2009;191:176–83.

[53] Carpinelli G, Celli G, Mocci S, Mottola F, Pilo F, Proto D. IEEE Trans Smart Grid 2013;4:985–95.

[54] Alarcon-Rodriguez A, Haesen E, Ault G, Driesen J, Belmans R. Multi-objective planning framework for stochastic and controllable distributed energy resources. IET Renew Power Gener 2009;3:227–38.

[55] Greenwood DM, Wade NS, Taylor PC, Panagiotis P, Heyward N, Mehtah P. 23rd international conference and exhibition on electricity distribution, Lyon, France, 2015.

[56] Greenwood DM, Taylor PC. IET generation. Transm Distrib 2014;8:2055–64.

[57] Tsekouras GJ, Hatziargyriou ND, Dialynas EN. IEEE Trans Power Syst 2007;22:1120–8.

[58] Strbac G. Energy Policy 2008;36:4419–26.

[59] Brown RE, Freeman LAA. Power Engineering Society summer meeting, 2001. Vancouver, BC, Canada: Power Engineering Society. vol. 2, p. 1013–1018.

[60] Wang P, Yi J, Lyons P, Taylor P, Miller D, Barker J, presented at the CIRED Workshop, Lisbon, Portugal, 2012.

[61] Seng LY, Taylor P. Innovative application of demand side management to power systems. Industrial and Information Systems, IEEE, First International Conference, Peradeniya, August 8–11, 2006, p. 185–189. http://ieeexplore.ieee.org/xpl/articleDetails.jsp?arnumber=4216581&filter=AND(p_Publication_Number:4216542).

[62] Stanojević V, Silva V, Pudjianto D, Strbac G, Lang P, MacLeman D. 20th international conference on electricity distribution, Prague, Czech Republic, 2009.
[63] Schroeder A. Appl Energy 2011;88:4700–12.
[64] Yi J, Wang P, Taylor PC, Lyons PF, Liang D, Brown S, Roberts D. Third IEEE PES international conference and exhibition on innovative smart grid technologies, Berlin, Germany, 2012.
[65] Tonkoski R, Lopes LAC, El-Fouly THM. IEEE Trans Sust Energy 2011;2:139–47.
[66] Mokhtari G, Nourbakhsh G, Ghosh A. IEEE Trans Power Syst 2013;28:4812–20.
[67] Tan KT, So PL, Chu YC, Chen MZQ. IEEE Trans Power Deliv 2013;28:704–13.
[68] Xin H, Lu Z, Qu Z, Gan D, Qi D. IET Control Theory Appl 2011;5:1617–29.
[69] Wang L, Liang DH, Crossland AF, Taylor PC, Jones D, Wade N. IEEE Trans Smart Grid 2015;. DOI: 10.1109/TSG.2015.2452579.
[70] Lyons PF, Wade NS, Jiang T, Taylor PC, Hashiesh F, Michel M, Miller D. Appl Energy 2015;137:677–91.
[71] N. S. Wade, K. Wang, M. Michel T. Willis, in Innovative Smart Grid Technologies Europe, Grenoble, France; 2013.
[72] Lang PD, Wade NS, Taylor PC, Jones PR, Larsson TG. 21st international conference on electricity distribution, Frankfurt, Germany, 2011.
[73] Lang P, Michel M, Wade N, Grunbaum R, Larsson T. IEEE Power and Energy Society general meeting, 2013, Vancouver, BC, Canada.
[74] Chang W-Y. ISRN Appl Math 2013;2013.
[75] Omar N, Monem MA, Firouz Y, Salminen J, Smekens J, Hegazy O, Gaulous H, Mulder G, Van den Bossche P, Coosemans T, Van Mierlo J. Appl Energy 2014;113:1575–85.
[76] Waag W, Käbitz S, Sauer DU. Appl Energy 2013;102:885–97.
[77] Patsios C, Wu B, Chatzinikolaou E, Rogers DJ, Wade N, Brandon NP, Taylor P. An integrated approach for the analysis and control of grid connected energy storage systems. J Energy Storage 2016;5:48–61.
[78] Wu B, Yufit V, Marinescu M, Offer GJ, Martinez-Botas RF, Brandon NP. J Power Sources 2013;243:544–54.
[79] Chaturvedi NA, Klein R, Christensen J, Ahmed J, Kojic A. IEEE Trans Control Syst 2010;30:49–68.
[80] Ramadass P, Haran B, Gomadam PM, White R, Popov BN. J Electrochem Soc 2004;151: A196–203.

Chapter 22

Off-Grid Energy Storage

Catalina Spataru*, Pierrick Bouffaron,†**
*Energy Institute, University College London, United Kingdom; **MINES ParisTech, PSL Research University, Centre de mathématiques appliquées, France; †Berkeley Energy & Climate Institute, University of California, Berkeley, United States of America

1 INTRODUCTION: THE CHALLENGES OF ENERGY STORAGE

Energy storage is one of the most promising options in the management of future power grids, as it can support discharge periods for standalone applications such as solar photovoltaics (PV) and wind turbines. The main key to a successful minigrid and microgrid is a reliable energy storage solution, including but not limited to batteries [1]. While mentions of large tied-grid energy storage technologies will be made, the chapter focuses on off-grid storage systems from the perspective of rural and island electrification, which means in the context of providing energy services in remote areas.

The electrical load of power systems varies significantly with both location and time. Whereas time dependence and magnitudes can vary appreciably with the context, location, weather, and time, diversified patterns of energy use are always present and can pose serious challenges for operators and consumers alike [2]. This is particularly true for off-grid systems and minigrids. In the last couple of years, renewable energy (RE) such as solar PV and wind has become a game-changer based on the rapid drop of hardware costs in the global market [3]. This is good news. RE is a key driver for a new, sustainable, energy future. Due to their intermittent nature, the rapid penetration of RE into traditional power systems has been challenging the day-to-day operation of traditional power grids by introducing another degree of forecast uncertainty [4]. This drawback in the operation of electric networks must be addressed to maintain power quality, reliability, and supply efficiency [5]. On the one hand, RE supply does not always coincide with peak demand. In large interconnected power systems, "peaker" plants can be used to make up for the gap between demand and shortage of RE supply, but dispatching resources collected from "peakers" can be economically and environmentally expensive. For minigrids and off-grid systems, energy storage technologies become a must when renewable penetration is high, especially with no backup diesel engine. On the other hand,

Storing Energy. http://dx.doi.org/10.1016/B978-0-12-803440-8.00022-1

RE sources injecting energy into the grid when the demand is low constrains operators to store, export, or lose the energy produced.

The obvious need for balancing renewable sources opened up new opportunities for energy storage technologies. Storage offers operators flexible assets in the management of power systems. In large power systems, it plays a major role in the alleviation of the peaking capacity crunch witnessed [6]. In smaller systems, storage plays two key roles: balancing the intermittency of RE sources and providing energy when those sources are not able to produce (e.g., during the night for solar resources). As RE penetration increases, energy storage technology installations around the world should pave the way for larger deployment [7] with possible great value, with the idea that multiple usage can be economic [8]. Despite this widely recognized necessity and the fast-changing electricity landscape worldwide the development and integration of energy storage technologies have been slowly rising in the last years. Three main challenges remain for the power sector:

1. *Technical challenges*. There is still uncertainty among stakeholders over the scale at which systems should be deployed [9]. Bulk power systems are deployed and managed centrally. They have long been the norm and the option favored by utilities. Alternatively, RE technologies become more competitive and broaden the technology mix [3]. For example, large-scale on-grid solar power capacity could become available at around \$1 W^{-1} (one dollar per watt) by 2020, down from more than \$8 W^{-1} in 2007 [10]. They bring new operation constraints and complexity and require the association of energy storage solutions to insure power quality and supply efficiency.
2. *Regulatory challenges*. The economic framework of energy storage is strongly dependent on favorable local regulatory conditions. To date, there is no universal answer regarding whether storage is profitable, how much value it adds to the system in terms of technical rules, ownership, and market regulation [11, 12]. Any storage valuation requires a clear regulatory framework.
3. *Economic challenges*. Innovative business models must be created to foster the deployment of energy storage technologies. A review is provided in [12] that shows energy storage can generate savings for grid systems under specific conditions. However, it is difficult to aggregate cumulative benefit streams and thus formulate feasible value propositions [13], especially in the case of off-grid systems where economic valuation is restricted. Furthermore, comprehensive and consistent assessments of cross-value chain storage have not yet been performed for many market situations.

2 WHY IS OFF-GRID ENERGY IMPORTANT?

Off-grid RE technologies are increasingly becoming a competitive solution for sustainable energy access in a range of remote locations, from individual homes, to minigrid level servicing 50–100 households, and to microgrids servicing islands [14–16]. In developing countries the dramatic uptake of mobile phones is a perfect example of how quickly decentralized services can develop on a commercial basis under the right conditions. The private sector plays a key role: it raises the prospect that private finance could also drive decentralized energy access for the poor [17]. This is a welcomed new trend. Since the 1950s the failure of many governments worldwide to achieve a significant electrification rate in many countries led to the rapid growth of decentralized power systems based mostly on fossil fuels. It is estimated that poor African households face recurring expenditures on fuels ranging between (10–25)% of their monthly income [15]. Any strong price rise due to oil volatility could wipe out the remaining income available for other services, or even for food. Today, RE technologies are becoming increasingly cheaper and, therefore, the most sustainable solution for energy access in numerous off-grid situations. Decentralized clean energy markets are fundamental in the development of local economies. Low-income households and microenterprises could benefit by reduced energy bills, increased net income, and improved health conditions [18,19]. Some clean energy markets in developing countries are already experiencing double-digit growth rates [20]. The growth rate of solar PV systems in Africa has grown by an average of 13% a year over the past decade. The local markets for off-grid RE lighting could experience 50% annual growth over the next few years [21].

What about energy storage technologies in this picture? While storage value has been identified in many cases, three use cases are essential when it comes to off-grid systems: power quality, power reliability, and balancing support. Indeed, energy storage can enable time shifting at the time of excess low-cost generation and the release of energy in times of peak demand [7]. Looking at the power transmission sector in large interconnected systems the balance between the supply and demand of electricity is mostly maintained by the market domain, which includes but is not limited to energy storage technologies and their usage. For smaller grids and off-grid the added value of energy storage goes further than just grid balance: power quality issues and power reliability are also addressed [17,22]. Power quality is the ability of the supplied electricity on the distribution grid to adhere to specified peak levels and standard voltage levels. Any change in the level (e.g., increasing, decreasing or random voltage) can impact loads and disturb frequency control. It might also lead to the disconnection of solar panel inverters in case of overvoltage, which implies a loss of available RE. One of the ways to avoid such a situation is to use storage and/or demand response schemes in accordance with load sensitivity. Power reliability is more easily understandable, since storage capacity is designed to handle adapted discharge periods and provide energy when no power source is available.

Insuring power quality and power reliability may become an increasingly complex task. Unlike the traditional power grids, future electricity networks will accommodate various kinds of electricity generation technologies as well as newly complex loads (e.g., loads with communication capabilities, such as electric vehicles) and allow for enhanced bidirectional energy exchanges [23]. In Western countries, soon to be followed by emerging economies, smart devices will soon be the norm and modify residential consumption patterns. Therefore, operators, utilities, and consumers alike currently face a new emerging paradigm: energy exchanges between stakeholders will become frequent, competitive, and key to the management of local power systems. When properly designed the inherent exchange mechanisms will show operational benefits for all parties [24].

3 BATTERY TECHNOLOGIES AND APPLICATIONS

3.1 Battery Technologies

We suggest looking at existing electrochemical energy storage (EES) technologies and most specifically those generally used or deemed to be used for off-grid, minigrid, and microgrid projects: lead acid (L/A) batteries, lithium-ion (Li-ion) batteries, sodium sulfur (NaS) batteries, and vanadium redox flow batteries (VRB). EES is indeed the most common storage option in off-grid projects, although a few hybrid storage systems have emerged during the past few years. Key parameters used to compare the types of batteries on the market are described below ([2,25,26]) and summarized in Table 22.1.

- energy storage capacity (kW h): the amount of energy that can be stored
- volumetric energy density (W h L^{-1}): the nominal storage energy per unit volume
- power density (W L^{-1}): the maximum available power per unit volume
- charge/discharge duration: the time needed for the storage to fully charge or discharge
- power output (MW): amount of power discharged within a typical discharge duration
- response time: the time needed for the storage to start providing power output
- lifetime: the number of cycles and/or years that a storage technology will continue to operate
- roundtrip efficiency (%): the ratio of energy discharged by the system to the energy required (including losses) to charge the system over each cycle
- capital cost ($ kW^{-1} or $ kW h^{-1}): the upfront investment costs of a storage technology per unit of power discharge ($ kW^{-1}) or energy storage capacity ($ kW h^{-1}).

3.1.1 Lead–Acid Batteries

First introduced in 1860, L/A batteries are the most common rechargeable large-format batteries available on the market today. Significant technological improvements have been made over the years [2], for example, the development of sealed L/A batteries, so that batteries are “spill proof” and can be used in any physical orientation: upright, on the side, or even upside down [27]. The main benefits of L/A batteries are their relatively low cost and ease of manufacture.

TABLE 22.1 Storage Technologies [2,6,25,26,33–36]

	L/A battery	Li-ion battery	NaS battery	VRB flow battery
Energy storage capacity/kW h	<100	<10	<100	20–50
Typical power output/MW	1–100	0.1–5	5	0.01–10
Energy density/W h L^{-1}	50–80	200–500	150–250	16–33
Power density/W L^{-1}	10–400	0	0	(2–8) h
Discharge duration	Hours	Minutes–hours	Hours	(2–8) h
Charge duration	Hours	Minutes–hours	Hours	<Seconds
Response time	<Seconds	Seconds	Milliseconds	
Lifetime/years	3–10	10–15	15	5–20+
Lifetime/cycles	500–800	2 000–3 000	4 000–40 000	1 500–15 000
Roundtrip efficiency/%	70–90	85–95	80–90	70–85
Capital cost per discharge/$ $(kW)^{-1}$	300–800	400–1 000	1 000–2 000	1 200–2 000
Capital cost per capacity/$ (kW h^{-1})	150–500	500–1 500	125–250	350–800
Power quality			✓	✓
Transient stability	✓			
Ancillary services				
Regulation		✓	✓	✓
Spinning reserves	✓	✓	✓	✓
Voltage control		✓	✓	✓
Energy services				
Arbitrage			✓	✓
Load following	✓	✓	✓	✓
firm capacity				
Congestion relief	✓	✓	✓	✓
Upgrade deferral	✓	✓	✓	✓
Advantages	Low-cost, high-recycled content	High efficiency, high energy density	High energy density, quick response, efficient cycles	Higher depth of discharge, high cycling tolerance
Disadvantages	Low energy density, large footprint, limited discharge depth	Cost prohibitive, overheats, limited discharge depth	Safety issues largely related to their high running temperature	Low energy density, low efficiency

The typical L/A battery is cheaper than any other competing kind of EES solution, with costs ranging from (50–250) $ per kW h^{-1}. Different types of L/A batteries can be used following end-usage, but two main applications can be found: (2) keeping the cell fully charged to maintain a constant output voltage (e.g., in UPS systems) and (2) load-leveling systems through periodic charging (e.g., in the context of rural electrification). The average L/A battery efficiency is close to 80%, and the cells can provide between 400–1500 complete cycles. Advanced L/A batteries have recently appeared on the market. While they present an enhanced lifecycle (4500 cycles would be possible), their average price is still three times that of a standard model. Self-discharge is about 2% per month for stationary usage [28].

3.1.2 Lithium-Ion Batteries

The electrochemical reaction of a Li-ion battery is based on charge transfer that occurs through ion intercalation rather than chemical reactions on the electrodes as in some other battery types. The electrode areas are demarcated by a porous separator such as polyethylene to allow lithium ion exchanges. In discharge mode, positive lithium ions move toward the cathode while electrons travel in the same direction via an external circuit to recombine with the lithium ions. The charging cycle consists of the reverse process to discharging. The costs of Li-ion cells are still high about \$850 kW^{-1} but the development of mass markets makes the technology more and more competitive. Efficiency is high (90–95)%, and 3000–4000 complete cycles are now possible, on average. Some advanced Li-ion technologies are able to offer up to 15 000 complete cycles. Self-discharge is about 5% per month.

Performance wise, Li-ion batteries have always been popular for their high specific energy, which is between (75–125) W h kg^{-1}. The major drawback with this battery technology is proneness to thermal runaway. Since Li-ion batteries have high charging and discharging capabilities, they can easily be overcharged. This inadvertently leads to temperature rise as the battery consistently generates more heat that it can dissipate. The consequences are the risk of leakage or, even worse, the possibility of an explosion. Another example of the weakness of Li-ion batteries is seen when the batteries are used outside their recognized temperature range, resulting in capacity fading due to self-discharging. For that reason, control circuitry is required to provide protection to Li-ion batteries, pushing the cost of an expensive battery to an even higher level. While Li-ion batteries are not commonly used in rural electrification projects the booming electric vehicle industry will rapidly provide for partially degraded, "retired," but still usable batteries in 2016 and beyond. These batteries can become the storage hubs for community-scale grids in the developing world [29].

3.1.3 Sodium–Sulfur Batteries

NaS technology is mature and industrialized. The only manufacturer is the Japanese company NGK Insulation Ltd. The NaS battery is made up of an anode made of molten sodium (Na) and a cathode made of molten sulfur (S). In most cases, NaS battery cells are designed in a tubular manner, in which the sodium electrode is usually contained in an interior cavity formed by the electrolyte. Typically, NaS batteries are characterized by their need and capability to operate under high temperature conditions between (300–350)°C. In this temperature range the ceramic material shows remarkable conductivity [30]. As a result the heat required to stimulate charging or discharge reactions is adequately supplied by the reactions themselves. This exothermic nature means that there is no need to have an external energy source. The energy and power density for NaS batteries are (100–175) W h kg^{-1} and (90–230) W kg^{-1}, respectively. The NaS battery has the advantages of high energy density, high efficiency of charge/discharge (89%) and long cycle life. The technology is mostly used in island contexts [31].

3.1.4 Vanadium Redox Flow Batteries

The other group of electrochemical cells that have liquid electrode reactants are the flow batteries, where energy is primarily stored in the active materials dissolved in the electrolytes that are stored externally [2,32]. The main advantages to these batteries are that the power and energy components are easily scalable and can be determined independently since electrolytes are stored externally.

The two major types of flow battery in early commercialization/demonstration stage are vanadium redox (VRB) (see chapter: Vanadium Redox Flow Batteries) and zinc–bromine batteries (ZnBr). Since each charge/discharge cycle returns the solutions to their initial states, the electrolytes can be used indefinitely, contributing to very long VRB lifetime, achieving over 10 000 cycles or above 10 years. The main drawback with VRB is their relatively low specific energy and energy density compared with other battery technologies. ZnBr is the primary alternative to VRB: its main advantage over VRB is igher specific energy; however, ZnBr batteries are slightly less mature compared with VRB technology.

Small projects have been deployed in rural Australia comprising 5 kW, 2 h systems to defer installing new transmission lines. For VRB, the case is similar with very few small-scale installations. The total accumulated installed capacity for flow batteries is about 3 MW worldwide [23]. A summary of the storage technologies of the various batteries is given in Table 22.1.

3.2 Battery Applications

EES systems have wide-application possibilities across the entire electric enterprise value chain and can be integrated at various levels into electricity systems: (1) upstream/downstream of a substation, (2) downstream of a medium-voltage/low-voltage (MV/LV) transformer, (3) nearby an RE producer connected to the LV grid, and (4) within or close to customer houses. EES thus has the inherent flexibility to be used as a generation, transmission, or RE integration asset as well as many combinations of these assets.

When considering distributed EES performing electricity arbitrage, transmission, and distribution infrastructural deferral, value stream aggregation is definitely one of the obstacles to its deployment. The work described in [37] introduces a distinction between "revenue" and "value" streams. For larger grids, utility-owned EES will indeed deliver a value stream with improved system efficiency yet associated cost savings are inaccessible to privately owned EES. The authors of [12] undertook an in-depth review of EES evaluations, demonstrating the net system benefits which storage technologies can bring. Recent studies in the United States evaluated EES at the grid scale and showed it facilitates increased renewable penetration. For example, [23] came up with a valuation tool to determine the value of EES for grid applications in California. In most scenarios, it was found that 50 MW batteries provide bulk storage and ancillary services with net benefit. Similarly, [38] demonstrates net savings due to EES on the island of Maui. Sandia estimates that more than 100 000 batteries are necessary to provide onsite power at US substations [39]. Finally, Viswanathan et al. [40] found that NaS or Li-ion batteries could be more cost effective than natural gas in supporting 15 GW of wind power in the US Northwest Power Pool (NPPW) system.

The breadth of available resources enables EES to provide a number of offgrid and tied-grid services, which can be broadly classified based on their timescale as power (short-duration) or energy (long-duration) services. The following list, compiled from [41] and [42], describes the most common offgrid and minigrid services provided by energy storage, with congestion relief and upgrade deferral being two further services that could be added (revenue stream for larger power systems). As already mentioned, three properties are key when it comes to minigrid, microgrid, and standalone power systems: power quality, power reliability, and balancing support:

- Power quality services support utilization of electric energy without interference or interruption, maintaining voltage levels within bounds.
- Transient stability services help to maintain synchronous operation of the grid when the system is subject to sudden (potentially large) disturbances.
- Regulation services correct short-term power imbalances that might affect system stability (generally frequency synchronization).
- Voltage control provides the ability to produce or absorb reactive power, and to maintain a specific voltage level.
- Energy arbitrage, which implies using power that is produced during off-peak hours to serve peak loads, that is, energy storage charges during off-peak times and discharges during peak times to provide load leveling/load shifting.
- Load following (balancing) which means adjusting power output as demand fluctuates to maintain power balance in the system.

4 DEALING WITH RENEWABLE VARIABILITY

RE sources are a key driver for a more sustainable energy ecosystem. This new paradigm is not without constraints: RE sources introduce some drawbacks in the operation of electric networks, which must be properly addressed to avoid deteriorating power quality, reliability, and supply efficiency [3]. In particular, one of the main issues of RE sources such as wind and solar energy is their unpredictability, which reduces the sound prevision of energy flows on networks. Indeed, the energy balance between load and generation has to be respected all the time, acting either on the injection of some additional flexible power plants (e.g., diesel engines) or the compensation offered by storage technologies [34], or both (this is the case for hybrid minigrids, for example).

Regarding solar PV, in recent years PV technologies have become much less expensive [3,43]. Solar PV is one of the most flexible RE available sources, in the sense that power outputs range from milliwatt to gigawatt and panels can be installed geographically almost everywhere (with inherent efficiency variations) [1]. Storage technologies are particularly useful in dealing with daily, monthly, and seasonal radiation changes, and this is particularly true for off-grid and minigrid contexts. Following the usage patterns of targeted areas, intermediate storage can also be used to cover load peaks. If nonelectrical energy storage systems—such as water tanks for a pumping system, or flywheels or hydrogen storage in specific locations and contexts—are sometimes a relevant solution, electrochemical storage technologies are the most common for off-grid installations [35]. As for wind energy, modern turbines can now supply inexpensive and relatively reliable energy in windy regions. Wind energy is available in the range of a few tens of watts to several kilowatts, which excludes this option for very low energy demand (i.e., less than 200 W h d^{-1}) [1]. The power of the wind is seldom sufficiently constant for a full system to work without energy storage. Therefore, it is now more and more common to combine wind and solar systems in so-called hybrid systems to increase reliability. What are the energy storage roles in this picture? Providing the energy reserve to bridge lulls in RE output is without any doubt the priority when it comes to off-grid and minigrid systems. Insuring power quality by smoothing out the ramping up and down of RE generation—and/or generator sets (gensets) for hybrid minigrids—is another key aspect. Finally, playing an arbitrage function in which green energy is stored when electricity prices are low and then dispatch it during peak demand is another possibility (the latter being particularly true when an optimization strategy has been put in place, which is less common for standalone installations). A single storage installation can potentially provide all or a subset of these services simultaneously [44].

High penetrations of RE sources are thus possible in remote areas, for minigrids, and microgrids. We can imagine having systems combining only intermittent RE systems and storage, but there are still technical challenges for systems larger than 200 kW [24]. Pilot projects attempt to accomplish wind/PV/storage hybrids with sophisticated control strategies, such as the El Hierro Island project in the Canaries [31,45].

5 THE EMERGENCE OF MINIGRIDS AND MICROGRIDS

The economies of remote areas and islands worldwide are often highly vulnerable to oil price volatility and supply disruptions. Islands especially are under stress as a result of environmental challenges such as climate change. Obviously, energy contexts and patterns for remote areas and islands in developing countries vary from those in developed countries, but lessons learned can be applied in both directions, especially under an energy storage perspective. Departing from the traditional centralized model, minigrids and microgrids have been very much in the air in recent years. There is a broad range of different areas where minigrids or microgrids could be useful, ranging from military bases, university campuses, to remote villages in Lao PDR [46] as well as tourist destinations in the Caribbean. All those areas are united by the common challenges of securing affordable and reliable energy. However, they would most probably require different energy policy approaches and a combination of RE and storage technology solutions [47,48].

Depending on the circumstances a minigrid or microgrid can operate either in parallel with the upstream grid (this is the case of most university campuses, for example) or in island mode. Energy sources that comprise the microgrid include technologies such as diesel generators, fuel cells, PV panels, and wind turbines associated with storage resources. Energy storage technologies play a key role in the operation of the system. For instance, to deal with the intermittence of RE sources, storage systems can be used to manage energy time shifting, provide ancillary services as well as power quality improvement. Because of the crucial role played by storage systems, their sizing is essential for assuring correct operation of the microgrid [24]. Sizing the storage system involves finding the optimal energy storage power and energy capacities with the aim of minimizing the operating costs. Unit commitment and economic dispatch problems must be solved: the unit commitment problem consists of determining the optimal schedule of generating units (and other resources) subject to the satisfaction of demand and other system operating constraints. Economic dispatch is a subroutine of the unit commitment problem whose aim is to locate optimal generator outputs such that the entire load may be supplied in the most economical way [49]. The capital cost of integrated storage technologies is to be considered, as we will see in the next section. The weight that this cost has on daily operation depends on the amortization period of the investment. For this reason the lifetime of the storage system has to be known. In the case of batteries the two main factors that affect the lifetime are the number of cycles and the state of charge at which the storage system operates.

Many minigrids or microgrids today are hybrid systems [31]. It is no secret that multisource power systems leverage the benefits of each technology to continuously insure the reliability of power. Hence, solar PV or wind power can be used to provide the bulk of the community's energy needs. Storage technologies can be used to bridge the energy supply during low production and high demand. A liquid fuel generator (or hydro, if available) can be switched on when the other sources are unavailable. Three examples of running minigrids integrating storage technologies, in various contexts, will be presented in the next section: (1) the island of Bonaire in The Netherlands; (2) the island of Miyakojima in Japan, and (3) the island of Eigg in Scotland.

6 ENERGY STORAGE IN ISLAND CONTEXTS

6.1 Island of Bonaire (The Netherlands)

In 2004 the main power plant on the island of Bonaire burned down. The long-term energy strategy of the island was renewed that day to move toward a 100% renewable electricity supply. In 2009 a hybrid wind–diesel facility was completed. 11 MW of wind power provides 50% of electricity demand, while a 15 MW diesel facility coupled with a 3 MW Li-ion battery provides the base power and backup to maintain power quality. The Li-ion batteries inject energy for 2 min to allow diesels to start if wind turbine output falls. The EES also provides a short-term dump load during system faults. Following the success of this strategy the plan of the Dutch government is to shift away from heavy dependence on diesel power and move toward use of wind resources on the islands [69].

6.2 Island of Miyakojima (Japan)

One of the largest of the Miyako Islands in Japan, Miyakojima has been expanding the use of RE sources since the early 1990s to reduce the use of and reliance on diesel (mostly solar PV systems, but also a fair number of wind turbines). In 1993, Mitsubishi Electric installed a 750 kWp (kilowatt-peak) PV with a 300 kW diesel hybrid in the framework of a project launched by the Japanese Ministry of Industry. In 2010, a 4 MWp PV (comprised of a 3 MWp solar farm and a 1 MWp total consumer side PV) and a 4 MW NaS grid stabilization battery system is expected to provide around 8% of the electricity needs of the island with expected emissions reduction of 4000 t of CO_2 [50]. In addition, 25 of the households with 4 kWp of consumer-side PV each have 8 kW h of Li-Ion battery storage. The project is up and running. PV power output with the storage capacities was analyzed in 2011: the output power can fluctuate up to 3 MW within minutes depending on weather conditions; by using storage systems this effect on the grid is minimized [51].

6.3 Island of Eigg (Scotland)

The island of Eigg is the second largest island of the Small Isles Archipelago in Scotland. A hybrid off-grid system was installed as follows: 119 kW of hydro power capacity using three turbines of 100, 10, and 9 kW; 24 kW of wind power capacity; 54 kW of solar PV capacity; and 160 kW of diesel generator capacity as a backup (total system installed capacity is about 357 kW). An additional battery bank and the associated inverter system are at the heart of the system: 48 V batteries with a total 4400 A h capacity provide enough storage from RE sources for delivery when the demand arises. Inverters control the frequency and voltage of the grid balancing the demand and supply and controlling power input and output to and from the batteries. The system load is distributed between 38 households and 5 commercial properties, connected through an 11 km long underground high-voltage distribution system. Not only are the residents of the island enjoying a reliable electricity supply but, more importantly, their carbon footprint has fallen considerably as 90% of their electricity comes from RE [52].

7 BRING CLEAN ENERGY TO THE POOR

According to the *World Energy Outlook*, there are still today about 1.4 billion people lacking access to electricity, some 85% in rural areas [53]. There is only a 21% electrification rate across the least developed countries (LDCs), with most electrified areas concentrated in urban areas [54]. Therefore, rural electrification is one of the largest markets for "off-grid" storage applications. Here lies one of the challenges of economic growth, since the link between energy access and development has been widely acknowledged [52,55]. Many reports discuss the benefit of job creation due to RE integration [56,57]. Notably, RE can undercut, on a level basis, the cost of traditional sources such as diesel, providing valuable savings for the entire value chain, from local governments to customers. Remote communities represent perfect testbeds to implement storage, grid integration strategies, and the management of high-penetration RE systems [46]. Despite double-digit percentage growth in some clean energy markets, the number of people without access to electricity in sub-Saharan Africa is expected to increase by 10%, from 585 million in 2009 to 645 million in 2030 under a business-as-usual scenario, as the rate of connections will not be able to keep pace with population growth [58]. Globally, over one billion people will remain without access to electricity by 2030. Clean energy, storage development, and access face a large number of interconnected information, institutional, behavioral, technical, and financial barriers.

An unfortunate trend is the design choices that are often made for rural electrification, especially on the combination of RE and batteries. Users tend to minimize the initial investment cost. In many cases the battery and the solar PV generators are designed too small for the task in hand, which leads to batteries not properly charged and systems which do not satisfy the demand of the user or the community [1]. The first consideration in designing any energy system is to insure that energy demand is limited to the fullest extent possible. This is important both in remote and in nonremote areas. However, a key difference in remote areas is that the value of managing energy use is greater due to the higher cost of generation and, therefore, the range of cost-effective measures is greater [59]. Properly characterized demand can more fully leverage the cost-saving benefits of RE technologies as well as storage technologies. Energy systems and storage systems such as batteries are no exception: they should be sized to meet specific load levels and optimized to provide power at expected levels.

Once demand is precisely defined, the question of how to integrate the economic valuation of storage technologies must be tackled. The leveled cost of electricity (LCOE) is often cited as a convenient summary measure of the overall competiveness of energy systems, giving precious information to validate a project [60,61]. It represents the per-kilowatt-hour cost (in real dollars) of building and operating a generating plant over an assumed financial life and duty cycle [48]. LCOE can be used to assess the additional system costs of standalone systems and minigrids. It was determined that adding batteries to an RE system can raise the system cost by up to (30–50)%, thus leading to a significantly higher LCOE [62]. Very often one of the assumptions made is that RE systems require storage and its costs should be included in the LCOE calculation. Such assumptions are not always true: numerous minigrids use storage technologies to improve the reliability and efficiency of combustion engines, insuring minimum energy input during the nights to turn the engines off [4]. When RE is also present (hybrid systems), storage will be used for balancing. Therefore, storage can be viewed here as part of a broader minigrid management strategy and not as an incremental cost of RE integration. On the other hand, off-grid, pico-scale PV (100 W or smaller), with or without attached energy storage, could have a much higher LCOE than a large diesel generator in an urban area, even with the skyrocketing cost of diesel fuel. Therefore, context is essential for LCOE calculation.

8 THE WAY FORWARD: COST–STRUCTURE EVOLUTION

Many current energy prospective scenarios today do not fully describe the challenges at stake in remote areas. By definition, these areas will most probably never be connected to transnational interconnected grids nor will they have access to the new natural gas-run power sources that would make energy "bridge" transitions away from coal and diesel feasible. Thus, the most practical and cost-effective energy future in islands, rural areas, and other remote areas is an RE future [44]. This is where energy storage will play a fundamental role and where relevant technology assessment of storage technologies must be tightly coupled with economic analysis that takes into account all of its potential revenue streams [46]. Indeed, while different revenue streams can potentially increase the value of storage technologies that have a large range of technical capabilities, most projects often focus on performing one particular service (e.g., generation capacity or time shifting). This is challenging both from a technical and a business perspective: how to dispatch storage according to different objective functions, and how to optimize the storage system for the investor as well as the user?

Commonly, storage technologies can compensate for generation capacity; bulk energy arbitrage; the provision of ancillary services and reductions in distribution access; and congestion charges. From the end-user and operator side, storage provides indirect financial benefits such as reductions in electricity rates, increases in reliability and power quality, and deferral of new capacity investments. Most importantly, storage can also significantly increase the revenue from RE sources (e.g., solar PV and wind turbines), for example, mitigating the risk of forecast uncertainty, allowing time shifting, playing the role of backup, etc. Naturally, the economic framework for RE is not as straightforward as this presentation of all revenue streams could suggest. In recent years, many projects failed because of market barriers, irrelevant technology choices, and other economic and regulatory barriers. For some time now the main argument offered to explain the low deployment of RE technologies was that the market would follow when "grid parity" (which means when RE capacity equalizes traditional power capacity) is reached [63]. Clearly, evidence shows this is not the case, even in remote areas where RE has often reached and surpassed grid parity. Although this is attributable in part to the noneconomic barriers we mentioned previously (design irrelevance, lack of skills to insure proper system operation, etc.), financing plays an important role too. The main challenges include the cost of capital (high interest rates due to higher risk in remote areas), the access and availability of capital (lower number of investors, less interest in remote areas), and the role of subsidies (conventional thermal generation is often subsidized at multiple levels: distribution, storage, direct pricing to customers). To face these challenges, public policies can help to spur new business models and accelerate the commercialization of RE and storage technologies. Alstone et al. [64] identify energy isolation barriers that remote areas keep experiencing as a result of three main dimensions of remoteness: geographic, of course, but also economic and political.

There is no universal answer on whether storage is a profitable investment or adds value to a system. Off-grid storage will be directly impacted by local regulation and codes, such as market design, technical rules for renewable energy scenario (RES) integration, ownership, and operation of storage designs. Based on previous experiences of market-driven programs launched by international organizations and NGOs to foster RE integration, public policies would have to focus priority on three fronts to enhance off-grid storage integration: (1) increasing affordability of storage technologies for the poor; (2) increasing access to financing for the poor; and (3) working at removing noneconomic barriers [65]. Low-income consumers do not have the luxury to reason in terms of LCOE as previously described: upfront costs remain the major bottleneck. Local governments may propose a set of methods to lower upfront costs, including reduction of balance-of-system costs (for a PV installation, balance of systems represents everything except the PV modules themselves, including mounting and racking components, inverters, permitting, etc.), the elimination of taxes and tariffs on clean energy devices, and the promotion of entrepreneurship through the new usage of energy services. Balance-of-system costs can represent 50% of a project's cost: opportunities to reduce it to about 25% have been reported, mainly based on the elimination of inefficiencies in business practices [43].

9 INTERNATIONAL EXAMPLES

Data reported in many different case studies are often incomplete, which can make it complex to benchmark, compare, and evaluate off-grid pilot projects involving storage technology precisely. The following examples contribute to the discovery of new insights about the implementation of storage systems in remote areas [51].

9.1 Developing a Microgrid with Racks of Lead–Acid Batteries: Akkan (Morocco)

An interesting case study is the hybrid microgrid based in Akkan (Morocco), fueled by solar PV, a diesel generator, and batteries. Akkan is an isolated Moroccan village made up of 35 households, a school, a mosque, a communal house, and public lighting. Before the microgrid project, access to energy services was limited to candles, kerosene, wood, and batteries. The new microgrid is composed of a 5.6 kWp PV installation, a 72 kW h battery bank, a 6 kW battery charge controller, a 7.2 kW inverter/rectifier, and an 8.2 kW backup diesel genset. Run in 2006 the users reported high satisfaction with the whole system and quickly requested greater electricity availability. The system was designed on basic residential energy needs (lighting, radios, TVs) as well as some light commercial needs. Each user has been required to sign a personalized contract for service supply. It is worth noting that the genset was not used at all during the year 2011, which seems to show that the project was designed appropriately.

9.2 Developing a National Policy for NaS Batteries: the Case of Japan

Japan is leading the market on battery storage systems, with over 200 utility applications based on NaS storage systems only [66]. NaS battery application in utility grids covers a range between (1.0–34) MW (over 200 MW h in total) worldwide with NGK Insulators Ltd. of Japan being the only manufacturer. On the policy side, Japan brought both technologies to a commercially stable level. The Ministry of Economy, Trade and Industry (METI) has set lofty ambitions aimed at performance enhancement and cost reduction through several programs. One of these priority programs is inclined toward renewable integration, as large-scale stationary batteries are developed to compensate for output fluctuation in wind turbines as well as solar farms, while another is targeted at smaller scale battery application as batteries are designed for residential dwellings for energy management purposes. The most recent was the unleashing of a 3 year subsidy program. This program covered stationary Li-ion battery energy storage, and it was valued at approximately ¥20 $\times$ 10^9 (£130.4 $\times$ 10^6). The Japanese legal framework clearly encourages legislation to support energy storage technology application at every consumption level: residential, commercial and industrial [67].

9.3 An Example of New Microgrid Project in the West

9.3.1 US Navy Smart Microgrid

The smart microgrid demonstration system consists of 150 kW of solar PV capacity and a 100 kW/400 kW h energy storage solution based on three equally sized vanadium flow batteries. The project hosted by the US Navy at its Mobile Utilities Support Equipment (MUSE) facility in Port Hueneme, California, aims to demonstrate how a solar PV system and battery storage disconnected from the grid can provide energy stability at a given time period. Four key attributes are: demand–charge management, load shifting, solar firming and ramp control, as well as island mode. Tests started in the summer of 2015.

9.3.2 New Microgrid Demonstration Project by General Electric and PowerSteam Inc. in Canada's Ontario Province

The project combines wind, natural gas, and solar generation with battery storage and electric vehicle charging. General Electric's (GE) microgrid control system determines when it is most economical to rely on resources within the microgrid and when it is better to buy power from the grid.

9.3.3 Microgrid at Ft. Bliss

The microgrid at Ft. Bliss is a great example of how RE can be integrated with the electric grid at a domestic military facility to provide increased reliability, security, and operational capabilities. The energy storage system provides a backup energy source in case of grid failure or intentional "islanding." (In intentional islanding the generator disconnects from the grid and forces the distributed generator to power the local circuit. This is often used as a power backup system for buildings that normally sell their excess power to the grid.)

9.4 Progress and Real Growth in Africa

As is well known, Africa has great potential for solar energy and is being increasingly exploited by African governments. Rwanda leads East Africa with almost 9 MW of newly installed grid-connected PV and South Africa is moving rapidly toward 1 GW of solar installations. In addition, there are several promising solar pilot projects at Kenyan tea farms and South African off-grid mining sites. Moreover, energy storage technologies offer great promise for Africa to solve electrical infrastructure challenges, can help toward more resilient and efficient grids, and defer expensive maintenance costs.

9.5 Off-grid Projects in Villages

Padre Cocha Village is a small Peruvian village on the Nanay River. In 2003 a PV/diesel/battery system was installed. It consists of two 14 kWp PV systems with a total capacity of 300 kW h d^{-1}, including 240 storage batteries of 375 A h and a single diesel generator of 128 kW [68]. Financial analysis of the system showed that revenues cover only 22% of total costs which includes first costs and only 59% of operational costs which includes maintenance, fuel, and equipment replacement. This makes the system unsustainable from the financial point of view. Even if one takes into account that PV costs have dropped considerably since 2003 when this system was installed, the conclusion is that lower capital costs would not resolve the mismatch between revenues and operational costs.

9.6 Off-Grid Projects on Islands

9.6.1 Haiti

After the devastating earthquake in Haiti a comprehensive hybrid power system was installed to power a complete hospital, with five Victron 24/5000/120 Quattros connected in parallel at a charity-run hospital in Cap-Haitian (Haiti). All six buildings of the hospital have their roofs filled with solar panels, which are connected to the outputs of the Quattro grid inverters. All excess solar power is used to charge the batteries. Only a small grid connection is available, with a capacity of a 100 A. When the required power is higher the Quattros supplement the grid with energy from the batteries. The output of the inverters is synchronized with the grid, which effectively adds power to the grid. When the load reduces the spare power is used to recharge the battery bank.

9.6.2 Alcatraz Island, San Francisco, California

In 2012 a system comprising a 400 kW (PV) and 400 kW (battery)was installed. It reduces approximately 80% of the island's carbon emissions. One of the biggest challenges faced was during the installation phase, which had to be carefully done: the solar array was placed on the roof of the prison in a flat configuration, while the inverters, battery rack, and generators were placed in the old generator room.

9.6.3 Apolima Island (Samoa)

One of the four islands that make up the nation of Samoa has since 2005/2006 installed a 100% renewable electricity system, consisting of PV and L/A battery storage. The PV system is 13.5 kW and provides electricity for 100 residents. The main challenge with this system was cost; this amounted to $223 500 or approximately ¢70 kW^{-1} (assuming a 15 year life, 4% discount rate, and 30% capacity factor). The basic issue is that being a rural community, where residents have very small incomes, recovery of these costs from users may not be realistic.

9.6.4 Bonaire (Venezuela)

Bonaire is an island near the coast of Venezuela, having a population of about 14 000 and a peak electricity demand of about 12 MW [69]. In 2004 the island's sole power plant burned down. This disaster prompted the government to take action and put forward a plan to develop 100% renewable sources for electricity on the island. A wind farm was installed producing 3500 kW h full load annually [70]. The system is stabilized by 3 MW from a battery storage backup system.

10 CONCLUSIONS

The chapter examines barriers to off-grid energy storage, providing a number of international examples. A couple of issues have been discussed. In rural communities, where residents have small incomes, to recover the costs directly from them is not realistic. For such locations and communities, there is a need for support from governments.

Following the examples and some of the financial analysis, it can be concluded that in such cases only a small percentage of total cost is initial costs, while the majority of the cost is for operational costs, which includes maintenance, fuel, and equipment replacement. From a financial point of view this is unsustainable. It has also been concluded than even lower capital costs would not resolve the mismatch between revenues and operational costs.

Despite the unsustainable financial cost the energy storage system provides a backup energy source in case of grid failure or intentional islanding, as well as great support to renewables integration in remote areas.

REFERENCES

[1] Adelman P. Existing markets for storage systems in off-grid applications. Electrochemical energy storage for renewable sources and grid balancing. Chapter 5. New York: Elsevier; 2015.

[2] Huggins RA. Energy storage. New York: Springer; 2010.

[3] Jordan PG. Global markets. Solar energy markets: an analysis of the global solar industry. Chapter 8. New York: Springer; 2014. p. 127–33.

[4] Han X, Ji T, Zhao Z, Zhang H. Economic evaluation of batteries planning in energy storage power stations for load shifting. Renew Energy 2015;78:643–7.

[5] Delfanti M, Falabretti D, Merlo M. Energy storage for PV power plant dispatching. Renew Energy 2015;80:61–72.

[6] DG ENER. The future role and challenges of energy storage. Working paper. Brussels, Belgium: European Commission, Directorate-General for Energy; 2013. https://ec.europa.eu/energy/sites/ener/files/energy_storage.pdf
[7] Grünewald PH, Cockerill TT, Contestabile M, Pearson PJ. The socio-technical transition of distributed electricity storage into future networks—system value and stakeholder views. Energy Policy 2012;41:815–21.
[8] Strbac G. Strategic assessment of the role and value of energy storage systems in the UK: low carbon energy future. Report. London: Carbon Trust; 2012.
[9] ERP. The future role for energy storage in the UK—executive summary and conclusions. Technology report. London: Energy Research Partnership; 2011.
[10] Dobbs R, Oppenheim J, Thompson F, Brinkman M, Zornes M. Meeting the world's energy materials, food and water needs. New York: McKinsey Global Institute; 2011. http://www.mckinsey.com/insights/energy_resources_materials/resource_revolution
[11] Eyer J, Corey G. Energy storage for the electricity grid: benefits and market potential assessment guide. Livermore, CA: Sandia National Laboratories; 2010.
[12] Zucker A, Hinchliffe T, Spisto A. Assessing storage value in energy markets: a literature review. JRC scientific and policy reports, report EUR 26056 EN. Luxembourg: European Commission; 2013.
[13] Taylor PG, Bolton R, Stone D, Upham P. Energy Policy 2013;63:230–43.
[14] Deichmann U, Meisner C, Murry S, Wheeler D. The economics of renewable energy expansion in rural sub-Saharan Africa. Policy research working paper 5193. Washington, DC: World Bank; 2010.
[15] Szabo S, Bodis K, Huld T, Moner-Girona M. Energy solutions in rural Africa: mapping electrification costs of distributed solar and diesel generation versus grid extension. Environ Res Lett 2011;6(3):034002 (pp9).
[16] UNEP. Financing renewable energy in developing countries—drivers and barriers for private sector finance in sub-Saharan Africa. Paris, France: UNEP Finance Initiative; 2012.
[17] Lillo P, Ferrer-Martí L, Boni A, Fernández-Baldor A. Energy Sust Dev 2014;25:17–26.
[18] Clemens E. Capacity development for scaling up decentralized energy access programmes: lessons from Nepal on its role, costs, and financing. New York: United Nations Development Program; 2010.
[19] Sovacool BK, Dhakal S, Gippner O, Bambawale MJ. Halting hydro: a review of the socio-technical barriers to hydroelectric power plants in Nepal. Energ Policy 2011;36(5):3468–76.
[20] Glemarec Y. Energy Policy 2012;47:87–93.
[21] IFC. Solar lighting for the base of the pyramid—overview of an emerging market. Washington, DC: International Finance Corporation; 2010.
[22] Foley AM, Gallachóir BP, Hur J, Baldick R, McKeogh EJ. Energy 2010;35:4522–30.
[23] EPRI. Cost-effectiveness of energy storage in California: application of the EPRI energy storage valuation tool to inform the California public utility. Commission proceeding R. 10-12-007. 3002001162 Technical Update, June. Palo Alto, CA: Electric Power Research Institute; 2013.
[24] Fossati JP, Galarza A, Martín-Villate A, Fontan L. Renew Energy 2015;77:539–49.
[25] Ecofys. Energy storage opportunities and challenges: a West Coast perspective white paper. Report, Contract EDF Renewable Energy with feedback. Utrecht, Netherlands: Ecofys Advisory Panel; 2014. http://www.ecofys.com/files/files/ecofys-2014-energy-storage-white-paper.pdf
[26] Castillo A, Gayme DF. Energy Convers Manage 2014;87:885–94.
[27] Rand DAJ, Moseley PT, Garche J, Parker CD. Valve-regulated lead–acid batteries. New York: Elsevier; 2004.

[28] Enos DG. Lead-acid batteries for medium- and large-scale energy storage. Advances in batteries for medium and large-scale energy storage types and applications. New York: Elsevier; 2015. 57–71.
[29] Ambrose H, Gershenson D, Gershenson A, Kammen D. Environ Res Lett 2014;9. 094004, 8.
[30] Koohi-Kamali S, Tyagi VV, Rahim NA, Panwar NL, Mokhlis H. Renew Sust Energy Rev 2013;25:135–65.
[31] Neves D, Silva CA, Connors S. Renew Sust Energy Rev 2014;31:935–46.
[32] Parformak PW. Energy storage for power grids and electric transportation: a technology assessment. CRS report 7-5700 for Congress. Washington, DC: Congressional Research Service; 2012.
[33] Kondoh J, Ishii I, Yamaguchi H, Murata A, Otani K, Sakuta K, et al. Energy Convers Manage 2000;41:1863–74.
[34] Gyuk I, Johnson M, Vetrano J, Lynn K, Parks W, Handa R. Grid energy storage. Washington, DC: US Department of Energy; 2013.
[35] Bradbury K. Energy storage technology review. Technical report. Durham, NC: Duke University; 2010.
[36] Simbolotti G, Kempener R. Electricity storage technology brief. Technical report technology policy brief E18. Bonn, Germany: IEA-ETSAP and IRENA; 2012.
[37] Bhatnagar D, Loose V. Evaluating utility procured electric energy storage resources: a perspective for state electric utility regulators. SAND2012-9422. Albuquerque, NM: Sandia National Laboratories; 2012.
[38] Ellison J, Bhatnagar D, Karlson B. Maui energy storage study. SAND2012-10314. Albuquerque, NM: Sandia National Laboratories; 2012.
[39] Eyer J, Corey G. Energy storage for the electricity grid: benefits and market potential assessment guide. Livermore CA: Sandia National Laboratories; 2010.
[40] Viswanathan V, Guo X, Tuffner F. Energy storage for power systems applications: a regional assessment for the northwest power pool (NWPP). PNNL-19300, vol. 19300. Washington, DC: Pacific Northwest National Laboratory, US Department of Energy; 2010.
[41] Delille G, François B, Malarange G, Fraisse J-L. Energy storage systems in distribution grids: new assets to upgrade distribution networks abilities. 20th conference on electricity distribution, Prague, Jun 8–11, 2009.
[42] Delille G. Contribution du stockage à la gestion avancée des systèmes électriques, approches organisationnelles et technico-économiques dans les réseaux de distribution. PhD thesis. Ecole Centrale de Lille; 2010.
[43] Bony L, Doig S, Hart C, Maurer E, Newman S. Achieving low cost solar PV. Industry workshop recommendations for near term balance of system cost reductions. Snowmass, CO: Rocky Mountain Institute; 2010.
[44] Lilienthal P. The role of storage in island power systems. Proceedings of the international renewable energy agency accelerated renewable energy deployment on islands with emphasis on the Pacific islands, Sydney, Australia, Oct 26–28, 2011.
[45] Gorona del Viento. http://www.goronadelviento.es/
[46] Martin S, Susanto J. Energ Sust Dev 2014;19:111–21.
[47] Simbolotti G, Kempener R. Electricity storage technology brief. Technical report, technology policy brief E18. Bonn, Germany: IEA-ETSAP and IRENA; 2012.
[48] Gershenson D. Increasing private capital investment into energy access: the case for mini-grid pooling facilities. Report for UNEP. Washington, DC: Crossboundary, Stanford University, and University of California Berkeley; 2015.
[49] Sioshansi R, Denholm P, Jenkin T. Energ Econ 2011;33:56–66.
[50] Datta M, Senjyu T, Yona A, Funabashi T, Kim C. IEEE Trans Energy Convers 2011;26:559–71.

[51] IEA-RETD. Renewable energies for remote areas and islands. Report with the involvement of Trama TecnoAmbiental, Meister Consultants Group, E3 Analytics, HOMER Energy. Paris, France: IEA Renewable Energy Technology Deployment; 2012.
[52] Chmiel Z, Bhattacharyya SC. Energy 2015;81:578–88.
[53] IEA. World energy outlook. Paris, France: Organisation for Economic Cooperation and Development/ International Energy Agency; 2014.
[54] Legros G, Havet I, Bruce N, Bonjour S. The energy access situation in developing countries: a review focusing on the least developed countries and sub-Saharan Africa. New York: United Nations Development Programme and World Health Organization; 2009.
[55] World Bank/AUSAID. One goal, two paths—achieving universal access to modern energy in East Asia and the Pacific. Washington, DC: World Bank/AUSAID; 2011.
[56] Kammen D, Kapadia K, Fripp M. Putting renewables to work: How many jobs can the clean energy industry generate? Berkeley, CA: Renewable and Appropriate Energy Laboratory, University of California; 2004.
[57] Renner M, Sweeney S, Kubit J. Green jobs: towards decent work in a sustainable, low carbon world. UNEP report. Washington, DC: Worldwatch Institute, Cornell University; 2008.
[58] UNDP. Transforming on-grid renewable energy markets. A review of UNDP-GEF support for feed-in tariffs and related price and market-access instruments. Special report. New York: UN Development Program; 2012.
[59] MAFA & Northern Economics. Alaska rural energy plan: initiatives for improving energy efficiency and reliability. Vol. I. Anchorage, AK: Alaska Energy Authority; 2004.
[60] WEC. World energy perspective: cost of energy technologies. Project with Bloomberg New Energy Finance. London: World Energy Council; 2013.
[61] NREL. Levelized cost of energy calculator. http://www.nrel.gov/analysis/tech_lcoe.html
[62] Rickerson W, Colson C. Renew Energy Focus 2007;8:50–2.
[63] Bronski P et al. The economics of load defection. RMI report Apr 2015. Snowmass, CO: Rocky Mountain Institute; 2015. http://utilityproject.org/wp-content/uploads/2015/04/2015-05_RMI-TheEconomicsOfLoadDefection-FullReport.pdf
[64] Alstone P, Gershenson D, Kammen DM. Nat Clim Change 2015;5:305–14. doi:10.1038/nclimate2512.
[65] Abdmouleh Z, Alammari RAM, Gastli A. Renew Sust Energy Rev 2015;45:249–62.
[66] Roberts BP. Sodium–sulfur (NaS) batteries for utility energy storage applications. IEEE Conference, Pittsburgh, PA, 2008.
[67] Esteban M, Zhang Q, Utama A. Areas and islands. Energ Policy 2012; 47:22–31.
[68] Moseley P. J Power Sources 2006;155:83–7.
[69] Bonaire Island (2015). http://www.edinenergy.org/bonaire.html
[70] Johnstone H. Power Eng Int 2010;18:1–5.

Part G

International Issues and the Politics of Introducing Renewable Energy Schemes

Chapter 23

Energy Storage Worldwide

Trevor Sweetnam, Catalina Spataru
Energy Institute, University College London, United Kingdom

1 INTRODUCTION: THE ENERGY STORAGE CHALLENGE

As the proportion of electricity supply derived from renewable sources with variable and uncontrollable output increases the storage of energy will become necessary to balance supply and demand. Energy storage is important for maintaining grid flexibility and grid stability, and is an important enabler of smart energy systems where all of the energy vectors and end uses within a society are linked, allowing the whole energy system to be optimized [1].

Energy storage within electricity systems is not a new concept, indeed pumped hydroelectric storage (PHES) has been used for bulk energy storage worldwide since before the advent of the "smart grid" and is an essential tool in managing the supply–demand balance. PHES currently accounts for almost 99% of worldwide storage; however, the construction of PHES stations is extremely large in scale involving disruptive engineering programs. Furthermore, suitable sites are limited in number and the planning and approval of new plants is a protracted process. Therefore, grid operators are increasingly seeking alternative forms of energy storage recognizing their advantages in terms of speed of deployment, flexibility of siting, and ability to manage macro- and micro-level grid issues and so on.

For alternative forms of storage to reach maturity, correct support structures are required. The chapter aims to move the debate on the development of energy storage forward by focusing and comparing the efforts of some of the leading nations. It begins with a general discussion of the barriers to energy storage development, followed by a discussion on developments in Japan, the United States, and Germany. Finally, we draw together some general lessons that may help to drive forward the development of energy storage.

2 BARRIERS TO DEVELOPMENT AND DEPLOYMENT

Barriers to the development and deployment of alternative energy storage technologies can be grouped into three principal categories: technological, market and regulatory, and strategic.

Storing Energy. http://dx.doi.org/10.1016/B978-0-12-803440-8.00023-3

At the technological level there are a number of barriers to deployment:

- *Capacity*. There is a need to increase the power and energy capacity of existing technologies as well as increasing their efficiency. This is not a simple task and in some cases requires the development of new materials and industrial processes. New battery technology, for example, can take 20 years to move from the laboratory to the market place [2].
- *Deployability*. For market penetration to grow technologies need to be made easier to deploy or new technologies must be made more compact. For example, there is a need to increase deployability at the small, decentralized scale; this is presently taking place in the field of battery technology where new packaged units are being brought to market for domestic applications [3]. At the other end of the market "containerised" battery units are now available [4].
- *Cost*. Cost is a major factor that needs to be addressed partly by development and partly by economies of scale and learning by doing.

There are also a number of market and regulatory issues that need to be addressed:

- Appropriate *market signals* are required to incentivize the building of storage capacity and the provision of storage services. This may involve providing short-term support that helps new entrants to compete and drive economies of scale.
- The true *value* of storage must be recognized. This means creating the process through which investors in storage can be rewarded for all the technical benefits that a unit delivers, for example, alleviating local constraints, managing grid frequency, and taking advantage of price fluctuations in the wholesale market [4].

Finally, a strategic framework is needed to allow for support to be deployed in a systematic way that bridges the technical, regulatory market, and political aspects of storage development.

The steps that three successful nations have taken in overcoming these barriers is discussed in the following section.

3 CASE STUDIES

3.1 The Situation in Japan

In Japan 15% of electricity is cycled through storage facilities [5]. While investment in the development of energy storage is small compared with Japan's investment in developing nuclear power the Japanese have an excellent record in spearheading technological trends in energy storage [6]. Since the Fukishima disaster, Japan has been facing particular constraints relating to the management of its grid, and storage has the potential to alleviate these issues.

Japan has been particularly strong in the area of battery storage with over 200 utility-scale sodium–sulfur (NaS) battery units operational and linked to the electricity grid [7]. There are also examples of the application of superconducting magnetic energy storage systems to curb instantaneous voltage drops caused by industrial units—this being a very new and specialized application of storage technology.

Continuous investment in battery development has led to battery storage technologies achieving a commercially stable level in Japan. Battery development has been driven by very specific technical performance (measured as power and energy densities) and cost targets set by the Ministry of Economy, Trade and Industry (METI) supported by five focused research and development programs.

METI has targeted a fivefold increase in energy density with 2.5 times current power density and a 95% cost reduction by 2030 [8]. To achieve these targets a series of research programs has been supported by METI in recent years. These have focused on:

- the development of lithium-ion batteries suitable for electric and hybrid vehicles;
- the development of large-scale stationary batteries for renewable energy integration using alternative battery chemistries;
- the development of small-scale batteries for energy management at the level of the individual home; and
- the development of new-generation battery technologies and materials.

As well as supporting battery R&D the Japanese have also used subsidies to support stationary lithium-ion battery storage, effectively reducing the cost of batteries for residential consumers by a third. This program is designed to support residential feed-in tariffs for solar generation while demonstration projects by the New Energy and Industrial Technology Development Organization (NEDO) focus on grid-scale applications. Japan has also run large programs to demonstrate residential fuel cells in an effort to drive down prices. Although the investment in battery and fuel cell technologies is significant, far more support has been given to research in the nuclear sector.

3.2 The Situation in the United States

The electricity sector in the United States is made up of a number of regional networks, each of which has individual drivers for energy storage. In California, for example, energy storage is used to deal with large peaks in PV generation and electricity demand to insure security of supply is maintained. Energy storage on the US grid totals approximately 23 GW with over 95% provided by PHES. This represents 18% of world storage making the United States a world leader.

The United States is arguably the most proactive country in terms of energy storage, and this is reflected in the intentions shown by US policy makers and the applications for new storage technologies. The US Department of Energy (DOE) had launched its Energy Storage Technology Program in 2009 primarily funded by the American Recovery and Reinvestment Act (ARRA) [9]. One primary objective was to improve the US electricity grid's flexibility, economic competitiveness, and the network's overall reliability and robustness. The DOE's aim is for energy storage technologies to make the transition from being an area of research to an attractive commercial proposition as quickly as possible. This industry is estimated to be worth between $(2–4) billion over the coming 20 years [10].

In the period up to 2010, ARRA provided $185 million in support for demonstration projects, valued at a total of $772 million. These demonstration projects addressed a range of areas including the use of battery storage for balancing wind generation and frequency regulation, as well as compressed air storage and other storage technologies.

In addition to ARRA the DOE further supports energy storage development via the Advanced Research Projects Agency–Energy (ARPA-E) program. ARPA-E provides support in four main areas:

- Advanced management and protection of energy storage devices (AMPED)
- Batteries for electrical energy storage in transportation (BEEST)
- Grid-scale rampable intermittent dispatchable storage (GRIDS)
- High-energy advanced thermal storage (HEATS).

GRIDS alone provided over $55 million in project funding for fiscal year 2010–11 [9]. The inclusion of advanced thermal storage in the ARPA-E portfolio is interesting, and a number of demonstration projects using ice thermal storage, mostly in universities and schools [11], have been carried out.

In tandem with national government activities such as DOE's Energy Storage Program, state governments have played a prominent role with their own activities. In California, several legislative mandates form the basis of which policies regarding energy storage are built upon. For example, the AB2514 statute requires publicly owned utilities to determine appropriate targets to procure energy storage systems by Dec. 31, 2016 [12]. This indirectly creates a regulatory focus for public utilities and sets the way to build an energy storage market in California.

Another driver behind the expanded deployment of energy storage in California is the Renewable Portfolio Standard (RPS). Under Senate Bill 107, California's investor-owned utilities were required to procure 20% of their electricity from renewable resources by 2010. The target was later increased to 33% by Dec. 2020. To achieve this the new law requires that utilities establish appropriate procurement targets to meet the 33% goal and be retained in subsequent years. The implication for energy storage is that the bill also requires investor-owned utilities to integrate renewable energy resources to the grid in a manner that would require the least additional transmission facilities [12].

The policy framework in California is only one of many examples that are set out to accommodate more energy storage deployment. In Texas, for example, Senate Bill 943 classifies specific energy storage equipment or facilities as generation assets, which directly means that these facilities are eligible to be interconnected to the grid, obtain transmission service, and sell electricity to the wholesale market. Judging by SNL's project map, California is arguably the leading state along with New York in terms of energy storage development.

Moves are also being made in the United States to make the market more supportive of storage by providing more recognition of its true value. The Federal Energy Regulatory Commission (FERC), the governmental agency that oversees the entire electricity market, realizing the importance of energy storage as a ramping source, has amended compensation practices for frequency regulation services and rewarded market operators based on their energy-ramping performances [9].

3.3 The Situation in Germany

The German government has invested heavily in renewable energy, particularly in residential solar PV but also in wind energy. These schemes have been so successful that excess generation is now a concern. In addition, Germany's abandonment of nuclear power after the Fukushima incident has caused further concerns regarding security of supply. To increase Germany's share of renewables, energy storage is seen as a means of eliminating flexible generation [13].

The first focus of German policy has been the expansion of pumped storage, with approximately 4.7 GW of new projects recently announced. To push forward the PHES vision the German Energy Act (EnWG) have offered exemptions to bulk storage facilities from grid access tariffs. This applies to any newly built storage or refurbished PHES scheme. EnWG has also brought moves to insure the eligibility of storage systems connection to the grid. The grid codes do not have any special requirements for storage systems to be connected to the grid, but the storage system must be able to meet the load as well as the generation requirement depending on its operation mode [14]. Evidently, large-scale, centralized energy storage is EnWG's immediate focus. Germany is fortunate that some of its neighbors—not least Norway and Sweden—have immense PHES resources that can be accessed with increased interconnections [13].

Compressed air energy storage (CAES) and advanced adiabatic CAES (AA-CAES) are the second area of focus. Although only one plant (in Huntdorf) is currently operational, salt caverns are being scoured out. and caverns that are currently used to store natural gas may provide further opportunities. Meanwhile an AA-CAES plant that requires no fossil fuel consumption during the gas expansion phase is under development in Saxony-Anhalt. Commercializing power-to-gas is a further area of focus, providing the potential to compensate for long periods of low wind output and seasonal variations in output and demand [13].

The German Renewable Sources Act (EEG) has also introduced a premium payment for residential PV producers with the condition that excess solar energy generated is consumed locally without being injected into the distribution grid [15]. The German government has also provided support for domestic energy storage to encourage self-consumption of solar generation (systems less than 30 kW at peak times). This scheme is a collaboration between the Federal Ministry for the Environment, Nature Conservation and Nuclear Safety and the state-owned KfW bank and provides soft loans and cash incentives (c. 30%) for battery purchases. Germany is aiming to lead the domestic storage market with a capacity of 2 GW h [16].

4 LESSONS FOR THE DEVELOPMENT OF STORAGE

A range of approaches can be seen within the three countries studied. Each has differing requirements for storage and each has brought forward a range of programs to support the storage technologies that respond to these requirements.

4.1 Overcoming Technological Barriers

Japan's approach could be seen as "battery centric," and it has been highly effective at targeting and achieving specific technical performance improvements. The United States has taken a less technology-specific approach, and has focused primarily on bringing technologies to the stage where they are ready for private investment. Germany's approach is to develop a number of technologies, and proven storage technologies are being developed to meet short-term aims while a suite of approaches across multiple scales is under development to meet longer term goals.

4.2 Market and Regulatory Developments

Regulatory frameworks should aim to create an level playing field for cross-border trading of electricity storage. They need to provide clear rules and responsibilities concerning technical modalities and financial conditions for energy storage. They must address barriers preventing the integration of storage into markets. The frameworks should be technology neutral, insuring fair competition between different technological solutions.

The development of a low-carbon electricity system, as set out in the EU 2050 roadmap, requires member states to work together to develop technologies, drive the necessary investments, and harmonize the different rules across the European energy markets.

Decisions to invest in the development of storage and the deployment of adequate capacity will depend on the evolution of the whole energy system. They are closely linked to other developments such as electricity superhighways together with large-scale deployment of renewable generation in the North Sea and North Africa, the growth of electric vehicles, and improvements to demand-side management.

The most important focus for regulatory reform is to break down the barriers to storage, capturing its true value to the grid. The business model for storage is often dependent on its ability to act in response to fluctuating wholesale market prices; managing system frequency; providing reactive power control; and relieving local constraints. In many jurisdictions there is a need to separate supply chain functions. For example, system operation and distribution network management mean that a storage operator has to deal with multiple market actors with conflicting concerns to secure a revenue stream.

Within Europe, balancing products are only exchangeable cross-border among member states to a very limited extent. Improved market conditions and regulations agreed at the EU level could spur a massive effort in technology development. The EU has suitable instruments, for example, the RTD framework programme, Horizon 2020, and the strategic energy plan, which have the potential to drive forward the storage agenda.

4.3 Strategic Framework

As mentioned previously, the strategic framework in each of the countries we have addressed responds to specific issues being faced by their respective energy sectors. The important similarity between these three leading countries is the presence of a strategic framework. This is the vital first step for the development of a storage market in any country.

Governments are rarely disposed to "picking winners" in terms of technologies or businesses. However, we consider that a successful strategic framework should clearly define the storage needs in terms of functionality (energy and power capacity, ramp rates, etc.). From this starting point the necessary support in terms of R&D funding and regulatory reform can be designed.

5 CONCLUSIONS

Having completed our examination of ongoing storage development in Japan, the United States, and Germany we can suggest the following steps to develop a thriving storage industry:

1. Put a strategic framework in place that defines the technical requirements for storage. The German case is a good example of a strategy that clearly defines short-term and long-term technology mixes across scales.
2. Define clear performance targets for R&D activity as the Japanese have done with batteries.
3. Work across scales to deliver a range of technologies from large centralized plants and highly distributed forms of storage.
4. Work across energy vectors; often the most effective forms of storage involve supplying energy across vectors, for example, ice storage.

5. Put in place market frameworks that allow storage, with its unique technical capabilities, to capture value across the energy supply chain. The question is whether this is better achieved by regulation or deregulation?
6. Aim for early commercial adoption so that development and commercial deployment is delivered by markets as early as possible, as the DOE is demonstrating.

Energy storage is sure to be a vital component of a future energy system where unpredictable generation is growing. The mix of technologies will be determined by the particulars of any energy system and is likely to involve a mix of scales and a mix of technologies including demand response.

REFERENCES

[1] Lund H. Renewable energy systems: a smart energy systems approach to the choice and modeling of 100% renewable solutions. Waltham, MA: Academic Press; 2014.
[2] Divya KC, Østergaard J. Electr Power Syst Res 2009;79:511–20.
[3] Tesla Motors. Tesla Powerwall. Palo Alto, CA: Tesla Motors, 2015. www.teslamotors.com/en_GB/powerwall.
[4] Cooper I, Heyward N, Papadopoulos P. Smarter network storage. SNS 1.12 Energy storage as an asset. London: UK Power Networks; 2015.
[5] Yang Z, Zhang J, Kinter-Meyer M, Lu X, Choi D, Lemmon J, Liu J. Chem Rev 2011;11:3577–613.
[6] Valentine SV. Energy Policy 2011;39:6842–54.
[7] Roberts BP. IEEE Trans Energy Convers 2008;11:658–64.
[8] Aki H. Energy storage research and development activities in Japan. 2011 IEEE Power and Energy Society General Meeting, 2011.
[9] EAC. Energy storage activities in the United States electricity grid. Washington, DC: Electricity Advisory Committee (DOE); 2011. http://www.energy.gov/sites/prod/files/oeprod/DocumentsandMedia/FINAL_DOE_Report-Storage_Activities_(2011)5-1-11.pdf
[10] EPRI. Electricity storage technology options: a white paper primer on applications, costs, and benefits. Palo Alto, CA: Electric Power Research Institute; 2010. http://www.besia.org.uk/public_html/uploads/publications_reports/BESIA - Epri_White_Paper_20130909162235.pdf
[11] Department of Energy, 2013. www.eesi.org/files/IssueBrief_Energy_Storage_080613.pdf
[12] CIEE. 2020 strategic analysis of energy storage technology. Portland, ME: Council on International Educational Exchange; 2011. http://uc-ciee.org/downloads/PIER_2020_Energy_StorageVisionProject_Final_Report_CIEE.pdf
[13] Auer J. State-of-the-art electricity storage systems: indispensable elements of the energy revolution. 2012. http://www.dbresearch.com/PROD/DBR_INTERNET_EN-PROD/PROD0000000000286166/State-of-the-art+electricity+storage+systems%3A+Indispensable+elements+of+the+energy+revolution.pdf
[14] DG ENER. The future role and challenges of energy storage. DG ENER working paper. Brussels, Belgium: EU Directorate General for Energy; 2011. https://ec.europa.eu/energy/sites/ener/files/energy_storage.pdf
[15] Nekrassov A, He X, Prestat B. Efficiency analysis of incentive mechanisms for energy storage integration into electrical systems. IEEE PES Trondheim PowerTech: the power of technology for a sustainable society. Power Tech 2011; 1–6, http://dx.doi.org/10.1109/PTC.2011.6019170.
[16] GTAI. Industry overview: the photovoltaic market in Germany. Berlin, Germany: Germany Trade and Invest; 2014. http://www.gtai.de/GTAI/Content/EN/Invest/_SharedDocs/Downloads/GTAI/Industry-overviews/the-photovoltaic-market-in-germany-en.pdf

Chapter 24

Storing Energy in China—An Overview

Haisheng Chen, Yujie Xu, Chang Liu, Fengjuan He, Shan Hu
Institute of Engineering Thermophysics, Chinese Academy of Sciences, Beijing, China

1 INTRODUCTION

Electrical energy storage (EES) refers to a process of converting electrical energy from a power network into a form that can be stored for converting back to electrical energy when needed [1–3]. Such a process enables electricity to be produced at times of either low demand, low generation cost, or from intermittent energy sources and to be used at times of high demand, high generation cost, or when no other generation is available [1–5]. EES has numerous applications including portable devices (mobile phones, laptops, toys, personal stereos, etc.), transport vehicles (electrical vehicles, yachts, autocycles, trains, etc.), and stationary energy resources [1–9]. This chapter concentrates on EES systems for stationary applications such as power generation, distribution and transition network, distributed energy resource, renewable energy, and industrial and commercial customers.

EES is currently enjoying somewhat of a renaissance, for a variety of reasons including changes in the worldwide utility regulatory environment; ever-increasing reliance on electricity in industry, commerce, and the home; power quality/quality-of-supply issues; the growth of renewables as a major new source of electricity supply; and all of these combined with ever increasing stringent environmental requirements [3,4,6]. These factors, combined with the rapidly accelerating rate of technological development in many emerging EES systems, with anticipated unit cost reductions, now make their practical application look very attractive on future timescales of only a few years. The governments of the United States [1,2,9,13–15], the European Union [3,6,10], Japan [10,16], and Australia [4] all have announced national programs on EES since the late 1990s. The anticipated storage level will boost energy by between (10–15)% in the United States and in European countries, and even higher in Japan in the near future [4,10] (as of 2015).

Although started later than other developed countries mentioned above, China has achieved much progress in research and application of EES. This

Storing Energy. http://dx.doi.org/10.1016/B978-0-12-803440-8.00024-5

chapter aims to review the current status of EES in China on both aspects of technology and development. As this book demonstrates, there are over 10 types of EES technologies in usage or under development at the present time. These include pumped hydroelectric storage (PHES) [11,12,17], compressed air energy storage (CAES) [18–22], flywheels [13,16,33,34], lead–acid batteries [23–27], lithium–ion batteries, sodium–sulfur batteries, flow batteries [3,4,6,13], fuel cells [24,28], solar fuel [4,29], superconducting magnetic energy storage (SMES) [30–32], cryogenic energy storage [33–43], and capacitor and supercapacitor storage [4,16]. Currently in China the first seven types of technologies have been in use, as large-scale, megawatt-scale facilities or as demonstration facilities. This chapter will focus on these seven types of EES technologies.

The chapter will include a discussion on the imperativeness and applications of EES technologies; technical characteristics, research, deployment, and the status of development of EES systems; and the prospects for EES technologies in China.

2 IMPERATIVENESS AND APPLICATIONS

EES is urgently needed by the conventional electricity generation industry [1–7] all over the world. Unlike any other successful commodities markets, conventional electricity generation industries have at present little or no storage component. Electricity transmission and distribution systems are operated for simple one-way transportation from remote and large power plants to consumers. This means that electricity must always be used precisely when it is produced. However, the demand for electricity varies considerably, daily and seasonally, and the maximum demand may only last for a few hours each day. This leads to inefficient, overdesigned, and expensive plants. In 2014 in China the average capacity utilization rate of power generation was only 49.8% and the average capacity utilization rate was below 55%. EES allows energy production to be decoupled from its supply. By having large-scale electricity storage capacity available (as shown in Fig. 24.1), system planners would need to build only

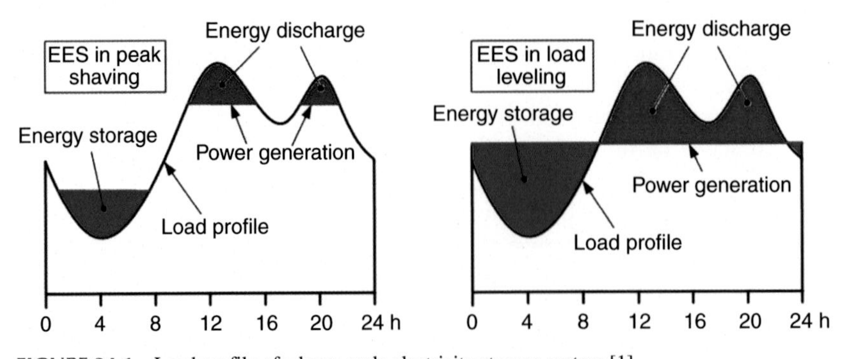

FIGURE 24.1 Load profile of a large-scale electricity storage system [1]

sufficient generating capacity to meet average electrical demand rather than peak demands [13]. Therefore, EES can provide substantial benefits including load following, peaking power, and standby reserve. Moreover, by providing spinning reserve and dispatched load, EES can increase the net efficiency of thermal power sources while reducing harmful emissions [44–46].

More importantly, EES systems are critical to intermittent renewable energy supply systems [2–7,49–53] such as solar photovoltaics and wind turbines. The penetration of renewable resources may displace significant amounts of energy produced by large conventional plant. It is expected that renewable energy will supply 16% of total electricity in China by year 2020. However, intermittency and noncontrollability are inherent characteristics of renewable energy-based electricity generation systems. Such disadvantages have become major hurdles to the extensive use of renewable energy. In 2014 in China the abandoned wind power ratio was 8.8% due to its intermittency and noncontrollability. A suitable EES could obviously provide an essential solution [51] in dealing with the intermittency of renewable sources and the unpredictability of their output as the surplus could be stored during periods when intermittent generation exceeds demand and then be used to cover periods when the load is greater than the generation.

Furthermore, EES is regarded as an imperative technology for the distributed energy resource (DER) system [2–7,32,47–49] in the near future. Deferred from the conventional power system, which has large, centralized units, DERs are usually installed at the distribution level, close to the place of utilization, and generate power typically in a small range from a few kilowatts to a few megawatts [47]. DER is regarded as a sustainable, efficient, reliable, and environmentally friendly alternative to conventional energy systems [47,48]. The energy resource system is undergoing a change to be a mixture of centralized and distributed subsystems with higher and higher penetration of DERs [48]. However, more drastic load fluctuation and emergent voltage drop are anticipated in DER systems due to smaller capacity and higher possibility of line faults than the conventional power system. EES is identified as a key solution to compensate for power flexibility and provide uninterruptible power supply in cases of instantaneous voltage drop for such distributed energy networks. It is expected that the installed capacity of DERs in China will be 50 GW in 2020 with about 10%, that is, 5 GW, being EES.

3 TECHNICAL AND DEVELOPMENT STATUS

3.1 Pumped Hydroelectric Storage

PHES is the most widely implemented large-scale form of EES. A PHES facility normally consists of (1) two reservoirs located at different elevations, (2) a system to pump water to the higher elevation, and (3) a turbine system to generate electricity when water is released to return to the lower reservoir.

The principle of PHES is that hydraulic potential energy is stored by pumping water from a lower reservoir to an elevated reservoir. During periods of high electricity demand, water is extracted through a turbine generator in a manner similar to traditional hydroelectric facilities. The amount of stored energy is proportional to the height difference between the two reservoirs and the volume of water stored. Some high-dam hydro plants have a storage capability and can be used as a PHES facility. Underground pumped storage, using flooded mineshafts or other cavities, are also technically possible. Open sea can also be used as the lower reservoir. A seawater pumped hydro plant was first built in Japan in 1999 (Yanbaru, 30 MW) [10].

PHES is a mature technology with large-volume, long storage period, high–efficiency, and a relatively low capital cost per unit energy. Owing to relatively small evaporation and penetration the storage period of PHES can be very long—typically hours to days and even up to years. Taking into account evaporation losses from the exposed water surface and conversion losses, c. (71–85)% of the electrical energy used to pump water into the elevated reservoir can be regained. The typical rating of PHES is about 1 GW [(100–3000) MW] and facilities continue to be installed worldwide at a rate of up to 5 GW a^{-1} (5000 MW per year). The rating of PHES is the highest of all the available EES systems. As a consequence, PHES is generally used for energy management, frequency control, and provision of reserve. According to data from an International Energy Agency (IEA) report in 2014 [54] the global installed capacity of PHES is about 140 GW which amounts to 99% of the total capacity of EES.

The major drawback of PHES is the scarcity of available sites for two large reservoirs and one or two dams. A long lead time (typically about 10 years); high costs (typically hundred millions to billions of US dollars) for construction; and environmental issues, (e.g., removing trees and vegetation from the large amounts of land prior to the reservoir being flooded) [45,46], are three other major constraints in the deployment of PHES.

Pump hydro is also widely used in China for peak shaving, peak loading, energy management, and renewable energy electricity. It now has the largest installed capacity in China over the other EES systems. The first commercial pump hydro in China was the Gangnan plant in Hebei province which was in operation in 1968 with a capacity of 11 MW. Up to 2014 the amount of stored energy in Chinese pumped hydro stations was about 22.1 GW, which is 99.2% of total installed capacity of EES. This amounts to about 1.7% of total generation capacity in China. Table 24.1 lists all the pumped hydro stations in operation in China. Total installed capacity is now third in the world (behind Japan and the United States). However, most of the stations are located in the southeast of China due to geological restrictions [1]. As a result other EES technologies are urgently needed in China.

TABLE 24.1 Pump Hydro Projects in China

Project name	System description/MW	Rated capacity/MW	In operation/MW	Under construction/MW
Shisanling	4 × 200	800	800	
Panjiakou	3 × 90	270	270	
Taishan	4 × 250	1 000	1 000	
Zhanghewan	4 × 250	1 000	1 000	
Miyun	2 × 11	22	22	
Gangnan	1 × 11	11	11	
Xilongchi	4 × 300	1 200	1 200	
Fengning	6 × 300	1 800		1 800
Huhhot	4 × 300	1 200	600	600
Tianhuangping	6 × 300	1 800	1 800	
Dongbai	4 × 300	1 200	1 200	
Xianghongdian	2 × 40	80	80	
Yixing	4 × 250	1 000	1 000	
Langyashan	4 × 150	600	600	
Shahe	2 × 50	100	100	
Xikou	2 × 40	80	80	
Xianyou	4 × 300	1 200	1 200	
Xiangshuijian	4 × 250	1 000	1 000	
Liyang		1 500		1 500
Xianju	4 × 375	1 500		1 500
Jixi	6 × 300	1 800		1 800
Huilong	2 × 60	120	120	
Tiantang	2 × 35	70	70	
Baoquan	4 × 300	1 200	1 200	
Lianxu	4 × 300	1 200	1 200	
Hongping	4 × 300	1 200		1 200
Heimifeng	4 × 300	1 200	1 200	
Baishan	2 × 150	300	300	
Pushihe	4 × 300	1 200	1 200	
Huanggou	4 × 300	1 200		1 200
Dunhua	4 × 350	1 400		1 400
Yangzuoyong		90	90	
Guangxu	8 × 300	2 400	2 400	
Huixu	8 × 300	2 400	2 400	
Qingyuan	4 × 320	1 280		1 280
Shenzhen	4 × 300	1 200		1 200
Qiongzhong	3 × 200	600		6 000
Total		36 223	22 143	14 080

3.2 Compressed Air Energy Storage

CAES is the only other commercially available technology (besides PHES) able to provide the energy (above 100 MW in a single unit) needed for large-scale energy storage. A CAES system involves aboveground and underground components that combine man-made technology and natural geological formations to accept, store, and dispatch energy through a series of thermodynamic cycles.

The major components of a CAES installation include five aboveground and one underground components:

1. A motor/generator that employs clutches to provide for alternate engagement to the compressor or turbine trains.
2. An air compressor that may require two or more stages, intercoolers, and aftercoolers, to achieve economy of compression and reduce the moisture content of compressed air.
3. A turbine train containing both high- and low-pressure turbines.
4. Equipment controls for operating the combustion turbine, compressor, and auxiliaries and to regulate and control changeover from generation mode to storage mode.
5. Auxiliary equipment consisting of fuel storage and handling, and mechanical and electrical systems for various heat exchangers required to support the operation of the facility.
6. An underground component which is mainly the cavity used for the storage of compressed air. The storage cavity can potentially be developed from any of three different categories of geological formations: underground rock caverns created by excavating comparatively hard and impervious rock formations; salt caverns created by solution- or dry-mining of salt formations; and porous media reservoirs made by water-bearing aquifers or depleted gas or oil fields (e.g., sandstone, fissured limestone). Aquifers in particular can be very attractive as storage media because the compressed air will displace water, setting up a constant pressure storage system while the pressure in the alternative systems will vary when adding or releasing air.

The principle of CAES is based on conventional gas turbine generation. CAES decouples the compression and expansion cycle of a conventional gas turbine into two separate processes and stores the energy in the form of elastic potential energy of compressed air. At times of low demand, energy is stored by compressing air in an air-tight space. To extract the stored energy, compressed air is drawn from the storage vessel, heated, and then expanded through a high-pressure turbine, which captures some of the energy in the compressed air. The air is then mixed with fuel and combusted with the exhaust gas expanded through a low-pressure turbine. Both the high- and low-pressure turbines are connected to a generator to produce electricity.

CAES has a relatively long storage period, low capital costs, and high efficiency. Typical ratings for a CAES system are in the range (50–300) MW, and currently manufacturers can create CAES machinery for facilities ranging from (1–350) MW. The rating is much higher than for storage technologies other than PHES. The storage period is also longer than for other storage methods since the losses are very small; a CAES system can be used to store energy for more than a year. A typical value of storage efficiency of CAES is in the range (40–75)%. Capital costs for CAES facilities vary depending on the type of underground storage but are typically in the range (\$300–660) kW^{-1}.

Similar to the situation with PHES, the major barrier to implementation of CAES is the reliance on favorable geological structures, and CAES is only economically feasible for power plants that are near to rock mines, salt caverns, aquifers, or depleted gas fields. In addition, in comparison with PHES facilities and other currently available energy storage systems, CAES is not an independent system, and each facility must be linked to a gas turbine plant. It cannot be used in conjunction with other types of power plants such as coal-fired, nuclear, wind turbine, or solar photovoltaic plants. More importantly, the combustion of fossil fuel leads to emission of contaminates such as nitrogen oxides and carbon oxides which render CAES less attractive [19,45,46]. Many improved CAES systems are proposed or under investigation, eg, small-scale CAES systems with fabricated small vessels; advanced adiabatic CAES (AA-CAES) systems with thermal energy storage (TES) [19,21]; and compressed air storage with humidification (CASH) [13,20].

Although CAES is a mature, commercially available energy storage technology, there are only two operating CAES systems in the world. One is in Huntorf (Germany), the other is in McIntosh, Alabama (United States). Currently in China, CAES is still under research and development and there is no CAES station in commercial operation. The largest CAES station is the 1.5 W demonstration project at the Institute of Engineering Thermophysics (Chinese Academy of Sciences) (IET-CAS), Beijing, and a 10 W project which is under construction at the same institute. Table 24.2 lists the CAES projects in China. The institutions working on CAES include Zhejiang University, Shandong University, Tsinghua University, IET-CAS, and the Datang Power Company. Among the projects, Shandong University, IET-CAS, and Datang Power Company have been supported by national high-tech programs. The project conducted by Datang Power Company has now been terminated due to lack of an air-tight cavern. Scientists at IET-CAS are also working on two 1.5 MW scale projects for industrial users; these two projects are not listed in Table 24.2 as they are only at the design stage.

3.3 Flywheel Energy Storage

Flywheels have been used for thousands of years to store energy. Energy is stored through the angular momentum of a spinning mass. During charge the flywheel is spun by a motor driven by electrical energy; during discharge the same motor acts as a generator, producing electricity from the rotational energy

TABLE 24.2 Compressed Air Energy Storage System in China

Project name	System description	Rated capacity	Status
Institute of Engineering Thermophysics (Chinese Academy of Sciences)	Advanced CAES based on supercritical air	1.5 MW	Demonstration
Institute of Engineering Thermophysics (Chinese Academy of Sciences)	Advanced CAES based on supercritical air	10 MW	Under construction
Zhejiang University	CAES for automobiles	~10 kW	Finished
Sandong University	CAES with screw turbine	~20 kW	Demonstration
Tsinghua University	CAES without combustion	500 kW	Demonstration
Datang Power Company	Conventional CAES	200 MW	Terminated

of the flywheel. The energy of a flywheel system is dependent on the size and speed of the rotor, and the power rating is dependent on the motor/generator.

A flywheel storage device consists of the following components:

1. A flywheel that spins at a very high velocity to achieve maximum storage of rotational kinetic energy within given constraints.
2. A containment system that provides a high vacuum environment of between (10^{-1}–10^{-5}) Pa [(between 10^{-6}–10^{-8}) atmosphere pressure] to minimize friction losses and protect the rotor assembly from external disturbances. The containment system can also absorb the energy of the exploding rotor and contain the debris within a defined volume envelope in a failure situation.
3. Bearing assemblies which provide a very low loss support mechanism for the flywheel rotor.
4. A power conversion and control system which is an integrated electrical apparatus that can operate either as a motor to turn the flywheel or as a generator to produce electrical power on demand using the energy stored in the flywheel.

The major advantage of the flywheel is the capability of several hundred thousand full charge–discharge cycles [1] which provides a much better lifecycle than many other EES systems. The efficiency of flywheels is high and typically in the range (90–95)%. Their application is principally high power/short duration, (e.g., 100 s of a kilowatt every 10 s). The most common power quality application is to provide a ride through of interruptions up to 15 s long or to bridge the shift from one power source to another. Such systems may often be implemented in a hybrid configuration with standby generators (e.g., diesel generators). Flywheels have also been used for demand reduction and energy recovery in electrically powered mass transit systems. Megawatt flywheels can also be used for reactive power support, spinning reserve, and voltage regulation by power quality–sensitive customers such as communications facilities and computer server centers; the duration could be up to tens of minutes using a magnetic levitation bearing. Urenco Power Technologies (UPT) has recently demonstrated the application of flywheels to the smoothing of the output of wind turbine systems and the associated stabilization of small-scale island power supply networks. The rail traction industry represents another significant and high added value application for flywheel storage, particularly for trackside voltage support. Such an application could well represent a significant growth area in the years ahead, with the prevalence of increasing numbers of larger and heavier trains being imposed on existing infrastructures.

Compared with other EES systems the major disadvantages of the flywheel system are short duration, relatively high frictional loss (windage), and low energy density which restrict the use of flywheel systems. Much of the current research and developmental effort in relation to flywheel energy storage systems is directed toward high-speed composite machines, running at 10 000 rpm and utilizing fabric composite materials technology [39]. Units have already been supplied on a commercial basis by UPT, and further systems are being developed by AFS-Trinity, Beacon Power, Piller, and others. The largest project in the world is the Stephentown Advanced Energy Storage project located in Stephentown, New York (United States). The project was constructed by Beacon Power and its scale is 20 MW/5 MW h with an efficiency of 97% for frequency regulation or a roundtrip efficiency of 85% for an overall charge/discharge cycle.

China started research and development into flywheels during the 1980s when the Institute of Electrical Engineering (Chinese Academy of Sciences) investigated a 10 kW prototype with a capacity of 10 W h and rotating speeds of (66–133) Hz (4000–8000) rpm. In China work on flywheels is currently being conducted at the Institute of Electrical Engineering (Chinese Academy of Sciences); Tsinghua University; Harbin Engineering University; Beihang University; China Electric Power Research Institute; and the Ying-Li Company. In 2008 the first flywheel system for industrial application was supplied by the China Electric Power Research Institute, which was installed in a hospital in Beijing with a capacity of 250 kW/15 s. Zhengjiang University and Harbin Engineering University have announced prototypes with power capacity in the region of 100 kW. Tsinghua University is now working on a high-speed system with rotation speeds of 300 Hz (18 000 rpm) and a capacity of 100 kW per unit. A 1 MW array system is under investigation by Tsinghua University which expects to install a demonstration system within 3 years. Table 24.3 gives a summary of current flywheel projects in China.

TABLE 24.3 Flywheel Projects in China

Project name	System description	Rated capacity	Status
China Electric Power Research Institute	Unit	250 kW/15 s	Completed
Institute of Electrical Engineering (Chinese Academy of Sciences)	Unit	10 kW/10 W h	Completed
Zhejiang University	Unit	~100 kW	Prototype
Harbin Engineering University	Unit	~100 kW	Prototype
Tsinghua University	Unit	100 kW/25 kW h	Demonstration
Ying-Li Company	Unit	30 kW/15 kW h	Demonstration
Tsinghua University	Array	1 MW	Demonstration

3.4 Lead–Acid Battery

The lead–acid battery, invented in 1859, is the oldest and most widely used rechargeable electrochemical device. It consists of (in the charged state) electrodes of lead metal (Pb) and lead oxide (PbO_2) in an electrolyte of about 37% (5.99 molar) sulfuric acid (H_2SO_4). In the discharged state both electrodes turn into lead sulfate ($PbSO_4$) and the electrolyte loses its dissolved sulfuric acid and becomes primarily water. The chemical reactions are (charged to discharged):

Anode (oxidation): $Pb(s) + SO_4^{2-}(aq) \leftrightarrow PbSO_4(s) + 2e^-$

Cathode (reduction): $PbO_2(s) + SO_4^{2-}(aq) + 4H^+ + 2e^- \leftrightarrow PbSO_4(s) + 2H_2O(l)$

The electrolyte is dilute sulfuric acid, which provides the sulfate ions for discharge reactions. There are several types of lead–acid battery: the flooded battery, which requires regular topping up with distilled water; the sealed maintenance-free battery, which has a gelled or absorbed electrolyte; and the valve-regulated lead–acid battery.

A lead–acid battery is a low-cost, (\$300–600) kW h^{-1}, highly reliable, efficient (70–90%), and popular storage choice for power quality, uninterrupted power supplies (UPS), and some spinning reserve applications. Its application for energy management, however, has been very limited due to its short lifecycle of between 500–1500 cycles and low energy density of (35–50) W h kg^{-1} due to the inherent high density of lead, which results in a high total mass for large energy storage requirements. The lead–acid battery also has poor low-temperature performance and therefore requires a thermal management system. Nevertheless, lead–acid batteries have been used in many commercial and large-scale energy management applications such as the 8.5 MW h h^{-1} system in the BEWAG Plant (Berlin, Germany), the 4 MW h h^{-1} ESCAR system at the Iberdrola Tech-

TABLE 24.4 Lead–Acid Battery Projects in China

Project name	System description	Rated capacity	Status
Institute of Electrical Engineering (Chinese Academy of Sciences)	Demonstration	100 kW/600 kW h	Completed
Hebei Branch, State Grid Company	Distributed energy resource system	80 kW/128 kW h	Completed
Dong-ao Island project, Zhuhai, Guangdong province	Island energy system	1 MW/2 MW h	Demonstration
Hu-xi Island project, Wenzhou, Zhejiang province	Island energy system	2 MW/4 MW h	Demonstration

nology Demonstration Center (Madrid, Spain), and the 14 MW h 1.5 h system in PREPA (Puerto Rico). The largest is the 40 MW h system in Chino, California (United States) which can work with a rated power of 10 MW for 4 h.

There are many companies and research institutions in China working on lead–acid batteries, such as Nandu, Fengfan, Suangdeng, Institute of Electrical Engineering, and the Chinese Academy of Sciences. There were 14 projects in operation in 2014 with a total capacity of 12.1 MW. Major projects include:

1. A 100 kW/600 kW h system being developed by the Institute of Electrical Engineering (Chinese Academy of Sciences) and installed in Zhangbei county in Hebei province.
2. A 1 MW/2 MW h project developed by Singye Solar Company in Zhuhai in Guangtong province.
3. A 2 MW/4 MW h project developed by the State Grid Company in Weizhou in Zhejiang province. The major lead–acid battery projects in China are listed in Table 24.4.

3.5 Sodium–Sulfur Battery

The sodium–sulfur battery was invented by the Ford Company in 1966. A NaS battery consists of liquid (molten) sulfur at the positive electrode and liquid (molten) sodium at the negative electrode with the active materials separated by a solid beta alumina ceramic electrolyte. The electrolyte allows only positive sodium ions to go through it and combine with the sulfur to form sodium polysulfides:

$$2Na + 4S = Na_2S_4$$

During discharge, positive Na^+ ions flow through the electrolyte and electrons flow in the external circuit of the battery; a potential of about 2.0 V is produced. This process is reversible as charging causes sodium polysulfides to release the positive sodium ions back through the electrolyte to recombine as elemental sodium. The battery must be kept at c. (300–350) °C to allow the reactions to process. NaS batteries have a typical lifecycle of about 2500 cycles. Their typical energy and power density are in the range (150–240) W h kg^{-1} and (90–230) W kg^{-1}, respectively.

NaS battery cells are efficient (75–90%) and have pulse power capability over six times their continuous rating (for 30 s). This attribute enables the NaS battery to be economically useful in combined power quality and peak-shaving applications. NaS battery technology has been installed at over 30 sites in China with a capacity of more than 316 MW/1896 MW h. The largest NaS installation is a 6 MW, 8 h unit for Tokyo Electric Power Company (TEPCO). Recently, Japan's NGK Insulators Ltd. has commissioned a NaS energy storage system of 8 MW/58 MW h at a Hitachi plant in Japan.

The major drawback of the NaS battery is that a heat source is required which uses the stored energy of the battery, partially reducing battery performance. Initial capital cost remains another issue (c. \$2000 kW^{-1} and c. \$250 kW h^{-1}), but it is expected to fall as manufacturing capacity expands.

The Shanghai Institute of Ceramics (Chinese Academy of Sciences) first started research on the NaS battery in the 1960s and the first 6 kW prototype for vans was in operation in 1977. In 2007 a unit with a capacity of 650 A h was successfully used for an EES operation. In 2011 the ShangHai NaS Company was established and its operation was based on the NaS technology developed by the Shanghai Institute of Ceramics (Chinese Academy of Sciences). The ShangHai NaS Company is now the leading company in China working on NaS technology. One 100 kW/800 kW h demonstration project, developed by the company, is in operation and a 1 MW project is under investigation. A demonstration project of capacity 200 kW is being operated by the Narui Company in Nanjing, Jiangsu province; it was supplied by the NGK Insulators Company (Table 24.5).

3.6 Lithium-Ion Battery

Lithium-ion batteries, first proposed in the 1960s, came into reality once Bell Labs developed a workable graphite anode to provide an alternative to lithium metal (lithium battery). The cathode in these batteries is a lithium metal oxide ($LiCoO_2$, $LiMO_2$, $LiNiO_2$, etc.) and the anode is made of graphitic carbon with a layer structure. The electrolyte is made up of lithium salts (such as $LiPF_6$) dissolved in organic carbonates. When the battery is being charged the lithium atoms in the cathode become ions and migrate through the electrolyte toward the carbon anode where they combine with external electrons and are deposited between carbon layers as lithium atoms. This process is reversed during discharge.

TABLE 24.5 Sodium–Sulfur Battery Projects in China

Project name	System description	Rated capacity	Status
Shanghai Institute of Ceramics (Chinese Academy of Sciences)	Used for vans	6 kW	Completed
Shanghai Institute of Ceramics (Chinese Academy of Sciences)	Unit	650 A h	Completed
ShangHai NaS Company	Stationary system	100 kW/800 kW h	Demonstration
ShangHai NaS Company	Stationary system	1 MW/2 MW h	Under Investigation

The first commercial lithium–ion batteries were produced by Sony in 1990. Since then, improved material developments have led to vast improvements in energy density terms from figures of (100–175) W h kg^{-1} and increased lifecycles as high as 20 000 cycles. The efficiency of Li-ion batteries is almost 100% which is another advantage over other batteries. While Li-ion batteries took over 50% of the small portable market in just a few years, there are some challenges for making large-scale Li-ion batteries. The main hurdle is the high cost (>$600 kW h^{-1}) due to special packaging and internal overcharge protection circuits. Many companies are working to reduce the manufacturing cost of Li-ion batteries to capture large-scale energy markets. The leading companies are A123 System, NEC, LG, Samsung, Tesla, and BYD. The first megawatt-scale Li-ion battery project was supplied by the A123 System Company and developed by AES in the United States. The largest project announced recently is the Elkins project in West Virginia in the United States with a scale of 32 MW.

China started research into Li-ion batteries in the 1980s, and during the period 2000–10 products for mobile phones and laptops experienced a rapid increase in development. The development of Li-ion batteries for large-scale stationary applications started in 2009 when BYD tested its 1 MW system. In 2013 the Chinese company Wanxiang Group purchased A123 System and became one of the leading companies in the area of Li-ion battery development. As of 2015, more than 30 projects have been announced using Li-ion batteries over the past 5 years with a total capacity of about 62.3 MW. Table 24.6 lists the major projects in China with scales greater than 1 MW. Of them the 20 MW/40 MW h system developed by BYD in 2014 is the largest project in China.

TABLE 24.6 Li-Ion Battery Projects in China

Project name	System description	Rated capacity	Status
BYD Project	Demonstration	1 MW/4 MW h	Completed
Southern Grid Project in Shenzhen, Guangdong province	Demonstration	4 MW/16 MW h	Demonstration
State Grid Project in Zhangbei, Hebei province	Demonstration	14 MW/56 MW h	Demonstration
BYD Project	Commercial usage	20 MW/40 MW h	Operation

3.7 Flow Battery

The flow battery is a form of battery in which electrolyte containing one or more dissolved electroactive species flows through a power cell/reactor in which chemical energy is converted to electricity. Additional electrolyte is stored externally, generally in tanks, and is usually pumped through the cell (or cells) of the reactor. The reaction is reversible allowing the battery to be charged, discharged, and recharged.

In contrast with conventional batteries, flow batteries store energy in the electrolyte solutions. Therefore, the power and energy ratings are independent, the storage capacity being determined by the quantity of electrolyte used and the power rating determined by the active area of the cell stack. Flow batteries can release energy continuously at a high rate of discharge for up to 10 h.

Three different electrolytes form the basis of existing designs of flow batteries currently in demonstration or in large-scale project development. These electrolytes are sodium bromide (NaBr) by Regenesys in the United Kingdom, vanadium bromide (VBr) by VRB Power Systems, Inc. In Canada, and zinc bromide (ZnBr) by ZBB Energy Corporation. In China only the vanadium redox battery (VRB) has been extensively developed. The following discussion will focus on this kind of flow battery.

The principle of VRB is that it stores energy by employing vanadium redox couples (V^{2+}/V^{3+} in the negative and V^{4+}/V^{5+} in the positive half-cells). These are stored in mild sulfuric acid solutions (electrolytes). During charge/discharge cycles, H^+ ions are exchanged between the two electrolyte tanks through the hydrogen–ion permeable polymer membrane. The reactions that occur in the battery during charging and discharging can be expressed simply by:

Positive electrode: $V^{4+} \leftrightarrow V^{5+}e^-$

Negative electrode: $V^{3+} + e \leftrightarrow V^{2+}$

Cell voltage is between 1.4 and 1.6 V. The net efficiency of this battery can be as high as 85%. Like other flow batteries the power and energy ratings of VRB are independent of each other.

VRBs are suitable for a wide range of energy storage applications for electricity utilities and industrial end-users. These include enhanced power quality, uninterruptible power supplies, peak shaving, increased security of supply and integration with renewable energy systems. The majority of development work has focused on stationary applications due to the relatively low energy density of VRBs.

The VRB was pioneered in the Australian University of New South Wales (UNSW) in the early 1980s. Australia's Pinnacle VRB bought the basic patents in 1998 and licensed them to Sumitomo Electric Industries (SEI) and VRB Power Systems. VRB storage of up to 500 kW, 10 h (5 MW h) has been installed in Japan by SEI for Kwansei Gakuin University and other institutions. VRBs have also been applied for power quality applications (3 MW, 1.5 s, solid–electrolyte interphase) for Tottori Sanyo Electric.

China started research and development on VRBs in 2000; the major institutions involved are Dalian Institute of Chemical Physics (Chinese Academy of Sciences), Tsinghua University, and China University of Geosciences. In 2008, Beijing Prudent Energy purchased VRB Power Systems, resulting in China becoming the leading country in VRB battery technology. In 2009 the first VRB demonstration project was installed in Tibet. The system was supplied by Dalian Rongke Power Company, used for the PV power system, and had a capacity of 5 kW/50 kW h. In 2012 Prudent provided a 2 MW system to China Electric Power Research Institute for the National Wind–Solar Energy Storage Demonstration Project which is the first megawatt-scale VRB system in China. In 2013 Dalian Rongke Power Company provided a 5 MW VRB system to the Wo-Niu-Shi Wind Power Farm in Liaoning province, which is the largest VRB system up to now. Overall, 8.2 MW of VRB flow battery systems have been installed in China up to 2014. The major flow battery projects in China are summarized in Table 24.7.

TABLE 24.7 Flow Battery Projects in China

Project name	System description	Rated capacity	Status
State Grid Project in Zhangbei, Hebei province	Demonstration	2 MW/8 MW h	Completed
JinFeng Project in Beijing	Microgrid	200 kW/800 kW h	Demonstration
Wo-Niu-Shi in Liaoning province	Wind power farm	5 MW/10 MW h	Operation

4 SUMMARY AND PROSPECTS

Fig. 24.2 gives a summary of installed capacity of energy storage systems in China up to 2014 [55]. As can be seen the installed capacity of pumped hydro is dominant and shares 99.2% of total capacity of EES. Li-ion battery, flow battery, and CAES systems are the other major EES technologies in China.

Research and development on EES in China has made great progress during the past (10–15) years. However, it is still below the leaders of EES in the world. As shown in Fig. 24.3, most EES technologies are under investigation or at the demonstration stage in China. In China, PHES and lead–acid battery storage are technically mature systems; CAES, NaS, Li-ion, and flow batteries have been technically developed and are in demonstration; and flywheel energy storage is still being developed.

It is expected that total installed EES in China will be 70 GW by the year 2020 which will be (4.0–5.0)% of total power generation capacity [55].

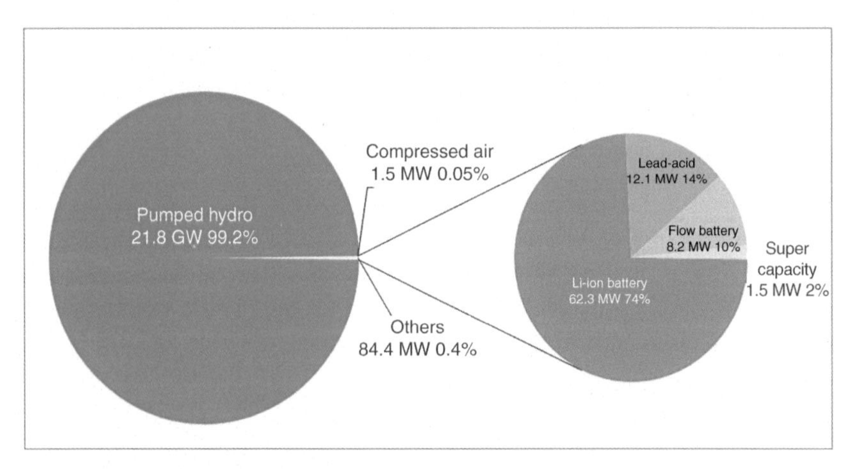

FIGURE 24.2 Total installed capacity of energy storage systems in China, 2014 [55]

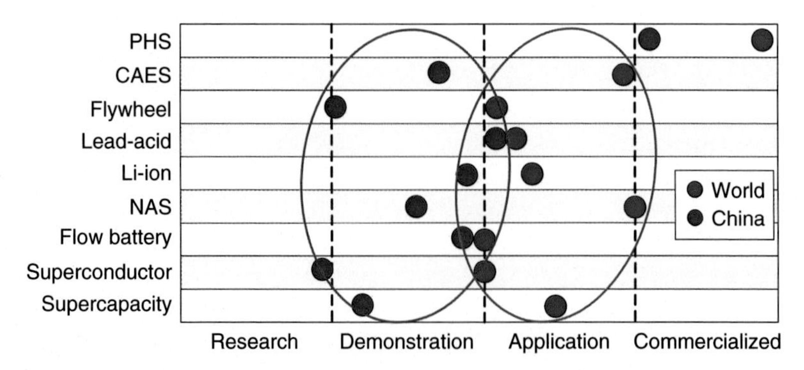

FIGURE 24.3 Comparison between China and the world's leading developments in EES.

Furthermore, total installed EES in China is expected to be 200 GW by 2050 [54]; this will equate to (10–15)% of total power generation capacity. The demand and market potential of China are enormous.

5 CONCLUSIONS AND REMARKS

The wide variety of available EES technologies and systems in China have been reviewed in terms of imperativeness, application, technical characteristics, research, and deployment. The conclusions drawn are the following:

1. EES is urgently needed by the conventional electricity generation industry, distributed energy resource system, and intermittent renewable energy supply systems. By utilizing EES the challenges facing the industry—such as raised volatility, reduced reliability, and stability—can be greatly diminished.
2. The applications of EES are numerous and various and cover the full spectrum, ranging from increased scale, generation and transmission–related systems, to distribution network and even customer/end-user needs. The various technologies provide three primary functions: energy management, bridging power and power quality, and reliability.
3. Research and development on EES have experienced fast and fruitful development during the past (10–15) years in China, although overall research and development is still below that of world leaders.
4. There is great potential for EES applications in China and this is summed up by the expected installed capacity of EES to be 70 GW in 2020. EES has a bright future in China.

ACKNOWLEDGMENT

This work was supported by Key Project of Chinese National Programs for Fundamental Research and Development (973 program) under grant No. 2015CB251300 and China Natural Science Foundation under grant No. 51522605.

REFERENCES

[1] Chen H, Cong TN, Yang W, Tan C, Li Y, Ding Y. Prog Nat Sci 2009;19:291–312.
[2] Alanne K, Saari A. Renew Sust Energ Rev 2006;10:539–58.
[3] Baker JN, Collinson A. Power Eng J 1999;6:107–12.
[4] Chen LG, Zheng JL, Sun FR. Energ Convers Manage 2003;44:2393–401.
[5] Cornelissen RL, Hirs GG. 1998; 39:1821–1826.
[6] Denholm P, Kulcinski GL. Energ Convers Manage 2004;45:2153–72.
[7] Denholm P. Renew Energ 2006;31:1355–70.
[8] Denholm P, Kulcinski GL, Holloway T. Environ Sci Technol 2005;39:1903–11.
[9] Denholm P, Holloway T. Environ Sci Technol 2005;39:9016–22.

[10] Ding Y. Experimental evaluation of heat transfer processes associated with a prototype cryogenic engine. Report. London: Highview Enterprises Ltd.; 2005.
[11] MIT. Energy and environment. Report. Cambridge, MA: MIT Laboratory for Energy and the Environment; July 2005. p. 1–24.
[12] Hands BA. Cryogenic engineering. London: Academic Press; 1986.
[13] Hersh DJ, Abrardo JM. Cryogenics 1977;17:381–444.
[14] Ishimoto J, Ono R. 2005; 45:304–316.
[15] Jacobsen RT, Penoncello SG, Lemmon EW. Thermodynamics properties of cryogenic fluids. New York: Plenum Press; 1997.
[16] Karpinski AP, Makovetskiy B, Russell SJ, Serenyi JR, Williams DC. 1999; 80:53–60.
[17] McClintock PVE. Low-temperature physics: an introduction for scientists and engineers. London: Blackie & Son Ltd; 1992.
[18] Kashem MA, Ledwich G. Electr Power Syst Res 2007;77:10–23.
[19] Kishimoto, K, Hasegawa, K, Asano, T. Mitsubishi Heavy Industries, Ltd. Technical Review 1998;35:117–120.
[20] Kluiters EC, Schmal D, Ter Veen WR, Posthumus K. J Power Sources 1999;80:261–4.
[21] Knowlen C, Herzberg A, Mattic AT. Automotive propulsion using liquid nitrogen. AIAA J 1994;1994:1994–3349.
[22] Knowlen C, Williams J, Mattick AT. Quasi-isothermal expansion engines for liquid nitrogen automotive propulsion, SAE International 1997, No 972649, doi:10.4271/972649.
[23] Knowlen C, Matick AT, Bruckner AP. High efficiency energy conversion systems for liquid nitrogen automobiles, SAE International 1998, No 981898, doi:10.4271/981898.
[24] Knowlen C, Mattick AT, Hertzbeg A. Ultra-low emission liquid nitrogen automobile, SAE International 1999, No 1999-01-2932 doi:10.4271/1999-01-2932.
[25] Kondoh J, Ishii I, Yamaguchi H, Murata A. Energ Convers Manage 2000;41:1863–74.
[26] Korpaas M, Holen AT, Hildrum R. Electr Power Energ Syst 2003;25:599–606.
[27] Koshizuka N, Ishikawa F, Nasu H. Physica C 2003; 386:444–450.
[28] Krane RJ. Int J Heat Mass Transf 1987;30(1):43–57.
[29] Lavrenchenko GK. Cryogenics 1993;33:1040–5.
[30] Manwell JF, Elkinton CN, Rogers AL. Renew Sust Energ Rev 2007;210–34.
[31] McLarnon FR, Cairns EJ. Energy storage. Annu Rev Energ 1989;14:241–71.
[32] Najjar SH, Zaamout MS. Energ Convers Manage 1998;39:1503–11.
[33] Nakaiwa M, Akiya T, Owa M, Tanaka Y. Energ Convers Manage 1996;37:295–301.
[34] Narinsky GB. Cryogenics 1992;32:167–72.
[35] Ordonez CA, Plummer MC. Energ Sources 1997;19:389–96.
[36] Ordonez CA. Energ Convers Manage 2000;41:331–41.
[37] Rosen MA. Energy 1999;24:167–82.
[38] Scott RB. Cryogenic engineering. Princeton, NJ: D. Van Nostrand Company, Inc.; 1959.
[39] Ratering-Schnitzler B, Harke R, Schroeder M, Stephanblome T, Kriegler U. J Power Sources 1997;67:173–7.
[40] Seliger B, Hanke-Rauschenbach R, Hannemann F. Sep Purif Technol 2006;49:136–48.
[41] Smith AR, Klosek J. Fuel Process Technol 2001;170:115–34.
[42] Sun S, Wu Y, Zhao R. Cryogenics 2001;41:231–7.
[43] Wang W, Chen L, Sun F, Wu C. Appl Therm Eng 2004;25:1097–113.
[44] Weisend II JG. Handbook of cryogenic engineering. London: Taylor & Francis; 1998.
[45] Williams J. AIAA paper 1997, No 1997-0017, doi:4271/970017.
[46] Williams J, Knowlen C, Mattick AT. AIAA paper 1997, No 1997-3168, doi:4271/973168.
[47] Wu Y, Peng S, Chen L, Xie H. Cryogenics 1992;32:300–3.

[48] van der Linden S. The commercial world of energy storage: a review of operating facilities (under construction or planned). Paper presented at the 1st annual conference of the Energy Storage Council, Houston, Texas, Mar 3, 2003.
[49] http://en.wikipeida.org/wiki/Hydroelectric_energy_storage
[50] Bueno C, Carta JA. Renew Sust Energ Rev 2006;10:312–40.
[51] Cavallow AJ. J Sol Energy 2005;123:387–9.
[52] Chalk SG, Miller JF. J Power Sources 2006;159:73–80.
[53] Steinfeld A, Meier A. Encyclopedia of Energy, Elsevier Inc., 2004;5:623–637.
[54] van der Hoeven M. Technology roadmap of energy storage. Paris: International Energy Agency; 2014.
[55] White paper on energy storage industry. Beijing, China: China Energy Storage Alliance; 2014. www.ESCexpo.cn.

Chapter 25

The Politics of Investing in Sustainable Energy Systems

Alan Owen, Leuserina Garniati
Centre for Understanding Sustainable Practice (CUSP), Robert Gordon University, Aberdeen, Scotland, United Kingdom

1 INTRODUCTION

In the chapter we argue that an entirely new economic and political model is required to address global energy issues in the 21st century.

Renewable energy systems are generally seen as beneficial to the environment, desirable for an advanced society, and/or an expensive indulgence that business cannot afford. The positions of many governments, corporations, and influential individuals (positive or negative) on renewable energy are often based on selective use, misuse, or ignorance of the facts regarding renewable energy and the technologies required to deliver it.

Access to energy resources is largely what is presently used to define an "advanced" nation as opposed to a "developing" nation and is therefore an entirely political component of human civilizations. The terms "advanced," "emerging," "developing," etc. are themselves political statements seemingly designed to determine the rightful place of any given nation as perceived by those nations who consider themselves to be "advanced". Since the "advanced" nations are largely in debt to the point of no return and whose economies depend entirely on the exploitation of dwindling resources, we would argue that "advanced" nationhood is not sustainable and therefore should not be held up as a desirable way forward. From this we suggest that for the purposes of energy consumption the terms under-developed nation, developing nation, developed nation and over-developed nation are more appropriate.

Noting that there has been little change in ranking between 2003–12 the top 35 nations jointly consume 65% of the world"s primary energy, the next 35 consume 22%, nations ranked from 70–105 consume 9%, and the lowest 35 consume 4%. Based on Table 25.1, if we assume (somewhat simplistically, it is recognized) that the average is about right, then the average is ranked at 47th in the world (in both 2003 and 2012). Assuming reasonable equity for social and political purposes then around 4.3 billion people need to improve their access

Storing Energy. http://dx.doi.org/10.1016/B978-0-12-803440-8.00025-7

TABLE 25.1 Examples of National Energy Consumption Per Capita (ca) Per Annum (a)

Example classification	Nation	Rank (2003)	kW h ca^{-1} a^{-1} (2003)	kW h ca^{-1} a^{-1} (2012)	Rank (2012)
Overdeveloped nations	Iceland	1	22 477	23 640	2
	Trinidad & Tobago	2	21 187	19 140	3
	Qatar	3	17 041	24 622	1
	Kuwait	4	16 248	14 177	4
	Luxembourg	5	11 107	9 755	8
Developed nations	Slovakia	35	4 367	4 162	35
	United Kingdom	36	4 332	3 960	37
	Ireland	37	4 284	3 877	38
	Libya	38	4 011	3 710	40
	Israel	39	4 001	4 099	36
Average		47	3 374	3 419	47
Developing nations	Indonesia	90	1 153	1 151	91
	Dominican Republic	91	1 118	979	97
	Ecuador	92	1 113	1 239	88
	Armenia	93	1 052	1 333	85
	Zimbabwe	94	1 017	930	103
Underdeveloped nations	Senegal	132	362	400	130
	Haiti	133	304	532	122
	Bangladesh	134	278	285	134
	Eritrea	135	188	173	135
	Afghanistan	136	119	No data	

World Bank (2003, 2012).

to energy resources, while 2.7 billion, that is, the developed and overdeveloped nations, need to reduce their consumption, some by significant amounts. Since fossil fuels are a diminishing resource and access to nuclear is a contentious issue, then the only mechanism for long-term energy equity is increased access to renewable energy.

We also argue that sustainability in its true sense should be used to describe sustainability of the human population; in effect, the Earth as a support system is ambivalent about the existence (or otherwise) of any one particular species, and human beings have no particular claim to exclusivity. Renewable energy is promoted as being part of the "Save the Earth" movement, but the Earth will continue to exist (and flourish) long after *Homo sapiens* has starved to death, dehydrated, or created some apocalyptic disaster that effectively removes the species completely or reduces its influence considerably. At present it is suggested that 1.5 Earths [1] are required to meet sustainably the existing resource demand and that 2 will be needed by 2030; obviously this is not possible and the politics of reducing demand by (50–70)% or reducing the human population by the same amount is not being addressed by anyone. Additionally, it is unlikely that the solution to this issue can be generated by using the same 200 year old investment model built on the European Industrial Revolution and colonial exploitation that has created the problem. For the human population to be sustainable, it must be able to provide sufficient resources for everyone—not excess for a few and poverty for the many. Now that the global human population has passed the 7 billion mark the availability of all resources becomes more acutely focused, and the inextricable bind between food, energy, and water must be respected.

Water and energy are needed to grow food, to drive the associated processing, and preservation activities, and energy is required to treat and transport water. The relationships and tradeoffs within this triangle of resources iterate that food, water, and energy are inextricably interdependent. While current policies and business models often treat water, energy, and food security separately, issues in one of these sectors must be addressed with the understanding of this interdependence and seek holistic solutions to address the water–energy–food nexus.

2 SUSTAINABLE ENERGY SYSTEMS POLICY AND POLITICS

Geospatial regions, nations, and districts have different requirements for energy due to variations in environmental, economic, social, and political constraints, which influence demand profiles and generation capacities. Hence, these constraints determine the policy development and implementation priorities for sustainable energy consumption, generation, and distribution.

Sustainable energy systems discussed in the following sections consist of:

1. Sustainable energy consumption, which is understood as energy conservation measures including energy use reduction and energy efficiency.
2. Sustainable energy generation, which is understood as renewable energy provision.
3. Sustainable energy distribution, which is understood as equal and secure access to energy resources.

A sustainable energy future sees the spread of energy services to reach disadvantaged populations, the practice of rational pricing strategy, and actions for structural reform to ensure facilitation and financing of technology transfer [2]. The social component of sustainable energy can be expanded to cover community involvement, affordability, social acceptability, lifestyles, and aesthetics [3].

Sustainability requires that the four legs—economic, political, social, and environmental—are considered to be equally important, thus sustainable energy resources must be economically viable, politically supported, socially equitable, and environmentally acceptable. We therefore define sustainable energy as any economically viable energy resource (not only electricity) that is not, in its lifecycle, a net contributor to climate change *and* does not have a substantially negative environmental or social impact (actual or potential). It follows from this that just because an energy resource can be defined as renewable does not make it necessarily sustainable. For example, palm oil, a significant component of biodiesel, is frequently grown on plantations created by clearing rainforest, displacing indigenous people and wildlife. The CO_2 impact of felling, clearing, and burning alone is more than the CO_2 emissions saved by adding a small percentage of palm oil to road fuels [4].

Sustainable energy requires a balanced composition between energy security, economic development, and environmental protection [5]. It encompasses energy systems, which are based on three core dimensions: energy security, social equity, and environmental impact mitigation [6]. A large component of this would be the incorporation of renewable energy into the existing energy mix, but it does not eliminate the efficient use of conventional sources to ensure sustainable energy security [7]. The term also takes into account the issues of creating an internal energy market and coordinating international collaboration, which in itself constitutes the efforts of efficiently managing energy consumption and energy distribution.

Meanwhile, policy is described as sets of decisions toward a long-term approach to a particular problem, which in governmental scopes are usually embodied in legislation and real, driven actions to achieve its objectives [8]. Therefore, for the purpose of this review, sustainable energy policy can be understood as sets of decisions, which encourage investments from private sectors, present clear business cases to its strategies, and are developed in a participatory, transparent, and accountable way, to achieve energy use, generation, and distribution practices, which are economically viable, environmentally responsible, and socially acceptable for the long term.

The available literature on sustainable energy policies has been found to be extensively documented surrounding economically developed countries and those countries emerging as so-called new industrialized countries (NICs). It is however the contrary for those that are considered countries with developing economies. These limited publications have been captured by an article which provides an overview of energy for sustainable development in developing nations [9], but policy types and sectors in the category are only briefly and broadly touched upon. Global reports on specifically applied policies on sustainable energy have also been produced by the International Energy Agency (IEA) [10] and the World Energy Council (WEC) [11], but as highlighted earlier for peer-reviewed literature, not all of these reports have specifically addressed those country groups with developing economies.

In many developing nations, due to financial and time constraints, policy formulation is often superseded by direct technology implementation without any robust, strategic planning. Lack of political foresight to support policy making creates unclear and insufficiently embedded organizational structures of implementing agents and procedures for appropriate technology utilization. The process then creates energy generation, distribution, and utilization systems, which are not thoroughly planned and often temporarily designed to quick-fix major problems. The process is also more negatively influenced by changes of people, roles, and positions. Politics should instead act as the persistent but subtle force in shaping ideal environments for:

- evaluations on agent of change and initiator of policy modification
- decisions on startup programs for modifying existing policies
- identification (or facilitation/creation if necessary) of timeframe for modifications of existing policies based on local political atmosphere and social readiness to accept changes
- ensuring process design's flexibility and robustness for transfer of knowledge and paradigm between governments.

3 IMPLICATIONS FOR INVESTMENT IN SUSTAINABLE ENERGY SYSTEMS

In the next decade the global food production system will come under increasing pressure, along with water scarcity and deforestation, due to exponential population growth [12]. Forest products contributed US$100 $\times$ 10^9 a^{-1} ($100 billion per year) to the global economy between 2003–07 and the value of nonwood forest products (mostly food) was estimated at US$18.5 $\times$ 10^9 ($18.5 billion) in 2005, yet approximately 40% of the world's natural forests have disappeared in the last 300 years [13] and are predicted to decline by 13% from 2005–30, mostly in South Asia and Africa [14]. Along with forest degradation the rate of natural ecosystem loss continues unabated [15]. While palm oil, rubber, and pulp-based economic activities provide people with short-term income, they also accelerate greenhouse gas emissions, increase air pollution, and harm the forests that Indonesians and the rest of the global population depend on. The signs of ecosystem breakdown and stress have illustrated how dependent livelihood and business operations are on the critical services these ecosystems provide as the decline in ecosystems is:

1. making natural resources scarcer and more expensive
2. increasing the costs of water
3. escalating the damage caused by invasive species to sectors including agriculture, fishing, and food production.

Therefore, by definition, the current use of the word "investment" must mean something different to future generations than to those of the last 50 years.

Investment in energy systems must be based on long-term peaceful societal benefit and not "exploitation as usual."

4 TECHNOLOGY SELECTION

Currently, technology selections are made based on the emphasis of efficiency, for example, technologies developed in industrialized countries are designed primarily for capital intensive and labor minimization [16]. However, societies with developing economies have different supply and demand requirements for renewable energy, thereby often creating a mismatch between the proposed technology to be implemented and the technology that will create optimum impact. Adapting complex and sophisticated technologies within a local context remains a challenge. On the other hand, the engineering capabilities of the indigenous communities to design, manufacture, install, operate, and maintain their own tailor-made technologies for their specific contexts are also still very limited, especially in the most vulnerable regions of the developing world. These two contradictory issues have become the precursor for addressing the needs and priorities of appropriate technology.

The most significant roles of investment should be for the development of appropriate technology in creating access to sustainable energy for developing nations to benefit their local population and, wherever possible, there should be universal access to knowledge of developing sustainable energy technologies for locally appropriate use. These technology systems must be developed with the desire to encourage independence, not just for financial gain, thus ensuring that the skills for developing technologies are available to all, and any restrictions due to economics/politics are removed. Access to energy, which is reliable and affordable, will potentially increase economic activities and thereby reduce the potential for social/political conflict in these vulnerable regions.

In summary, sustainable energy technical resources have outlined issues related to energy in societies with developing economies in a way that may not be immediately obvious to a Western, educated, industrialized, rich, and democratic (WEIRD) mindset. It is clear from field experience that simply transferring complex technology is of little help without indigenous skillsets being developed to support subsequent service life. Substantial indigenous wisdom exists which can be used if the external fieldworker takes the time to engage and form constructive relationships with local communities.

Correct timing and accurate actor identification, encapsulated as the politics of policy making in the sector, are also prerequisites to ensuring policy goals in sustainable energy access are practiced at ground levels. Politics as the fourth component of sustainable energy systems in practice needs to understand how technology works; how devices are manufactured, operated, and maintained; and how engineering systems affect its social/natural surroundings. It needs to value how community functions and interact with its social/natural surroundings.

Politics needs to be aware of how the natural environment behaves, dictates limiting factors, and sets boundaries to constraints.

5 TRANSITION

The economic and political challenges of transitioning to a sustainable energy future require that making money in large quantities is no longer the motivating factor and recognition of the fact that sustainable energy supplies are themselves a potential point of conflict. As an example, it may be seen that using renewable energy technology to replace diesel generation in remote areas is a good thing to do; it reduces CO_2 emissions, reduces energy costs to the local community, and reduces dependency on dwindling fossil fuels. However, what is often not recognized is that the diesel fuel currently used in that location is part of an existing supply chain and will represent a number of business investments, individual livelihoods, and income-generating practices. If diesel generation is removed or reduced then a conflict will arise between the existing energy provider and the new one. If the existing energy provider is to become the new energy provider, then who will provide the additional investment, training, and product support—not to mention writing off the existing supply chain infrastructure?

Equally, many companies and individuals in the United Kingdom have invested, with the support of government-funded feed-in tariffs, in wind, solar, and other forms of renewable energy technology. However, due to peak-loading needs, inequity between availability and demand, and existing grid capacity constraints, these energy supplies are often curtailed or, put more simply, not used but still paid for. If renewable energy sources are for operational purposes unable to supplant the existing systems, how is the transition to sustainable energy systems to be made?

6 GLOBAL IMPLICATIONS

It is unfortunate that most of the world's influential governments are tied to, and funded by, large corporate interests who will fight any attempt to change their commercial and financial models which are based on linear exploitation, that is, resource extraction, application of energy, product manufacture, sale, use, and disposal. This linear approach ensures constant demand for new products but fails to take into account a finite and dwindling supply of resources to meet that demand. Continued linear exploitation of resources will only undermine our long-term survival; there is no magic solution that will allow us to continue our current usage patterns. In short, capitalism in its current form is not sustainable, but the highly indebted nature of overdeveloped nations and the need to service that debt mean that it is a fairground ride that we cannot get off.

Already mass economic migration between Africa and Europe is creating additional economic pressures that cannot be withstood. Lack of human equity across the globe, much of which is ultimately traceable to access to energy, will only exacerbate the situation, while the stripping of the planet for the purposes of monetizing natural resources only damages our own life support systems.

Having said that, it is difficult to envisage a global system that does not recognize the need for a business case; every operation must make business sense, otherwise it fails the economically viable test of sustainability. The primary requirement is that simply operating energy systems as a means of profit is not sufficient.

7 THE CIRCULAR ECONOMY

As a potential solution to the linear economy the concept of a circular economy attempts to manage our economic activity in the same characteristic style of the circulatory systems of the global environment. It requires that demand for resources, including energy, are reduced and that everything should be designed for reuse and recycling—effectively zero waste. A report from the Ellen MacArthur Foundation estimates [17] that $\$2 \times 10^{12}$ (2 trillion USD) is wasted and that this is potentially a new economic opportunity. However, referring back to the remote diesel example given earlier, if the recycling market is potentially worth $\$2 \times 10^{12}$ (2 trillion USD) then that must, under current circumstances, be able to replace $\$2 \times 10^{12}$ (2 trillion USD) of existing resource consumption, that is, an excellent plan in terms of global resource use, but someone's existing investment is going to be threatened. The report's key findings suggest that:

- Household food waste: an income stream of $\$1.5 \times 10^9$ ($1.5 billion) could be generated annually in the United Kingdom alone for municipalities and investors by collecting household food waste and processing it to generate biogas and return nutrients to agricultural soils.
- Textiles: revenue of $\$1975\ t^{-1}$ (USD 1975 per tonne) of clothing could be generated in the United Kingdom if collected, remade, and sold at current prices, comfortably outweighing the cost of $680 (USD 680) required to collect and sort each tonne.
- Packaging: a cost reduction of $20\%\ hL^{-1}$ (20 percent per hectoliter) of beer sold to consumers would be possible across all markets by shifting from disposable to reusable glass bottles, which would lower the cost of packaging, processing, and distribution.

It is estimated that 75% of our global energy resource is used for raw material production and processing contributing significantly to CO_2 emissions. In many developing nations manpower is used where overdeveloped nations would simply apply energy-intensive equipment. For example, at Soekarno Hatta Airport in Jakarta (Indonesia), the authors observed 6 men pushing a line of some 50–60 trolleys back to the baggage carousels, while on the same trip one man drove an electric "tractor" at Schiphol Airport, Amsterdam towing fewer than 10 trolleys.

One of the biggest factors in energy use and, therefore, energy reduction is global transportation of goods, many of which are exported from a country while similar goods are being imported into the same country from elsewhere or domestic suppliers are unable to compete. As an example, one tonne of shipping from Australia or New Zealand requires around 10 MW h of energy to reach the United Kingom, yet UK producers such as farmers often find it difficult to sell into their own domestic markets. It is however difficult to make an investment case for not using energy as part of global trade.

Therefore, in addition to a circular economy our economic model should take into account the need to reduce globalization and increase localization, which often leads to cries of protectionism from those nations wishing to export into the local economy and is in direct opposition to the present investment mantra.

8 CONCLUSIONS

The authors apologize for the somewhat gloomy energy investment prospectus outlined in the chapter and for raising more questions than we can answer. However, it is clear to us that "investment" in the accepted sense of the word is no longer appropriate for the situation that the human population finds itself in. For sustainable energy and the renewable energy systems that must be part of such a concept a new model of investment is required, one that places the emphasis on community benefit and equity rather than just maximizing profit for a few, but this is a very difficult political argument to make.

The inequality of access to energy resources between underdeveloped, developing, developed, and overdeveloped nations must be addressed, and quickly, to avoid significant global movement of people trying to extricate themselves from where they are to where they perceive to be a better place, even if it is a false dawn built on borrowed money.

The biggest single economic and energy-related challenge is to persuade the top 25% of energy-consuming nations that they are not entitled to 65% of the world 's primary energy supply no matter what their relative economic advantage. Not only must these nations reduce their consumption, they must support the lowest 50% of energy-consuming nations to improve their access to energy beyond the existing figure of 13%.

If sustainability of the human race is considered to be important, then the politics of investing in renewable energy systems is not (whether we like it or not) best served by the post–Industrial Revolution capitalist model. Similar sentiments have been expressed in Ref. [18].

REFERENCES

[1] WWF. Living planet. Report. Gland, Switzerland: World Wide Fund for Nature; 2015.
[2] Saha PC. Energy Policy 2003;31:1051–9.
[3] Rosen MA. Sustainability 2009;1:55–80.

[4] ZSL. Indonesia. London: Zoological Society of London; 2015.
[5] International Energy Agency, Energy efficiency, policies, and measures, http://www.iea.org/textbase/pm/?mode=pm&action=result
[6] World Energy Council. Policies for the future: 2011 assessment of country energy and climate policies. London: World Energy Council; 2011.
[7] European Union. Green paper—a European strategy for sustainable, competitive, and secure energy. Brussels, Belgium: Commission of the European Community; 2006.
[8] Food and Agricultural Organisation of the United Nations, Definition of policy, http://www.fao.org/wairdocs/ILRI/x5499E/x5499e03.htm
[9] Kaygusuz K. Renew Sust Energy Rev 2012;16:1116–26.
[10] International Energy Agency, Renewable energy policy considerations for deploying renewables. Paris: International Energy Agency; 2011.
[11] World Energy Council, Energy efficiency policies and measures. Energy efficiency: a recipe for success. London: World Energy Council; 2010. http://www.wec-policies.enerdata.eu/
[12] KPMG International. Expect the unexpected: building business value in a changing world. KPMG International; 2012.
[13] Food and Agriculture Organization . Global forest resources assessment 2005, main report: progress towards sustainable forest management. FAO Forestry Paper 147. Rome, 2006.
[14] KPMG International. Expect the unexpected: building business value in a changing world. KPMG International; 2012.
[15] Butchart SHM, Walpole M, Collen B, van Strien A. Science 2010;328:1164–8.
[16] Kaygusuz K. Renew Sust Energy Rev 2012;16:1116–26.
[17] http://www.ellenmacarthurfoundation.org/business/reports/ce2013
[18] Meadows DH, Randers J, Meadows D. Limits to growth: the 30-year update. White River Junction, VT: Chelsea Green Publishing Company; 2004.

Subject Index

A

B

C

D

G

H

M

N

O

P

V

W

Y

Z

Printed in the United States
By Bookmasters